FORCED INDUCTION PERFORMANCE TUNING

FORCED INDUCTION PERFORMANCE TUNING

A. GRAHAM BELL

Haynes Publishing

First published by Haynes Publishing 2002

British Library Cataloguing in Publication Data
A catalogue record for this book is available from the British Library

ISBN 1 85960 691 1

Library of Congress catalog card no. 2002103351

Haynes Publishing, Sparkford,
Yeovil, Somerset, BA22 7JJ, UK.

Tel: 01963 442030 Fax: 01963 440001
Int. tel: +44 1963 442030 Int. fax: +44 1963 440001

E-mail: sales@haynes-manuals.co.uk
Web site: www.haynes.co.uk

Haynes North America, Inc.
861 Lawrence Drive, Newbury Park,
California 91320 USA

Printed and bound in Great Britain by J. H. Haynes & Co., Ltd

Contents

Preface

This book has been written on the premise that even the most intimidating subject can be made understandable if it's properly explained. Forced induction in its various forms has been around for a long time, and has been developed and put to use by ordinary people like you. It is not a technology that defies comprehension. I wasn't born with this knowledge. I built my first supercharged engine 35 years ago and I'm still learning. When something doesn't work out as planned I do more research and figure it out, and I talk to other people and look at what they are doing. I have committed those solutions to print in this book, explaining the principles involved as simply and as clearly as possible.

Forced induction would appear to have much in common with war. Journalists and ad men tend to hold up power as some sort of benevolent and virtuous force totally devoid of any hint of an evil side. The reality, though, can be harshly different. I first saw a turbo race car in the mid-1960s. It looked the part, sitting low to the ground and riding on suitably wide wheels, even if it was only a Ford Anglia. But, hey, under the bonnet was a Lotus Cortina twin-cam engine, which in itself was a pretty trick in that era. However, it was what was hanging off the side of that engine that grabbed everyone's imagination – a big turbo!

How did this car fare out on the race circuit? Well, after a year of trying it was still running close to last against Minis with one third the power. The driver had the necessary talent, and there was no shortage of money to pour into the project. What was lacking was an understanding of turbo matching and how to correct the fuelling and ignition to enable the engine to rapidly respond to driver inputs.

At that time tuners struggled with clockwork ignition and carburettors, and turbos that were matched to truck diesels. Today it should be much easier. Car manufacturers have produced a vast array of vehicles in performance levels from mild to almost wild in both turbocharged and supercharged configurations. They have poured millions into developing the hardware to make forced induction work, and at the same time electronic engine management systems have been refined, enabling drivers to exercise delicate control of even very powerful engines. Yet, in spite of these advances, frequently I see modified forced induction engines that simply do not work. Certainly these engines produce loads of power, but because all that power is not easily accessible and manageable it is useless in most environments.

The aim of this book is to tell the story of forced induction as it should be told. I don't want to delude you either, for that matter, into thinking that the forced induction route is the highway to driving nirvana or horsepower heaven. There are many traps along the way but this book will give you a broad understanding of all aspects of turbocharging, supercharging and nitrous oxide injection so that you can stay clear of those traps. Also, you will be empowered to work your way through developmental problems and go on to achieve the level of performance that you require from your road or competition engine.

A. Graham Bell
Maitland
New South Wales
Australia

Acknowledgements

The author wishes to express his sincere thanks to the following people and organisations for their help in providing information and illustrations, or for opening their doors and letting him loose with a camera to enable him to provide a more graphic insight into the subject of forced induction than could have been expressed by mere words.

Brad Host and Ewan Wasiak at The Toy Shop, Cardiff
Ian and Ryan Benson at Cardiff Wreckers
Steve Cox at Quickfit Motorsport, Hornsby
Sam at Hornsby Prestige and Performance
Con and Vic Elfes at C&V Performance, Waterloo
BMW Motorsport
Renault Sport
Ford Motorsport
Fiat/Lancia Auto Abarth
Cosworth Engineering
SAAB
Porsche
Eaton
Volkswagen
KKK
IHI
Jaguar
ZF
Mazda
Opcon Autorotor
Opel
Audi
National Motor Museum

Chapter 1

An introduction to forced induction

In the past, if we wanted more performance there were two relatively easy options available; either we went to our motor dealer and traded up to a car with a larger engine or we went to a performance tuner and had our stock engine worked over with free-flow inlet and exhaust and a modified cylinder head and cam. In both situations we were in effect saying, perhaps sometimes unknowingly, 'I want to get more air into the engine.'

All engines are in effect just sophisticated air pumps with a few add-on decorations to make them appealing to the eye. The amount of power the engine produces is determined by how much fuel it burns. Maximum power is produced when an engine burns 1 pound of petrol (gasoline) for every 13 pounds of air. Hence, when we get an engine to consume more air the potential exists to introduce more fuel into the cylinders, the end result being a rise in horsepower.

Those who chose to trade up to a car with a larger engine were basically spending a lot of money to obtain larger cylinder capacity – bigger lungs. Those bigger lungs, while offering the potential for more power, have a down side. Bigger cylinders frequently mean engine weight increases and possibly also car weight increases. More fuel is burned but performance is lost because of that weight increase. On top of that, internal friction in the engine also goes up, again with a performance and fuel consumption penalty. Consequently a 50% increase in engine capacity will perhaps see only a 40% performance increase at best.

The hot-up modification route also gives more power, but not by increasing cylinder capacity. Rather this method makes the engine breathe more efficiently, enabling it to draw more air into small lungs. However, this increase in breathing efficiency comes at a price. Usually the engine no longer works so well at lower revs and it is only in the upper portion of the power range where more power is available.

Forced induction is yet another method of getting more air into an engine. We can supercharge the engine using either mechanical or chemical means. The supercharger may be a belt- or gear-driven air pump, usually called a blower or a supercharger. Many

For some the route to more power is by means of big engine capacity, and even 351cu in (5.7 litres) may be insufficient!

Others see improved breathing as the answer. Consequently, they follow the conventional engine tuning route of exhaust headers and multiple carburettors. Plus, in this instance, the stock 5-port cast-iron head has been replaced by an alloy 7-port crossflow unit.

The alternative path is some form of forced induction.

different designs exist, as you will discover in the following chapter. Another type of supercharger is driven by the engine's hot exhaust gases. In this instance the supercharger is called a turbocharger, or more likely just a turbo. Whatever the design, the objective of the supercharger is to force more air into the cylinders so that the engine performs as if it were considerably larger.

Fig. 1.1. The four-stroke engine cycle. ***Induction stroke*** *– as the piston descends, the inlet valve opens fully and the cylinder is filled with air and vaporised fuel.* ***Compression stroke*** *– the piston rises, and part way up the inlet valve closes, allowing the ascending piston to compress the air/fuel mixture. As the piston nears top dead centre (TDC) the spark plug fires to ignite this mixture.* ***Power stroke*** *– the rapid expansion of the burning mixture produces pressure which drives the piston down. Around two thirds down the exhaust valve begins opening.* ***Exhaust stroke*** *– as the piston again ascends to TDC, the burned gases are expelled past the open exhaust valve. Before the piston actually reaches TDC, the inlet valve again begins to lift off its seat to allow in the fresh air/fuel charge when the piston descends on the next induction stroke.*

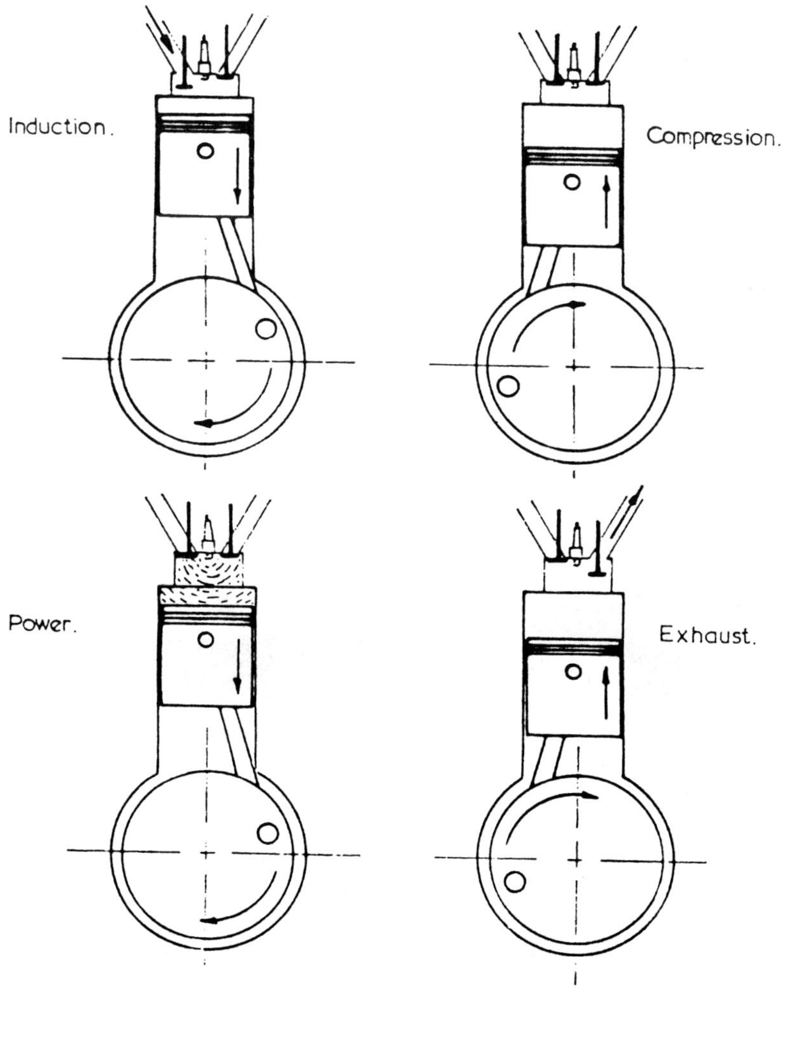

At sea level normal air pressure is about 14.7psi (pounds per square inch). Therefore, within an engine when a piston moves down in its cylinder it leaves in its wake an empty space, a vacuum. Because the inlet valve is open at this time, air at 14.7psi rushes in to fill the void created by the descending piston. In some very high performance engines and many race engines the cylinder will completely fill with air. We call this 100% volumetric efficiency (VE), because the air density or concentration level within the cylinder is the same as that outside the engine, i.e. cylinder pressure equals ambient pressure (*Fig. 1.1*).

With any form of supercharging the aim is to raise the air pressure in the inlet manifold to well above atmospheric pressure. If we were to add boost of 14.7psi the potential exists to force twice as much air into the cylinder before the inlet valve closes. In this case the engine would produce twice as much power, or to view the situation from another perspective the engine would perform like it was twice as big. That is why in 1987, just before the end of the Formula 1 turbo era, tiny 1.5-litre 4-cylinder BMW and 6-cylinder Honda engines produced over 1200hp when operating at 60psi boost during qualifying laps. In fact, insiders at BMW say 1500hp for qualifying in 1987 was closer to the true figure, but that is just a guess as their race dyno couldn't measure past 1280hp. However, apparently the special qualifying engines far exceeded that figure. By 1988 the rules limited boost to 2.5bar (36.3psi) and fuel loads were pegged at 150 litres. This meant power levels never exceeded 700hp under race conditions.

Apart from supercharging we can greatly magnify the amount of air an engine receives by chemical means. Many substances in either liquid or gaseous form contain oxygen as a component. You may be aware that nitromethane, for example, is 53% oxygen by weight. This fuel enables big inefficient 8-litre dragster engines to produce in excess of 1500hp without the aid of a blower. With a blower, power levels are near 6000hp.

Less well known oxygen bearing liquids are nitropropane, hydrogen peroxide and nitrogen peroxide. Nitropropane has some limited use, often as a 'cheater fuel' in motor sport. Hydrogen peroxide may prove to be of some value in the future, but its extremely high latent heat of evaporation has precluded its use up to this time. Nitrogen peroxide is quite unstable so is of little value outside of the laboratory.

This leads us to laughing gas, more properly called dinitrogen monoxide, or more commonly known simply as nitrous or nitrous oxide. As discussed in Chapter 15 nitrous is stored in a gas cylinder in liquid state at around 800psi at normal ambient temperature. When released from its pressure container it converts from liquid to vapour with an oxygen content of just over 36% by weight. On the surface this would seem to suggest that a one third power increase could be expected. In small doses that is about the minimum power increase. At maximum sustainable concentration levels the power increase will be at least double that, i.e. 72%. This means 2-litre engines making 500hp, while monster 10-litre Pro Street drag engines produce well over 2000hp.

Please don't read into my comments on the two previously mentioned conventional methods that I am advocating forced induction as the superior method for enhanced engine performance. Such is not the case. Forced induction is simply another route to more horsepower and, as you will discover in later chapters of this book, just as inept hot-up tuning can produce extremely disappointing performance results so it is with the forced induction route. In fact, there are far fewer true forced induction experts in the world and because, unfortunately, the pseudo-experts blow a disproportionate amount of hot air the potential for disappointment is indeed great. However, this and other good books on the subject, plus a working knowledge of the general principles of engine operation and tuning, will enable you to know exactly what you should want and expect from your planned modifications, and will arm you with the questions to ask to test out advertising claims.

Remember, knowledge is power. Do take the time to learn. Start out with small aspirations. Learn what you can from that experience before moving on to more complex projects. Too often I see young people, their pockets loaded with money and their heads loaded with ideas, but with no real understanding of why they want their desired level of performance, what they will do with it once they've got it and how irritating, impractical or even downright dangerous it will be for day-to-day street use.

I can't cover *everything* you need to know about engines in this book. I don't know it *all* either, and I fully intend to keep on learning years into the future. Consequently, I recommend that you read this book along with another of my books, either *Four-Stroke Performance Tuning* or *Modern Engine Tuning*. If you want to expand your knowledge and understanding to the maximum, and if you are after the very highest levels of performance for the road or for competition, go for *Four-Stroke Performance Tuning*. For those who are not into the science of getting power from an engine, but rather want to quickly pick up the basics of engine tuning and modification, with the emphasis on street-able horsepower, then *Modern Engine Tuning* is for you.

Chapter 2

Learning from the past

Probably because we live in a high pressure must-have-it-now society, many live by the maxim 'forget about where you have been, just concentrate on where you are going.' Basically, this is a pretty stupid philosophy, particularly when it comes to technology. I find that a regular glance into the rear-view mirror keeps me out of a lot of trouble. Without doing that we would regularly waste considerable time and resources reinventing the wheel. A better maxim is 'learn from other peoples' mistakes and successes'. That means not wasting time and money making the same errors others have made, or running into the same developmental cul-de-sacs. It also means having a good careful look at what others who are successful have done or are doing. Ask why they did it that way, what are the pros and cons? If you simply focus on getting the ball, from directly in front of the goal, across the goal line you will not succeed, particularly when tuning problems come up, as undoubtedly they will. Rather you must thoroughly learn every move, and every counter move for that matter, in moving the ball from the middle of the field right up and across the goal line. That way, when things don't work out as planned, you will be able to figure out why, and then move on by working out a solution.

A word for looking into the past is history. Many are bored by it, but the history of forced induction is fascinating, exciting and stimulating. It will also help you develop a healthy respect for what was achieved in the past under pretty difficult circumstances, and with that under your belt you will be far less dismissive of so called 'old technology' and you will be less likely to gullibly believe the line that new technology is always superior.

The history of supercharging dates back to the nineteenth century. Less than 20 years after the first successful four stroke internal combustion engine, Rudolf Diesel, in 1896, introduced his supercharger design. The idea of using exhaust gas to drive the supercharger followed within 10 years when Dr Alfred Büchi filed a patent for a design which proved to be the forerunner to the turbocharger.

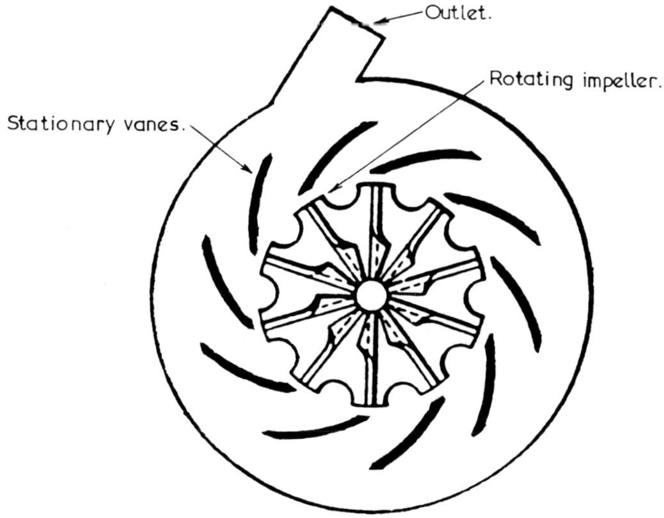

Fig. 2.1. The centrifugal blower.

BLOWN RACE CARS

The first blowers, as I prefer to call mechanically driven superchargers, were a centrifugal type (*Fig. 2.1*) driven via gears off the front of the crankshaft or by a belt off the flywheel at around five times engine speed. By 1910, race cars which were required to hold together over rugged 200-mile road races regularly exceeded 100mph.

Not long after that the world was at war. Initially, Roots-type positive displacement blowers (*Fig. 2.2*) with a history dating back to 1848 were pressed into service on fighter planes, but were soon ditched in favour of the more efficient centrifugal type. Rolls-

Fig. 2.2. The Roots positive displacement blower.

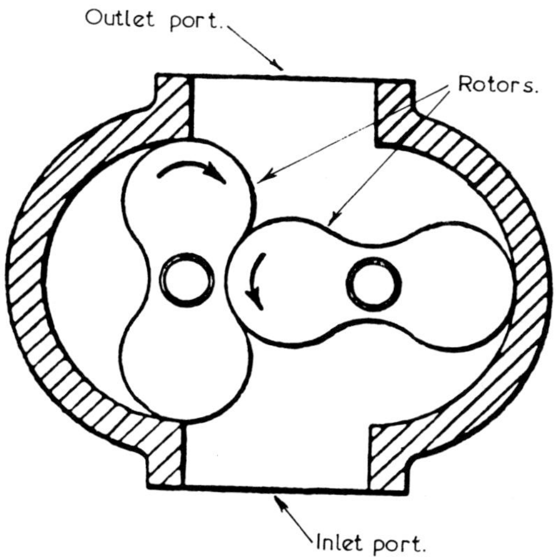

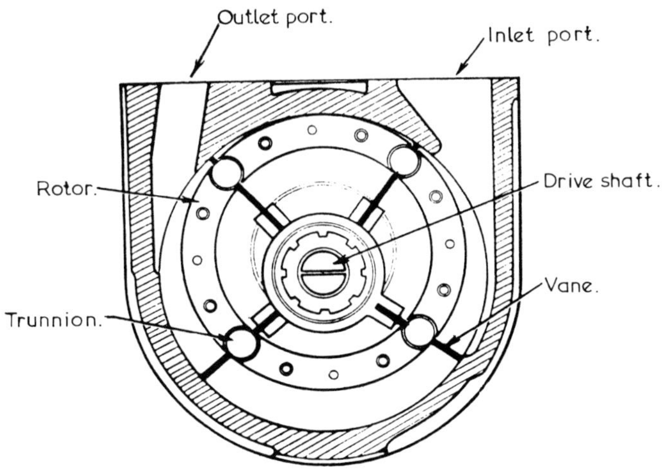

Fig. 2.3. The vane blower.

Royce continued development of the latter for aeroplane engine applications and eventually their design incorporated a complex 3-speed gearbox drive coupled to a dual stage blower.

In 1921 Mercedes supercharged a 1.5-litre 4-cylinder twin cam sports car for the Targa Florio competition. They chose a Roots-type blower operating at 7psi boost. Of main interest, though, was the fact that, by means of a cone type clutch, the blower engaged only at full throttle.

The first Grand Prix racer to wear a blower was the 2-litre straight-eight Fiat in 1923. It used a vane type blower (*Fig. 2.3*) and, like Mercedes, they chose to employ part-time drive. However, this decision may have been in the interests of keeping the blower functional for the full duration of a race, as this design is prone to severe lubrication problems, and these problems were not overcome until many decades later. At any rate, Fiat soon ditched their Wittig vane blowers and swapped to the Roots type, but they introduced a new factor into the equation, namely charge air cooling.

Within a year or two, other Grand Prix engines also sported Roots blowers, notably Sunbeam and Alfa Romeo. The straight-eight Alfa reportedly developed 135hp at a then extremely fast 5500rpm!

The Europeans weren't the only innovators. In the USA the 1924 Duesenberg, which won the famous Indianapolis 500, had its 2-litre straight-eight engine pressurised by a centrifugal blower spinning at 40,000rpm at an engine speed of 5000rpm. What was unusual was the position of the carburettor and the type of fuel. Whereas all previous designs had blown through the carburettor, they instead chose to put the carb in front of the blower and suck through it. Running on methanol, this produced greatly improved blower efficiency by cooling the air. In Roots blowers the fuel improves the seal between the rotors themselves and between the rotors and the blower case. The result being more boost at lower blower speed, which in itself cuts the amount of heat put into the intake charge.

Back in Europe, and in particular Germany, Grand Prix engine development was continuing apace. Just prior to 1940 both Mercedes and Auto Union were locked in combat for race supremacy. The 6-litre V16 Auto Union was reportedly putting out

Right: The 2 litre 4-cylinder Mercedes which competed in the 1924 Targa Floria, with Roots blower driven off the snout of the crankshaft through a throttle linked clutch.

Below: This 1936 rear-engined V16 Auto Union Grand Prix racer used a Roots blower drawing through a huge carburettor.

520hp at 5000rpm. This was using 26psi boost and at the time a very high 9.2:1 compression ratio on methanol. The Mercedes contender, like Auto Union, also used a two-stage Roots blower but Mercedes were running a higher 38psi boost to produce 646hp.

Alfa Romeo also continued to campaign blown 1500cc race engines. Before World War II they were running engines at 12psi boost, but after the war development continued and by 1951, in their last season of Grand Prix racing, the same engine was operating at 43psi boost and producing 425hp at 9300rpm.

The era of blown Grand Prix engines closed in 1951 with the unsuccessful 1500cc V16 BRM. It was powerful but fragile racing alongside the 4.5-litre naturally aspirated Ferrari. With an extreme boost of 72psi from a dual stage Rolls-Royce centrifugal blower, a 7.5:1 compression ratio and producing 600hp at 12,000rpm it was bound to be unreliable. Some readers may be interested in a few more of its specs: bore and stroke 55.56mm x 38.5mm, 31.75mm inlet valves, cam timing 55°/70° inlet and 70°/46° exhaust (99.75° lobe separation, 2.25° advanced).

BLOWN DRAG CARS

The history of blowers in competition, however, didn't end with the demise of the V16 BRM Grand Prix engine. In the USA a new form of competition, 1/4-mile drag racing was attracting attention. Pretty soon racers were bolting on Roots style blowers sourced from two stroke GM diesels. One GM engine line had an individual cylinder capacity of 71 cubic inches. The smallest 71 series engine had three cylinders, hence the designation 3-71. The blower from the 6-cylinder GM, the 6-71 was soon a favourite. Later tuners found that blower too small. They were having to drive it faster and faster to get sufficient air flow, so they swapped to the bigger 8-71 blower.

For 50 years drag racing has been the traditional home for Roots blowers.

Top Fuel drag engine specifications	
90º V8 Bore and stroke:	106.4mm x 114.3mm (8.13 litre)
Typical power output:	5500—6000hp at 7500rpm 4700—5000lb/ft torque at 6000rpm Maximum engine speed 8300rpm
Piston:	forged flat top with intake valve relief, weight 755gm staggered compression ratio from 5.8:1 to 7.5:1 piston pin 330g weight, 39.5mm dia x 83.8mm long x 13 mm bore piston rings top — Dykes section 0.028in end gap second — ductile moly faced 0.024in end gap oil — double rail expander
Cylinder head:	solid billet aluminium (no water jacket) inlet valve 60.3mm dia. titanium exhaust valve 50.2mm dia. Inconnel inlet port 58mm x 56mm wide exhaust port 38mm x 57mm wide head gasket 2.36mm copper plus 'O' rings each head secured by 17 x 9/16in studs tensioned to 125lb/ft
Valve train:	roller cam 9310 steel, 112º lobe centres, advanced 2º inlet lobe timing 61º/101º, 302º duration at 0.050in inlet valve lift 0.720in with 1.57:1 rocker ratio exhaust lobe timing 102º/54º, 296º duration at 0.050in exhaust valve lift 0.660in with 1.52:1 rocker ratio
Bottom end:	block — aluminium wet liner liners — cast ductile iron 4.5mm thick walls crankshaft — billet steel 31kg weight con rods — 7075 T6 forged aluminium, length varies around 177mm depending on compression ratio, weight 900gm
Fuel system:	blower — Roots type 14-71 with Teflon tipped rotors, 45psi maximum boost, driven 26% overdrive inlet — 3 x 127mm dia butterflies injectors — 10 nozzles into the blower 8 nozzles into intake manifold 16 nozzles into cylinder head pump — 2 each flowing 235 litre/min at 500psi fuel — 100% nitromethane at 1.3:1 air/fuel ratio
Ignition system:	magneto — dual, each generating 800 volt/44 amp primary output ignition advance — locked at 50º spark plugs — 2 per cylinder with 1.1mm gap plug wires — 8mm suppression
Exhaust:	8 pipes 64mm dia x 460mm long
Lubrication system:	sump — 11 litre wet type oil pump — 85 litre/min at 200psi oil — 70 weight

By now the favoured 426cu in Chrysler Hemi engines were making over 2000hp and 'Jimmy blowers', as GM blowers are fondly called, were being turned close to 15,000rpm, an incredible speed considering they were originally designed to be geared up to run a max of 5200rpm on GM diesels. Naturally, those high blower speeds weren't being achieved on stock blowers. Blower builders initially began rebuilding stock 'Jimmys' to more exacting tolerances. Then, as blower science developed, these former blower builders became innovators. By 1970 they were gutting the stock blowers and replacing the rotors and gears with parts of their own design. This allowed the high rotor speeds and the high boost pressures, which at times exceeded 25psi.

After another decade the innovators were designing and making their own blowers, patterned after the Jimmy Roots type. Boost pressures were now over 30psi and dragsters wore blowers with names like Mooneyham, with a size designation of 14-71. With these larger blowers, rotor speeds were reduced to under 11,500rpm and power climbed to 2500hp.

By 1995, 500cu in drag engine power had risen to almost 6000hp at 7500rpm. At launch these engines quickly get to over 8000rpm, but interestingly (because of Teflon rotor seals and other advances) the same blowers, 14-71s, with their 19-inch long rotors are pumping more pressure, over 40psi, at lower rotor speed, typically up to 10,500rpm.

ROOTS BLOWERS ON ROAD CARS

Naturally, all this development was not going unnoticed by blower manufacturers around the world. In the 1960s Eldred Norman in Australia, and Shorrock in the UK, turned their attention to overcoming the original heat and lubrication problems which Fiat had encountered 40 years earlier with the same vane-type blower. However, most attention was directed towards the Roots type, with particular emphasis on improving its efficiency so that it didn't put so much heat into the intake air charge. Also, to make it a viable option on road cars, the blower's noise level had to be reduced. Consequently, during the last two decades of the twentieth century a number of car makers, including some prestigious brand names, released production saloons (sedans) and coupés with Roots-type blowers strapped to the engine. The list includes Fiat/Lancia, Ford (USA), GM's Buick, Oldsmobile and Holden divisions, Toyota, Mazda, Jaguar, Aston Martin and Mercedes.

The majority of these car makers chose blowers made by Eaton who, with finances provided by a couple of the auto makers (primarily Ford), set about a major redesign of the Roots blower in the early 1980s. The three lobe rotors were twisted, screw like, through 60°. The rotors were also given a special coating which, along with improved manufacturing tolerances, meant that rotor clearance was able to come down to 0.2mm. This greatly reduced leakage through the rotors, improving pumping efficiency. The improved tolerances and material specs allowed rotor speeds to increase safely to 12,000rpm in most models, while the M92 developed for Ford can run up to 15,600rpm.

Along with these improvements Eaton also added a bypass valve (a simple butterfly valve) at the blower inlet. This allows the engine to draw air directly from the air filter, bypassing the blower at idle and at cruise when a vacuum exists in the inlet manifold. In turn, this bypass gave car makers the option of employing an electro-magnetic clutch

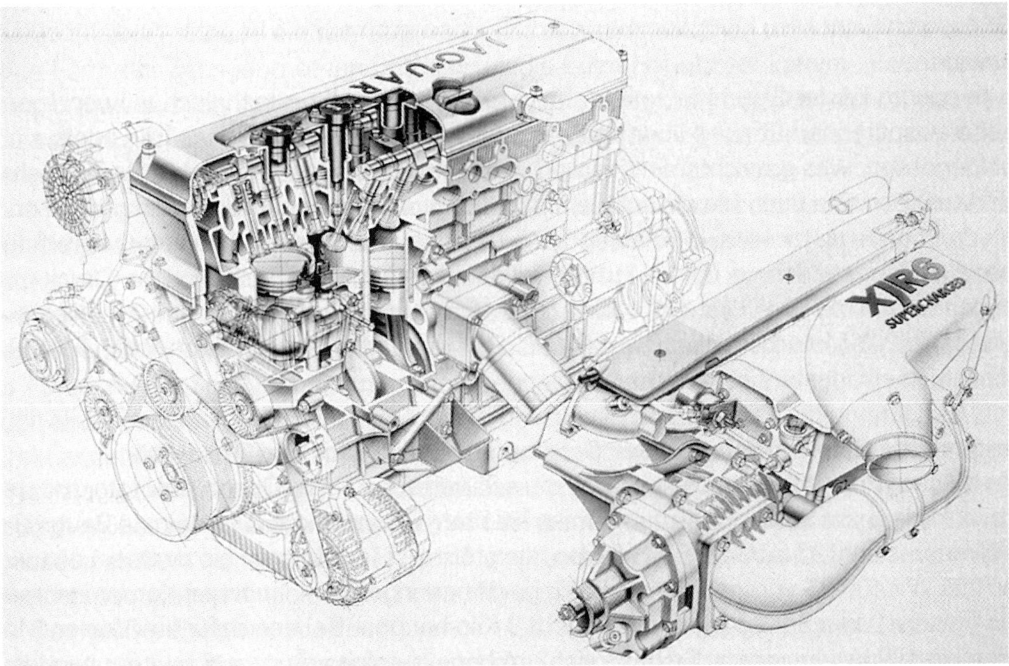

Above: Luxury car makers like Jaguar use Roots blowers in their high performance models.

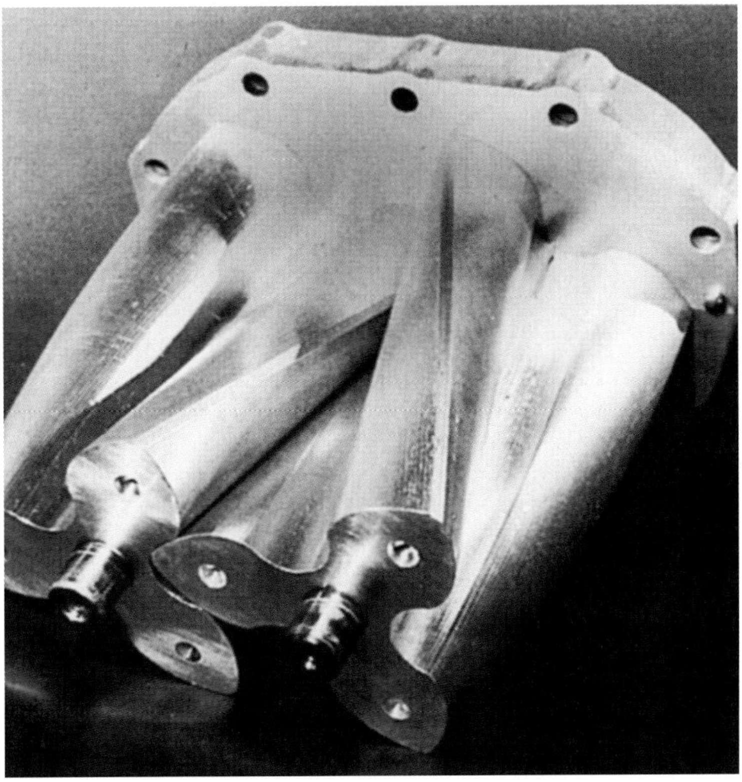

Right: Roots blowers are produced with a variety of rotor designs. The twisted 3-lobe design developed by Eaton reduces noise and charge pulsations.

(like an air-conditioner electro-magnetic drive) to disconnect the blower when extra power wasn't required. Toyota had long used this type of drive, and Mercedes followed the same path with their C230 model in 1996. *Fig. 2.4* illustrates how the Eaton blower at very low boost, only 7.3psi, raises the power of the Mercedes. Note how the power increases right from idle speed. This is a characteristic of all mechanically-driven superchargers, with the centrifugal type being the one exception.

CENTRIFUGAL BLOWERS

During this era of refining the Roots blower, two other blower designs were also attracting their fair share of interest. In the USA two distinctive camps were emerging; one group favoured the big, heavy, noisy, inefficient GM diesel blower, while the other camp was looking for something smaller, lighter and more efficient. What they settled on was the centrifugal blower. The compressor half of a turbocharger is a centrifugal compressor, so with all the developmental attention and dollars that had been poured into turbos it was quite simple to copy existing turbo compressor technology and come up with a good compressor design. Then it was just a matter of adding a suitable drive to spin the blower at extremely high speed (usually about 7 to 15 times engine

Fig. 2.4. The Roots blower provides a power boost from very low engine speeds.

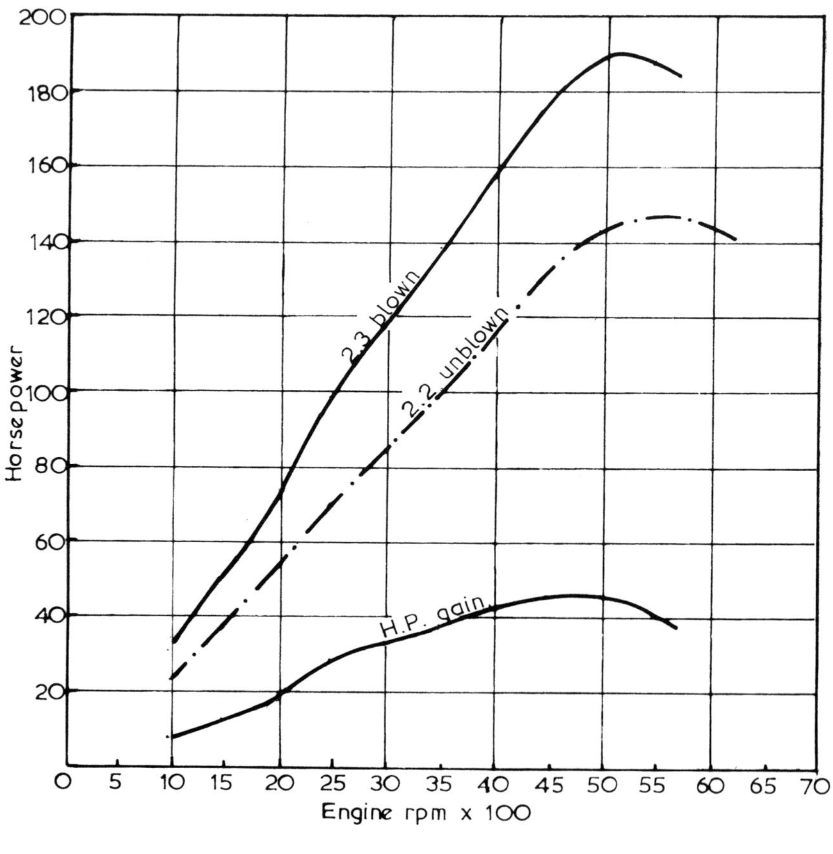

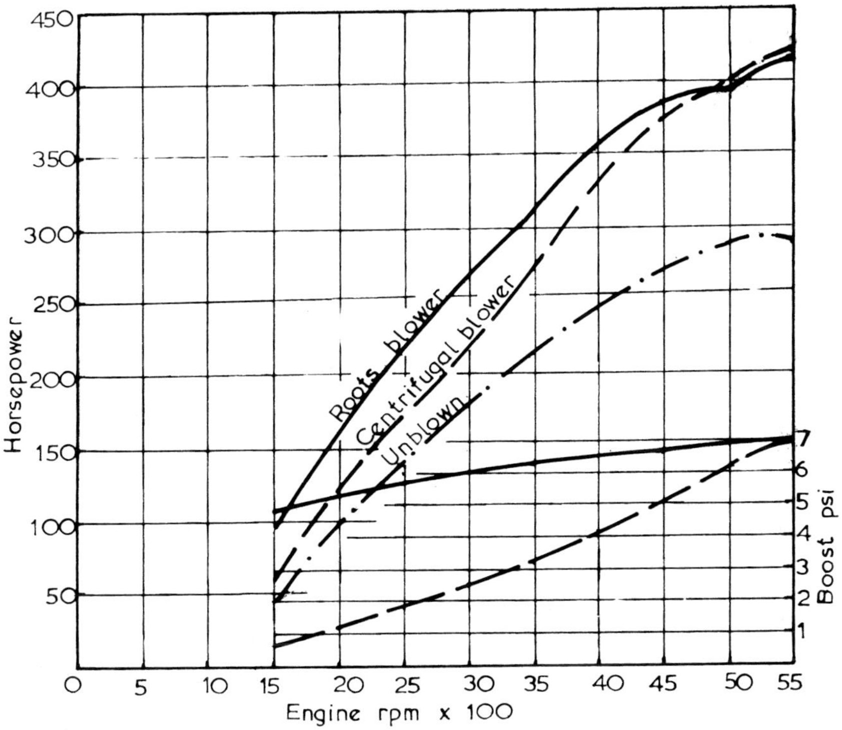

Fig. 2.5. The centrifugal blower is more efficient at higher speeds, but gives lower boost and smaller power gains at lower speeds.

speed). Brand names like Paxton, Vortech and Powerdyne became well respected in this field.

Figure 2.5 illustrates what some consider to be a drawback with centrifugal blowers. Compared to the Roots type there is only limited boost, and power increases at lower rpm. However, some see this as a virtue because large power increases at low revs can greatly increase transmission wear, and maintaining traction can be a problem with larger engines like the 5-litre Ford Windsor used in this test. Of course, by spinning the blower faster, higher boost levels could be achieved at lower rpm, but then the boost would be too high for normal pump petrol further up the rev range. Fitting a dump valve (blow-off valve) to bleed off high boost pressure is impractical. It makes no sense to unnecessarily put extra heat into the intake air and burn extra fuel driving the compressor, only to dump excess pressure.

A viable solution to this problem would, of course, be a variable drive to spin the compressor very fast at low engine rpm to produce more boost and then progressively reduce the compressor drive ratio at higher engines speeds to avoid excessive boost pressures. If you have been paying attention to this history lesson you will recall Rolls-Royce achieved something similar during World War I by means of a 3-speed gearbox. Later, in the 1950s, Paxton used a variable speed V-belt drive referenced to the throttle in an attempt to overcome this problem. More recently the German company ZF, best known for its gearboxes, produced a complex electromagnetic 2.5:1 variable speed drive.

Left: Centrifugal blower with ZF drive and KKK compressor.

Below: This cutaway of an Opcon Autorotor screw blower shows the basic design of the Lysholm principle. Air enters through ports in the blower end-plate (far right) and exits through the two triangular shaped ports in the blower case.

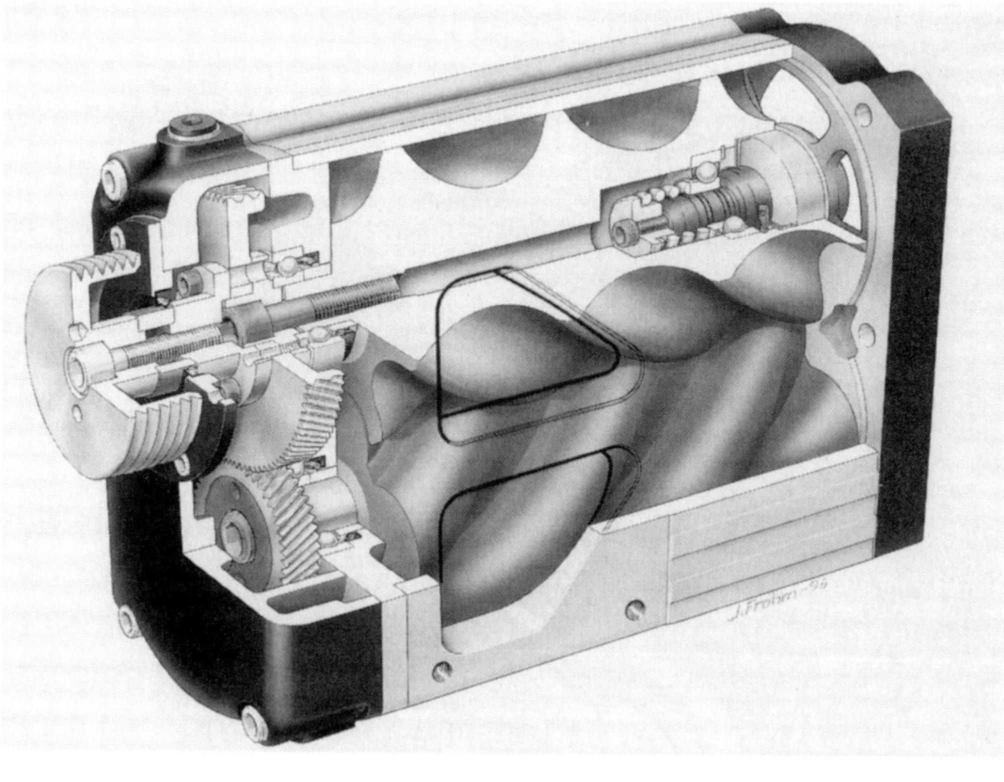

LYSHOLM SCREW BLOWER

While in the USA attention was mainly fixed on the centrifugal blower, in Europe various screw-type blowers were being investigated as a high-efficiency alternative. Compared to most other supercharger types, with family trees dating back 100 to 150 years, the screw-type blower is the new kid on the block. On the outside the screw-type looks like a Roots blower, but inside the two rotors are twisted into a complex screw pattern, more properly called a helix.

The idea of helixes within an enclosed chamber goes back to 1784 when Cameron, an employee of James Watt the father of the steam engine, patented the helix engine. Others adopted the idea of using a helix or helixes as a compressor but it wasn't until 1934 when Swedish engineer A. Lysholm patented the Lysholm screw supercharger that the idea was shown to be practical. Even so, at that time it was only suitable in industrial applications because oil was used to provide a seal between the helixes. However, in time a number of companies invested in machining the two rotors to much tighter tolerances to allow the rotors to run dry, at a clearance of less than 0.002mm, thereby allowing the screw type blower to be used in automotive applications. Fleming Thermodynamics of Scotland manufacture, and also license companies in other parts of the world to make, the well-known Sprintex; in Sweden Opcon manufacture the Autorotor, PSI Corporation in the USA and also the Whipple brand are well regarded. The Japanese company IHI, well known for its turbochargers, also make the Lysholm screw blower for the first production car ever fitted with this type of blower, the Mazda/Eunos 2.3-litre Miller cycle V6.

The screw-type blower appears to be the supercharger for the future. Its biggest hurdle at present is cost. However, the design has much in its favour. It has an efficiency of 70–85% (*Fig. 2.6*), which far exceeds the 35–60% of the Roots, and it is well in front

Fig. 2.6. The Lysholm screw blower is very efficient over a broad range of pressure ratios and blower revs.

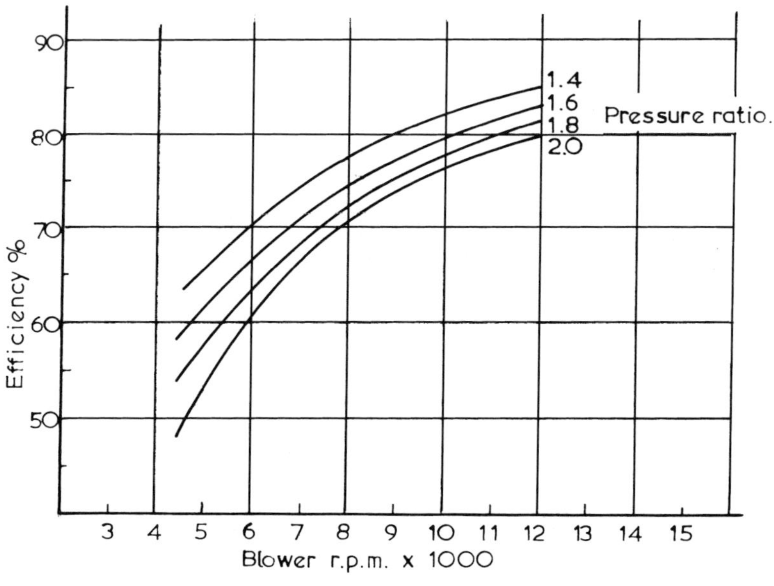

of centrifugal blowers too. Another plus is, unlike the Roots, it provides non-pulsing flow. It is lighter and smaller than the Roots and uses a lot less power to drive. However, I believe the major factor in the Lysholm's favour is its ability to produce substantial boost at low speed just like the Roots, but additionally it continues to provide boost at very high efficiency at high speed, substantially the same as a centrifugal blower or turbocharger. In effect, it crosses the boundary by giving the best of both worlds. It was banned from Top Fuel (nitromethane) class drag racing in the early '90s but PSI Corporation's biggest screw blowers are running at the front of the 500cu in 3000hp Top Alcohol class, and I do not expect that they will be challenged by an other class of supercharger.

Fig. 2.7 illustrates the power increase when a Whipple 1600A Lysholm screw blower was fitted to a stock 4-litre Ford. Even without any form of intercooling or exhaust modifications, the blower produced a healthy 41% jump in maximum power with boost kept below 6psi.

OTHER BLOWERS

Apart from the more common blower types already mentioned there are some others which will probably be abandoned, just as the vane-type was two decades ago. The German turbo manufacturer KKK, for example, produced the promising rotating piston Ro-charger, but not many have shown any firm interest in it. Similarly with the Comprex pressure wave blower developed by Brown Boveri who, incidentally, were probably the first to mass-produce turbochargers way back in 1923. Now called ABB (ASEA Brown Boveri), their Comprex blower was tried without any real success by Ferrari in 1981 during the Formula 1 turbo era. Since then it has been relegated to use on small diesels. Finally, there is Volkswagen's own blower, the G-lader spiral blower. It is based on principles originally patented by a Frenchman, L. Creux, nearly 100 years ago in 1905. VW used a version with 40mm wide spirals, the G40, on their small 1.3-litre Polo

Fig. 2.7. The screw blower provides impressive performance gains at all engine speeds.

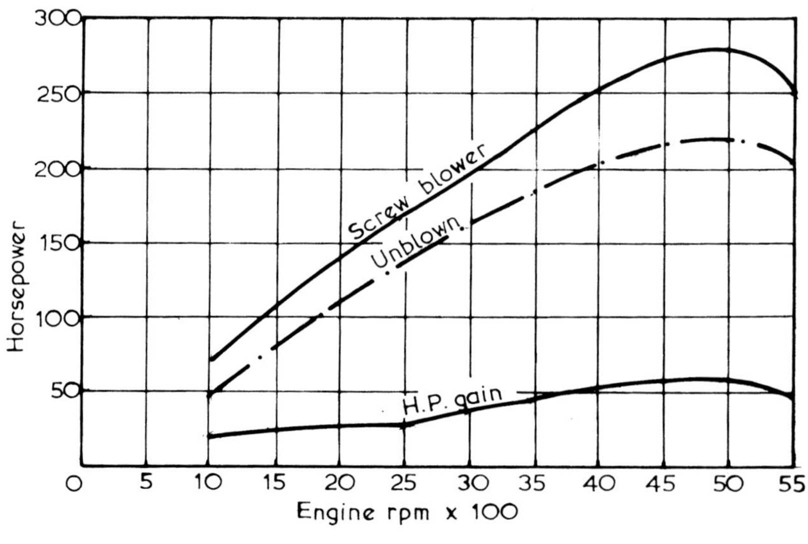

The Volkswagen G-lader blower showing the two case halves (top) that are fixed. The G-spiral (lower left) is oscillated within the case by the coupled motion shafts.

models and the larger G60, with 60mm wide spirals, on the 1.8-litre Golf with good success during the '80s and '90s but, unfortunately, interest in the concept has waned. This in spite of the fact that the G-lader is somewhat more efficient than the Roots at 55–68% and makes good levels of boost at low rpm. It would appear that a worldwide romance with turbocharging may have killed it.

Chapter 3

Turbocharging past and present

As mentioned near the beginning of the previous chapter, the history of the turbo goes back to 1905. Then, in 1917, Rateau experimented with turbos on French aircraft. A year later in the USA the General Electric Company were involved in the same pioneering work. By 1923 Brown Boveri were building the first exhaust gas driven superchargers, called turbochargers, as a regular production item. In 1925 Büchi discovered that dividing exhaust gas pulsations into split manifolds improved turbo efficiency.

Early turbo developments were mainly being directed towards diesel engines as the much higher exhaust gas temperatures of petrol engines quickly led to turbo failure. However, General Electric persisted and, by the beginning of World War II, famous American war aircraft such as the B17 Flying Fortress, the Liberator P38, the Lightning B24 and Thunderbolt P47 were successfully operating GE turbos at temperatures up to 800°C at turbine speeds of 22,000rpm. As jet engines began to take on a greater role in military aircraft some felt that it would lead to the end of piston engines and turbos. In fact the reverse happened. The development of high-tech metals that were both fatigue and temperature resistant as well as lightweight, for use in jet engines, opened the door to rapid advances in turbo development. Using similar materials and production processes enabled turbo manufacturers to make good advances in reliability as a first priority, and later to reduce spool-up time (lag) and enhance air flow efficiency.

TURBOS ON THE OVALS AND ROAD CIRCUITS

In 1952 the first turbocharged race car entered the Indianapolis 500, with a diesel engine! However, in the next decade turbo power came to the fore, and for more than 30 years, until an organisational change took place, the famous 'brickyard' reverberated to the sound of turbocharged engines – with the Offenhauser and the Cosworth Ford possibly being the best remembered. Prior to the imposition of boost limits the 2.61-litre

4-cylinder Offys in 1978 were producing over 800hp on methanol with an 8.1:1 compression ratio on boost levels of up to 45psi without intercooling. By the late '80s 2.65-litre V8 Porsche engines were producing 700hp at only 8.7psi boost on 9.5:1 compression ratio. A decade later the boost limit was lowered even further to only 5.5psi but, even so, engines such as that developed by Honda were producing in excess of 900hp. A few years later similar engines were approaching those outputs at even lower boost levels, 3.5psi on road courses and ovals and 3psi on the super speedways, still without intercooling being permitted. Clearly there is much more to high hp levels than merely turning up boost pressure! (Towards the end of 2001 boost was further reduced to just 2psi to cut car speed.)

Back in the early 1970s, Porsche introduced turbo power into sports car racing, first in the Can-Am series and later in Le Mans type racing. The mighty 5.4-litre 917 model produced 1200hp in 1974. However, what really led to rapid developments in both the design of turbos and their actual application was what took place in Formula 1 Grand Prix racing following Renault's appearance with a turbo car in 1977. That early attempt wasn't too successful but, even so, a little single turbo 1.5-litre V6 making 500hp was enough to make some sit up and take notice. Renault discovered big problems needing to be solved. Those same problems are still with us today. Charge cooling and engine cooling were the priority, but turbo lag, fuelling and ignition advance control were also major issues. By 1981, Ferrari with its own V6 engine and Toleman with a 4-cylinder Brian Hart engine entered the fray. In 1982 the BMW 4-cylinder contender appeared, followed by Alfa Romeo's V8, Honda's V6 and the TAG Porsche V6 in 1983. Latecomers were the V6 Motori Moderni in 1985, the V6 Cosworth Ford in 1986 and in 1987 Alfa made another attempt with a twin turbo 4.

The high power outputs of this era are legendary, with claims that most F1 teams were running in race trim in 1984 with 750hp at their disposal when needed. However, during qualifying when there was no requirement to restrict fuel consumption and very high boost levels of up to 65psi could be ulitised, even if only for a few seconds, power levels of around 1200hp are reported in 1987 for those teams using BMW, Honda and

Porsche V8 Indy engine.

The flat-12 Porsche 917 engine featured driver-adjustable boost control and blow-off valves in the inlet ducts to dump boost and keep up turbo speed whenever the throttle was quickly closed.

Ferrari power. On legal fuel the long stroke RF15 Renault qualifying engine was making the same power a year earlier.

However, rather than salivating and speculating over the power levels, it is much more relevant to look at some of the technical aspects of these F1 engines. The table opposite provides general specifications, but it is other details which give us greater insight into the challenge of making high hp not just for a few seconds, but for around the two hours of a Grand Prix event.

FORMULA 1 ENGINE DEVELOPMENT

For example, some teams tried alloy engine blocks, then ditched them when they found the cylinder walls bowing under pressure. Cylinder distortion caused gas blow-by, resulting in lost performance, and when pistons and piston rings lose contact with the cylinder wall they cannot shed their heat loads and are either damaged or destroyed. More often than not the latter. Renault used cast-iron blocks for qualifying, with the boost

Formula 1 engine specifications[1]

	Alfa Romeo	[2]Alfa Romeo	[3]BMW	[4]Ferrari	[5]Hart	[6]Renault	[7]Renault	TAG Porsche
Engine type	90° V8	in line 4	in line 4	120° V6	in line 4	90° V6	90° V6	80° V6
Bore & stroke (mm)	74.0 x 43.5	92.3 x 56	89.2 x 60.0	81.0 x 48.4	88.0 x 61.5	86.0 x 43.0	80.5 x 49	82.0 x 47.3
Stroke:bore	0.588	0.607	0.673	0.598	0.699	0.5	0.609	0.577
Compression ratio	7:1	7.2:1	6.7:1	7:1	6.8:1	7:1	7.5:1	7:1
Included valve angle	25°	25°	40°	38°	40°	22°	22°	30°
Horsepower (@ 1.7–2.0 bar boost)	745 @ 11,300rpm	885 @ 10,500rpm	840 @ 9,500rpm	870 @ 10,800rpm	710 @ 10,000rpm	850 @ 11,000rpm	870 @ 10,650rpm	860 @ 11,000rpm
Bmep @ max hp	571.4psi	730.5psi	766.3psi	698.2psi	615.3psi	669.7psi	708.0psi	677.6psi
Turbocharger	2 x KKK	2 x Garrett	1 x KKK	2 x KKK	1 x Garrett	2 x KKK	2 x KKK	2 x KKK

Other details:

[1]1988 rules limited boost to 2.5 bar and 150 litres of fuel. Previously, teams used up to 3.6 bar with 195 litre fuel limit.

[2]Exhaust ports divided to feed both turbos. Maximum torque 470lb/ft @ 8500rpm (bmep 775.3psi). Maximum engine speed 11,500rpm.

[3]Production car block (cast iron). 35.8mm inlet and 30.2mm exhaust valves. Titanium con rods 153.6mm long. Up to 1500hp in qualifying.

[4]Earlier engines used 6.5:1 compression ratio.

[5]Compression ratio varied between 6:1 and 7:1 over the years.

[6]Cast-iron block for qualifying.

[7]Later long stroke engine, designated RF15. Around 1200hp in qualifying on legal fuel.

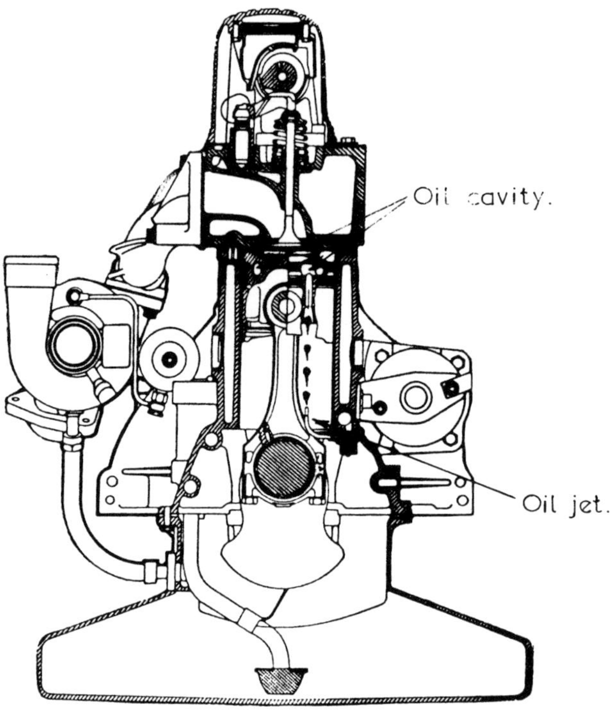

Fig. 3.1. Piston cooling by means of oil spray into the piston crown.

wound right up, and lighter alloy blocks for racing. The two most powerful engines, the Honda and the BMW, both used cast-iron blocks. TAG Porsche took a different approach. Their all-alloy engines used 5-ring pistons. Undoubtedly the two additional rings reduced gas leakage down the Nikasil cylinder liners, and the additional ring face area in contact with the cylinder wall provided a physically larger heat transfer path.

Another approach to piston cooling is to use the lubricating oil. The most common method by far is to tap a series of spray nozzles into the main oil gallery so as to direct a continuous spray of engine oil to the underside of the piston crown (*Fig. 3.1*).

OVERCOMING HEAD GASKET PROBLEMS

At elevated power and boost levels, head gasket integrity is always a worry. If a head gasket is properly clamped between the head and block, usually only one thing will destroy the seal and gasket – detonation. Unfortunately, when you are running at the very edge, detonation is a part of life. Brian Hart's F1 engines got round the problem by eliminating the head gasket! Hart followed the lead set by Offenhauser decades earlier by casting the head as part of the block. Incidentally, Porsche followed a similar path by welding the head to the barrel on their sports car race engines.

BMW accepted that gasket failure at times would be inevitable, so they took steps to enable their engines to keep running. What they did was to mill tiny slots in the deck of the block from close to the edge of the cylinder walls to the outside edge of the block. Therefore, in the event of head gasket leakage, the hot combustion gases from the

When BMW went racing in Formula 1 using a production car block, they machined these slots in the deck to provide an escape route for combustion gases in the event of head gasket 'O' ring failure.

offending cylinder would be vented to the atmosphere. Without the slots, those hot gases would blow into the water jacket and the engine would soon be destroyed by coolant loss. The alternative is that these gases would blow across to the adjoining cylinder. With two cylinders gone the engine would be right down on power, and soon those adjoining cylinders would be detonating and the engine would be destroyed.

Unlike BMW, who were using a production car engine block (suitably seasoned in a curing oven and stiffened) with block to head water passages, other engine designers chose the dry deck route. A dry deck means that there are no water passage connections between the face of the block and the face of the head. If a designer chooses for some reason to flow the coolant from the head to the block, or vice versa, there will be external plumbing connections. Consequently, a dry deck, apart from making a head and block much stiffer, and thereby less likely to suffer distortion which may cause a head gasket to blow, does not permit gas leakage into the cooling system. Of course, if a gasket does fail there is still the possibility of it blowing between cylinders and of subsequent engine failure.

SPECIAL FUELS DEVELOPED

Another way to beat head gasket failure is to eliminate engine detonation. With that in mind, Ferrari engineers and their fuel supplier Agip developed a system of impacting water into racing fuel. Why not use a simple water injection system?, you ask. Water injection will suppress detonation both by lowering the combustion flame temperature and taking the edge off combustion flame speed. However, because it is not possible to obtain a state where at the beginning of every combustion event the water is evenly dispersed throughout the inducted air/fuel mixture, you are faced with a dilemma. If you play it safe and ensure every part of the combustion space has sufficient water droplets, other areas will be oversaturated and combustion will progress slowly, consequently power will be down. If you go for high hp on every combustion stroke and reduce the quantity of water injected, you run the risk that there will not be an adequate quantity of water vapour in one or more areas of the combustion chamber. Those water-lean-areas then have the potential to detonate, and when they do it will trigger simultaneous detonation of all the air/fuel mixture.

By developing a process which enabled a tiny water droplet to be encapsulated in a larger fuel droplet, Ferrari were able to get the best of both worlds – high power levels and freedom from detonation. Another benefit, according to some engineers, was that as combustion proceeded the water droplets would burst, breaking the fuel droplets into smaller more combustible particles. At any rate, at the time, it was clear Ferrari had a significant power advantage, and the trick fuel technology may have been a major contributor.

Soon, though, the regulations changed. Refuelling during the race was no longer allowed and fuel capacity was progressively reduced to 250 litres, then 220, then 195 and finally 150 litres. Obviously the teams could not waste space with water (water was counted in as part of the fuel load) so the race was on to obtain more power on less fuel, and avoid detonation. The rules limit the fuel to a maximum of 102 octane as measured by the Motor method. After much experimentation, Honda found that the best fuel comprised 90% toluol with the remaining 10% consisting of various substances which

The extremely successful TAG Porsche Formula 1 engine.

would contribute to detonation suppression and fuel economy, but importantly they had to lower the octane level down to 102MON. The resulting fuel was heavy, with a specific gravity of over 0.85. Avgas, by comparison, is only 0.69 so, theoretically, 150 litres of this stuff was like having 185 litres of Avgas, plus true detonation resistance more like a 118 octane brew.

Another way to keep detonation at bay and ensure maximum charge density is to employ air charge cooling. Quite simply, compressing air adds heat and even a very efficient compressor, such as is used on a turbocharger, adds considerable heat. Consequently, all the F1 turbo cars used an elaborate system of intercoolers to take away much of that heat added by the compressors. The Honda, Renault and TAG Porsche engines used two air-to-air intercoolers plus a pair of water-to-air intercoolers.

OVERCOMING TURBO LAG

Turbo lag was always going to be a problem. Big power requires big turbos and it takes much more exhaust energy to get a big heavy turbine wheel up to high rpm rapidly to make good boost. Before the need to restrict fuel consumption arose it was common to see F1 cars accelerating out of corners with plumes of black smoke coming from the exhaust. Some assumed they were running rich to cool the piston and combustion chamber. The truth, in fact, was more closely related to the anti-lag systems being employed.

The joke in the pits was that some teams had developed a five stroke engine. This expression was borrowed from the earlier pre-World War II blower Grand Prix car era. In its most basic form the engine management system was programmed to dump a lot more fuel into the engine when the driver lifted and braked for a corner. Simultaneously, the ignition would be retarded enabling pretty much unburned mixture to pass into the turbo to be ignited by the extremely hot 1000ºC turbine wheel. The resulting burn created lots of black smoke, but it rapidly accelerated the turbo so that as the car clipped the apex of the corner boost was already building and at the exit the engine was at full boost. Another approach, tried by Renault, which produced a similar outcome was to place a fuel injector directly into the exhaust manifold.

Ferrari, though, experimented from a different angle. They adopted a system whereby the inlet and exhaust manifolds were joined via a bypass passage containing a valve which opened as the throttle was closed. Therefore, any time the driver lifted, compressed air from the inlet manifold rushed into the turbo. With excess fuel available, plus a red hot turbine wheel and now air being dumped into the turbo, enormous gas flow was created to quickly accelerate the turbo to full boost. Drivers loved the throttle response with almost a total absence of throttle lag. However, turbine materials weren't up to the excessive temperatures generated by such violent afterburn and a string of turbo failures saw Ferrari investigate other ways of minimising turbo lag.

Soon after commencing its F1 engine development programme, Renault discovered that one large turbo took too long to spool up, and it began trialing a system based on two smaller turbos with less inertia. What made the situation even more difficult for Renault was that their engine was actually a destroked version of their winning 2-litre F2 race engine. With such a short stroke the engine was lacking in low rpm torque and, with turbo lag added, the problem was enormous. Then came the infamous afterburn fifth stroke era, but for the 1986 season Renault moved to a new engine with smaller bore and a longer stroke.

BMW, though, stuck with a single turbo. Being based on a production block, their engine already had a relatively long stroke in F1 engine terms, so they had nowhere near so much trouble with lag. From the beginning they raced with split pulse turbine housings which turbo pioneer Büchi discovered nearly 60 years earlier provided superior response and spool-up times.

Towards the end of the F1 turbo era, turbo manufacturers Garrett, KKK and IHI were all able to supply variable geometry turbines that could survive in the very hostile petrol engine racing environment. As explained later, variable geometry turbines help engine performance all the way through the engine rev range. At low rpm they function like a small responsive turbo, but as engine speed increases the geometry changes allowing the turbine to flow like a much larger unit. The result is more air flow at low rpm and little exhaust restriction at maximum revs. Previously this technology had been restricted to diesels with their much lower exhaust gas temperatures. However, the IHI turbos on the Honda engine, with turbine wheels made of silicon nitride could survive turbine speeds of 160,000rpm and exhaust gas temperatures of up to 1170ºC. To gain some indication of how much air a pair of RX6 turbos could pump into the little Honda, consider that the turbine diameter was 62.9mm while the compressor wheel was 82mm diameter.

RALLY ENGINE DEVELOPMENT

When considering turbo competition engines it is easy to focus on the heady days of development within the exclusive confines of the F1 club and neglect the contribution made on the world rally scene. Turbo pioneers SAAB led the charge with their works car producing 270hp in 1981. Renault, though, were carving up the field in tarmac rallies with their R5 car using their vast Formula 1 engine knowledge. Running just 20psi boost the 1.4-litre long stroke (76mm x 77mm) 2-valve pushrod engine made 300hp at 6800rpm in the 1983 Tour de Course. A year later, with engine capacity at 1.5 litres (77mm x 82mm), the hp climbed to 350hp at 6500rpm with 26.7psi boost, but significantly torque rose from 246lb/ft at 5000rpm to 324.8lb/ft at the same rpm.

The stock Renault R5 engine. Note the well formed exhaust manifold, and how a tank on the air-to-air intercooler is utilised as the inlet manifold.

Renault R5 specifications		
	Road spec	Maxi 5 tarmac rally spec
Bore and stroke	76mm x 77mm	77mm x 82mm
Capacity	1397cc	1527cc
Compression ratio	7.0:1	6.7:1
HP	160 at 6000rpm	350 at 6500rpm
Torque	154.5lb/ft at 3250rpm	324.8lb/ft at 5000rpm
Valve timing	28°/52° inlet 66°/14° exhaust (cam advanced 7°, lobe separation 109°)	57°/63° inlet 83°/37° exhaust (cam advanced 10°, lobe separation 103°)
Ignition advance		28° at 5000rpm 34° at 6500rpm
Supercharging	Garrett T2 turbo	Garrett T31/T04 turbo
Maximum boost	12.5psi	26.85psi
Intercooling	1 air-to-air	1 water-to-air plus water injection
Exhaust manifold	4-2 (split pulse turbo)	4-2 (split pulse turbo) 32mm id pipes x 500mm long

This later engine used what Renault called DPV (Dispositif Pre-rotation Variable) – a fancy name for what was basically a variable nozzle in the compressor inlet. Mechanically linked to the throttle, it changed the compressor inlet area and the direction at which the inflowing air hit the compressor wheel, according to throttle position. Therefore, when the driver lifted off the throttle the DPV closed more rapidly than the throttle ensuring the turbo kept spinning fast in vacuum-like conditions. Then, as the throttle was opened, the DPV controlled both the air flow inlet area and direction to minimise compressor wheel speed loss. Consequently, Renault's F1 engine supremo, Jean Pierre Boudy (the man responsible for pneumatic valve control) could confidently say the drivers 'don't know what turbo lag is,' even though these engines were running relatively large Garrett turbos using T31 turbines and T04 compressors.

In 1985, a year later, Boudy went over to Peugeot taking the DPV system with him. Soon, though, the DPV was electronically controlled by the engine management system, allowing even larger turbos to be used. The Peugeot 205 in E2 guise made 435hp with maximum torque of 383lb/ft running 23psi boost and 1775cc engine size.

Water injection figured prominently during Boudy's time at both Renault and Peugeot. The water injection was programmed to trigger at very high boost, 22psi on the Formula 1 V6 and 20psi on the 1.5-litre Maxi 5 rally engine, delivering fine water particles (fog) at the intercooler outlet. Only very small quantities of water were injected; during race circuit testing of the Maxi 5, water consumption was in the region of 4 litres per 100km, and on tarmac rally stages this figure dropped to 2–3 litres per 100km.

Perhaps the best remembered cars from this era were the brutish Group B Audi four-wheel-drive monsters. Running 5-cylinder 2144cc engines, the works drivers had 400hp at their disposal in 1982. When the engine design was later changed to twin-cam 4-valve, power rose to 500hp. These later twin-cam engines were further developed for other motorsport programmes in the USA – IMSA sports racing and the Pike's Peak hill climb. The hill climb spec engine in 1987 was 2133cc (80mm x 84.9mm). Running 43psi boost, power was 720hp at 9500rpm with maximum torque of 428lb/ft at 7200rpm. The inlet system used three staged fuel injectors per cylinder. One set of injectors sprayed fuel onto the back of the inlet valves. The second set was located about midway between the valve seat and the face of the intake trumpets, while the third set was fixed to the air box lid and sprayed down the throat of the trumpets. The pistons were oil cooled, as were the exhaust valve seats.

Group B was killed off in 1986 when the sport's regulators deemed horsepower levels were too dangerous. However, before that time manufacturers had given a clue to where developments were headed.

Ford produced its RS200 mid-engine 4WD car, and to gain homologation had to build a number of road cars. While there was nothing especially innovative about the engines in these road spec cars, remember they were developed with input from Cosworth. At the time the Cosworth DFX turbo was the engine to have in Indy cars, so how the road cars developed 230hp from only 1.8 litres with 11.4psi boost and mild cams is of interest.

Ford RS200 specifications

	Road spec	Rally spec
Bore and stroke	86mm x 77.6mm	
Capacity	1803cc	
Compression ratio	8.2:1	7.2:1
HP	250 at 6000rpm	420 at 7000rpm
Torque (lb/ft)	214 at 4500rpm	354 at 6000rpm
Valve timing and lift	20º/56º 0.358in in 56º/20º 0.358in ex 0º advance, 108º lobe separation	
Ignition	Mapped digital	
Fuel injection	Mapped digital 2 injectors per cylinder	
Supercharging	Garrett T3/T04 turbo	
Maximum boost	10.9psi	17.4psi
Intercooling	1 air-to-air	
Exhaust diameter	80mm	100mm

LINKING A TURBO AND BLOWER

The situation was similar with the Lancia Delta S4. Like Ford, they too produced the required number of detuned road specials. However, their development exercise started from a base engine which was already blown with a Roots-type Volumex supercharger, and now they were adding a turbo as well!

As you will have gathered from earlier pages, the Roots-type blower produces good boost at lower engine speeds but gets pretty inefficient at higher rpm. Turbos are just the opposite, with big efficient turbos unable to produce any boost up to about one third

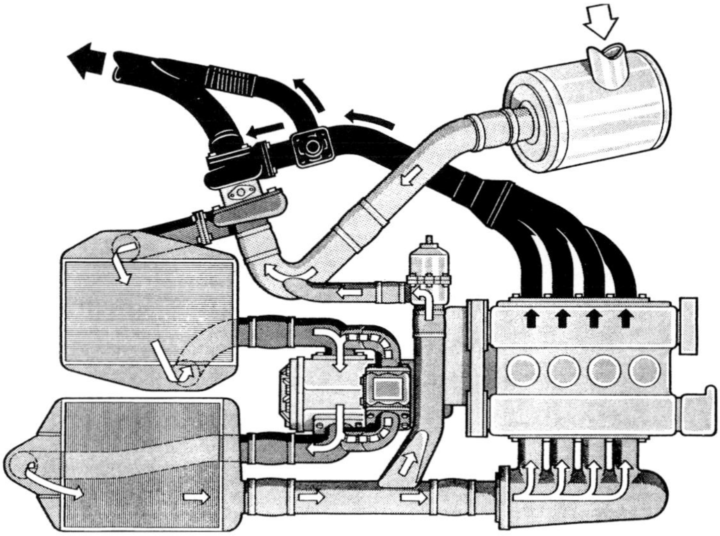

Above: A schematic diagram of the complex Lancia Delta S4 combined turbo and Roots-type blower supercharging arrangement.

Below: This cutaway drawing shows where Lancia managed to fit everything on the Delta S4.

maximum engine speed and huge amounts of boost at the top end of the power band. Lancia's aim was to combine both characteristics to produce good power and crisp throttle response at all engine speeds.

Naturally enough, linking both technologies into a workable package was quite a feat, and getting one's head around the complex and tortuous intake system is seemingly no less of one! Basically, the air flow path is as follows: the air flows from a high-mounted air filter down through the turbo compressor and on through the first of two air-to-air intercoolers. From there the flow path can take one of two routes, or even both routes, depending on turbo and blower outputs. Before the turbo begins to provide boost, the flow path continues on from the first intercooler through the second intercooler and the inlet manifold into the engine.

However, as engine speed increases and turbo boost begins to rise, a bypass valve at the blower progressively opens, bypassing some flow from the turbo directly to the second intercooler and into the engine. The bypass opens further as turbo boost rises, thereby progressively reducing flow through the blower. Finally, when fully open, all output from the turbo compressor bypasses the blower. With the blower now merely freewheeling, power loss to it is minimal at higher engine revs.

Lancia Delta S4 specifications		
	Road spec	Rally spec
Bore and stroke	88.5mm x 71.5mm	
Capacity	1759cc	
Compression ratio	7.6:1	7.0:1
HP	250 at 6750rpm	450 at 8000rpm
Torque	214lb/ft at 4500rpm	325lb/ft at 5000rpm
Valve timing	40°/68° inlet 66°/18° exhaust (advanced 5°, 109° lobe separation)	
Ignition	Mapped digital	
Fuel injection	Mapped digital	
Supercharging	Roots-type blower & KKK type K26 turbo	
Maximum boost	12.9psi	25.9psi
Intercooling	2 air-to-air	

SEQUENTIAL TURBOCHARGING

The next proposal for a Group S rally engine from Fiat/Lancia's engine tuning arm, Abarth, involved what Fiat in 1987 called 'modular turbocharging', but is now known as sequential turbocharging. Abarth quoted figures of 600hp at 8000rpm and 396lb/ft torque from the 4-cylinder 1759cc engine.

As well as an unconventional (at the time) twin turbo arrangement, Abarth also deviated from what most considered normal practice when it came to arranging ports and valves in the twin cam 4-valve head. Since inception, with few exceptions, all exhaust valves have been in line down one side of the head and operated by a single camshaft with eight exhaust lobes. The inlet side was similar, with all eight valves and inlet lobes arranged down the opposite side.

Abarth figured differently. Following the pattern of a previously unsuccessful layout BRM had earlier tried with its Formula 1 engines in the 1960s, the valves on one side were arranged inlet-exhaust-inlet-exhaust, etc., while on the other side the order was reversed to exhaust-inlet, etc. The set-up was right for a small engine with four exhaust ports on each side supplying exhaust flow to their respective turbo, while the eight inlet ports were in the valley between the two cams. Apart from making for easier plumbing to the two turbos the design also served to even out temperatures around the combustion chamber. Additionally, exhaust valve and valve seat temperatures are not such a worry (*Fig. 3.2*).

The conventional layout of both exhaust valves on one side of the combustion chamber creates a compound problem. First, it is more difficult to maintain the bulk of coolant flow down one side of the head to prevent head warpage and even out temperatures around the combustion chamber. Second, the thin peak of metal between the two exhaust valves is vulnerable to overheating and buckling. Likewise for the valve seats and the edges of the exhaust valves bordering that area, which leads to valve warping and burning. Of even greater concern, though, the heat in that area will heat the inlet charge and bring on detonation (*Fig. 3.3*).

While the Abarth sequential turbo system was at the time unconventional, it is by no means unique. Two years earlier, Porsche showed off its Group B 959 sports racer with a

Fig. 3.2. The Abarth twin turbo engine featured a cylinder head with diametrically opposed inlet and exhaust valves.

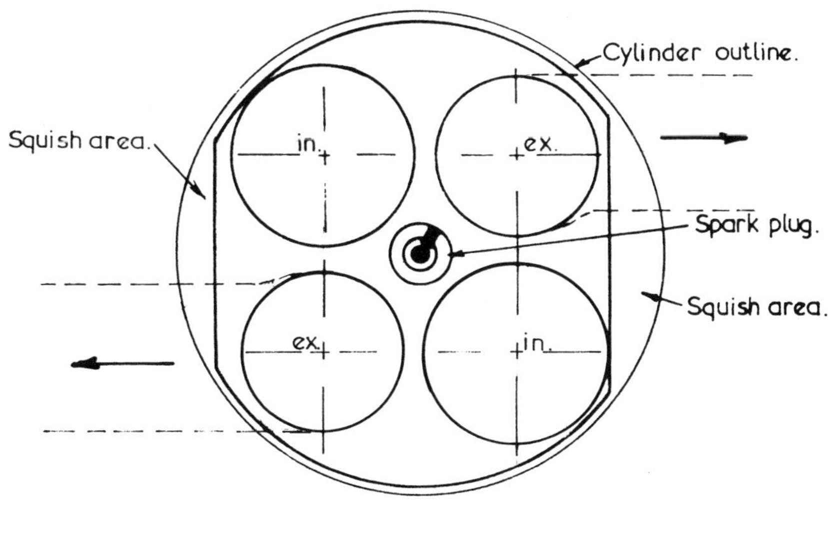

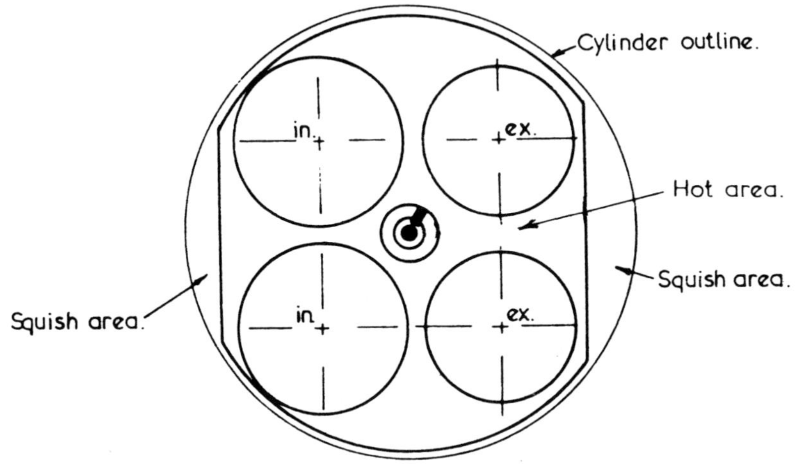

Fig. 3.3. In a conventional 4-valve combustion chamber the area between the two exhaust valves is vulnerable to overheating. Some manufacturers limit exhaust valve size so as to keep the valves well separated. In their 5-cylinder turbo competition engines Audi machined oil galleries around the exhaust valve seats to allow circulating engine oil to carry away heat.

Fig. 3.4. The Porsche 959 power curves illustrate how a well executed sequential turbo arrangement provides good power at lower rpm without limiting top end power.

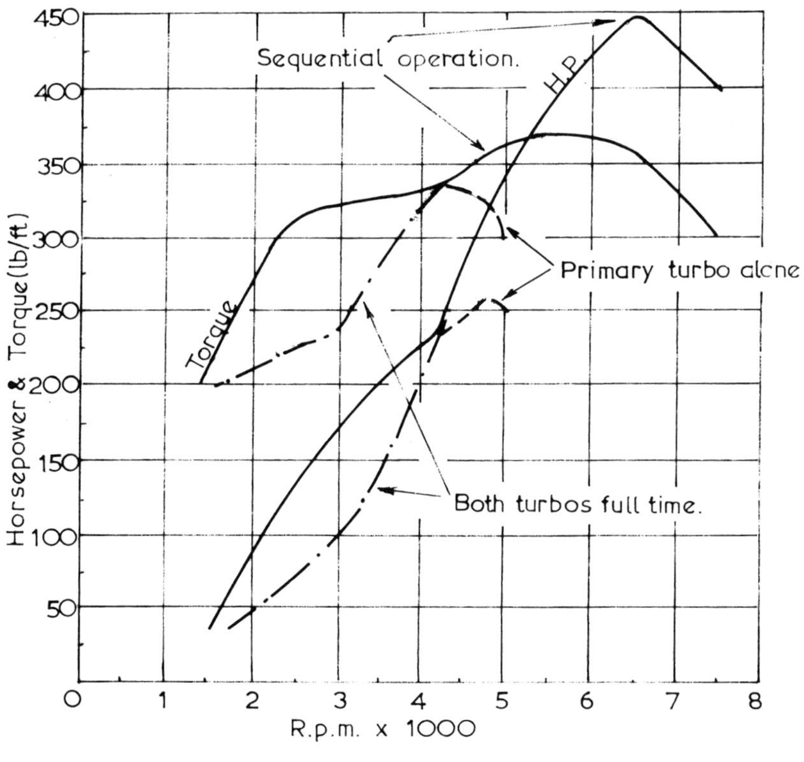

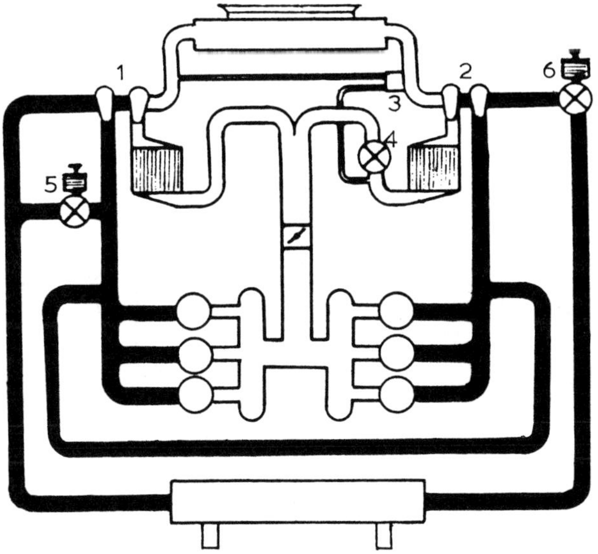

Fig. 3.5. Porsche 959 sequential turbo arrangement. The components illustrated are: 1. primary turbo; 2. secondary turbo; 3. vent valve; 4. strangler valve; 5. wastegate; 6. exhaust blocking valve.

similar set-up, and later Mazda and Toyota released production sports cars, the RX-7 and Supra, with sequential turbos as stock.

The idea behind the sequential turbo system is quite simple. A small turbo provides boost at low speed, then at higher speed exhaust flow is progressively diverted to a big turbo to maintain a high boost level and not restrict exhaust flow. *Figure. 3.4* illustrates the effect on the road version of the 2.85-litre Porsche 959 running 8.3:1 compression ratio and a maximum of 14.3psi boost. The small turbo, while providing good throttle response, severely limits maximum power and wastes fuel. On the other hand, running two turbos full time results in sluggish performance up to nearly 3500rpm, which is 45% of maximum engine speed.

While the sequential turbo concept is simple, the actual application is anything but, as *Fig. 3.5* illustrates. Basically, at low speed all exhaust flow is to the smaller left side turbo, and its wastegate is closed. Exhaust gas can back up to the right side turbo, but it can't flow through the turbine because of a valve, which is like a wastegate, blocking the exit pipe. Likewise on the inlet side, an IMPCO-like strangler valve prevents inlet charge flow reaching the compressor. (This is important because in the next phase that lack of inlet back pressure will allow the turbo to spin up to speed more rapidly.)

As engine speed and exhaust flow increases, the small turbo's compressor flow and boost will begin to peak (*Fig. 3.6*). In a normal system the wastegate would be actuated at this time, but in a sequential set-up the car's computer (ECU) will signal the secondary turbo's turbine exit valve to begin opening. This action permits flow through the right side turbo and allows it to spool up. However, with the inlet strangler valve still closed and the compressor vent valve (blow-off valve) open, it does not supply any air to the engine. All flow from the compressor goes round in a closed loop.

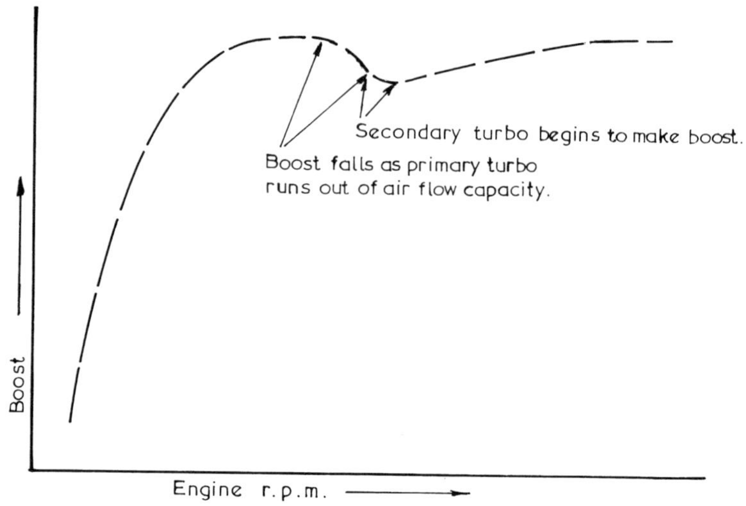

Fig. 3.6. Sequential turbo engine boost curve. As engine speed increases, the small primary turbo runs out of air flow capacity, so boost begins to drop. Then, as exhaust flow increases, the larger secondary turbo begins to make its contribution, and boost rises again as both turbos fill the engine's air flow needs.

Further increases in engine rpm and exhaust flow signals the ECU to fully open the secondary turbo's exhaust blocker valve and simultaneously close the inlet vent valve. The in-line strangler valve now opens as boost pressure builds up and the secondary turbo's compressor begins to supply air flow to the engine. As flow from this compressor continues to rise, boost pressure will again peak. The ECU receives the signal and pulses the wastegate to avoid an overboost situation. Should the wastegate fail, the ECU on the Porsche 959 will then activate a fuel cut to avoid dangerous boost levels.

ANTI-LAG DEVELOPMENT

After the decision to ban the very powerful Group B rally cars, steps were taken to limit horsepower to 300hp. This was achieved by reducing the diameter of the compressor's air intake, effectively limiting how much air the turbo can drag in and compress. While these mandatory 34mm restrictors did cut the power, a serious downside in the form of massive turbo lag arose. Overcoming the problem required very careful turbo matching, new turbo design technologies and effective anti-lag systems which did not compromise turbo reliability.

The early attempts of Formula 1 engineers at anti-lag controls were soon curtailed by fuel load regulations and turbo failures. By 1993, after nearly a decade of development, turbo technology and especially turbine material technology, had moved on. Now the turbine housing and wheel, made of more exotic materials, could survive at considerably higher temperatures. Previously, anything over about 950° spelled the death of the turbo, but in the early 90s the critical temperature had risen to nearly 1100°C. (Today, those with unlimited resources can obtain turbos which will live for short spells at up to 1250°C.) This, in turn, opened the possibility of utilising fuel afterburn in the turbo to keep it spinning at high speed any time the driver lifted off the throttle.

The workings of a typical anti-lag system are shown in *Fig. 3.7*. There are different ways of doing it, and driver preferences also influence the choice of method. Some drivers like an aggressive system with instant high boost the moment they get back on the throttle. Others prefer a softer response, particularly for mud or marbles road surfaces, so the system is tailored accordingly using a combination of hardware and ECU programming.

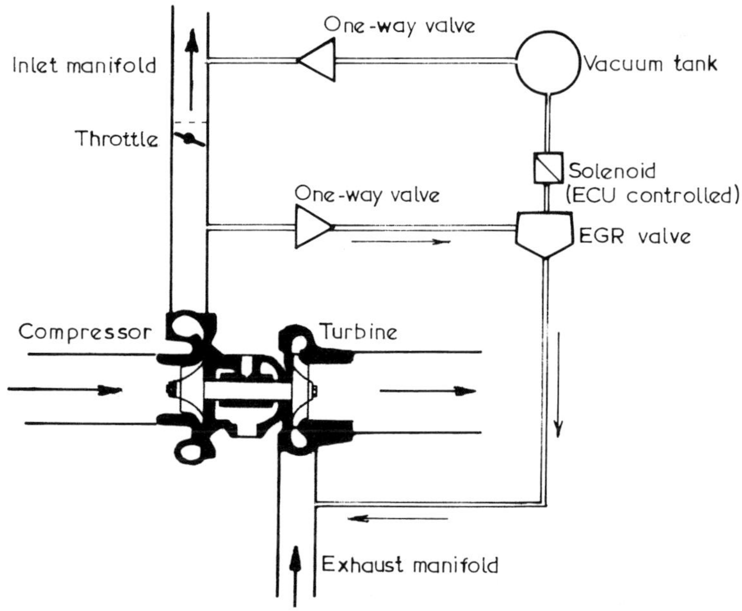

Above: Fig. 3.7. Anti-lag system works like a jet turbine, with air under pressure being fed into a fuel-rich and extremely hot exhaust manifold.

Right: EGR valve used to divert inlet air into the turbine for anti-lag operation.

A basic system, when activated, will instruct the ECU to increase fuelling by 15% at any time it senses a trailing throttle (less than $^1/_4$ open) at engine speeds over 2000rpm. The ECU will then take all spark advance out of the ignition (the ignition will revert to the base timing, often about 10° before TDC) and it will selectively cut spark duration to one cylinder at a time by 75% (a complete cut could cause plug fouling), allowing fuel to pass into the exhaust manifold unburned.

At the same time a suitably modified EGR (exhaust gas recirculation) valve opens in response to an ECU-controlled solenoid opening a valve between the EGR valve and a vacuum reservoir. With the EGR valve open, air under pressure from the compressor side of the turbo is forced into the exhaust manifold. Here, close to the hot turbine inlet, the unburned fuel is ignited by this blast of pure air.

This combustion, much like that occurring in a jet turbine, creates heat and pressure sufficient to spool the turbo to produce in excess of 20psi boost with the throttle still closed. As the throttle begins to open the boost quickly drops to less than 8psi. Once past $^1/_4$ throttle, the ECU discontinues anti-lag operation, and normal exhaust flow takes over to accelerate the turbo up to full boost of about 28psi at low rpm (the inlet restrictor causes the engine to run out of air at higher rpm so the boost falls back).

Subaru WRX specifications

	Road spec	Rally spec
Bore and stroke	92mm x 75mm	
Capacity	1994cc	
Compression ratio	8.5:1	9.2:1
HP	208 at 5800rpm	312 at 6400rpm
Torque (lb/ft)	199 at 4000rpm	431 at 3800rpm
Supercharging	IHI TD05H	IHI RB52
Maximum boost	11.6psi	28psi
Intercooling	1 air-to-air above engine	1 air-to-air in front of radiator with water spray of 30 litres per 40km special stage

CAR MAKERS ADOPT TURBOCHARGING

With all the focus on what has been achieved with competition power units, it is very easy to overlook or even be dismissive of the accomplishments of motor vehicle manufacturers when they turned their attention to producing turbo road cars. To do so would be short sighted – particularly if our prime interest is in street performance. The world's car makers invest millions getting what they perceive to be the right mix of performance, fuel economy and reliability, so there is much we can learn from them when we take the time to scrutinise what they have done, and where they have moved on to.

The first production turbo car came from General Motor's Oldsmobile division. It hit the streets in 1962 called the F85 Jetfire Turbo Rocket! The engine was from GM's Buick division – a 215cu in (3.5-litre) all-alloy V8 (later sold to British Leyland for use in Rovers, Range Rovers and the Australian P76). With a high compression ratio of 10.25:1 and a maximum boost of 6psi reached at 2200rpm it increased the power of the standard engine from 185hp to 215hp, but more importantly the torque figure improved by 74lb/ft at 3200rpm. The Garrett turbo featured a 61mm diameter turbine wheel made of nickel steel with stellite and cobalt added. The 63.5mm compressor impeller was die-cast alloy and the turbo shaft ran in plain bearings.

The carburettor was a special side-draught Rochester and, to prevent detonation, a water injection system was included. To avoid icing up it actually used a 50/50 water and alcohol mix, described by Old's as 'Rocket Fuel'. In hindsight, there may have been some type of special upper cylinder lubricant in the Old's Rocket Fuel because one of the big problems which gave GM executives a good deal of heartburn was cylinder bore corrosion, most probably caused by car owners not wanting to buy the expensive Old's brew but instead substituting a plain town water (frequently from underground supplies and so containing many minerals) and heating alcohol mix. The corrosive effects of alcohol and water on piston rings and cylinder walls are well documented, but even today the evidence is frequently ignored.

The water injection system was triggered to begin functioning as the boost approached 1psi. To avert an overboost situation in the event the wastegate stuck closed, a diaphragm-operated butterfly between the carby and compressor inlet would close, strangling airflow.

Within a few months, GM's Chevrolet division released the much more successful Corvair Monza Spyder. It had a 2.3-litre air-cooled flat six which in stock form made 90hp at 4400rpm. Adding a Garrett turbocharger providing 7–8psi boost raised this to 150hp at 4400rpm, with a torque figure of 214lb/ft at 3200rpm. This large power gain is as a result of a much larger turbo being used than on the Olds, even though the Corvair engine was only two thirds the size. The turbine wheel grew to 75.5mm and the compressor impeller diameter was increased to 76.2mm. With a few more modifications, power rose to 180hp in 1965.

To help the engine cope with increased mechanical and heat stresses, Chev engineers 'toughened up' the base Corvair engine in a number of areas. Performance oriented valve springs and engine bearings were added. Specifications of the crankshaft, con rods and pistons were revised in a number of key areas. The exhaust valve material was changed to Nimonic steel. The ignition advance mechanism was changed to keep the advance at 24° all the way to 3800rpm without boost. Once 2psi boost was reached, the pressure retard canister pulled the spark back 9° to only 15° before TDC. As the engine accelerated past 4000rpm it could handle more spark lead, so the centrifugal advance mechanism progressively added 12° advance by 4500rpm.

A decade later, in 1973, Porsche began development of the air-cooled flat six engine in its 911 model using lessons learned from racing the brutish 1200hp 917 Can-Am sports cars. Early 1975 production began with an engine that had progressively grown from 2 litre to 3 litre (95mm x 70.4mm). Without intercooling, a compression ratio of 6.5:1 and maximum boost of 11.4psi, the engine produced 260hp at 5500rpm and 253lb/ft torque. In 1978 engine capacity was increased to 3.3 litre (97mm x 74.4mm), and a modest air-to-air intercooler was added which allowed the compression ratio to be increased slightly to 8:1. With the same maximum boost and cam as the 3-litre turbo, the power increased to 300hp at 5500rpm and maximum torque was now 304lb/ft. Without intercooling, the charge temperature was 125°C at full boost, and Porsche say adding the intercooler provided a 5% performance gain.

The main change in the two turbo engines in comparison with the standard 911 was the move to much milder cams with shorter duration, less valve lift and less overlap. The 210hp 2.7-litre used relatively radical cams with timing and lift figures of 64°/76° and 0.457in inlet, 64°/44° and 0.406in exhaust, and 108° overlap. The 200hp 3-litre engine was toned down to make it acceptable in the USA. Valve lift remained the same, but the timing was shortened to 24°/76° inlet, 66°/26° exhaust, with overlap of 50°. The 3-litre and 3.3-litre turbo cams were softened even further to 22°/62° and 0.382in inlet, 52°/20° and 0.350in exhaust, with overlap of 42° (Table 3.1).

Table 3.1. Porsche 911 cam characteristics

	2.7-litre	3.0-litre	Turbo
Valve timing and lift			
Inlet	64°/76° 0.457in	24°/76° 0.457in	22°/62° 0.382in
Exhaust	64°/44° 0.406in	66°/26° 0.406in	50°/20° 0.350in
Duration			
Inlet	320°	280°	264°
Exhaust	288°	272°	252°
Duration at 0.050in lift			
Inlet	262°	232°	209°
Exhaust	232°	220°	200°
Overlap	108°	50°	42°
Overlap at 0.050in lift	40°	1°	–11°
Valve lift at TDC			
Inlet	0.208in	0.051in	0.032in
Exhaust	0.120in	0.051in	0.032in
Lobe separation angle	98°	113°	108°

At the same time as the 3.3-litre 911 appeared SAAB also debuted its first turbo car, as did GM's Buick division. However, after a relatively small start during the '70s the following decade saw the majority of the world's auto manufacturers jump onto the turbo performance bandwagon.

Not to be outdone, Japanese motorcycle manufacturers (Honda, Yamaha, Suzuki and Kawasaki) also wanted the euphoria and hype, which was part and parcel of the word turbo, associated with their products. The air-cooled 738cc 4-cylinder Kawasaki was technologically advanced with digital sequential fuel injection and a knock sensor. Operating at 10.5psi boost, it produced 96hp at 9500rpm and 63.3lb/ft torque at 6500rpm (measured at the countershaft; equivalent power at the crankshaft would be about 10% higher). Very impressive figures for a 2-valve engine with very mild cams. Whereas the stock cams provided inlet and exhaust valve events of 50°/80° and 74°/42° respectively, with 92° overlap and 0.335in lift, the turbo figures are 34°/64° and 56°/24° (105.5° lobe separation, 0.5° advance) plus only 58° overlap and 0.295in lift. Additionally, the turbo had 1mm smaller inlet valves (33mm) and 2mm smaller exhaust valves (28mm), and correspondingly smaller ports.

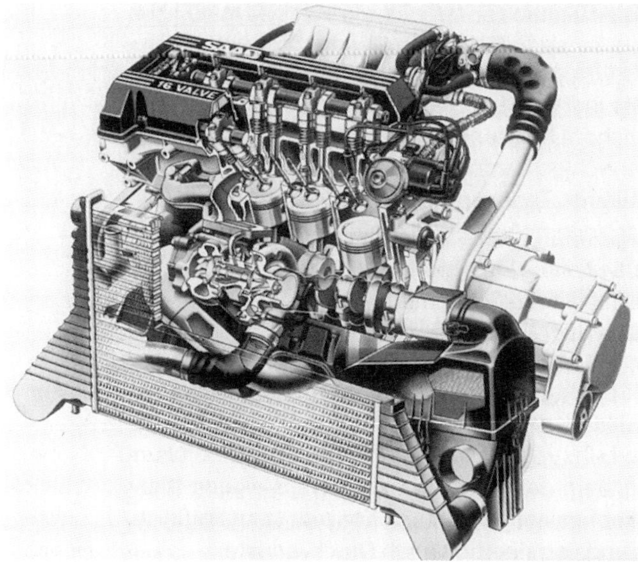

SAAB were early pioneers of turbocharging.

MAKING THE TURBO CAR USER-FRIENDLY

Most attention, though, should be focused on car makers because from that arena sprang the truly remarkable developments which in isolation did not look to be significant, but when bundled with other innovations enabled the next generation of turbo engines to make a giant leap forward. Most of these developments did not make for more powerful engines, rather they made turbo cars more fun to drive because engine responsiveness and turbo reliability rose. Consequently the majority of turbo cars in the late '90s had well integrated turbo packages. They no longer felt as though they had been hotted up by a tuner simply looking at obtaining big impressive horsepower numbers, at the expense of massive doses of lag and dismal performance in real day-to-day driving conditions.

In hindsight, Buick probably started this trend in 1978 with their 3.8-litre V6 Regal. The stock engine developed an almost unbelievably poor 105hp at 4300rpm. That base engine was turbocharged to wave the flag for Buick doing pace car duty during the 1976 Indy 500. Using production cylinder block, crank and rods the engine produced 315hp with boost of 22psi. That encouraged Buick to set about developing a turbo V6 for production with the objective of providing V8 performance and V6 fuel economy, along with minimal lag.

Right away this demanded that the stock 8:1 compression ratio (standard fuel is 91RON unleaded) be retained. Obviously, if detonation was to be avoided and excellent fuel economy maintained, precise spark control was of the essence. A distributor employing a mechanical advance mechanism and vacuum/pressure advance/retard controls would not make the grade. The solution was to use a stock type distributor with mechanical and vacuum advance arrangements along with electronic spark retard. In operation, a knock sensor located on the thermostat housing on the inlet manifold sends a signal to the electronic control centre whenever engine detonation is detected. The control centre in turn sends a delayed command signal to the ignition module which then retards the ignition firing point. Varying degrees of retard are possible up to the maximum limit programmed into the electronic unit.

Then, to give the engine more low rpm 'grunt' and take away any impression of turbo lag, a new cam was developed. The exhaust lobe retained the stock engine's 277° duration and 0.366in lift, but the inlet lobe had the lift reduced from 0.383 to 0.323in. The inlet opening angle wasn't altered, so overlap was unchanged. However, the closing point arrived 24° earlier with duration reduced from 270° down to 246°.

The final result was 165hp at 4000rpm and 265lb/ft at 2800rpm, with maximum boost of 8.9psi and no intercooling. Development continued, and in 1986 engine performance rose to 235hp at 4400rpm, with maximum torque of 330lb/ft at 2800rpm. The later engine employed full electronic engine management with sequential fuel injection (the original engine used a 4-barrel carb) and electronic wastegate control; the latter permitting overboost of 15psi for 2 seconds before pulling back to 13psi. Also contributing to the higher performance was an air-to-air intercooler with crank-mounted auxiliary fan, which reduced intake charge temperature by up to 52°C.

The 1979 release by Porsche of its model 924 turbo introduced us to a new device carrying a Porsche patent. Porsche described the fitting nestled within the compressor

Fig. 3.8. Porsche 924 turbo 'pop-off' valve opens when the boost pressure spike in concert with manifold vacuum drawing against the diaphragm (both as a result of the throttle being suddenly closed) overcomes spring pressure and forces the valve open, allowing boost to escape into the compressor inlet.

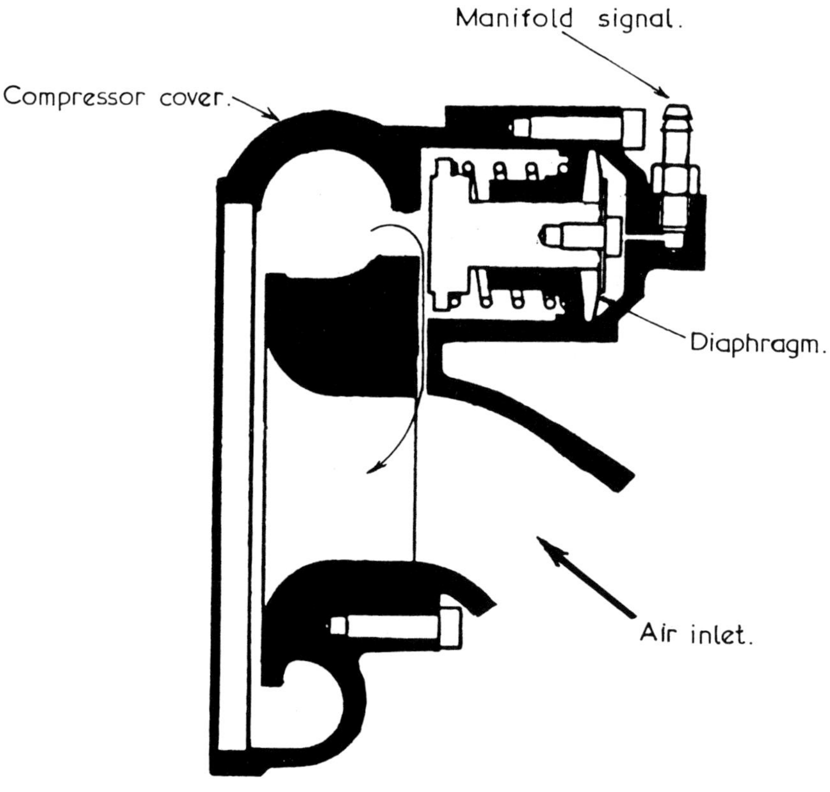

housing of the KKK turbo as a 'pop-off' valve. Developed as a result of sports racing experience, the valve opened when the driver lifted off the throttle, instantly dropping pressure in the compressor to atmospheric. This allowed the turbo to keep spinning at high revs with minimal resistance. Next, when the driver got back on the throttle the exhaust gases would begin accelerating a turbine already spinning at high rpm, and so reduce turbo lag (*Fig. 3.8*).

1981 saw SAAB add the letters APC to the description of its revised approach to turbocharging. SAAB had found that a solid car plus a small turbo engine and a low 7.2:1 compression ratio equals lethargic off-boost performance. SAAB claimed APC (Automatic Performance Control) was the solution. The plan was for the turbo engine to return to the stock engine's compression ratio of 8.5:1 and ignition timing. APC would permit what was thought at the time to be pretty radical thinking.

Off boost, the engine should offer the same level of performance as the stock engine. Then, under boost conditions, whenever the knock sensor detected detonation, that signal would go to the electronic control centre. From there a signal is sent to the wastegate solenoid, which pulses at a rate of 12 times per second, to open the wastegate and reduce boost until detonation ceased.

Under different circumstances, though, the engine might tolerate considerably more boost. For example, the ambient air temperature and the engine operating temperature could both be low or the car may have been filled with 98 octane fuel rather than 95RON. Now the boost levels would be allowed to increase by means of the wastegate solenoid control keeping the wastegate closed to a higher inlet manifold pressure level.

The Porsche 944 turbo in 1985 advanced SAAB's control system to the next level. In addition to normal wastegate control Porsche introduced an 'overboost' facility. For brief periods maximum boost was permitted to increase by 10% to improve acceleration. Additionally, the knock sensor was connected to Bosch's DME (Digital Motor Electronics) ignition computer which could advance or retard the ignition timing of individual cylinders to provide higher levels of performance and, if necessary, detonation protection. For example, if all four cylinders were knocking, the wastegate would lower boost pressure, but if only one cylinder was detonating, the timing would be retarded to that particular cylinder. In a two and two situation the computer would analyse a number of inputs. Based on those inputs it either could leave the boost level alone and retard the two detonators, or reduce the boost and advance the spark to the two 'cool' cylinders, thereby maintaining the best possible performance outcome.

Other features contributed to this 2.5-litre 2-valve engine producing 217hp at 5800rpm and 243lb/ft of torque at 3500rpm. (In 1989 this increased to 274hp and 258lb/ft.) The intercooler dropped intake charge temperature from 130ºC to 50ºC under ideal conditions. To keep heat in the exhaust gases for quicker turbo spool up, thick ceramic liners were fitted in the exhaust ports (this also assists cooling by preventing heat transfer to the head) and the exhaust headers were double-walled stainless tubes. For reliability the turbo centre was water cooled. Another important advance was the upgrade from mechanical to electronic fuel injection. While the move from carburettors to mechanical injection significantly improved fuelling and turbo engine drivability, the next step to electronic fuel injection on the 944 brought even greater benefits. Fuelling precision was further enhanced with a corresponding increase in throttle responsiveness and reduced turbo lag. Also, when the DME system retarded the ignition the turbo could now be protected from excessive exhaust temperatures by increased fuelling.

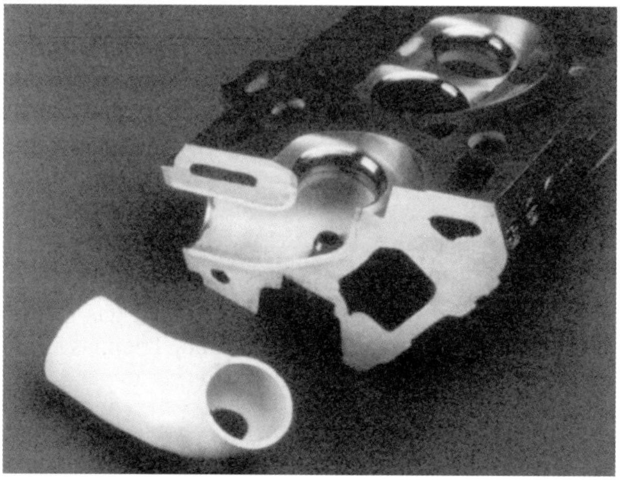

Porsche 944 turbo had exhaust port inserts for quicker turbo response.

TURBOCHARGER DEVELOPMENTS

Early 1987 saw the release of Buick's revised turbo V6 in the Regal GNX. Its Garrett turbo pointed to the direction future turbocharger progress would take to be rid of turbo lag. By substituting a lightweight aluminium oxide ceramic turbine wheel, typically about half the weight of the conventional alloy steel wheel, Garrett found that the rotating inertia could be reduced by as much as 40%. Field testing proved this reduced turbo lag by more than 30% (*Fig. 3.9*).

Fig. 3.9. Ceramic turbine wheel with less rotating inertia is able to accelerate more rapidly, building boost sooner and reducing lag.

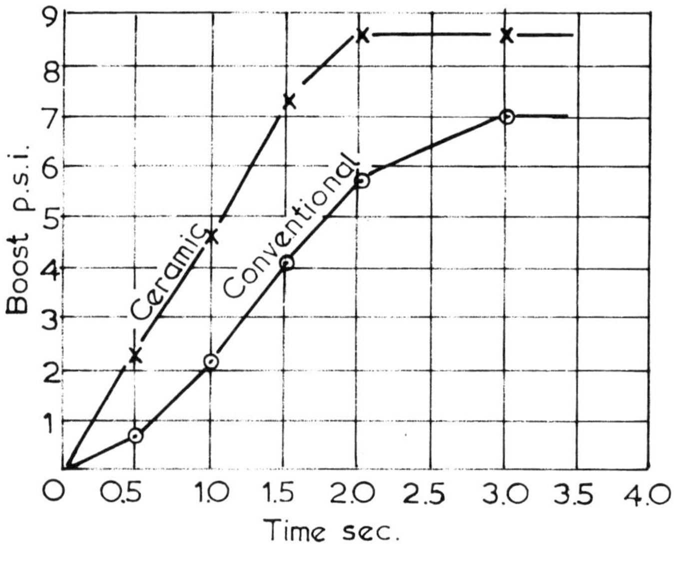

Of perhaps greater importance was the use of VNT (Variable Nozzle Turbine) technology. This technology had been used successfully on diesel engines, but petrol engines with their much higher exhaust temperatures were found to cause the VNT mechanism to seize up. Now the breakthrough had arrived and with it came the ability to make the turbine side perform as if it were much smaller for better low speed response, but as engine speed rose it could be adjusted to work like a big turbo with corresponding power and fuel economy improvements.

The Garrett VNT mechanism incorporates multiple vanes that function much like adjustable louvres. At low speed the engine's electronic boost control system moves these vanes into a closed position, forcing exhaust gas to flow through a much smaller passage (nozzle) to rapidly accelerate the turbo. Then, as engine speed and boost builds, the vanes are progressively opened, so preventing the turbo from spinning faster and exceeding maximum boost pressure. In this mode it functions much more effectively than a wastegate, and as such in some installations makes the wastegate redundant.

The other development later that year was considerably less important to those concerned purely with turbocharged road cars, but for high performance tuners it was significant. When Ford released a batch of special RS500 Sierra Cosworths for motorsport certification purposes their engines were given a number of upgrades. Among the upgrades was the move to two sets of fuel injectors. One set was mounted in the usual location firing fuel at the back of the inlet valve. These primary injectors provided fuel sufficient for normal operation, but they were too small to flow enough fuel at very high power levels. Larger injectors were not the answer because fuel flow would now be too high off boost at cruise. The solution was to mount a second row of injectors in the airbox lid, arranged so as to spray fuel into the mouth of the internal cast trumpets. These were staged to supply the engine's additional fuel requirements at high air flow levels. Being mounted over the inlet trumpets, fuel issuing from them pulled heat out of the inlet tracts and enhanced charge density.

Prior to 1990 a number of manufacturers had released exotic low volume performance cars outfitted with two turbos. However, in 1990 the situation changed when both Nissan and GM's European divisions released regular production cars with two turbos. The Opel Lotus Omega (Vauxhall Lotus Carlton) was timed at 282kph (175mph), the highest speed recorded for a 4-door production saloon (sedan). Its straight six 3.6-litre engine developed 377hp at 5200rpm and massive torque of 419lb/ft at 4200rpm with a compression ratio of 8.2:1, maximum boost of 10.2psi and a water-to-air intercooler. The Nissan 3 litre V6, by comparison, was a kitten with a more modest, and affordable, 280hp at 6400rpm and 274lb/ft torque at 3600rpm with 8.5:1 compression ratio, air-to-air intercooler and – another first for production turbo cars – variable valve timing. This latter feature, at low speeds, ensures more trapped cylinder pressure with a corresponding performance dividend, and at higher rpm the valve overlap period is extended to enhance power output.

The reasoning behind the twin turbo system is quite simple – a big turbo takes a long time to spool up to speed so turbo lag becomes an issue. Also, in Vee style and boxer engines a single turbo invariably has to be located off to one side. This, in turn, leads to a long crossover exhaust from one side of the engine to the turbo. That long pipe loses a lot of heat energy that should have been put to better use accelerating the turbo. Obviously that heat has to go somewhere and it continues to make problems by adding to the total underbonnet heat load, which ultimately stresses the cooling system and puts heat into the inlet charge.

Two smaller turbos can be mounted up close to the exhaust ports, minimising heat loss and maximising exhaust energy. Smaller turbos have a lower moment of inertia so they spin up to boost more rapidly, thereby minimising lag. Additionally, turbine wheels accelerate better when hit by distinct and powerful exhaust pulses. More cylinders into a common exhaust sees exhaust flow activity even itself out and pulse intensity being diluted. Fewer cylinders exhausting into a single turbo spaces the pulses further apart, but each pulse has a greater intensity peak to give the turbine wheel a good solid bang, so it accelerates faster, producing more boost more rapidly.

During all these years the internal design of turbos was being improved so that smaller housings and smaller turbine and compressor wheels flowed more air and exhaust gas with greater efficiency. The number of blades on wheels was changed, as were the size and shape of the blades. These changes weren't so visible and, even when they were pointed out, an untrained eye would be hard pressed to spot any differences.

A more visible change in the early 1990s was a move from plain bearings to ball bearings to support the shaft carrying the turbine and compressor wheels. IHI proved that special ball bearings could withstand tremendous rotational speeds (160,000rpm) and heat (1170ºC) in its RX6 turbos fitted to the Formula 1 Honda. Consequently, it was not surprising to see both Mazda and Subaru equip some of their engines with IHI ball bearing turbos. Nissan, too, were quick to adapt ball bearings into their Garrett licensed turbos.

Many scoff at the idea of ball bearing turbos spinning more freely. However, Nissan found a 22% reduction in spool-up time to only 5.8psi boost pressure. *Fig. 3.10* illustrates the amount of power consumed to spin a turbo. While the power differences do not appear large, remember that when the throttle is lifted exhaust flow almost stops. Therefore, anything which aids the turbine and compressor wheels to maintain their rotational speed will improve acceleration response.

Fig. 3.10. Ball bearings have reduced friction, so less exhaust energy is wasted to frictional losses as the turbo accelerates up to speed. Also, when the throttle is closed there is less drag to slow the turbo down, so boost builds more rapidly when the throttle is reopened.

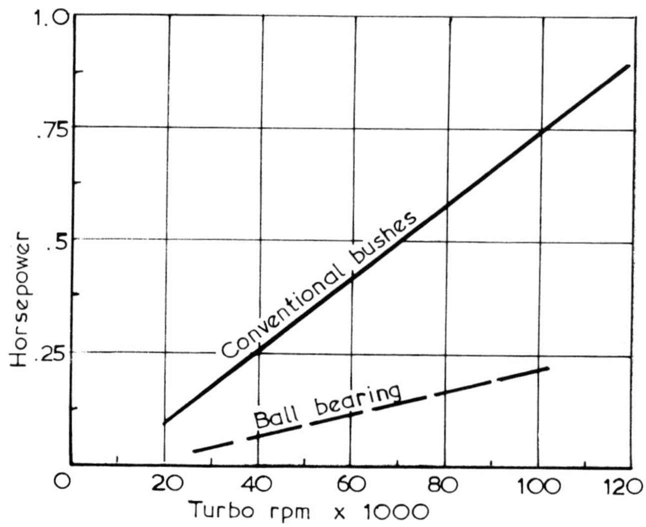

Chapter 4

Nitrous oxide up until today

In comparison to the other forms of forced induction nitrous oxide injection is relatively unknown. It was first described in military aircraft engineering circles as 'nitrous oxide supercharging'. That was in the last years of World War II. However, the history of dinitrogen monoxide (N_2O) goes back to 1772 when it was identified by Joseph Priestley. A half century later its usefulness as an anaesthetic was realised, but it was well over a century later before well-known British internal combustion engine experimenter Sir Harry Ricardo briefly noted in his extensive writings the effects when nitrous oxide was released into the carburettor airway of a running engine.

Following the war, interest in high-performance piston engine military aircraft pretty well died, and at the same time interest and accumulated knowledge in 'nitrous oxide supercharging' also faded into obscurity. Obviously, returning pilots and aircraft engineers may have dabbled briefly with nitrous oxide injection in various forms of motor sport, but its first recorded usage is attributed to American stock car racer and engine guru Smokey Yunick around 1956. The crude system on Smokey's Hudson Hornet stock car consisted of a gas cylinder of nitrous oxide tied down under the driver's seat with a length of hose running into the mouth of the carburettor. After wrecking a few engines he figured how to regulate nitrous flow and carb jetting. Then, during qualifying, a carburettor with appropriate rich jetting would be fitted, the nitrous hose secured in the air intake. During the hot lap the driver would reach under the seat and turn the gas valve open exactly four turns, virtually guaranteeing that Smokey's car started at the front of the pack. It took some time, but eventually others figured that his car's impressive turn of speed was attributable to something additional to Yunick's considerable engine tuning nous. During 1958 NASCAR rules changed and nitrous disappeared from stock car racing – for a time!

NITROUS BROUGHT OUT OF THE CLOSET

By 1962 in California, Ron Hammel was tuning engines based on Sir Harry Ricardo's writings. Hammel worked for a company with an engine dyno, so during the next few years he was able to accurately document the effects of nitrous on engine power. Ricardo hadn't written much on the subject, so Hammel was really pioneering the measured use of nitrous oxide to enhance car engine performance. Finally, in 1969, he had accumulated the required knowledge and had developed the necessary delivery systems to confidently install nitrous systems on competition engines (*Fig. 4.1*).

35 years ago Ron Hammel was pioneering nitrous oxide injection kits.

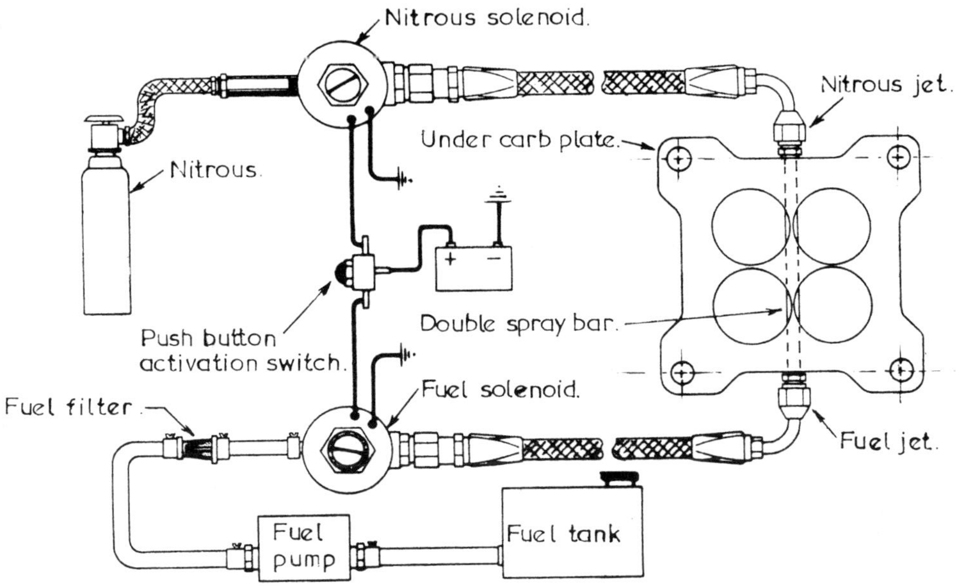

Fig. 4.1. Early nitrous systems delivered nitrous and additional fuel through spray bars inserted in a plate fitted between a 4-barrel carburettor and the inlet manifold.

Initially the systems were hidden or disguised as fire extinguishers in numerous drag boats and drag cars, and again in NASCAR stock car racing. Obviously, a few pounds of gas enhances power for only a short time (Ricardo had found that 5lb of liquid nitrous will provide 100hp for 1 minute) so drag racing appeared to be its natural home. However, as Smokey Yunick had discovered much earlier, laughing gas is a useful performance booster in circle track racing too, particularly for a hot qualifying lap, or to transform overtaking attempts into actual overtakes. As more racers got 'on the bottle', power greed led to larger nitrous doses being used, which in turn led to numerous engine failures. By 1976 the less well funded teams began haemorrhaging on all fronts, so they decided to blow the whistle on the use of nitrous oxide.

This had two outcomes. First, and most important, it brought nitrous oxide out of the closet. Everyone now had equal opportunity to be fully informed about the use of nitrous for instant hp. They could learn how to use the stuff without breaking engines and they now knew where they could purchase nitrous equipment and gas supplies. Second, motorsport's rule makers were put on notice that gas was out there and cheaters were using it. They could no longer ignore its use. Either they would have to outlaw nitrous and strictly police the rules, or they could permit nitrous usage within a set of rules.

Subsequently nitrous oxide was banned from most categories of motorsport outside of drag racing and unlimited hydroplane boat racing. Within drag racing it is permitted in a number of classes, but it was in unofficial late night drag racing (frequently for money) on the streets and industrial estates of the USA that gas usage has blossomed in the past two decades. In that environment a mildly modified pleasant-to-drive 450hp grocery getter can be transformed quite easily into a 600hp match racer by night.

The traditional home for nitrous oxide has been in drag racing. Hidden systems like on this engine (an 8-nozzle arrangement which adds over 300hp) are favoured for illegal match racing.

Fig. 4.2. Nitrous provides impressive power gains, even at low revs.

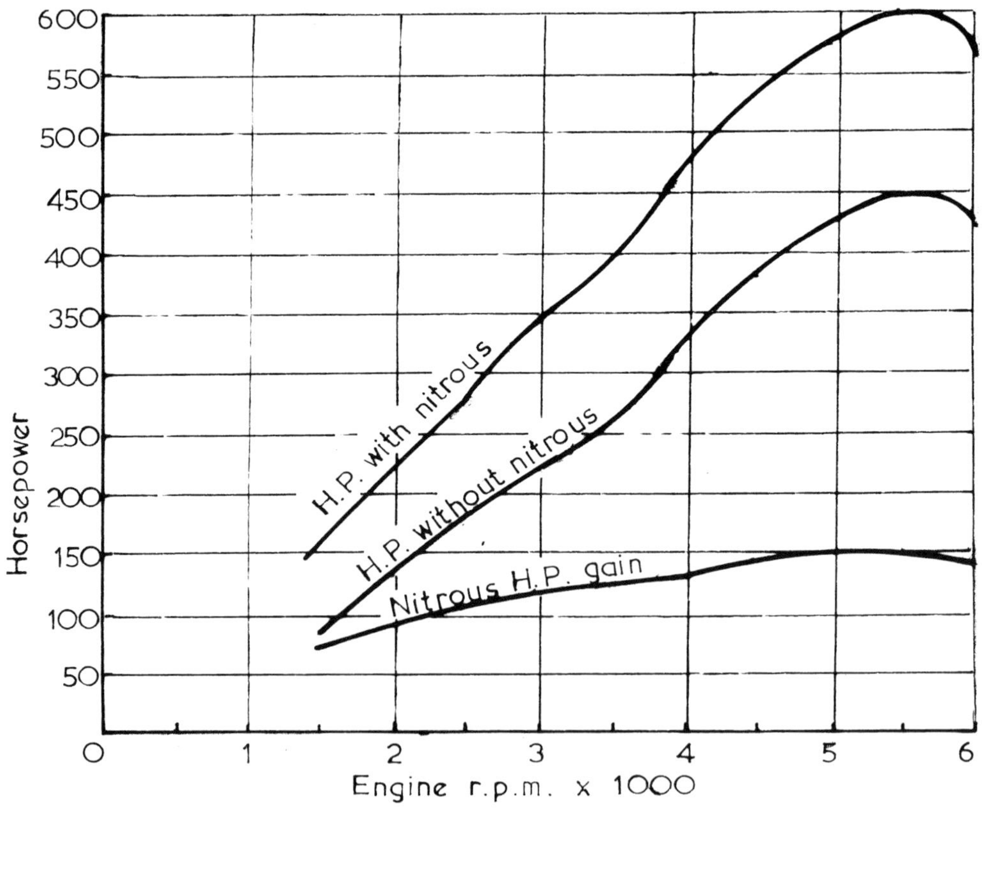

A ONE-THIRD POWER GAIN

A one-third power gain may not sound too impressive in comparison with some other horsepower enhancement methods but, remember, this sort of gain is available through virtually the entire rpm range of the engine, from idle to maximum revs! *Figure. 4.2* is a typical N_2O power curve on the dyno. However, when actually installed in a car with street tyres a system like this would have to be toned down somewhat by using a two stage system. In lower gears the ECU would be programmed to introduce a smaller dose of nitrous to avoid uncontrollable wheelspin. Only in higher gears, or when wheel speed sensors detected that wheelspin was minimal, would the full dose second stage of N_2O be released.

Because of the huge amounts of wheelspin, or the slam-in-the-back sensation when the accelerator is floored, the horsepower gains of nitrous tended to be grossly overstated in earlier times. In truth, many N_2O kit manufacturers did not have engine test facilities so they frequently guesstimated power gains, or at best timed the car over a $^1/_4$ mile and calculated horsepower from tables based on those speeds. Consequently, kits that supposedly provided a 100hp increase often gave more like 60–70 extra horsepower.

OVERCOMING ENGINE FAILURE

Engine failures with the high power kits were fairly frequent in the past. The main problem was poor distribution of N_2O from cylinder to cylinder. Remember, most V8s at that time were stuck with a 4-barrel carb at best, and with the nitrous being introduced at the base of the carb there was no guaranteeing that the correct proportions of nitrous and extra fuel reached every cylinder. Lean cylinders soon detonated, destroying pistons, and even the cylinder wall and valves in extreme cases.

There was also a considerable lack of understanding how gas bottle temperature and pressure, and gas bottle and gas line location affected nitrous flow rates. For example, on a cool night (4ºC) gas pressure will be 520psi, but when the day temperature rises to 21ºC nitrous pressure goes all the way to 760psi. Clearly, at higher pressures a calibrated orifice (a jet) will flow considerably more N_2O. Therefore, if the fuel to gas calibrations were worked out to suit night-time match racing and the guy ran at a sanctioned drag event the following day, nitrous flow would increase by virtue of the 46% pressure rise, probably melting down the engine because of mixture leanness.

Bottle and nitrous linc location can have a reverse effect. Nitrous systems are calibrated on the premise that liquid, not gaseous, N_2O is injected. Clearly, if the

Fig. 4.3. Nitrous cylinder is fixed as shown to ensure liquid nitrous is delivered.

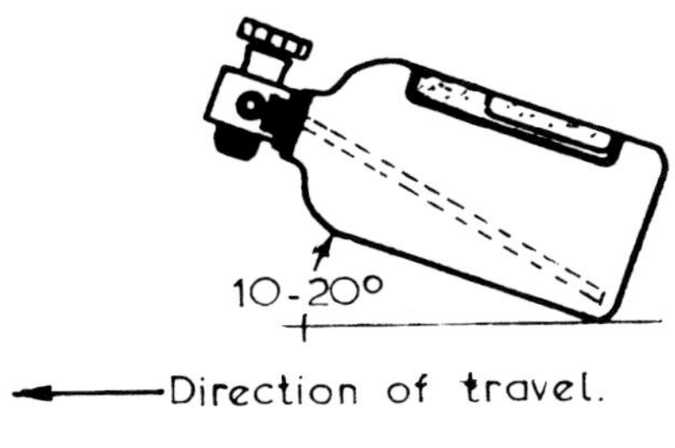

nitrous cylinder isn't correctly oriented, or if the cylinder is free to roll around in the boot (trunk), the pick-up tube could be surrounded by gaseous N_2O rather than liquid (*Fig. 4.3*). Likewise for the nitrous line if it is routed close to the exhaust and hot engine parts.

In time, kit manufacturers provided better fitting instructions to overcome these basic installation problems. Maintaining a fairly constant bottle temperature wasn't so easy. Hot water bottles and blankets have been tried, but the best arrangement was an electronically controlled heater to keep the bottle at 25ºC ±2ºC.

OTHER USES

Other uses for N_2O were also tried, but without much success. Apart from traditional applications of providing more oxygen for the engine, there are probably only two other legitimate uses. One is to assist turbo spool up by massively increasing exhaust gas volume and velocity at lower engine rpm. This makes the engine perform as if it is supercharged by a Roots-type blower and a turbocharger working together, such is the low speed power improvement. An option with nitrous assist is to opt for a much larger and more efficient exhaust turbine on the turbo to provide a greater top-end power boost, plus superior cruise fuel economy because of reduced exhaust restriction. Some have also injected the nitrous right into the compressor, to directly impinge on the impeller blades, supposedly providing some extra force to speed up the turbo. However, the theoretical benefits were never realised in practice.

The other possible application for N_2O is as an intercooler chiller. In horsepower heroes competitions the winner is the fellow, or girl, whose car makes the highest horsepower on a wheel dyno. Such a closed environment is not especially helpful. The air heats up fairly quickly and oxygen content pretty rapidly falls. One way competitors running air-to-air intercoolers get around this, in competitions which permit the practice, is to blast the intercooler with nitrous directly from the bottle. This not only chills the intercooler, but savvy individuals also ensure that the air intake is appropriately located so ensuring that the engine is drawing in air with an oxygen content closer to a normal 23%. (You probably learned in physics classes that air contains 21% oxygen. By volume that figure is correct, but by weight air is 23.3% oxygen.)

THE POWER GAINS POSSIBLE TODAY

In 1953, in his last book, *The High Speed Internal Combustion Engine*, Ricardo noted that liquid nitrous oxide 'permitted a large increase in power, at least 40 per cent.' That figure holds true pretty well to this day. Solid, well-built stock engines will accept a nitrous assisted power increase up to about 35% when everything has been done correctly, and fuel and N_2O calibration and spark advance are precisely controlled.

Competition engines with heavy duty components will live at 40% power boost levels as Ricardo stated. However, in pure drag race engines, which are only required to barely survive one ¼-mile trip, the power increase can be more like 50–80%. The monster 650cu in Pro Street engines make 1100–1250hp without nitrous. On the first stage of nitrous this rises 400–500hp and when the second stage is added the power rises another 400–500hp!

Pro Street drag engine specification	
90º V8 Bore and stroke:	117.48mm x 122.43mm (10.62 litres)
Typical power output:	1850–2200hp at 7500rpm 1700–1800lb/ft at 5500rpm
Maximum engine speed:	8500rpm
Pistons:	forged 3 ring with top ring 8–10mm down 15:1 compression ratio
Cylinder heads:	Dart Big Chief aluminium, CNC ported inlet valve 61mm exhaust valve 48.3mm head gasket – solid copper plus 'O' rings
Valve train:	roller cam 9310 steel, 114º lobe centres inlet lobe timing 48º/96º, 289º duration at 0.050in inlet valve lift 0.910in with 1.77 rocker ratio exhaust lobe timing 108º/60º, 308º duration at 0.050in exhaust valve lift 0.810in with 1.77 rocker ratio shaft mounted roller rockers
Bottom end:	block – cast iron GM Bow Tie crankshaft – 4340 steel con rods – 7075 T6 forged aluminium, 170.2mm long
Exhaust:	stepped headers and twin exhausts and silencers (mufflers) header pipes 60.3mm diameter to 63.5mm diameter x 850mm long, into 127mm diameter collector

Chapter 5

Forced induction theory and principles

Most people hate the idea of getting into involved physics and associated maths. Like it or loathe it, we are surrounded by physical laws that affect us from day to day. Some of those laws we can pretty much ignore, but others constantly remind us of their presence. Take the law of gravity or the day/night cycle as the earth rotates around the sun. Now, we don't need to know the science behind all this to live, but having some knowledge of their workings sure helps us establish a pattern which, in turn, enables us to work with, and use, things to our advantage.

Forced induction is no different. If we can train ourselves to 'see' those patterns established by physical laws, we can plan accordingly to good effect. However, if for some reason we have to work against a natural law, we can make provision to ensure that our engine survives.

UNDERSTANDING AIR DENSITY

The very first principle that we have to come to grips with is that air density is not constant. At sea level air is 35% more dense than at 10,000ft altitude. Therefore, it is often said that the air is thinner at high altitude as the gas molecules are spaced further apart. Consequently air weighs less at altitude. The figures are 0.0765lb per cubic foot at sea level and 0.0565lb/ft^3 at 10,000ft.

Taking this principle a step further, it should follow that a 3-litre engine will fill its cylinders with 3 litres of air at 10,000ft just as it does at sea level. However, at altitude, because those 3 litres of air contain less oxygen (about 26% less), the engine will produce substantially less power. In fact, a rule of thumb is that hp will decrease 3% for each 1000ft.

However, altitude is not the only factor influencing air density. Heat also expands air, consequently at sea level 40°C air is less dense than air at 5°C. Therefore, our 3-litre engine will make less power on hot air than cold air. Another rule of thumb: you lose 1% hp for each 7°C increase in air temperature.

Standard atmosphere chart									
Altitude		Air pressure			Temperature		Density		Relative density
ft	m	in Hg	psi	mB	°F	°C	lb/ft3	kg/m³	
Sea level		29.92	14.7	1013	59.0	15.0	0.00765	1.225	1.000
1000	305	28.86	14.2	977	55.4	13.0	0.00742	1.189	0.997
2000	610	27.82	13.7	942	51.9	11.1	0.00722	1.156	0.993
3000	914	26.81	13.2	908	48.3	9.1	0.00714	1.144	0.989
4000	1219	25.84	12.7	875	44.7	7.1	0.00680	1.090	0.986
5000	1524	24.90	12.2	843	41.2	5.1	0.00660	1.058	0.982
6000	1829	23.98	11.8	812	37.6	3.1	0.00639	1.024	0.979
7000	2134	23.09	11.3	782	34.1	1.2	0.00682	1.093	0.975
8000	2438	22.23	10.9	753	30.5	-0.8	0.00601	0.962	0.972
9000	2743	21.39	10.5	724	26.9	-2.8	0.00582	0.933	0.969
10,000	3048	20.58	10.1	697	23.4	-4.8	0.00565	0.905	0.965
11,000	3353	19.80	9.7	670	19.8	-6.8	0.00547	0.876	0.962
12,000	3658	19.03	9.3	644	16.2	-8.8	0.00530	0.849	0.958
13,000	3962	18.30	9.0	620	12.7	-10.7	0.00513	0.822	0.954
14,000	4267	17.58	8.6	595	9.1	-12.7	0.00498	0.797	0.951
15,000	4572	16.89	8.3	572	5.6	-14.7	0.00483	0.773	0.947

The jacket on the turbine housing and the wrapping on the turbo dump pipe is intended to keep as much heat as possible away from the inlet tract, so as to increase charge density. The air filter is fed cool air from a large bonnet scoop.

Pressure exerted by humidity (millibar)											
Temperature		Relative humidity %									
°C	°F	10	20	30	40	50	60	70	80	90	100
10	50	1	2	4	5	6	7	9	10	11	12
20	68	2	5	7	9	12	14	15	19	21	23
25	77	3	6	10	13	16	19	22	25	29	32
30	86	4	8	13	17	21	25	30	34	38	42
35	95	6	11	17	22	28	34	39	45	51	56
40	104	7	15	22	30	37	44	52	59	56	74
45	113	10	19	29	38	48	57	67	77	86	96
50	122	12	25	37	49	62	74	86	99	111	123

Humidity affects air differently. High humidity, high moisture content in the air, causes a barometer to read higher. However, rather than being more dense the air actually has less density – it is simply the combined weight of the air plus the water that appears to make the air more dense. Generally this is not a large consideration. However, high humidity at elevated temperatures does reduce the oxygen content enough to reduce power. For example 65% humidity and 35°C (95°F) at sea level reduces the oxygen content to a level similar to that at 1000ft elevation.

Do you see a pattern emerging from the preceding four paragraphs? That is, an engine will make more power at sea level on dense air which is cold and dry. Conversely, power decreases with increases in altitude, air temperature and humidity.

AIR DENSITY WITHIN THE ENGINE

The next principle that we need to establish is that two 3-litre engines operating in the same atmospheric conditions can have different hp outputs because the air is more dense in the cylinders of one than the other. Even though air quality started out at the same level, things change on the passage into the cylinders. One engine has everything working against it. It is drawing in air which has been heated by the radiator and exhaust manifold. Its inlet manifold receives no cooling air flow, but rather hot air that has passed through the radiator. Internally this engine is very hot. The water temperature is regulated at 95°C (203°F) so the air flowing into the cylinders collects heat from the inlet port walls and expands on its way through. The ignition timing is retarded (in itself this reduces power) which raises the temperature of the exhaust valve and that area of the combustion chamber, as well as the piston crown. Consequently, as the intake air discharges into the cylinders it expands rapidly and the cylinder volume is quickly filled by what is now 'thin' air.

However, this engine's problems don't stop here. The exhaust system is quite restrictive in itself, but additionally the CAT has overheated and the internal honeycomb-like ceramic material has melted, blocking many of the gas flow passages. This is

Right: The restrictive stock air box has been replaced by an aftermarket air filter featuring an impressive bellmouth fitted onto the inlet duct. Unfortunately, the engine will now be breathing hot low-density air, especially as the turbo is only inches away (top right).

Below: A melted ceramic core inside a CAT will block exhaust flow.

causing exhaust back pressure which is not allowing the cylinders to completely empty of exhaust gas. These remaining gases mingle with the fresh inlet fuel/air mixture and upset combustion, but prior to that event, because their temperature can easily exceed 800ºC, they quickly heat the inlet mixture, causing it to rapidly expand and block further inlet flow. Consequently the cylinders are filled with a thin, low density, inlet charge.

On the inlet side the story isn't good either. The air filter is blocked with dust and the air filter box doesn't flow well. It has been designed to keep induction noise very low by using various ribs, baffles, resonators and tortuous passages. The final outcome at full throttle is that the engine has a vacuum reading of 2in Hg (mercury), measured just before the throttle body. The vacuum reading, ideally, should be zero, so that we can take advantage of a full 14.7psi air pressure to push air into the cylinders. A vacuum reading of 2in Hg is almost 1psi, so this engine has only 13.7psi working for it to fill the cylinders.

Further on in the inlet tract there are more restrictions. Gaskets overlap a number of inlet passages, but the main concern is the massive build up of carbon on the back of the inlet valves. This is blocking flow into the cylinders until the valves are lifted nearly 50% off their seats. Obviously all these restrictions work to further lower air density in the cylinders.

The other engine, by comparison, has everything working in its favour to ensure maximum charge density in the cylinders and maximum hp. The inlet system is free of restrictions. The air intake is at the front of the car collecting air at ambient temperature. The water temperature is regulated so as not to exceed 85°C, thereby minimising the amount of heat the air will gather in the inlet passages. The exhaust is a free-flow item with zero back-pressure to minimise the amount of exhaust gas left behind in the cylinders. The ignition timing has been optimised for best power. As a side benefit this lowers exhaust gas temperature, so any residual gas puts less heat into the inlet charge. Also, the exhaust valve and piston run cooler, which helps mixture density as well.

The pattern here is that cylinder charge density is influenced by numerous factors which the engine builder or tuner can control. If the engine is 'plugged up' on the inlet side and/or the exhaust side cylinder charge density will be lowered. Also, if the engine or individual components are excessively hot because of high coolant temperatures or incorrect spark timing (excessive advance will also overheat the piston and combustion chamber, not just retarded ignition) cylinder charge density will be lowered. On the other hand, an engine which breathes freely on both the inlet and exhaust side, and is operating

Left: Carbon build-up on inlet valves blocks air flow and massively lowers air density in the cylinders.

Below: Stock valves, lightly polished and backcut, help cylinder filling.

at the ideal coolant temperature and ignition advance, will have a cylinder charge density closer to atmospheric air density.

BEWARE OF CRUTCHES

However, there is another factor to be considered. At this point we have two 3-litre engines, one filled with thin air the other filled with air closer to atmospheric density. Is it possible that at the point of ignition the charge densities could be made almost equal? Obviously, if we squeeze the low density charge into a small combustion space using a high compression ratio the density can be increased, but remember this will be limited by available fuel octane and the low density charge is much hotter (this is not always the case) so more likely to detonate anyway. Perhaps if we were extremely conservative and set the compression ratio way too low on the good engine, at certain engine rpm the compressed charge densities could possibly be almost equalised; but the problem really hasn't changed – fiddling the compression ratio, using some crutch, usually does not compensate for low charge density.

In other words, we must do all we can to avoid low charge density. Otherwise adding any form of forced induction could become just another crutch. Sure, the engine may run much better, but nothing close to its true potential will be realised. Some people are unconcerned at paying full price for half a cup of coffee – I am not one of them!

By now it should be clear that it is high charge density that makes power and, apart from poor engine breathing characteristics, the big enemy of charge density is heat. Any time heat is added to the air, charge density goes down.

HOW COMPRESSOR EFFICIENCY AFFECTS AIR DENSITY

Unfortunately, compressing air to increase charge density adds heat to it. The more you compress it the hotter it becomes. Some compressors are more efficient and so add less heat. However, all compressors add heat, and all compressors are less efficient under certain conditions so they add considerably more heat at that time. (Technically some types of superchargers are not compressors, but in reality the end result is the same.) Consequently, increasing pressure in the inlet manifold from normal air pressure of 14.7psi to double that pressure, 29.4psi, by means of 14.7psi supercharge boost, will not double the charge density in the cylinders, nor will it double the engine's power output.

Pressure Definitions	
Gauge pressure:	The pressure you read off a normal pressure gauge. This is the pressure above atmospheric pressure. Usually simply called the boost pressure when referring to the inlet manifold pressure in supercharged engines.
Absolute pressure:	The pressure above a perfect vacuum. Thus, at sea level, normal air pressure is 14.7psi absolute. Adding supercharge boost of 5psi would then give a reading of 19.7psi absolute.

To work out how much power increase we should be anticipating in this example of a boost pressure of 14.7psi at sea level, requires a series of simple calculations.

First, we need to know at what pressure ratio the compressor is operating. Use the formula:

$$Pr = \frac{Bp + Ap}{Ap}$$

where Bp = boost pressure psi (gauge pressure)
Ap = air pressure psi (barometric pressure or, if you don't know the pressure, by assuming it is 14.7psi at sea level and subtracting 0.5psi for each 1000ft above sea level will get you close.)

In the above example the pressure ratio is 2.0:

$$Pr = \frac{14.7 + 14.7}{14.7} = \frac{29.4}{14.7} = 2.0$$

The next step involves determining how much heat will be added to the inlet air charge by compression. Ideally, we need access to a compressor map such as *Fig. 5.1* to be able to accurately read off the compressor efficiency. The compressor on this particular turbocharger has a maximum efficiency of 71%, ranging down to below 62%. At maximum hp revs our engine will require compressor air flow of 426cfm (cubic feet per minute). With a pressure ratio of 2.0 we see that at peak power the compressor will be working at 70% efficiency.

Fig. 5.1. Compressor map illustrates that with the compressor pumping 426cfm at a pressure ratio of 2.0, the compressor is about 70% efficient.

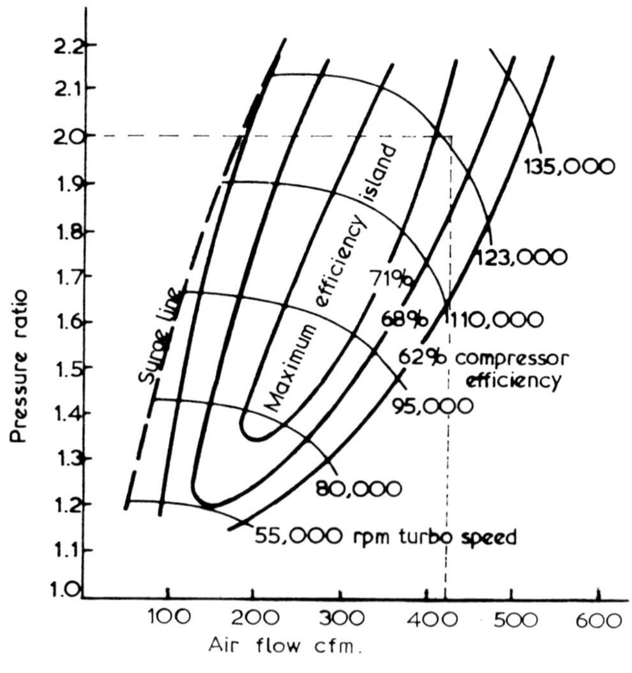

Table 5.1. Factors for calculating compressor outlet temperatures

Pr	F	Pr	F	Pr	F
1.35	0.089	2.25	0.258	3.3	0.402
1.4	0.100	2.3	0.266	3.4	0.414
1.45	0.110	2.35	0.274	3.5	0.426
1.50	0.120	2.4	0.281	3.6	0.437
1.55	0.130	2.45	0.289	3.7	0.448
1.6	0.142	2.5	0.296	3.8	0.459
1.65	0.152	2.55	0.303	3.9	0.470
1.7	0.162	2.6	0.311	4.0	0.480
1.75	0.172	2.65	0.318	4.2	0.501
1.8	0.181	2.7	0.325	4.4	0.521
1.85	0.192	2.75	0.331	4.6	0.540
1.9	0.199	2.8	0.338	4.8	0.559
1.95	0.208	2.85	0.345	5.0	0.577
2.0	0.217	2.9	0.352	5.3	0.603
2.05	0.225	2.95	0.358	5.6	0.628
2.1	0.234	3.0	0.365	5.9	0.653
2.15	0.242	3.1	0.377	6.4	0.691
2.2	0.250	3.2	0.390	7.0	0.735

The formula to calculate the compressor outlet air temperature rise is:

$$Tr = \frac{F \times (At + 273) \times 100}{E}$$

where F = factor from Table 5.1
At = ambient temp ºC (for ºF change the formula to At + 460)
E = compressor efficiency %

Hence, if the air temperature is 25ºC and the turbo air inlet is located so as to draw in air at ambient temperature, then the compressor outlet temperature rise will be:

F = 0.217
At = 25ºC
E = 70%

$$\text{Therefore } Tr = \frac{0.217 \times (25 + 273) \times 100}{70} = \frac{0.217 \times 298 \times 100}{70} = 92.38°C$$

To find the actual discharge air temperature (Dt) at the compressor outlet we have to add Tr and At:

$$Dt = 92.38 + 25 = 117.38°C$$

Obviously, air heated to in excess of 117ºC (242°F) will suffer a considerable decline in density. So, even though it has been compressed to double the air pressure, its density will be nothing like double the density of ambient air with a temperature of 25ºC (77°F).

To calculate how the density of this hot (117.38°C) air boosted to 14.7psi compares with the density of ambient air at 25°C we use the formula:

$$Dr = \frac{(At + 273) \times Pr}{(Dt + 273)}$$

where Dr = compressor density ratio
At = ambient temp °C (for °F change the formula to At +460)
Dt = compressor discharge temp °C (for °F change formula to Dt + 460)
Pr = compressor ratio

$$\text{Therefore } Dr = \frac{(25 + 273) \times 2.0}{(117.38 + 273)} = \frac{298 \times 2.0}{390.38} = 1.53$$

So, in this example turbo boost raises the air density in the cylinders by a factor of 1.53, or 53%. This indicates that the engine should produce very close to 53% more power with 14.7psi boost (assuming the engine will tolerate that level of boost and suitable high octane fuel is used). Therefore, a stock output of 200hp will increase to 200 x 1.53 = 306hp.

Now that you have waded through these calculations and understand how we arrived at the various figures, I have a confession to make. We can obtain an approximation of

Fig. 5.2. Pressure ratio to compressor discharge temperature graph illustrates the importance of compressor efficiency, and the importance of matching the compressor to the engine so that it is operating at maximum efficiency.

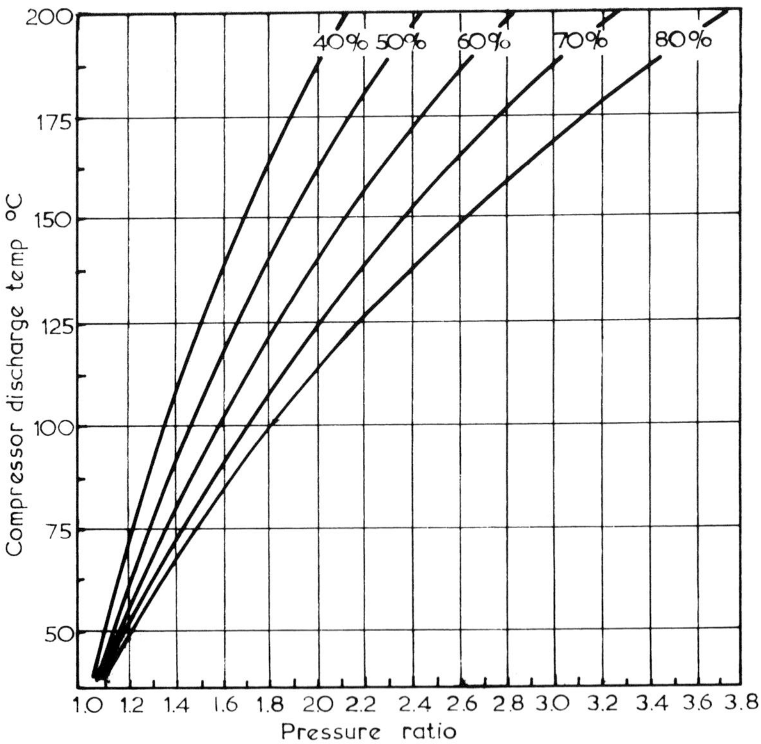

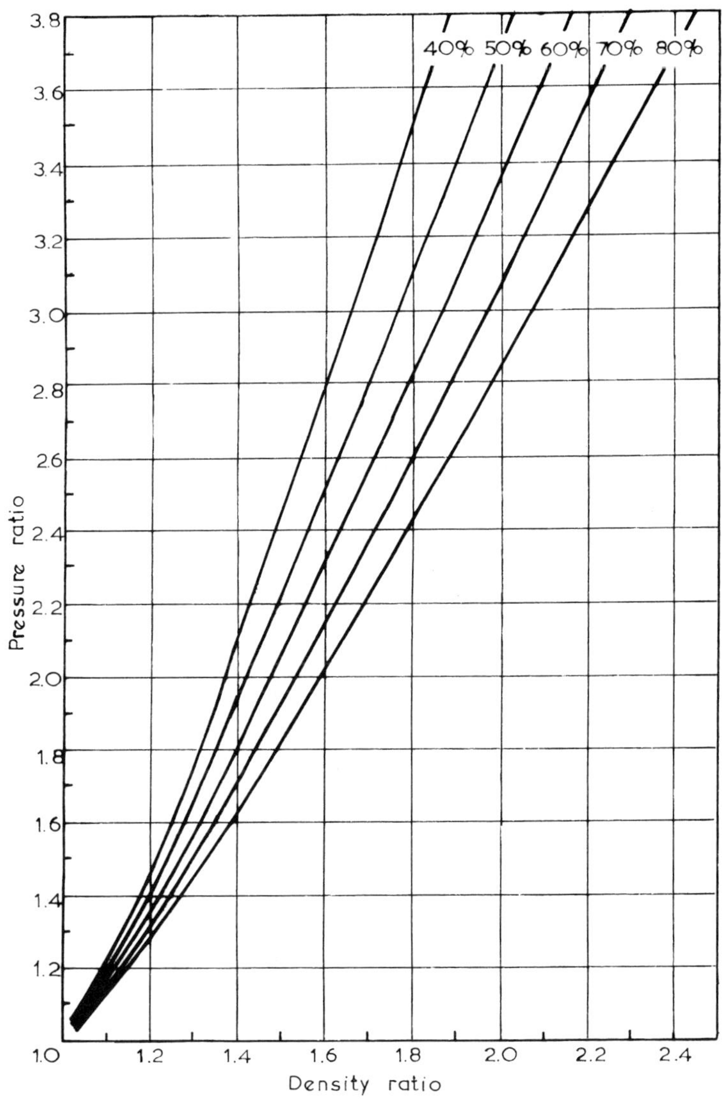

Fig. 5.3. Pressure ratio to density ratio graph illustrates how compressor efficiency affects the density of air entering the engine.

the compressor discharge temperature and the pressure ratio from *Figs. 5.2* and *5.3*. The results are not completely accurate but they are useful for quick comparisons. Also, both vividly demonstrate how important it is that we operate a compressor in a high-efficiency zone so as to minimise compressor discharge temps and maximise the compressor density ratio.

In Chapter 10 we will consider how we can improve charge density and lower the charge air temperature before it enters the cylinders by means of charge air cooling – intercooling. However, even with the benefit of very efficient intercooling, say in the 75–85% range (many stock intercoolers are quite inefficient), we cannot arbitrarily choose a supercharge boost figure and expect that the engine will provide good overall performance at that boost level.

MORE BOOST MAY NOT BE BETTER

Some have the very misguided belief that if a little is good, then more has to be better. As a consequence they figure that, since the engine is a well-engineered solid unit, a boost increase from the factory 8psi figure to 11psi must be better. Maybe it will be better, but let's consider some scenarios which could point to improved performance in some areas and a reduction in others, or even an overall decline in performance (unless the engine is fed an expensive fuel cocktail).

As we have already established, more compressor boost pressure means more inlet charge temperature. This may not be a problem on icy nights, but how about in the middle of summer? Will the engine run better then? Don't expect the knock sensor to compensate. Many engines respond to knock sensor signals by retarding the spark timing and, as I will demonstrate later, more boost and less spark advance often spell less hp.

This increase in inlet charge temperature is often a compounding problem. You can lose two ways: first, from more pressure putting extra heat into the inlet charge and, second, at the elevated pressure ratio at which the supercharger is now operating, its pumping efficiency may easily drop 10%, which puts another bundle of heat into the inlet charge.

We will spend considerably more time on ignition timing and combustion in Chapter 17, but we will touch on the subject now because it will help you better come to grips with the related topic of compression ratio when, shortly, we get to that.

Starting the fire late – that is all retarded spark is – prevents cylinder pressure reaching a peak too close to top dead centre (TDC). Most engines perform best overall

All compressors have a certain air flow capacity. Spinning them faster by increasing the boost may only increase charge temperature and actually reduce charge density.

when cylinder pressure peaks at 12–14° after TDC and performance is usually acceptable with a pressure peak as late as 18–20° after TDC. At this point, piston acceleration away from TDC is relatively slow, so at the time when combustion pressure is peaking the combustion space is not rapidly increasing in volume. Consequently the pressure holds at a high level and gives the crankshaft a good shove, thereby making high hp.

DETONATION AND PRE-IGNITION

However, when the charge temperature is excessive, what should have been a progressive burn near TDC becomes a violent explosion – detonation. The combustion space is small, piston acceleration is slow, so piston, rings, head gasket, valves, spark plug, cylinder walls, bearings and crankshaft cop a good hammering. Enter the saviour for many – spark retard. Let me make this clear, optimum timing is just that – it is the best. Any deviation from that hurts performance, but, and this is why spark retard is a crutch for so many, it can save engines from being wrecked. However, it too can be a self-perpetuating problem, even an engine destroyer, as I will show you later.

Let's consider the positive aspect, if you can really call it that, of spark retard. The fire is started late to ensure the cylinder pressure peak is reached well past 20° after TDC. At the time peak combustion pressure is reached, the piston has begun more speedy acceleration; thereby rapidly enlarging the combustion space. Consequently, cylinder pressure quickly tapers off as the fire chases the piston down the cylinder, thereby warding off detonation. With the plateauing of cylinder pressure, combustion slows markedly, causing a decline in cylinder pressure. Less pressure equals less push on the crankshaft which in turn equals less hp.

However, combustion slows so much that as the exhaust valve opens, the fire is still going, consuming fuel. This increases exhaust valve temperatures as the fire races out past the valve/s into the exhaust manifold. The exhaust side of the combustion chamber becomes hotter too, as does the piston crown. Soon the inlet valve opens to admit the fresh fuel/air charge, but all these hot parts quickly heat the inflowing charge. It quickly expands, filling the cylinder with a low density charge that will result in a poor power outcome at best.

At worst, combustion has continued so late on the down stroke that the piston crown, exhaust valve and combustion chamber have had insufficient time to cool before the fresh charge enters the cylinder. This retained heat is enough to get combustion started long before the plug fires. The intense heat resulting from pre-ignition is much like a blow torch and melts through the crown of the piston.

HOW SPARK RETARD AFFECTS POWER OUTPUT

Rather than getting into the real evils of spark retard, let's have a closer look at its effect on power output in a best-case scenario. The guinea pig in this case is a Nissan 200 SX. Because it is a road car the engine retained its stock 8.5:1 compression ratio as a reasonable compromise to providing acceptable off boost performance and fuel efficiency. Several modifications were performed to maximise flow into and out of the engine, and the turbo itself was revised in a number of areas. Finally, on an engine dyno the ignition timing/boost pressure relationship was looked at to provide best overall

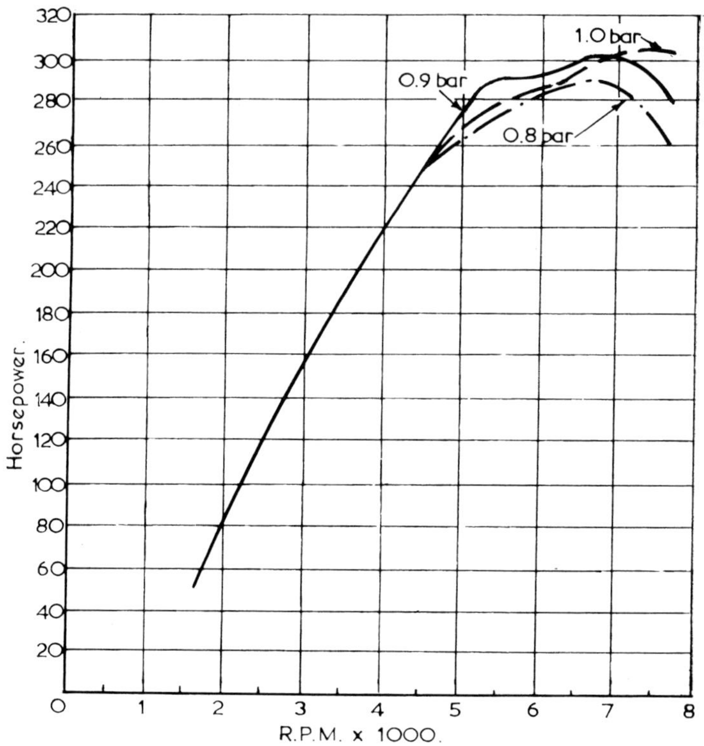

Fig. 5.4. Additional boost may enable a higher power output at close to maximum engine revs, but in the most important area of the power band output could be down because of the need for spark retard to keep the engine free of detonation.

performance rather than peak hp or best high rpm power while operating on readily available 95/96RON unleaded fuel, rather than 98 octane unleaded which was not available in all areas where this car would be driven.

Many tests were carried out to find a best-performance target area. Once identified, tests were concentrated in that area to provide the optimum outcome. *Figure 5.4* illustrates the degree of power loss or gain over a very small boost range, 0.8 to 1.0bar (about 11.6–14.5psi). Clearly 0.9bar boost produced the best peak power over the broadest range. At 1.0bar pressure, spark retard to prevent detonation lost power over a range of about 2500rpm, but gained power near the rev limit. Below 4500rpm virtually identical outputs were recorded with suitable manipulation of spark lead.

COMPRESSION RATIO vs. BOOST

Now let's move on to the compression ratio/boost pressure balancing act. Over the years, various mathematical formulas have been suggested as a basis for calculating the equivalent compression ratio when matched with a particular boost pressure. They don't work. First, they do not take into account intake charge temperature which can vary by a huge amount for identical boost pressures because of a wide range of supercharger efficiencies (45% for Roots blowers up to nearly 80% for centrifugal compressors), and intercooler efficiency (0% with no intercooler up to 85% with a very efficient intercooler, or even over 100% for ocean race engines using sea water intercooling). Second, they do not make any allowance for the fact that different types of superchargers develop manifold pressure at various engine rpm. Clearly, even moderate boost of say 5psi at 1800rpm with a Roots blower will require a lower compression ratio than an engine with a turbo which does not reach 5psi until 6000rpm.

Third, they do not allow for the fact that engines are different. Alloy heads, 4-valve combustion chambers and long duration cams usually indicate the engine will tolerate a higher compression ratio. Cast-iron heads, 2-valve chambers, short cam duration, carburettors or mechanical fuel injection and mechanical advance distributor point to a lower compression ratio requirement. Fourth, they do not give consideration to vehicle differences such as weight, aerodynamics, overall gearing and spacing between gears. A heavy vehicle shaped like a brick, with a small engine, with lazy gearing and huge gaps between gears demands a lower compression ratio. On the other hand, a light vehicle such as a motorcycle with a large engine, low gearing which sees the engine easily bump the rev limiter in top gear and carries six close gears, will work with higher compression.

With those generalities out of the way we can be a little more specific. However, now we have to give consideration to the vehicle's purpose – where it will be used. Will it be a daily driver that spends most of its life in the city or on twisting country roads? Perhaps it will see most use on unrestricted freeways and race circuits for weekend track days. Maybe it will be a pure competition vehicle.

If it is a general use road car there are a number of factors to consider. What type of fuel is most widely available and how much octane can you afford for every tank refill? Generally the choice will be 95 octane by the Research method (95RON), but if you live in a land where 91RON is the regular fuel and 95RON is considerably more expensive, you may decide on the affordable stuff.

The next factor is fuel economy. You may scoff at the idea now, but if you travel 1000km each week you will want to squeeze more kilometres out of every litre. A higher compression ratio means better economy. For example, reducing the CR from 9:1 to 8:1 will result in about 4–5% more fuel usage at cruise.

If you are going the turbo route you must also make driveability a priority. A low compression turbo engine is not fun most of the time. It may be acceptable if you spend all your time on unrestricted freeways, but for general driving you will enjoy your car a lot more if the CR is at least 8.5:1, and preferably in the low to mid-9s if 98RON fuel is available and affordable.

SAAB'S VARIABLE COMPRESSION ENGINE

A better solution would be some way of giving the engine a variable compression ratio. At cruise the almost closed throttle blade strangles air flow, so air density in the cylinders is quite low. In this situation we would achieve improved fuel economy and throttle response with a CR above 12:1. Conversely, on boost, elevated performance is usually found at CRs below 7.5:1.

Twenty years back, SAAB engineer Per Gillbrand reckoned the same way and, from his sketches, SAAB powertrain guru Torsten Anderlind oversaw production of the SVC engine as a development platform for Gillbrand's ideas. As I write this, the experimental engine has continuously variable CR from a low of 8.0:1 up to 14.0:1, courtesy of an

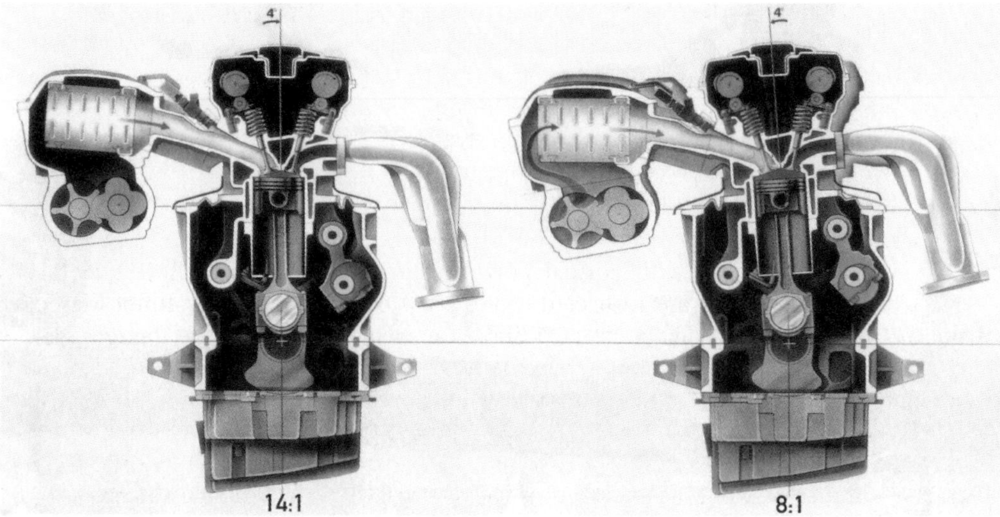

SAAB Variable Compression engine. The cylinders are attached to the head, and the combustion chamber size is varied by a head-lifting eccentric (exhaust side of block) tilting the head/cylinder mono-unit supported by a pivot shaft (inlet side of block), a maximum of 4º. A flexible 'bellows' seals crankcase to the cylinder/head.

ingenious method of tilting the cylinder head up to a maximum of 4° so as to alter combustion chamber volume.

With supercharging provided by a screw blower at up to 2.8 bar absolute (26psi boost), the 1.6-litre 5-cylinder engine (68mm bore x 88mm stroke) currently produces a promising 225hp at 6000rpm and a hefty torque peak of 225lb/ft at 4000rpm. Even more impressive, however, is that torque is already 133lb/ft at only 1000rpm, and at 6500rpm is holding at 177lb/ft (219hp). SAAB state that, if needed, over 300hp and 370lb/ft torque would be easily achievable.

COMPETITION RULES AND FUEL SPECIFICATION

Competition engines are another matter. Often rules will dictate the direction which one will have to take. Many rally competitions stipulate a 34mm turbo inlet restrictor. This means that once past max torque revs charge density is rapidly decreasing, and the only way to reverse the situation is to use a high CR above 9:1. The situation is similar at Le Mans with the 2001 direct injection 3.6-litre Audi V8 operating on a CR just over 12:1, with boost of 1.67bar absolute (9.5psi boost) and two 32.4mm restrictors as per the rule book. (The previous year, with normal port injection, the CR was a little over 11:1.) CART Champ car engines are restricted to very low boost. In 2000 the Ilmor Mercedes engine operated at 14:1 with boost of 4.88psi on methanol. Generally, though, we are desirous of high boost pressures to make big hp, and that means low compression ratios of probably less than 7.8:1 and perhaps as low as 6.5:1.

To give some guidance in this matter I have prepared a graph, *Fig. 5.5*, which will be useful if you keep some disclaimers in mind. The compression ratio/boost relationship illustrated will be reasonably safe for modern 4-valve engines operating with digital electronic injection and ignition on 98RON unleaded with intercooling efficiency as shown. The graph specifically applies to turbos and centrifugal blowers. Blowers making good boost at low rpm will require a CR up to 1.0 lower, while alcohol competition engines could benefit from a compression ratio about 1.0 higher.

If you use fuel with a lower or higher octane rating, the CR/boost relationship changes. Usually you can reckon on an octane decrease of 2.0 requiring a boost reduction of at least 1psi, and vice versa. Therefore, a change to 95RON or 91RON will require boost decreases of 1.5 and 3.5psi respectively. On the other hand, switching to 100RON or Avgas 100/130 will permit increases of 1psi and about 5psi.

CALCULATING THE COMPRESSION RATIO

Another point to note is that I am discussing real, not imaginary, compression ratios. Some manufacturers work to quite close tolerances, so the CR of their engines usually

Fig. 5.5. This graph shows what boost levels should be acceptable for a modern 4-valve engine operating on 98 RON fuel.

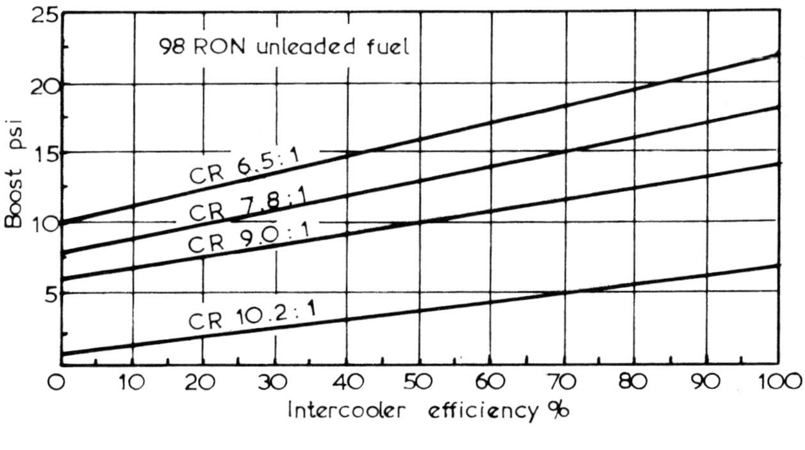

do not vary more than plus or minus 0.15 either side of what is stated in advertising literature. Unfortunately, most vehicles for the masses show much wider variations. Many are no closer than plus or minus 0.3, and I have seen worse than that! Consequently an engine which you think is 9.5:1 may in fact be 9.8:1. So, instead of figuring boost pressure from the 9.0:1 line on the graph you should, in fact, be using the 10.2:1 line. (Note that I do not advocate using CRs over about 9.5:1. The bottom line on the graph is there to show you what boost pressures are suitable if you choose to supercharge a stock high-compression engine.)

The actual compression ratio is the relationship between the total volume of the cylinder, head gasket and combustion chamber, with the piston at BDC (bottom dead centre) and the volume contained in the space between the piston crown, head gasket and combustion chamber at TDC (top dead centre). This is expressed by the formula:

$$CR = \frac{CV + CCV}{CCV}$$

where CV = cylinder volume
CCV = combustion chamber volume

Naturally, CV is the engine capacity in cc divided by the number of cylinders. CCV, however, is not so easy. This volume is made up of the combustion chamber volume, plus the volume that remains above the piston when the piston is at TDC, plus the volume caused by the thickness of the head gasket, plus the volume of the dish if dished pistons are used, or minus the amount displaced by high top pistons if these are fitted.

When we know what compression ratio we want, then we can calculate what the volume CCV should be to give that ratio, using the formula:

$$CCV = \frac{CV}{CR - 1}$$

Assuming the engine has 500cc cylinders and we want a compression ratio of 9.2:1 then:

$$CCV = \frac{500}{9.2 - 1} = \frac{500}{8.2} = 61cc$$

To find what the actual value of CCV is in our engine we first must physically measure the volume of the combustion chamber using a burette filled with liquid paraffin or water (*Fig. 5.6*). Incidentally, the combustion chambers should have been equalised in volume so that, for road engines, there is not more than a 0.1 change in compression ratio from cylinder to cylinder. This means that if the pistons are the same level below the top of the block at TDC, then an engine with 300cc cylinders will have a variation of no more 0.5cc between the largest and smallest combustion chamber. For a 500cc cylinder the spread between the largest and smallest should not exceed 0.8cc, and for 700cc cylinders the spread is 1cc. However, for competition engines we must be much more exacting. Personally I work to equalise the chambers within 0.1cc, except in engines which have a history of detonating a cylinder (usually because of coolant flow problems). Then I would take some compression out of the detonator by means of modifications to both the chamber and piston crown.

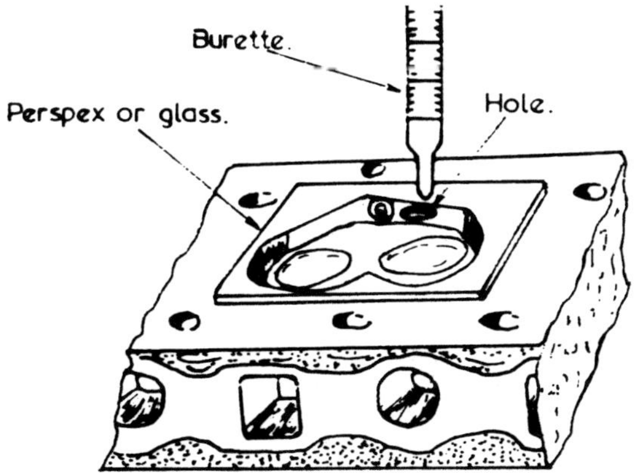

Fig. 5.6. Measuring combustion chamber volume.

If the engine has pistons which are either dished or high top, or have valve cut-outs, then the increase or decrease that they are causing has to be measured. For instance, if the cylinder bore is 90mm in diameter and the piston crown at the edge is 6mm from the block deck, using the formula:

$$V = \frac{\pi D^2 \times H}{4000}$$

where $\pi = 3.1416$
D = bore diameter in mm
H = height between piston crown edge and block deck in mm

this volume should be:

$$V = \frac{\pi \times 90^2 \times 6}{4000} = 38.2\text{cc}$$

If, on measuring the volume with a burette, it is found to be 27.2cc (*Fig. 5.7*), the lump on top of the piston decreases CCV by 11cc (38.2 – 27.2 = 11).

If the piston in this example had a dished top and the volume, as measured by the burette, was 52.7cc, then the dish volume would be 14.5cc, increasing CCV by 14.5cc (52.7 – 38.2 = 14.5).

Fig. 5.7. Measuring high-top piston displacement.

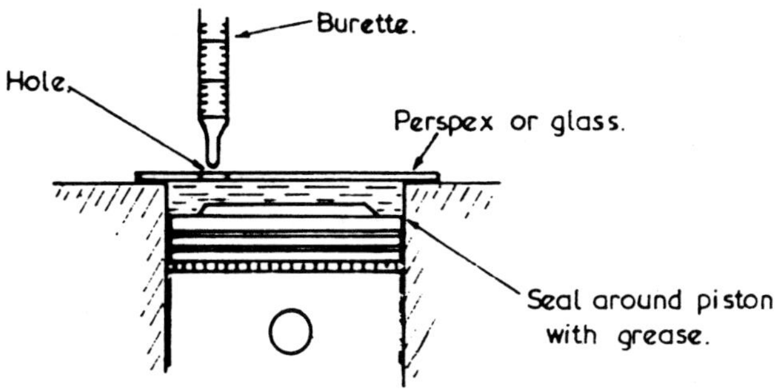

This piston for a 2-valve combustion chamber has an offset bowl, located under the spark plug, to provide a generous squish area opposite the spark plug.

The above formula is also used to find the volume of the head gasket by measuring the thickness of a used, so compressed, gasket. If the pistons are flat tops and they are below the block deck at TDC, the same formula is used to find this volume as well.

DECK CLEARANCE AND SQUISH IN ROAD ENGINES

To assist in equalising compression and combustion pressures, the crown of every piston should rise to the same point in each cylinder. This is called the deck height or deck clearance (*Fig. 5.8*). In road engines I like to keep the squish clearance between 0.035in and a maximum of 0.045in to assist combustion and cylinder scavenging. Hence, if the compressed thickness of the head gasket is 0.030in then the deck clearance required will be 0.005–0.015in. Usually, to achieve this figure the top of the block will have to be milled. Note, in a road engine, that it is not essential for all pistons to rise to the same height in each cylinder; a variation of 0.005in is acceptable. The reason I like to build a good deal of squish into a road engine is that this causes the piston to squish the fuel/air

Fig. 5.8. A tight squish clearance of 0.035in to 0.045in helps combustion in road engines. The squish clearance is the sum of the compressed thickness of the head gasket and the piston deck clearance.

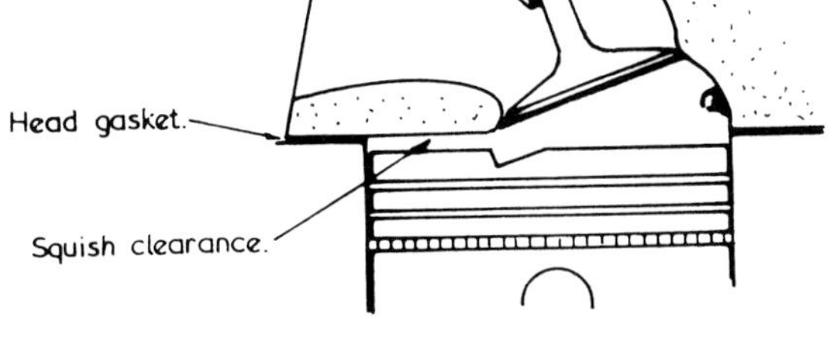

charge from the edges of the cylinder towards the spark plug. The fast-moving gases meet the spark plug and quickly carry the combustion flame to the extremity of the combustion chamber, thereby reducing the risk of detonation. This opens the way to use a higher compression ratio to enhance off boost performance and fuel economy.

With the passing of time, more benefits of building squish into the engine have come to light. The mixture being purged across the combustion chamber from the squish areas homogenises the fuel/air mixture more thoroughly and also mixes any residual exhaust gas still present with the fuel charge. This serves to speed up combustion by preventing stale gas pockets from forming. Such pockets slow down, and in some instances can prevent, flame propagation. With improvements in combustion quality, performance is improved, particularly at cruise. Consequently the engine is markedly more responsive and fuel consumption falls.

Turbulence caused by the squish effect also serves to enhance heat transfer at the spark-initiated flame front. Without proper heat transfer, jets of flame would tend to shoot out toward the edges of the combustion chamber, prematurely heating the surrounding gases to start off the cycle leading to detonation.

Whenever dished or reverse-dome pistons are fitted to lower the compression, the outline of the dish should ideally be a mirror image of the outline of the combustion chamber. This means that the piston has a squish surface which exactly matches the cylinder head squish area. Unfortunately, pistons machined to mirror the combustion chamber tend to be much more expensive because of the intricate machine operations involved. Hence, car manufacturers tend to use lower cost pistons with a central dish and narrow squish band or an offset dish which places a large squish area over to one side of the piston crown. When the piston squish band is as wide as the cylinder head area the required squish is produced. However, the squish band on some pistons is much narrower than the cylinder head squish area, and some tuners, when lowering the compression ratio of high boost turbo engines, also machine the squish band too narrow.

SQUISH IN RACE ENGINES

The situation is different with competition engines. Here, low speed combustion quality and cruise economy are not usually a concern. What we are after is the best hp over a range of about 35–45% engine revs at the top of the power band, usually on throttle openings wider than 60% open. Under such circumstances mixture density is quite high, so there is a corresponding increase in mixture turbulence as the piston rises to TDC on the compression stroke. Therefore, the squish area can be reduced or the squish clearance can be increased in many race engines without any power loss. Perhaps there may be a power gain.

I have experienced engines on the dyno that sounded good and sharp with wonderful throttle response, but with only average power output. Frequently opening up the squish clearance would see a nice power gain and much smoother running. The conclusion I reached was that tight squish was causing some sort of combustion instability, or possibly the very rapid compression as the piston squeezed the gases between the squish faces meant that the gases became the equivalent of solid matter, so in effect it was like the piston banging into the head.

While you digest that last thought, keep in mind that engines literally can go solid, or 'hydraulic', when supercharged. The problem is more often noted in nitro-burning engines running pistons which are not dished, but I have also seen it in engines operating on very rich alcohol mixtures and also when crude water injection systems have been employed. What happens is that as the piston accelerates up the cylinder it collects all the unvaporised fuel or water. Violent wind across the piston crown disperses this liquid away from the centre of the piston out towards the edges. If the clearance between piston and head is sufficient, no harm is done as the piston comes up to top dead centre. If it is too close, the engine goes solid as the piston attempts to compress the liquid against the head. There is no time for this liquid to be gently squeezed out of the squish gap into the combustion chamber as some suppose. Nor does fuel or water puddled into one area on the piston crown get 'flattened' or dispersed across the entire squish area. Basically, the damage will range from distorted pistons through to a totally destroyed engine.

In two-stroke engines the squish areas are extremely tight, down to 0.020in, but best combustion often results from squish areas which diverge towards the centre of the combustion chamber. Therefore, while the squish clearance may be only 0.025in at the very edge of the piston, it may open up to 0.032in about 16mm from the edge.

I have experimented with the same idea on four-strokes and, while it had the desired effect, it did not appear to be superior to simply opening up the squish clearance, or reducing the squish area. Usually the clearance has to be opened up to around 0.065in to 0.080in. This can be achieved in many ways. By far the simplest method is to use a thick solid copper head gasket. These are expensive but can be reused over and over. Of course, the thicker gasket also lowers the compression ratio, but usually with a competition engine this is to our advantage. With motorcycle engines it is even easier as you can raise the barrel a similar amount (up to 0.080in if you wish to bring the compression down as well as open up the squish) using either a copper or an aluminium spacer.

PISTON CROWN THICKNESS, RING POSITION AND ROD LENGTH

The piston crown can also be machined. However, there are a couple of things to keep in mind. Obviously, taking material off the crown reduces its thickness, thereby weakening the piston. Personally, on competition engines which are expected to live 1000km between rebuilds, I like piston crown thickness to be at least 0.3in up to 0.35in in bore sizes over about 84mm. Such thick crowns, in addition to being very strong, act as a heat sink, so release their heat load to the piston rings more slowly. This aids ring life over the long run, but even short term the extra metal can protect the rings from severe overheating that could seriously compromise their performance. Because of this latter problem I don't like the top ring to be less than 8mm from the top of the piston, and, in fact, 9.5mm is better in very high boost engines. Clearly, milling the piston to open up the squish could move the top ring too close to the top of the piston and expose it to excessive heat. Therefore, a better solution is to order special pistons with a lower compression height to get the correct squish clearance (and/or lower the compression ratio), plus specify lower top ring placement to move it away from so much heat.

Another alternative is to use shorter con rods. In a naturally aspirated engine we usually want to increase rod length for superior performance, but forced induction engines are different. Here, within reason (don't go to a rod length to stroke ratio of less than 1.5:1), a shorter rod can be used to open up the squish clearance and/or lower the compression ratio without losing hp. In fact, because short rods accelerate the piston more rapidly away from TDC they can help in keeping away from detonation.

Chapter 6

What you need to know about turbochargers

When you look at a turbo, or even hold one in your hand, it appears to be a relatively simple device. In fact, in comparison with all the wires, hoses, relays, sensors, etc. adorning the modern engine, the basic mechanical aspects of it are simple. However, just as a simple cam lobe dictates the basic characteristics or nature of the naturally aspirated engine, so also with the turbo on a supercharged engine. Similarly, just as an error in cam choice can make a vehicle extremely unpleasant to operate, whether on the road or the race track, the very same holds true with turbo selection.

The first wave of engines given the 'hot-up' treatment by tuners simply bolting on a turbocharger were a real disappointment. When the throttle was pressed down at speeds below 3000–4000rpm nothing happened; it seemed as if the engine was running with all of the plugs fouled. Then, after gaining some boost, the vehicle would rocket off with an incredible surge of power which was most disconcerting and intimidating. Controlling a turbo car on a twisting road was a nightmare. That was 35 years ago when turbocharging first became fashionable on race cars. Tuners didn't know much about this new technology and tried to transfer what little knowledge had been gained on the race circuit to the road. It didn't work. Two decades later, many car makers were involved in their own development projects which eventually led to the design of smaller, lighter turbos and compatible engine management systems that have virtually eliminated the dreaded 'turbo lag'. The majority of new generation turbo cars are pleasant to drive, but with a few ill-conceived modifications they can be quickly transformed into the volatile beasts of former times. Therefore, before adding a turbocharger, or before undertaking any modifications on a factory turbo car, a thorough knowledge of the subject is an asset if driveability problems are to be avoided.

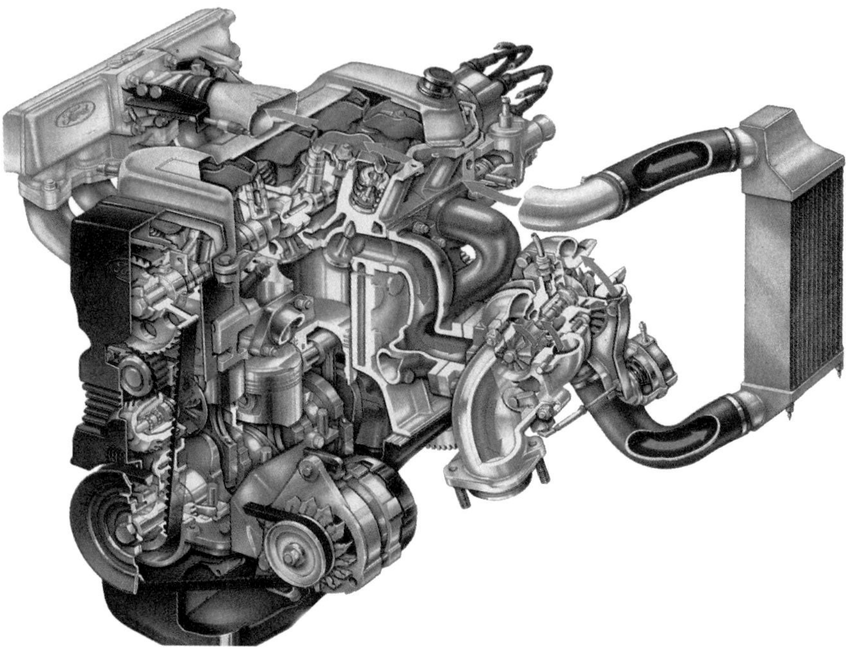

The Ford RS Escort produced 132hp on 7.1psi boost.

Fig. 6.1. From this graph, power increase at various levels of boost can be estimated.

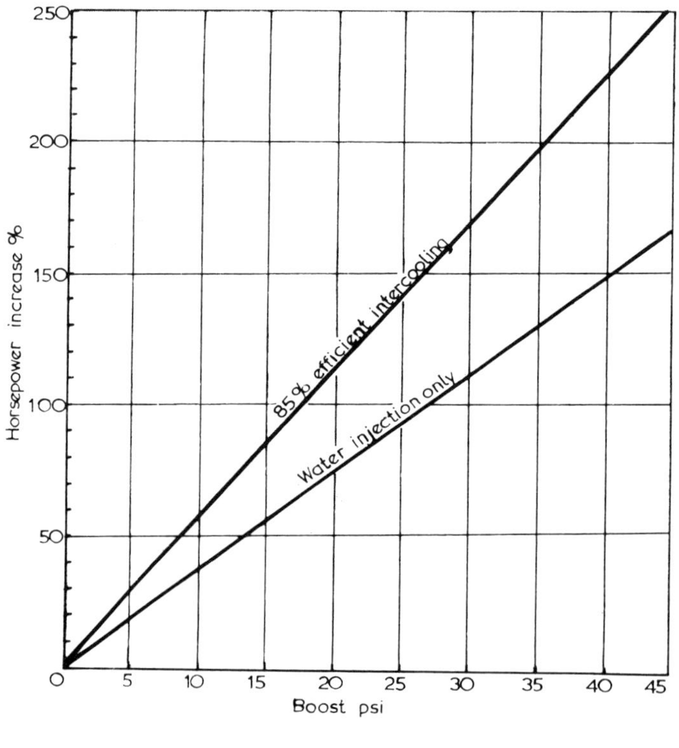

HOW TURBOCHARGERS BOOST POWER OUTPUT

There is nothing magical or hard to comprehend about why a turbo greatly increases the potency of an internal combustion engine. Basically, the power an engine produces is a product of the amount of fuel/air mixture it is able to breathe in and then burn. A turbocharger is a type of supercharger or compressor that forcibly crams into the cylinders far more mixture than the engine could draw in in naturally-aspirated form. At 100% volumetric efficiency (VE) and a manifold vacuum level of zero, a 500cc cylinder will be filled with 500cc of air/fuel mixture of the same density as the atmosphere, because of the effects of normal air pressure (14.7psi at sea level). Theoretically, if the inlet boost is 7.35psi above atmospheric the cylinder will fill with mixture which is 50% more dense, so the engine will perform as if it were 50% bigger.

You know, of course, that compressing air adds heat which lowers the theoretical power increase. However, with efficient charge cooling we can pull much of the heat out of the air before it enters the engine. This, then, gives the potential to achieve a power increase in the order of 40% from 7.35psi boost (*Fig. 6.1*).

The turbo is made up of a centrifugal compressor, which is part of the induction system, connected by a common shaft to a turbine which is joined to the exhaust system. The engine's exhaust gases then drive the turbine, making use of energy that would otherwise go to waste. The turbine, in turn, spins the impeller to force fuel/air mixture into the engine. Certainly, the concept is very simple, but things become a lot more complex when translated into the real world. The turbocharger itself is an expensive,

Fig. 6.2. The turbo turbine wheel is driven by the engine's exhaust gases. Being connected to a common shaft, the compressor impeller forces the air charge into the engine under pressure. When full boost is reached, the wastegate opens, bypassing some exhaust flow around the turbine.

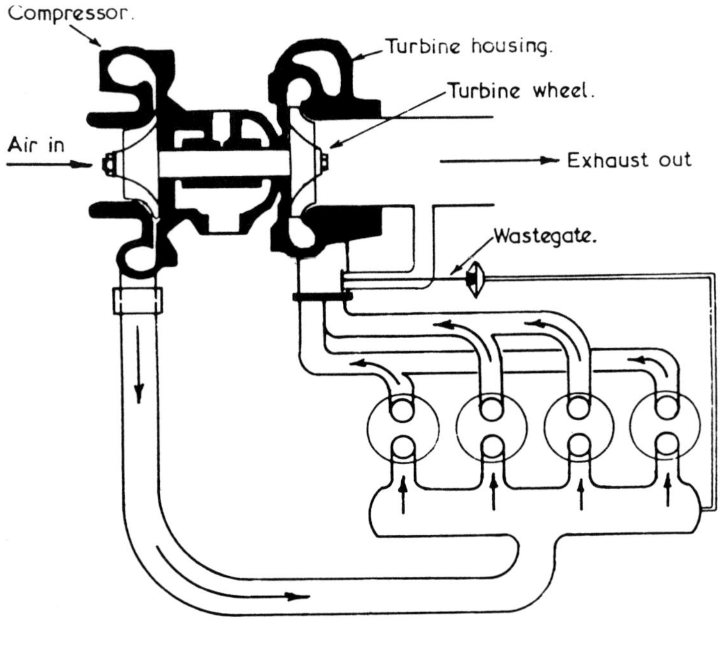

highly-engineered, precision item. On the exhaust side it has to be able to withstand enormous stresses from operating temperatures which may exceed 1000°C (1832°F), while at the same time the turbine is spinning at 100,000rpm and more. Additionally, the precise shape and dimensions of the turbine and impeller, and their separate housings, need carefully to be tailored to give the required throttle response, driveability and power (*Fig. 6.2*).

Aside from this there are the problems arising out of a fundamental characteristic of a centrifugal type compressor; namely that flow through it rises in proportion to the square of its rotational speed. Hence, at lower engine rpm there is little or no boost, and at higher rpm and full throttle, boost is excessive. It is in the resolving of this dilemma of how to get low-speed performance up without unduly restricting high rpm power, and perhaps causing excessive high-speed fuel consumption, in which the real science of turbocharging lies.

BOOSTING LOW-SPEED PERFORMANCE

Looking first at the low-speed component of this problem it should be obvious that modifications which reduce the low rpm performance of a naturally aspirated engine will have a like effect on a turbo engine. In fact, the decline in performance and economy will be greater in the turbo engine because of the compression ratio being 1 to 1.5 points lower in street machines, and perhaps as many as 6 points lower in competition engines. Therefore, big ports and cams are definitely not for turbo cars. We will look at this in a lot more detail in Chapter 14, but for now it is sufficient to say that for road use you will probably be happiest with the standard factory cam or the standard factory cam out of a non-turbo engine.

In a twin-cam engine, increasing the valve overlap period of the stock cams by means of adjustable sprockets will frequently yield impressive power gains, providing the exhaust has been modified to reduce back-pressure. Usually about 3° or 4° advance on the inlet cam and 4–6° retard of the exhaust cam will do the trick. In fact, a number of successful race engines compete with retimed stock cams. Even very powerful race engines use what would be classed as quite mild cams in a naturally aspirated (NA) engine. For example, the 4-cylinder 2.6-litre Offy Indy engines made 800hp using inlet and exhaust cams with only 254° and 244° duration measured at 0.050in lift respectively, and 0.425in lift. The V8 2.6-litre Foyt (Ford DOHC) engine from the same era ran even softer cams. The inlet measuring 242° duration and 0.390in lift, and the exhaust cam 232° duration and 0.380in lift.

TURBINE SIZE INFLUENCES PERFORMANCE

The other factor in low-speed performance is the actual boost of the turbo at lower rpm. Here, the size of the exhaust turbine and the dimensions of its housing play a major role. Naturally, a large diameter turbine, which will flow a lot of exhaust gas without causing much restriction at high rpm, is very heavy so it will need a large exhaust flow to get it spinning up to speed to create boost. Because the exhaust flow is low at lower rpm with the throttle closed, the turbine will be rotating slowly. Opening the throttle increases flow but it takes time to accelerate the heavy turbine up to speed to produce boost. It is this time-lag between opening the throttle and actually getting response which is called 'turbo lag'. A small turbine, on the other hand, has less inertia so it spools up to speed to produce boost more quickly. However, if it is too small it plugs up the exhaust at higher rpm, and by restricting exhaust flow reduces power and increases fuel consumption. Clearly, a compromise must be struck, with the majority of manufacturers choosing comparatively small turbines to enhance performance, particularly from 1800 to 4000 rpm. Most readers of this book are probably more interested in something beyond factory turbo performance levels but, even so, for road cars we should be selecting a turbine size that will have the compressor making boost by at least one third maximum hp revs.

A quick guide as to the performance potential of a turbine is to take a look at the size of the exducer bore (*Fig. 6.3*). While we cannot read a whole lot into this it should be obvious that a turbine with an overly large exducer bore will not be able to drive the compressor fast enough to make good mid-range boost. Conversely, a tight bore will strangle top-end performance. Therefore, generally speaking, we need to be looking for an exducer bore size in the range indicated in *Fig. 6.4* to achieve the horsepower output which we are striving for.

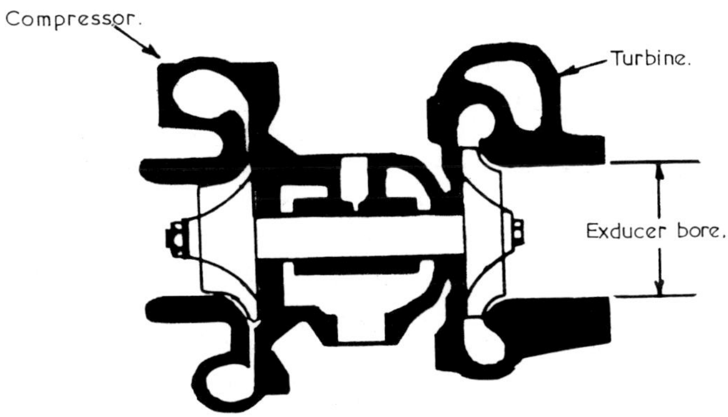

Fig. 6.3. The exducer bore diameter hints at a turbine's likely potential.

Fig. 6.4. To achieve a particular power level, the turbine exducer bore size will have to fall somewhere between these curves.

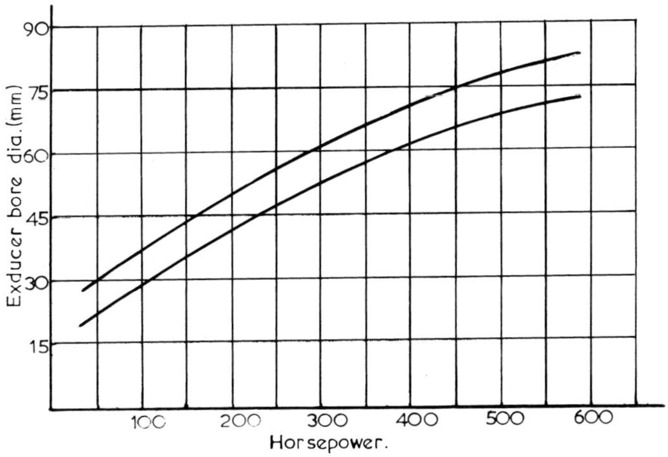

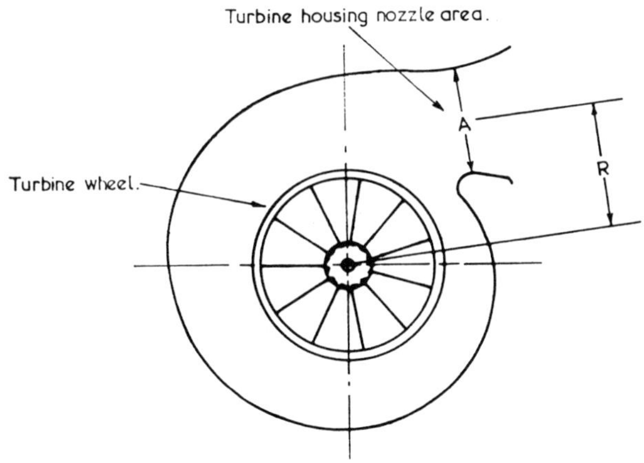

Fig. 6.5. The turbine housing A/F ratio is determined by dividing the area of the turbine nozzle A by the radius R from the centre of the turbine axle to the centroid of the housing throat.

TURBINE A/R RATIO AND HOUSING SELECTION

While the exducer bore size gives us some idea of the turbine's potential, the really critical dimension is what is called the A/R ratio. Within a single turbine size group several turbine housings will be available in a number of A/R ratios to fine tune the turbine to best suit our performance requirements.

As illustrated in Fig. 6.5, the A/R refers to the ratio between the area of the turbine nozzle A, and the radius from the centre of the turbine axle to the centroid of the housing

Fig. 6.6. Small A/R turbine housings make more boost at lower engine rpm but reduce maximum power because of exhaust flow restriction.

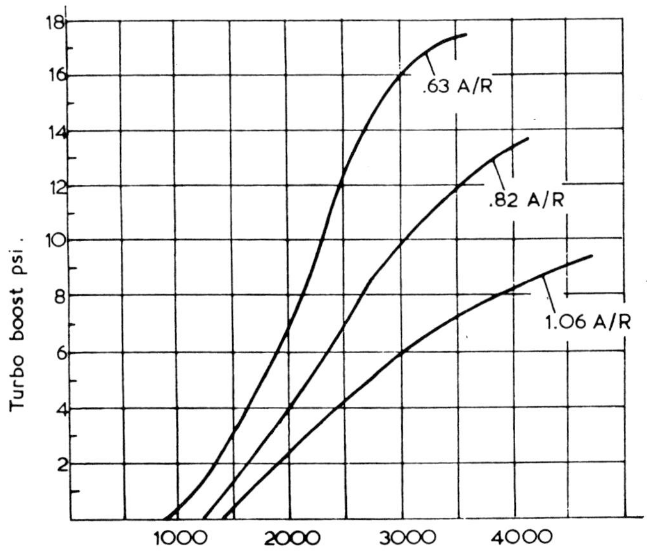

throat R. The smaller the A/R, the smaller the passage (nozzle) and the faster the turbo spins for a given exhaust flow. This is good for low-speed boost but a housing which is too small causes excessive back pressure and heat. A large turbine housing A/R flows the exhaust gas without causing so much restriction at higher rpm. However, because of the turbine speed being reduced at lower revs, low speed boost may be poor. Again, a compromise is necessary, with car manufacturers generally opting to get either low-speed or mid-range boost up at the expense of maximum power and economy.

Turbocharger manufacturers do have turbine performance maps to guide them in turbine selection, but the proven method is 'how the car feels'. While recording the boost levels with the engine running on a dyno clearly indicates how soon a specific number A/R housing makes boost (*Fig. 6.6*) and hp numbers indicate the power output, these numbers are of much more concern only in competition engine development. Of far greater importance for road vehicles is how the engine responds in actual day-to-day driving. Clearly, if it feels sluggish the A/R is too large for the available exhaust flow. On the other hand, a car that is very 'jumpy' and difficult to drive smoothly at lower engine rpm, and which seems to run out of power in the upper rev range, probably has a housing with an A/R too small.

FACTORS INFLUENCING COMPRESSOR SELECTION

On the inlet or compressor side of the turbo, selecting the correct impeller and housing is considerably easier as we can do the hard yards on a calculator and compressor maps. Here, the main point to keep in mind is that a compressor is more efficient, adding less

The turbocharger is a visually unimpressive device that can quite easily double engine power when care is exercised to carefully match compressor flow to engine air flow requirements.

heat to the air, at a particular combination of boost pressure and air flow. The aim, then, should be to choose a compressor size with maximum efficiency over what we deem to be the most valuable part of the engine's rev range. Obviously that will vary considerably depending on the application. For example, an engine tuned to operate on a circuit consisting mostly of very long straights would benefit from a compressor that is very efficient over a narrow range of perhaps only 2000rpm right at the top of the power band. On the other hand, the optimum compressor for a road engine will be one that delivers flow with least heat somewhere around mid-range revs, over as broad a range as possible with the recognition that at higher air flow rates the compressor will add considerably more heat to the inlet charge.

Unlike on the turbine side, the size of the compressor does not have much influence on turbo lag or when the compressor will begin to make boost. The major concern with compressor selection is that it will flow the quantity of air we want, at the pressure we want, and in so doing add a minimum of heat. Large flow rates obviously demand larger impeller diameters but, as this is the lightest part of the turbo's rotating assembly, any size increase has a minimal effect on total inertia. Also, while a large compressor wheel does experience more leakage past the blades at low rpm the engine speed at which the turbo begins to make boost is really more a function of turbine rpm which in turn is dictated by exhaust flow and turbine geometry.

Whereas turbine size selection is really a matter of trial and error, plus a good dose of experience thrown in, compressor size selection is made with the aid of a calculator and turbo manufacturer's compressor maps. When compressor maps are not available, try to find other engines in a similar state of tune as yours and copy their compressor size.

CALCULATING ENGINE AIR FLOW

In the previous chapter we looked at calculations to determine the pressure ratio and the density ratio. Later, we will see how intercooling will help us get heat out of the compressed intake air, and this will enable us to get the density ratio back closer to the pressure ratio. However, before we can use any of those figures to aid in compressor selection we must figure out the rate of air flow into our engine. The rate of air flow into an engine is generally expressed as so many 'cfm', which means cubic feet per minute. While some compressor maps use cfm others use pounds of air per minute (lb/min). Most compressor maps are corrected for 85°F/28.4in Hg or 20°C/981mB air conditions; so, to convert from cfm to lb/min, multiply by 0.07.

The basic calculation to find the air flow into the compressor is:

$$\text{CFM} = \frac{\text{L x rpm x VE x Pr}}{5660}$$

where L = engine capacity in litres
rpm = maximum engine speed
VE = engine volumetric efficiency
Stock 2-valve = 85%
Stock 4-valve = 90%
Street modified = 93%
Competition = 105%
Pr = pressure ratio (see Chapter 5)

Therefore, a lightly modified 4-valve 2-litre engine which is usually run up to 7200rpm, operating at maximum boost of 11.4psi (Pr = $\frac{14.7 + 11.4}{14.7}$ = 1.78) will have air flow into the compressor of:

$$\text{CFM} = \frac{2 \times 7200 \times 93 \times 1.78}{5660}$$

$$= 421.2 \text{ (29.5lb/min)}$$

COMPRESSOR SELECTION USING A COMPRESSOR PERFORMANCE MAP

Looking at the first compressor map, (*Fig. 6.7*), we see that our plotted point of 29.5lb/min air flow at a pressure ratio of 1.78 puts us off the right side of the map at something less than 62% (most likely about 58%) compressor efficiency. The plot on the second map (*Fig. 6.8*) appears to be more promising as we are just into the 73%

Fig. 6.7. At the air flow required, this compressor will be operating in a low efficiency zone.

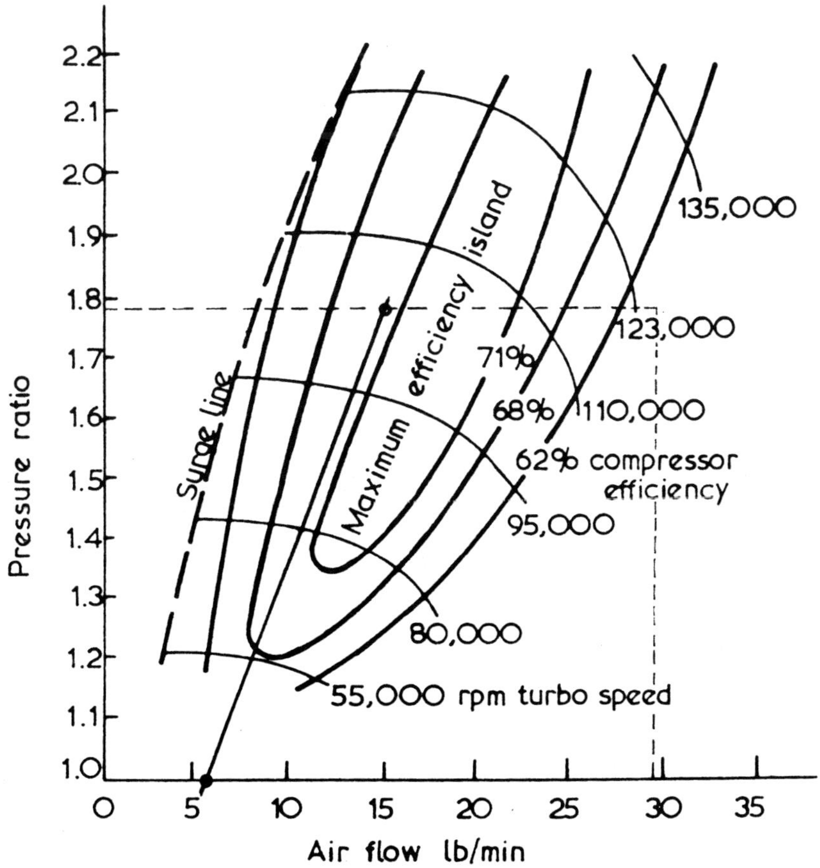

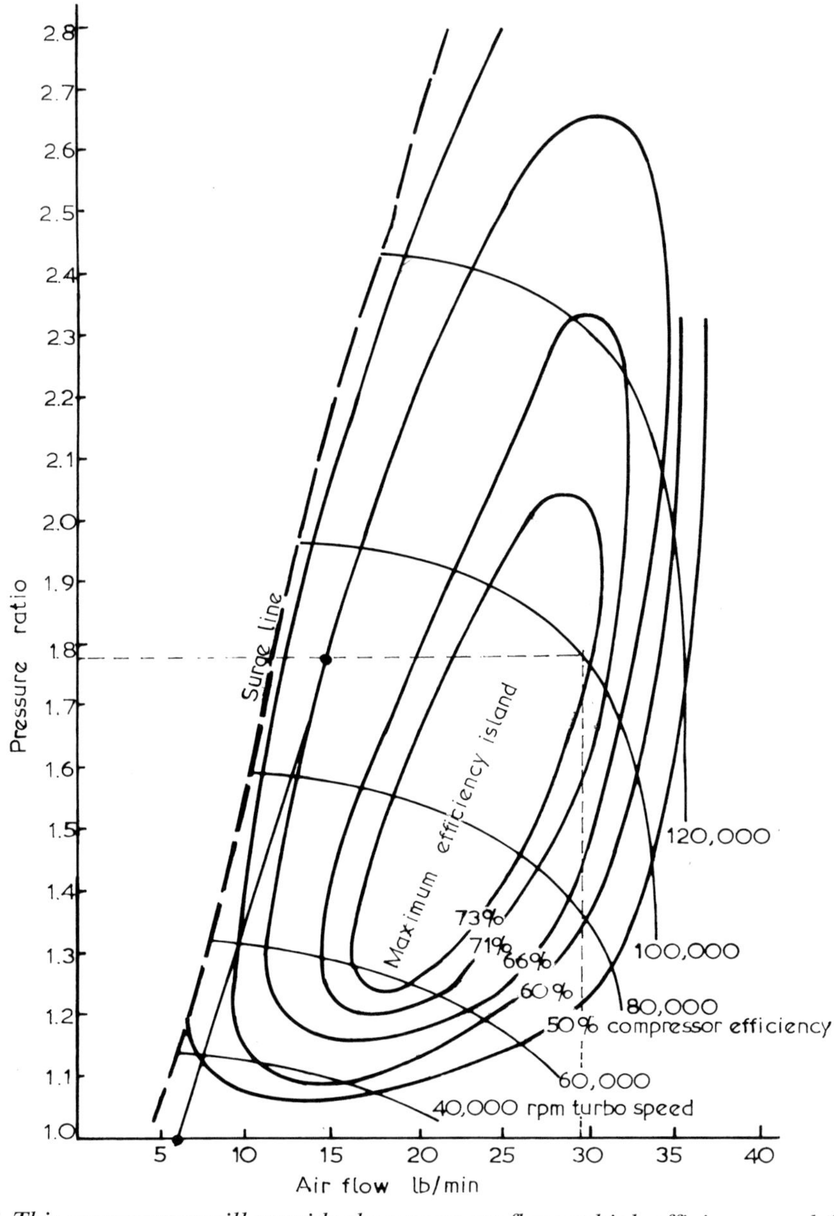

Fig. 6.8. This compressor will provide the necessary flow at high efficiency, and the 50% to 25% air flow plot line is away from the surge line at sea level, but at altitude it could be in trouble.

maximum efficiency island. At this point, if these were the only compressors we had a choice of, the second compressor, which provides the best efficiency by a clear margin of something like 15%, would be the better choice.

However, there is another plot we must make, and that is to make sure we don't ever run the compressor into surge. This can be calculated, but as the installed manifold volume and intercooler efficiency can cloud the result it is just as productive to use another method which has proven itself to be a reasonable guide even if not entirely accurate.

To check that we are away from the surge line we must plot two more points on each compressor map. Because most turbo installations make full boost by 50% maximum engine rpm we simply plot the air flow at these revs, which naturally enough is half the maximum air flow (421cfm or 29.5lb/min), 211cfm or 14.8lb/min against our maximum pressure ratio of 1.78. Next, we calculate 20% air flow (84.2cfm or 5.9lb/min) and plot that point on the bottom axis (Pr 1.0). When we draw a line between these two points it should always fall completely to the right of the surge line, not crossing it anywhere.

Having plotted those points and drawn the lines, we can see that both compressors appear to be safe. One passes easily, but our preferred compressor is quite close to the surge line at the lower point. At sea level there shouldn't be any problem but at altitude we would have to keep our ears open for the telltale sound of surge; a pop or backfire out of the compressor inlet. At high turbo rpm, surge can lead to impeller blades breaking with serious consequences for the turbo itself and, possibly, the intercooler and the engine. While surge damage to this extent is not common, it does happen.

FACTORS IN DECIDING ON ONE OR MULTIPLE TURBOS

When planning a turbo system, an early consideration should be: Will I use one or multiple turbos? For some the answer is easy – they will try to avoid multiple turbos every time. Their reasoning is that one turbo means less cost initially and half the cost for maintenance and repairs in the future. It is easier to find space in the engine bay for one turbo. Also, they are not convinced that the spool-up time for one big turbo receiving full exhaust flow can be any worse than for smaller, multiple turbos driven by divided exhaust flow.

In general, those reasons are all quite sound. Therefore, I find it hard to justify two turbos on 4- or 5-cylinder road engines. However, in some forms of competition the choice is not so clear cut, unless race regulations demand a single turbo. For road-going 6-cylinder engines it is probably a 50/50 call. I will explain why later, but unless space was a serious problem I would go two turbos every time. Likewise for competition 6s, with one exception; in drag racing one big turbo is usually not a disadvantage in comparison with twin turbo installations. However, when we get to V8, V10 and V12 engines the choice should always be two turbos.

Of course, for some the choice of two turbos is purely a cosmetic thing. Oh, they may justify their decision to those who question their action with some sort of pseudo-science but in reality they like the look and the hype associated with twin turbos. However, in the above instances where I have indicated a preference for two turbos my motive is pure, as you will see.

Cost is a major concern in any turbo installation. A single large turbo that is suitable will be difficult to find second-hand and, because there are less about, your choice is limited. Often the only option will be a very expensive new turbo. However, going the twin-turbo route can open up a virtually unlimited supply of cheap used turbos. Most of these turbos will be off Japanese or European engines in the 1.8- to 2.3-litre range. Being a popular size, these turbos have considerably more technology invested in them. Obviously you won't have access to compressor maps but you can usually get excellent

A view of the twin turbo Nissan Skyline GT-R straight-six that few ever get to study.

The usual view of the Nissan RB26 engine.

results simply by comparing relative engine sizes and power outputs. For example, a pair of turbos off identical 1.8-litre engines producing 175hp will work well on a 2.6- to 3.5-litre engine that you are modifying to produce 320–380hp.

Turbos respond better to pulsing exhaust flow. However, when many cylinders dump into a common manifold the individual exhaust pulses are smoothed out. When we run two turbos we effectively spread the pulses further apart and increase their amplitude. Therefore, when the turbo is spooling up to speed rather than receiving a constant push it will get a series of vigorous shoves that help it accelerate faster. It is for this reason that twin-turbos frequently manifest reduced lag rather than that they have smaller low-inertia turbine wheels.

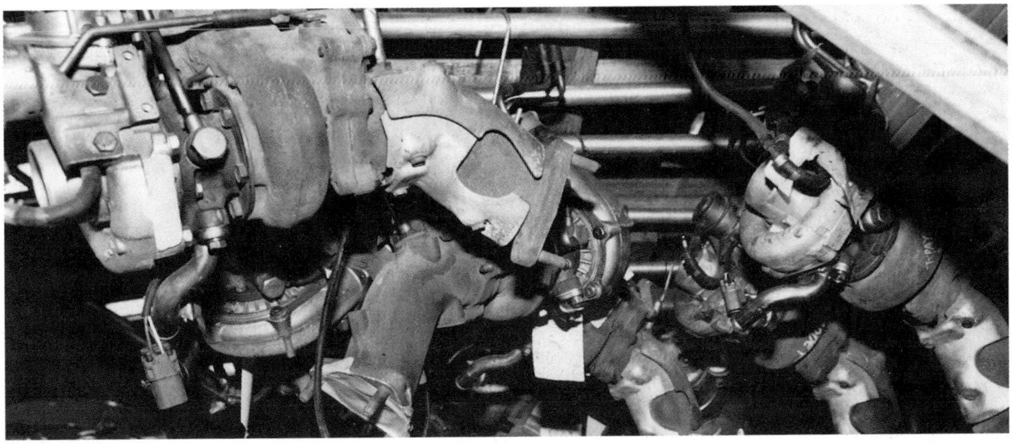

Top: The plentiful supply of cheap second-hand turbos from small engines makes the twin-turbo route very attractive.

Middle: Internal wastegate size is fixed, limiting how much exhaust flow can be diverted away from driving the turbine faster and thus spinning the compressor at inefficient boost/air flow rates.

Bottom: Twin-turbos are nearly always the best option on Vee and flat engines.

Because of the steep prices being demanded for external wastegates, many turbo installations have to make do with an internal wastegate. A single wastegate will only bypass so much exhaust flow around the turbine. So, as it approaches its flow limit more flow is forced through the turbine. This increases turbine and compressor rpm, so boost and inlet heat rises, and it may rise much higher than we desire and higher than the engine can withstand.

Also, this very high velocity flow from the wastegate will have been disrupting flow out of the turbine and impairing performance from the time the engine began to accelerate hard. We will discuss this in more detail in Chapter 16, but for now it is sufficient if you accept that anything we can do to reduce flow turbulence out of the turbine will provide a handsome performance reward. Therefore, when we halve wastegate flow we reduce considerably disruptive flow turbulence.

This brings us to actual exhaust design improvements when using two turbos. In V-type and horizontally opposed (flat) cylinder layouts it is easy to visualise how superior exhaust flow will be into twin-turbos. However, the same is equally true with straight-sixes. The other aspect, of course, is exhaust flow out of the turbine. Generally we will be hard pressed to find space down the side of the engine for anything larger than a 3in dump pipe. With two turbos, even small 2.5in dump pipes provide 37% more flow area to help turbulent turbine flows to more quickly settle, while 2.75in pipes provide a 67% increase.

Splitting the exhausts also means less problems with exhaust maintenance. The manifold between cylinder head and turbo is shorter so there is less metal expansion and contraction going on, reducing metal fatigue. Also, with exhaust flow divided, each turbo will run a touch cooler, which ultimately improves their service life.

HYBRID TURBOS

The question of hybrid turbos should also be discussed here. Some people refer to any modified turbo as a 'hybrid'. Therefore, they see a turbo which has had its stock 270° thrust washers replaced by 360° thrusts as a hybrid. Others would say that replacing the stock turbine wheel with a special Martinit 247 alloy wheel (good for 1300°C (2372°F) and costing what you would pay for two turbos) is making it a hybrid turbo. To me a hybrid is a turbo that has been machined and fitted with non-standard parts to increase flow, usually on the compressor side. And it most definitely is a hybrid if it comprises parts mixed and matched off two or more turbos.

Frequently the stock turbo is adequate. However, if you want better than stock power you often have to look at options which actually increase air flow out of the turbo compressor. Simply pushing up the boost pressure usually won't accomplish much in the way of performance, but some people are impressed with the boost gauge needle swinging around to a bigger number. If such makes them feel good, well so be it. The reason for the higher boost number and zilch extra power is that the compressor has reached its flow limit. At the higher boost, turbine speed increases because it is experiencing more exhaust flow (raising boost reduces bypass volume through the wastegate), therefore the compressor spins faster. All this additional churning heats the inlet charge causing the air to expand and, bingo, up goes the boost pressure. Recall the air density stuff we covered in the last chapter.

Compressor impeller trim

In each basic compressor wheel diameter a selection of wheels with different blade widths is available. However, rather than talking about blade width these distinctions are referred to as compressor trim. In technical terms trim is defined as the ratio:

$$\left\{\frac{Di}{Dt}\right\}^2$$

where Di and Dt are the inducer and tip diameters respectively, of the compressor wheel.

The way to obtain more compressor flow, apart from the more expensive alternative of getting a new turbo, is to fit a bigger compressor impeller. The larger wheel is not normally bigger in diameter, but rather has a greater blade width (*Fig. 6.9*). Obviously, for the new impeller to match the compressor housing its internal size has to be enlarged by careful machining. After the new rotating assembly is balanced, the turbo is reassembled and ready for use, with the potential to provide a 15% performance rise.

On the turbine side, flow performance can be enhanced by back-cutting the turbine wheel. This modification greatly benefits top-end power in similar manner to a large A/R ratio turbine housing. The down side is a slight increase in turbo lag.

A real hybrid, though, goes much further; so much so that the assembled turbo loses its original identity and can easily be relabelled by the brains behind it. For example, one very potent unit is based around components from three turbos. The Hitachi core, compressor housing and turbine wheel come from a Mazda RX7 13B rotary engine. The turbine housing, suitably machined to suit the Hitachi turbine wheel, is from a twin-cam 3-litre V6 Nissan. Finally, the compressor impeller is a Garrett TO4E that matches a suitably machined RX7 compressor cover. All up, a combination that will just about double the hp of most stock 2-litre turbo engines to over 320hp.

Fig. 6.9. Compressor impellers of the same diameter come in a variety of blade sizes.

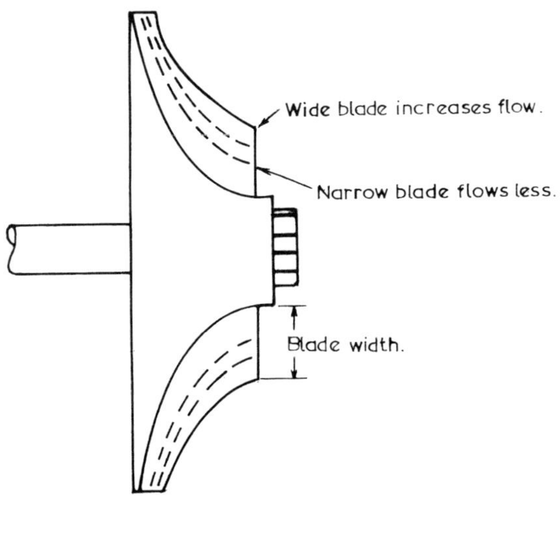

CERAMIC TURBINE WHEEL BOOST LIMITS AND PROVIDING WET PROTECTION

While the turbo's basic physical dimensions prescribe its performance potential, how long it does its stuff is largely up to you. The modern ceramic turbine wheel does wonders in reducing turbo lag, but if you raise the boost above 14.5–15.0psi it's bye-bye ceramic wheel. Why? To build more boost requires more energy or power. Consequently, the turbine wheel has to exert greater force (think of it as torque or leverage) on the common shaft it shares with the compressor impeller. In turn this places higher stresses on the ceramic wheel in two vital areas. First, where the blades attach to the wheel and, second, in the centre where the hub bonds onto the shaft. Both areas have a finite load limit and once exceeded the ceramic wheel will break that bond, or even shatter or break apart.

Another sudden-death problem is if the turbine housing meets with a good dousing. If it is a factory turbo installation the car manufacturer will have splash-proofed the turbo, either by locating it out of harm's way, or by means of appropriate splash shields. Therefore, before removing or modifying any splash guards to improve air flow, or for whatever reason, take care that it will not expose the turbo to water damage. Also, don't expect factory splash protection to be sufficient if the car will be rallied across creeks or water obstacles. The factory do not expect their products to be fording streams over a foot deep at 50kph. You may get away with the red hot (even white hot) turbine housing getting a ducking a few times, but eventually it will stress and crack. It goes without saying that non-factory installations will require some planning to protect the turbo from water damage.

TURBO LUBRICATION AND OIL RETURN

Turbo lubrication is a major consideration if you are installing a turbo on a non-turbo engine, or if you are changing the location of the turbo on a turbo engine. Turbos don't like high oil pressure, anything over about 65psi is excessive. This could overload the seals and give a telltale smoky exhaust. Fit a restrictor to bring the pressure down, but take care not to limit the flow excessively. At idle the minimum flow required is just under 0.5 litre per minute and at normal engine revs it should be around 2 litres per minute or more for plain bearing turbos. Ball bearing turbos require less flow, with about 1.5 litres per minute the maximum recommended at normal engine rpm. However, it is always wise to check on exact lubrication requirements with the turbo manufacturer. Usually it will be possible to take the turbo oil feed from an existing tapping in the main oil gallery, such as the oil pressure gauge tapping. Personally, I prefer a hard steel feed line with appropriate bracing to avoid vibration cracking. Aircraft type metal-braided flexible line also works a treat. Just be careful not to let it vibrate against or rub soft materials like aluminium as it will wear a hole right through.

With turbochargers the oil return arrangement is actually more critical than the feed system. The things to keep in mind here are that you are pretty much relying on gravity to move the oil from the turbo core to the sump, and after passing through the bearings the oil has been whipped up. Right away this should alert you to use a large bore return line which is free of acute bends and goes downhill all the way. A smooth bore $^7/_{16}$in id hard line is about the minimum I like to see used here, but certainly $^1/_2$in id rubber hose

Oil supply to turbo bearings via a pipe in the centre is vital for long turbo life. Of equal importance is the oil return line back to the sump.

(heat and oil resistant grade) is all right, and with this there is no need for expensive end fittings. To minimise flow impediments at both ends of the return line, do ensure that the turbo core is clocked with the oil outlet somewhere between the 5 o'clock and 7 o'clock position and that the sump end-fitting is well above normal oil level. If the turbo is mounted at a suitable height the return can also enter the engine block via a bespoke tapping or a redundant dipstick or breather fitting.

In some installations the turbo is mounted too low to permit a simple oil return. There are a couple of options in such cases. You can choose to fit an Aerodyne (Aerocharger) turbo which carries its own supply of special synthetic jet turbine oil (Mobil SHC630) and, as such, is not connected into the engine lube system, or you can use some sort of scavenge system which can be a belt drive or electric pump that draws oil out of a small sump attached to the turbo, and then returns it to the engine sump.

TURBO CARE BASICS

Everyone who has much to do with turbochargers regularly hears horror stories of turbo failure. Mostly the enthusiast blames his turbo or his oil for such early failure, often at less than 10,000 miles service. More often than not the driver is to blame, although there is some truth to oil-related failures, which is why I recommend using a 100% synthetic oil and changing it along with the filter at the intervals outlined in Chapter 20, but in essence this is just the start of turbo care.

Today, most non-competition turbos have a water-cooled bearing arrangement to increase the reliability and life of the bearings. Unfortunately, this has lulled many into a false sense of security as they feel the turbo no longer needs to 'idle down' before the engine is stopped. This just isn't so; after a hard run which has put a huge amount of heat into the turbo it must be allowed time to cool for a period of about three minutes. Actually I don't advocate allowing the engine to idle for that period. I would rather see it run at about 1500rpm or a little higher to promote a good flow of oil through the turbo bearings. At idle the oil flow will not be sufficient, which could allow the heat in the turbine to carbonise the oil. If the oil becomes carbonised it blocks the minute clearance between the shaft and bearings. When the engine is next operated, oil flow through the bearings is reduced and without adequate lubrication the bearings soon fail. However, this may not be the limit of the problem as the shaft may also be wrecked; the turbine and impeller wheels may touch their respective housings, damaging the turbo beyond economical repair.

Some enthusiasts do the right thing in allowing the turbo to cool before stopping the engine, but just before switch-off they give the throttle a few quick jabs to, in their words, 'clear the plugs!' Such action may spin the turbo up to 30,000rpm and then with oil flow cut, when the engine is stopped, the bearings soon run dry and suffer premature wear.

A similar problem occurs right after starting the engine. It can take up to 30 seconds for oil to flow to the turbo, so don't quickly rev the engine as this spins the turbine to high speed at a time when the bearings have only minimal lubricant in reserve. In similar vein, when a turbo is refitted after being off the engine for any time, or if a reconditioned unit is being fitted, do prime it with engine oil before attaching the oil feed line. When this isn't done the bearings are damaged from the time the engine is fired up. Likewise, if the engine has not been started for some time, as during winter, it is wise if the turbo or oil line can be reached easily to prime it with oil before starting the engine.

Some advocate the use of a pressurised pre-oiler system to supposedly provide turbo protection. These systems do some good in quickly getting oil to engine bearings and valve gear at start up, but it is nonsense to suggest they in any way extend turbo bearing life. If the engine loses oil pressure because of oil surge resulting from violent cornering, braking or acceleration, the pressurised auxiliary oiler will lube the turbo, but as it is also providing life fluid to the entire engine it is drawing a rather long bow to suggest that such are needed to extend turbo life. If you do as I recommend you can forget such a system, unless you have an oil surge problem.

Whenever a turbocharger is removed from the engine, place it in a plastic bag so as to prevent foreign matter entering the turbine or compressor, or the oilway. Also, cover the inlet and exhaust ducts on the engine so that no nuts or washers, etc. can get into them and later find their way into the turbo. By following these simple precautions expensive damage is easily avoided.

TURBO RECONDITIONING

In normal usage a turbocharger should require servicing only about every 30,000 to 50,000 miles. If the vehicle spends a good deal of time idling at lights or at high speed, the service interval could be more frequent. Usually, a service kit is available containing the necessary bearings, seals, bolts, etc. Turbos are of very simple construction and quite easy to service if approached with proper consideration for cleanliness of oilways and bearing surfaces, and if it is remembered that the rotating assembly has to spin at speeds in excess of 100,000rpm. Some manufacturers do not recommend that servicing be done except by authorised agents with the necessary, expensive, service equipment to rebalance the turbine/impeller assembly, as imbalance may cause undue noise. However, I have never experienced any problem when the compressor wheel and turbine wheel/shaft is undamaged, and consequently reusable. Simply marking the position of the compressor impeller on the shaft with a dab of paint during disassembly and then on reassembly accurately lining up the paint mark on the impeller with the paint mark on

Damage like this can usually be avoided with correct turbo matching (to avoid over-speeding or surge), adequate lubrication and care to keep things from getting into the turbo inlet.

the end of the shaft overcomes any imbalance/noise problems. Obviously, if either the impeller or the turbine wheel/shaft are replaced because of damage or wear, the complete assembly will require balancing.

During disassembly, carefully hold the centre housing/bearing assembly in a vice by the irregularly-shaped nut on the turbine end of the shaft. By the way, the nut is ground that shape to balance the assembly. Because there isn't much area for the vice to grip on to, be very careful when removing the other nut which holds the impeller onto the shaft. If the assembly slips you will wreck the turbine blades. When the blades are bent don't attempt to straighten them otherwise they will eventually break off. If it is an impeller blade that goes it could finish up inside the engine and do damage, or the impeller may explode and come through the compressor housing. Expert opinion should be sought when the blades are bent. When minor, the assembly may be reusable, but it is best to have this determined by someone experienced.

There are two other points of which to be mindful. Before dismantling the turbo, mark the positions (called the clocking) of the compressor housing and the turbine housing on the centre core assembly. This will save you a lot of assembly hassles later. Also, take care to follow carefully the torque settings recommended by the manufacturer when reassembling the unit. If the manufacturer says that a particular nut is to be tensioned to what may appear to be a small value, such as 10ft/lb, follow the instructions and do not rely on your own judgement. Possibly you will not have a tension wrench which is accurate at anything less than about 30ft/lb, so borrow or buy a suitable wrench.

Chapter 7

Turbo boost control

There are several good reasons why we have to limit turbo boost pressure. The prime reason is to prevent detonation (see Chapter 17) and, as we discussed in the previous chapter, to preserve ceramic turbine wheels. Others might also cite the need to prevent engine blow up, but clearly another couple of psi won't cause cylinder head bolts to stretch, pistons to collapse, con rods and crankshafts to break. What really causes this type of damage is usually either uncontrolled combustion (pre-ignition or detonation), or excessive engine rpm and high hp.

Another good reason to control boost is for the purpose of traction control. With very high boost, torque levels at the driving wheels can be so excessive as to make control of wheelspin impossible, particularly in the lower gears on wet and greasy roads. Limiting boost in such circumstances makes for safer and more rapid forward movement. In similar vein, boost control might be for the sake of transmission component reliability.

The fourth reason for boost control is to keep the turbo compressor working at maximum efficiency to prevent undue heating of the inlet air charge. Heat in the inlet charge reduces engine power and can induce detonation. A cooler air charge will allow more ignition advance, with accompanying power, driveability and economy benefits.

The final reason for boost control is because some competition rules limit boost to a stipulated figure. This is usually to even up the competition. Unfortunately, the cheaters know how to get around this one and legally run at a slighter higher psi than the rules permit.

EARLY BOOST CONTROLS

Years ago, boost control for road engines was more a type of engine asphyxiation than true boost control. Usually the exhaust was severely restricted using small bore mufflers or washers in the tail pipe. Obviously, high exhaust back-pressure after the turbo prevents the free flow of exhaust gas through the turbine, effectively preventing it

Cosworth CART Champ car engine with large blow-off valve mounted on plenum, as required by race regulations to limit boost.

spinning any faster. With turbine rpm capped, the compressor air flow rate and boost was also limited. The high back-pressure also meant the cylinders retained a lot of exhaust gas which, in turn, limited how much air/fuel mixture could be forced in to make power. This, in turn, led to less exhaust gas volume to spin the turbine, and so on and on.

The other strangulation method was applied on the inlet side. Many crude techniques were practised, with the most usual being a small bore carburettor. Today, of course, some competition organisers require inlet restrictors to limit horsepower outputs but, with special compressor design and massive intercooling, very high boost is possible at lower engine speeds where engine air flow requirements are lower.

Years ago some competition engines also used these crude boost limiting systems but in general more 'active' controls were used; either an inlet manifold blow-off valve or

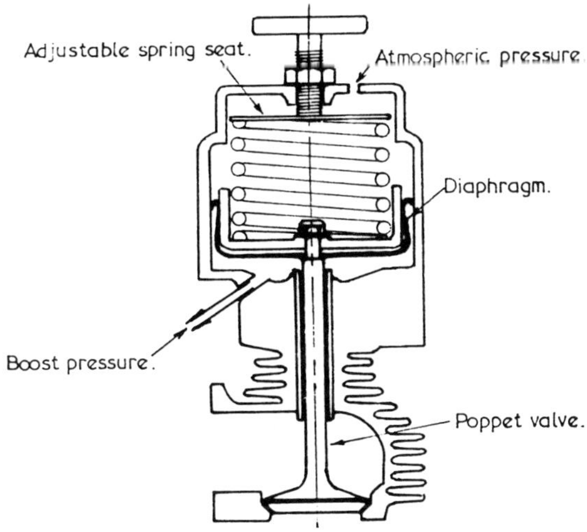

Fig. 7.1. Adjustable external wastegate. Boost pressure against the diaphragm raises the poppet valve, allowing some exhaust gas to bypass the exhaust turbine, thus limiting turbo speed and boost.

more likely an exhaust wastegate (*Fig. 7.1*, see also *Fig. 6.2*). Early blow-off valves were often radiator pressure caps. They did not work very well, being erratic in operation. In time, some car manufacturers developed more sophisticated valves, but these were really a last line of defence device to dump excess boost in the event of exhaust wastegate failure rather than a true boost control valve.

INLET VENT VALVES

Today, virtually all experienced turbo tuners scoff at the very suggestion of using any sort of inlet system vent to provide boost control, maintaining that an exhaust wastegate is the only way to go. However, I know of one outfit building very high performance road engines (200hp per litre) and competition engines (280hp per litre) who would never dream of using anything but an inlet dump boost control arrangement. Such a system is a fraction of the price of a big external wastegate, but these fellows insist money is not a motive. Simply, their system is more responsive and accurate. They and their clients have numerous trophies to back up their claim, so we should not be quick to dismiss this method of boost control.

First off, I think we have to remember that these are top hp engines running big turbos but, even so, company policy is to make their turbo cars respond like powerful naturally-aspirated cars. This means a linear power delivery rather than the big power surge that is typical of top-end hp turbo engines. Often two ordinary blow-off valves (dump valves), which would normally be employed to dump boost when the throttle is lifted, are used to provide the necessary venting characteristics. Because dump valves frequently leak, they are modified with stiffer springs. Actual valve control is not based on boost pressure alone, but rather a solenoid valve in the signal line responds to commands from a MoTeC ECU (electronic control unit – engine management computer). Additionally, the system is MAP sensor based.

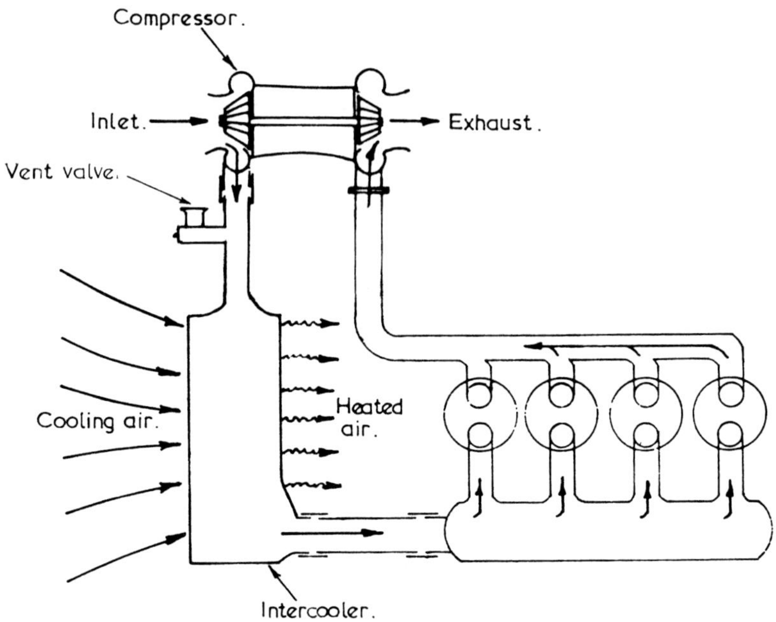

Fig. 7.2. A vent valve located before the intercooler improves intercooler efficiency.

Vent valve location has a bearing on performance. If placed after the intercooler, intercooler efficiency falls as it has then to work overtime taking heat out of compressed air, some of which is later dumped. Therefore, the place for vent valves is somewhere between compressor outlet and intercooler inlet. In this way the charge cooler will only be cooling that portion of compressed air that is actually going to be used by the engine (*Fig. 7.2*).

AFFECT ON AIR/FUEL RATIO

Something to keep in mind here while we are on the subject of dumping air from the inlet system, whether it be for the purpose of boost control or for the more usual function of maintaining compressor rpm and avoiding compressor damage when the throttle is closed at boost, is what effect this will have on the air/fuel ratio. If fuelling is metered according to inputs from a speed density MAP sensor based system, dumping air back to the atmosphere will not alter the A/F ratio. However, in any system employing some kind of air-flow meter, remember that once air flows through that meter it has been accounted for. Fuel delivery will then be calculated by the ECU, based primarily on the signal received from that source. Therefore, any time air is released to the atmosphere fuelling will go rich.

Some do not view this as a consideration, and to some extent in competition engines that may be OK, but only if the blow-off valve is employed to dump boost on overun. In such a situation the engine will be over-fuelled on overun when the throttle is closed. This excess fuel may, in fact, pass into the exhaust and burn, helping turbine speed and ensuring spectacular flames from the exhaust. However, in competitions requiring CATs, all that afterburn will overheat the core of the CAT, causing the ceramic structure to

A small dump valve plumbed into the inlet duct protects the turbo impeller wheel and helps the turbo maintain speed when the throttle is suddenly closed. These valves can also be used to control boost pressure, but should be located before the intercooler when used for that purpose.

melt, severely restricting gas flow. Obviously, the same applies to road vehicles, and there is the added concern of increased fuel consumption.

Therefore whenever you use any type of air metering system air exhausted from the blow off valve is usually best recirculated back into the air inlet system. The re-entry must be somewhere between the air-flow meter and turbo inlet, not before the air-flow meter.

Right away you can appreciate why a combination of air-flow metering and boost control by means of dump valves is not smart engineering. When these valves open as maximum boost is reached the exhausted air is extremely hot. Recirculating hot air into the compressor reduces its efficiency and raises inlet temperatures even more. As you know this will restrict power and could send the engine into detonation if the engine is operated at full boost for extended periods. Consequently if you want to control boost by means of blow off valves you should use a MAP sensor (speed density) based engine management system. (For those wondering why it is okay to recirculate air dumped on overun remember this is only for a very limited time, often less than a second. Also the engine is making zero or minimal boost when the throttle reopens and only a small volume of hot air is inducted.)

EXHAUST WASTEGATE

As mentioned earlier, the more frequently used way to control boost is by means of an exhaust bypass – a wastegate. Originally the only type available was the expensive external type illustrated earlier, but in time turbo manufacturers developed the much cheaper to manufacture and very compact internal wastegate such as we find on all stock turbo cars today. Both designs control boost by diverting a portion of exhaust flow away from the turbo turbine wheel at any time the preset inlet boost pressure is attained.

By controlling exhaust flow volume through the turbine, turbine wheel rpm is restrained, which in turn limits compressor speed and boost.

On the surface, the way wastegates work appears fairly straightforward, but there is a problem. In operation the ordinary wastegate is held closed by spring preload and is opened by boost pressure opposing that spring pressure as it pushes against a diaphragm. However, the spring does not simply collapse and allow the wastegate to snap open upon reaching the required boost pressure. Rather, boost pressure pushing against the spring forces the spring to progressively back off (called wastegate creep), allowing the wastegate to begin breaking open at around 50–65% boost. This means that long before maximum manifold pressure is achieved the wastegate has begun to leak exhaust energy which should have been utilised to aid the turbine to more quickly spool up to maximum boost.

Some don't see this as a problem, stating that they are struggling to maintain traction in the lower gears anyway. However, it is in the higher gears where turbo lag is more apparent, and at that time we do not want to reduce exhaust volume through the turbine because of premature wastegate activation. Therefore, on some competition engines, particularly those fitted in four-wheel-drive rally cars, hydraulic wastegates are used. Rather than boost pressure to open the wastegate, pressurised oil is used. In times past, hydraulic wastegate control was largely an expensive novelty, but more recently, with the electronic engine management system directly controlling the hydraulic valve that directs hydraulic pressure to the wastegate, wastegate operation is more in sync with what is actually required. This permits more rapid and precise wastegate operation, which reduces turbo lag and, when necessary, greatly improves traction control.

The Audi Quattro utilised a large external wastegate which directed exhaust bypass flow back into the exhaust dump pipe well away from the turbo.

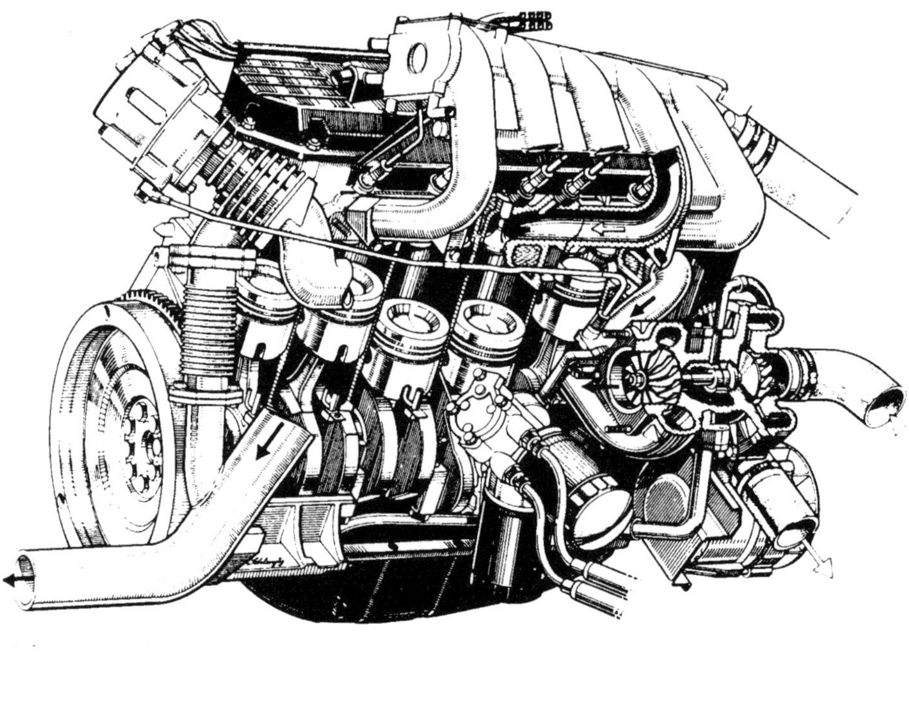

GAUGE PRESSURE ACTUATOR AND ABSOLUTE PRESSURE ACTUATOR

Looking back to *Fig. 7.1* you will note that this wastegate is connected so as to receive boost pressure at the lower port while the spring side of the diaphragm is vented to the atmosphere. (Note that in some wastegates the wastegate valve moves in the opposite direction to open. In wastegates of this type everything is reversed, with the boost port at the top, followed by the diaphragm, then the spring and the vent to atmosphere near the bottom.) In technical terms this arrangement is referred to as a gauge pressure sensor or actuator. Some, though, will call it a differential pressure sensor (or actuator), but this is not strictly correct unless the vent to atmosphere is connected back to the turbo compressor inlet. In theory, pressure at the turbo inlet should be the same as atmospheric. However, air flow restrictions from ductwork, air filter and air box can drop the pressure well below atmospheric in many road-going installations.

In nearly every automotive application the gauge pressure actuator will provide the necessary wastegate response. However, in some very unusual environments (for a motor vehicle) there are three other types of actuators which may be more suitable. I will only discuss one type, called an absolute pressure sensor (or actuator). Normally used in aeroplanes, this actuator has the advantage of limiting boost pressure to the absolute value (see Chapter 5) regardless of altitude or barometric pressure. Therefore, in motorsport competition where huge changes in altitude are experienced – Pike's Peak hill climb, for example – an absolute pressure sensor may prove more suitable.

The physical difference between the two is that whereas the gauge pressure sensor is vented to the atmosphere, the absolute pressure type is sealed. In fact, the upper chamber is evacuated to a high vacuum (as close to absolute zero pressure as practical).

SOURCING THE BOOST SIGNAL

The next consideration is the source of the actuator pressure signal. It is important to recognise that boost pressure is not uniform at all points between the compressor outlet and the inlet valve. Pressure variations occur for a number of reasons. Anything that restricts air flow will show a higher pressure before the restriction. The intercooler not only causes some flow restriction, but in cooling the air reduces its pressure. Venturi type devices don't necessarily restrict flow, nonetheless large pressure variations will be evident over the length of the venturi with the lowest pressure point being slightly after where the venturi walls begin to diverge.

Looking at stock turbo installations you will note that the signal line connects between the compressor outlet and the wastegate actuator. This is considered to be a safe location. The signal line is short, so there is rapid communication between the turbo outlet and the wastegate. Therefore, air in the signal line is less likely to compress and allow the turbo to overboost and possibly send a blast of very hot air (depending on compressor efficiency) into the intercooler. If the intercooler is none too efficient, that short blast of hot air will retard its ability to bring charge temperature down to normal values for quite some time because of heat soak in the intercooler. In turn, the high charge temperature could see reduced performance (because of lower charge density and ignition retard in response to knock sensor signals) or even eventual engine damage from detonation.

When and in what manner this internal wastegate opens will be determined by the type of actuator fitted and where the boost signal is sourced.

The down side of this rapid communication to the wastegate is that the wastegate begins to crack open extremely early, often at only 50% boost. As previously explained, this stifles how quickly the turbo comes up to full boost, which reduces performance particularly in the higher gears.

A compromise location for the actuator signal hose is somewhere after the intercooler and before the throttle plate(s). The pressure here is lower than at the turbo outlet, consequently in a stock installation moving the signal pick-up to this location will enable a higher actual boost pressure, as the engine sees it, without having to fiddle with wastegate actuator springs or rods. Also the wastegate will delay cracking open. However, for there to be a true overall power gain the intercooler must be large enough to pull the added heat from the inlet air.

Another possible spot for the signal pick-up is in the inlet manifold after the throttle plate. This won't work in engine installations using a separate throttle for each cylinder. Usually, sensing boost from this location has little in its favour. However, some engines do exhibit slightly superior turbo response times with the sensor line connecting here, so if intercooler capacity isn't a worry it could be worth a try.

It is worth noting at this point that there can be good reason, in competition classes with regulated boost limits (often 15psi), to thoroughly investigate the entire inlet system to find where boost is the lowest (or modify it using a suitable venturi). Then, at that exact point tap into your system with a suitable fitting as specified in the regulations where tech inspectors can check that your turbo is operating at legal boost levels.

Note that the turbo signal line does not need to be, and often should not be, connected to the same fitting. Rather, it must be located so as to provide best performance and best turbo response. Once you find the most advantageous spot for the sensor line, adjust the wastegate to bring boost pressure to the legal level at the checking port. However, make sure that running at maximum legal boost does indeed provide best performance at each competition venue. In warmer weather, or where full throttle is used for longer periods, do not be surprised if less boost is, in fact, superior in classes requiring stock turbos and stock intercoolers.

ADJUSTING BOOST PRESSURE

Some find the idea of fingertip boost control quite appealing. For example, to save on fuel costs for day-to-day use they might want to purchase cheaper low-octane fuel that allows for an upper boost limit of, say, 8psi. Weekend track days are a different story and on some exotic high-octane brew the engine is sweet at 11psi.

There are a number of avenues available to fool the wastegate. As you have just read, the signal hose pick-up location can give quite different pressure levels. Therefore, the signal at the compressor outlet might provide the 8psi signal, and shifting the signal line to the inlet manifold might raise the boost pressure to 11psi (measured at the turbo outlet) for track days.

External wastegates, like the one illustrated in *Fig. 7.1*, often have an adjuster to alter the preload on the internal spring. The full range of adjustment is usually about 4–5psi, from the adjuster fully backed off to fully turned in. However, with better types of external wastegates they can be dismantled and have stiffer springs, or an inner spring, installed. This isn't exactly fingertip boost control, but we will get on to alternatives shortly.

Internal wastegates, at first glance, do not appear to permit easily switched boost control. The boost can be changed by altering the length of the actuator rod (the rod connecting the actuator to the wastegate flap). If the rod isn't an adjustable type it is a simple matter to unbolt the actuator, saw the rod in half and thread it with a suitable thread such as ¼in NC. Using an internally threaded coupling sleeve and jam nut, the actuator can then be adjusted. Note that, when sawing and threading the rod, care must be taken to hold the rod securely to avoid damage to the actuator diaphragm.

An even simpler method, which allows for either a boost increase or quick boost changes, is to add a helper spring. I have seen tuners weld or screw a suitable lug to the actuator where they attach one end of the spring, and attach the other end to the wastegate flap pivot. Personally I don't like this idea, as welding or drilling easily ruins the actuator diaphragm. I much prefer to find a convenient bolt or screw in the general area of the actuator and fix a suitable lug under that. When a spring is stretched between this lug and the wastegate flap the boost pressure is increased. Unhooking the spring lowers boost to stock, so this is a convenient method of obtaining two boost settings (*Fig. 7.3*).

Fig. 7.3. A helper spring attached to an internal wastegate actuator.

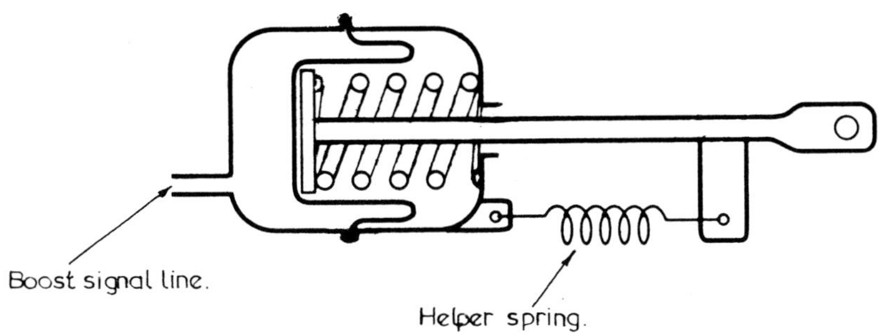

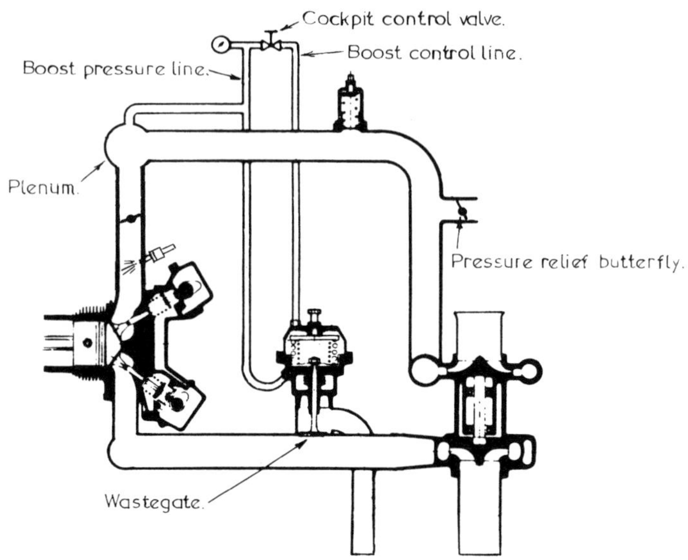

Fig. 7.4. The Porsche 917 racer had a cockpit control valve which permitted boost to be easily adjusted.

The arrangement Porsche used in their works racers 30 years ago we would today call dial-a-boost. Looking at *Fig. 7.4* you can see that the concept provides almost infinite boost control, and was for that reason originally used in the Porsche dyno cell as a means to quickly test engines for output and durability at various boost pressures. During a race when a Porsche 917 driver required more power he simply turned the knob of his cockpit mounted pressure regulator. This action then allowed some boost pressure to flow into the wastegate spring chamber area, effectively adding to the spring's preload pressure. For example, if the control knob was turned to permit 3psi pressure into the spring chamber, boost pressure increased from 20psi to 23psi and power rose 100hp to 1,200hp.

A variation on the Porsche dial-a-boost method has the pressure regulator venting to the atmosphere (the wastegate spring chamber is also vented to the atmosphere). In effect, the pressure regulator becomes a variable bleed valve which can leak a controlled amount of signal line pressure, thereby fooling the wastegate to increase boost pressure. For instance, if normal boost was 11psi, opening the valve to create a 4psi leak would increase inlet pressure to 15psi.

An inexpensive alternative which fairly quickly provides boost pressure adjustments also works on the signal bleed principle. The trick here is to insert a short length of hard plastic tube in a convenient place in the signal line and then perforate it with a number of bleed holes. Since only a small leak is required it is best to puncture the tube using a 0.040in needle which has been heated so as to melt a hole in the plastic. Begin with just one bleed hole and add additional holes, as needed, to achieve the desired boost level. There is no reason why you cannot use one piece of plastic with say one bleed hole for regular running and another piece with more holes for weekend competition. Just remember that whenever you use any type of signal line bleed that fuel must always be delivered into the inlet tract downstream from the signal line connection. If that isn't possible, then you will have to use some other method of boost control.

ELECTRONIC BOOST CONTROL

By far the best method of boost control is by means of electronics actually controlling signal line bleed. The first system of this type was the SAAB APC system introduced in 1981. In the signal line SAAB installed a solenoid valve which was controlled by the engine management system. The pulsing of this valve determined how much signal line pressure would be bled off. If the knock sensor did not detect any detonation the open/closed time ratio for the valve was changed to bleed off more signal pressure, so as to increase manifold boost. However, at the onset of knock the valve's open/closed duty cycle is adjusted, reducing signal bleed and thereby lowering boost pressure (*Fig. 7.5*). For some of their racers Porsche employed a pair of fuel injectors which were pulsed to bleed-off boost in the signal line. Spark retard could have been employed to protect the engine from detonation but, as pointed out in Chapter 5, that course should always be a last line of defence. Forced induction engines invariably perform better with a reduction in boost, rather than cutting spark advance, to keep the engine away from detonation.

Right away you can see how, with today's more powerful engine management electronics, such a solenoid valve (which I prefer to call a boost modulator) can have a much wider range of duties than simply keeping the engine away from detonation. For example, if traction is a problem the electronics can be programmed to reduce the

Fig. 7.5 The SAAB APC electronic boost control.

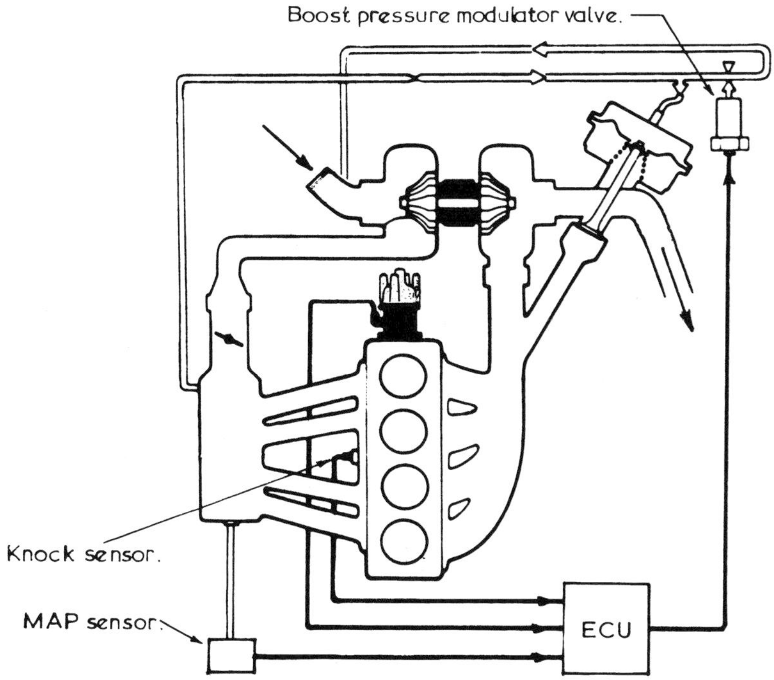

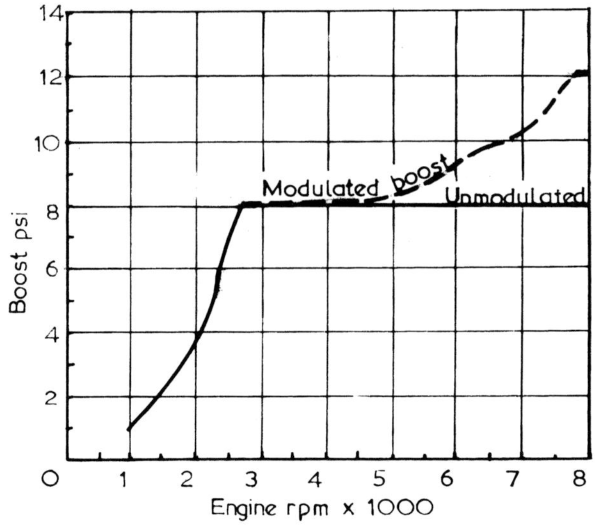

Fig. 7.6. Without electronic boost modulation, boost may have to be kept low to avoid detonation or transmission damage at lower rpm, thus also limiting top-end power.

amount of signal bleed at the boost modulator, so cutting boost psi at specific engine rpm in each gear, and then progressively restoring boost at higher rpm when engine torque begins to fall away. Alternatively an 'active' system using wheel speed sensors to provide feedback can be used. On the surface, such a system would appear to be superior and much less time consuming to programme. However, in practice it will often be found that totally eliminating wheelspin actually takes away driver feel and slows the car. For example, some wheelspin at launch improves acceleration and decreases driveline loads. On gravel about 15% wheelspin (some drivers like more) helps the tyres clear loose surface rubbish and dig into the harder base material.

An electronic boost modulator can help engine power levels by tricking the wastegate to give more boost at the top of the engine's rev range. Usually as an engine approaches max hp revs its volumetric efficiency begins to fall quite rapidly. This means charge density is now also falling, so if the turbo compressor and intercooler are working efficiently more boost can be applied at high rpm without the engine running into detonation (*Fig. 7.6*). A similar programme would also be employed if any part of the transmission/drive train assembly are borderline on failing because of excessive torque loads. As the torque falls away once past maximum torque revs, boost is progressively increased to raise torque back close to the safe maximum.

The overboost facility is also made possible by the boost modulator. Porsche, with their 944T, were first out of the blocks with this idea which allowed a 10% boost increase for a few seconds to improve acceleration. Since that time other manufacturers have followed suit, and also some race organisers in classes running a restricted sealed engine formula have programmed the electronics to permit a certain number of overboost periods (usually about five or six) of up to 15 seconds in each race to encourage more overtaking and force drivers to plan a race strategy.

Another benefit available from an electronic boost modulator is more precise wastegate operation. Rather than the wastegate being permitted to begin cracking open at as little as 50% boost, we can programme the management system to have the modulator bleeding off nearly all the boost signal and only closing just an instant before maximum boost is attained. This keeps the wastegate closed until almost full boost is reached and provides significantly improved acceleration in the higher gears.

There is some danger with this as there is always the temptation to delay wastegate operation too long. This causes problems similar to a near square camshaft lobe acting on an inlet or exhaust valve. Rather than opening progressively, the valve is slammed open, overshoots then slams back onto the seat. It will continue bouncing like that until signal line pressure and spring preload take over and balance out the uncontrolled oscillations. What generally is much superior is a progressive response whereby the amount of signal bleed-off trails actual boost pressure by 25–30% as soon as 70–75% of max boost is attained. Then by 90–95% of max boost full signal strength should be communicated to the wastegate. Therefore, if full boost is 12psi, then at something like 8.5–9psi boost signal line bleed should be cut back to allow a 6–6.7psi signal through, and at around 11psi the wastegate should be feeling the full 11psi signal.

OVERBOOST PROTECTION DEVICES

Something we have to keep in mind is that most car manufacturers don't trust our common sense. So, to keep us from hurting the engine, they invariably fit some sort of overboost safety device. The idea is that just perhaps the wastegate might seize in the closed position and with no boost gauge to warn of the danger we will drive the engine to destruction.

As mentioned at the beginning of this chapter, some car makers in the past fitted nothing more than a sophisticated radiator cap to pop open and vent excess boost. In time, though, manufacturers changed over to electronic safety overrides which cut either the fuel or the ignition. This creates a problem for the modifier as boost increases could be stymied by such a device coming into operation. There are overboost defencers on the market to defeat these electronic overrides, or when the ECU is being reprogrammed or rechipped the cut can be eliminated or modified.

A word of warning for those employing a fuel cut. When the fuel cut comes into operation, listen to the message it is giving you and get your foot off the accelerator quickly. Some fools don't and pay the penalty, wrecking the engine. With the fuel supply cutting off and on for a prolonged period the engine goes lean, then pistons and valves overheat.

INTERNAL WASTEGATE FLOW PROBLEMS

Wastegate control is not our only wastegate performance concern. A major factor in turbo engine performance is maximising exhaust flow energy to drive the turbine wheel up to speed as fast as possible. A wastegate always disrupts exhaust flow, so we have to investigate ways of minimising that impediment. Usually, integral wastegates are much worse in this respect than external types, but as you will see when we get to Chapter 16, poor plumbing of external wastegates can muck up exhaust flow into and out of the turbine pretty horribly too. At this point, though, we only want to look at internal wastegates so that you can see what flow problems you should be on the lookout for, and if perhaps an external wastegate may be a better option.

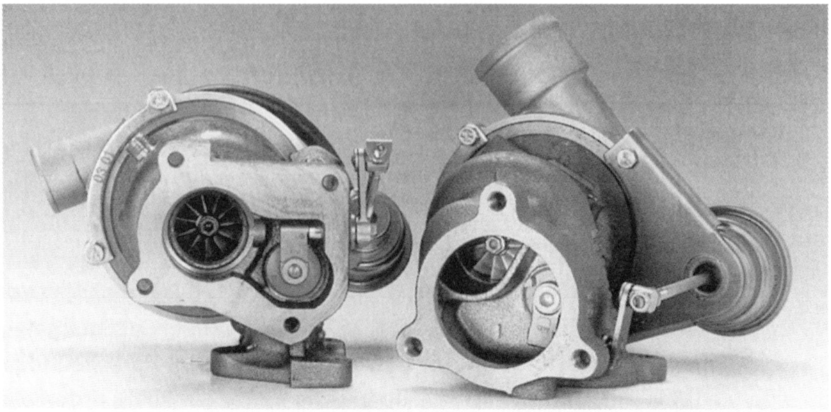

These two turbos have vastly different exhaust turbine housing designs. The turbo on the left has parallel flow paths from both the turbine and wastegate. However, the other one, because of wastegate location and the zigzag dump arrangement, causes massive flow turbulence as the gases exit the turbine.

Usually there isn't a lot we can do to help exhaust flow into the turbine. Depending on the design, sometimes we can streamline the web dividing flow between the turbine passage and the wastegate bypass passage. Also we can bellmouth as much as possible the wastegate passage entry. This provides two benefits, first in reducing flow turbulence through the wastegate, which can reach back into the turbine entry and disrupt turbine flow, and second, it increases the actual flow potential of the wastegate. If the flow volume is insufficient the wastegate will not be able to keep boost pressure controlled and the engine could have several pounds more boost than you want.

On the exit side of the turbo, have a good careful look at both the turbine and wastegate flow paths. Some designs leave a lot to be desired in this area. For some the problem is that wastegate flow is steered to cut across flow from the exducer bore (*Fig. 7.7*). Ideally we want the two flow paths to run parallel to each other. For others the

Fig. 7.7. Wastegate flow cutting across flow from the turbine wheel upsets exhaust flow and restricts turbo performance.

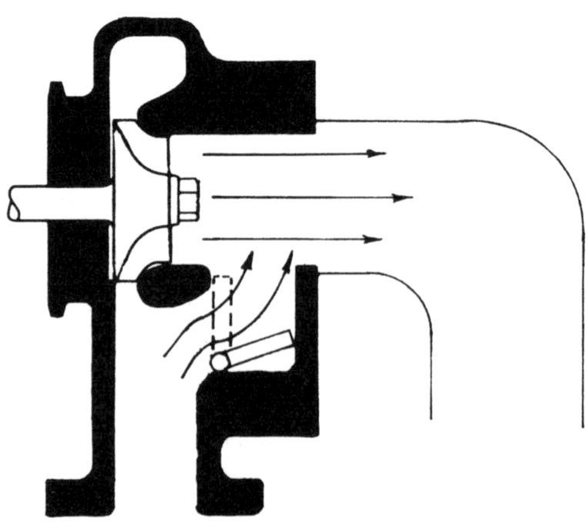

problem is that the exducer flow path looks like an obstacle course. In some turbos the flow path steps sideways one third the width of the exducer bore, forcing the gas to flow in an acute Z pattern. In others you will see obstacles such as the shaft the wastegate flap pivots on disrupting flow.

Again there isn't much you can do to easily fix such things up on the exit side. Car manufacturers often fit a special custom design turbine housing simply to speed up operations on the assembly line, or to make service operations easier for mechanics with hands the size of King Kong. Therefore, with suitable exhaust system modifications, which you will want to do anyway, it is quite likely that an ordinary (non-custom) turbine housing, providing superior flow, will fit with a few adjustments.

EXTERNAL WASTEGATE SELECTION

The alternative is, of course, an external wastegate. Fine if you have the money and sufficient space around the engine. When choosing such an item there are a number of things you should be looking at, apart from the price tag. First off, you have to consider how big, and if it is big enough for this engine will it be suitable for your next engine if you intend not to sell the wastegate with the engine. Another consideration is adjustability. Is there a simple adjusting screw in the top, or will it always be necessary to change the internal spring? Take a look at how far the diaphragm chamber is from the exhaust flow passage, how much cooling fin area there is, and what the main body material is. The more heat the diaphragm and spring are exposed to, the sooner they fail. Therefore, a stainless steel body is superior to cast steel. More fins are better. Also, choose one with more space between the diaphragm and bypass passage. Finally, measure the valve diameter and the diaphragm chamber diameter. The wastegate with the larger ratio is generally better for consistent, stable performance.

How you orient and cool the wastegate will also influence its ability to provide

Wastegate spring and diaphragm chamber protected from exhaust and turbo heat by heat shielding on BMW Formula 1 engine.

stable, consistent boost levels. If you refer back to *Fig. 7.1* you will note that the wastegate operates by lifting the valve up from off its seat. Exhaust flow and pressure must work in harmony with that action. Therefore, exhaust pressure should always push against the flat face of the valve, with flow entering the bypass passage around the valve face and exiting at the open end. If the direction of flow was reversed, exhaust pressure would be working against the action of the diaphragm. A duct directing cool air onto the wastegate is always helpful in extending the service life of the diaphragm and spring. In fact, with proper cooling, cheaper wastegates can be assisted to provide adequate levels of performance.

YOU NEED AN ACCURATE BOOST GAUGE

Your car may not have a boost gauge and you may never intend to fit one, nonetheless anyone who works with turbos should have an accurate boost gauge with a large easy-to-read scale. When you purchase a gauge, make sure that the supplier understands that you will be having it checked for accuracy and jumpy erratic readings, and if it fails on either test you will be looking for a replacement. Unfortunately, inaccuracy and the needle sticking or jumping are common problems with boost gauges. You can probably save yourself a fair bit of heartburn if when you go looking for a suitable gauge you go armed with a 25ml or 50ml hypodermic syringe and a short length of rubber tube. Connect the syringe, with the plunger pulled back, to the gauge. Now gently push the plunger in. Does the needle move smoothly without jerking or sticking at any point? If so, then that is a good start, but just to be a little more sure disconnect the hose from the syringe and, with the hose lightly clamped between your fingers, slowly release the pressure from the gauge, again noting if the needle jerks or falls smoothly. Repeat these tests until you are sure the needle does not jerk or stick either going up or down. Then make your purchase. Next take the gauge to a turbo specialist with an accurate calibrated large scale boost gauge and have him compare your gauge against his workshop gauge. If your gauge is consistently close, consider yourself very fortunate. If it is erratic or sticks at a particular pressure, get it replaced.

Chapter 8

Turbo anti-lag strategy

This is one area where supercharging by means of a blower has it all over a turbocharger. While turbo lag is now nothing like the problem it used to be, in part thanks to electronic fuel injection, electronic engine management and advancing turbo technology, turbo engines still experience lag to some degree. In standard road tune most current turbo cars are pretty much lag free, although there are a few notable exceptions. However, as soon as we move past the stock, or close to stock, situation we open the way for our tuning efforts to be overshadowed by turbo lag. The way I see it, it seems that many people chase big, impressive sounding hp numbers with little understanding that bottom-end power is going to fall right away and, even with some boost registering, throttle response could be quite sluggish. Consequently, on your average race circuit or tarmac road rally it is usually reckoned that a turbo car requires about 30% more power to put it on equal terms with an otherwise identical naturally aspirated racer, and on most hill climbs and tight race circuits the situation is even worse. Add to that the fact that a turbo car requires more driver skill, and because of the need for more intense concentration driver fatigue must also be factored in.

If we are building a competition engine or a powerful road engine there is no getting away from lag. However, the choices you make in four vital areas will have a large bearing on the degree of lag with which you have to contend. Those four areas are:

* Engine hardware choices
* Fine-tuning fuel and spark requirements
* Gearbox ratios and overall gearing
* Turbo sizing and technology choices

Back in Chapter 3 we took a look at the anti-lag system often employed on rally cars, so we won't go over that again here other than to remind you that because of the high exhaust and turbo temps involved it is not for use on road cars. Even in competition it is really only viable for those who can afford suitable turbos with extremely expensive turbine wheels and housings, and a relatively short service life between turbo rebuilds.

A big turbo on a small engine is usually a recipe for sluggish performance at low revs and in normal day-to-day driving.

A split pulse turbo spools up faster by making better use of exhaust pulses at lower engine speeds.

UTILISING THE SOFT REV LIMITER

A far less aggressive system (which works extremely well for drag starts and which, with some practice to co-ordinate foot movement, will also work in other environments) operates around a simple soft rev limiter activated by means of the clutch pedal. The rev limiter will be adjusted to limit revs to something like 4000–5000rpm, depending on where the turbo makes boost and the engine operating speed. Consequently, when the clutch is depressed the throttle can be floored and the rev limiter will go into action cutting spark to one spark plug at a time to hold engine revs at the rpm set. With spark alternately being withheld from the plugs, unburned fuel passes into the exhaust along with the oxygen from that cylinder. Once in the exhaust manifold the hot exhaust gases ignite that air/fuel mix to drive the turbine. Because spark timing isn't retarded with this arrangement, the exhaust gases are cooler. Hence afterburn combustion is relatively slow and incomplete, so turbo temperatures do not climb to dangerous levels.

Each time the clutch is depressed at engine rpm higher than that set in the rev limiter the turbo runs in afterburn mode. Therefore the turbo is able to maintain its speed during gear changes or from a standing start. It is not the afterburn alone that does this, but rather can it also be partly attributed to the fact that the throttle is open during this period. Therefore, air being pumped from the compressor is free to course into the engine, rather than backing up against a closed throttle plate and then dragging down compressor wheel speed. So effective is this system that some drivers are dusting off old technology turbos with heavy, large diameter turbine wheels. Their claim is that while being slower to spool up, the reverse is also true. Because of the flywheel effect they also resist slowing down, so during open throttle gear changes, either up or down the gearbox, the turbo can better maintain its speed and keep the boost higher.

Apart from the obvious benefit to acceleration, when you can hold the throttle open between gear changes, there is another advantage. Many drivers aren't great at the heel-and-toe technique of down changing into corners. With this arrangement you don't have to worry about blipping the throttle after you get the clutch in. You simply hold the throttle open with your heel, hit the clutch and down change. The rev limiter will hold the engine from over revving when the clutch is depressed.

This system does not have to be built around a soft rev limiter. Some ECUs, including piggy back ECUs, have double rev limiter circuits built in, one to actually limit maximum rpm and the other to function as a clutch (or switch) activated anti-lag or power shift system rev limiter.

ENGINE HARDWARE CHOICES

Engine hardware choices will be dealt with in detail in subsequent chapters. However, at this point it is prudent to consider a very basic truism applying to all turbo engines: any modification which takes away low-speed torque from a conventional engine will have the same effect on a turbo engine, but because the turbo engine has considerably more top-end power and a lower compression ratio coupled with reduced throttle response, the low-speed sluggishness is comparatively much worse. Therefore, quite small increases in cam timing and lift, inlet and exhaust port diameter, and also exhaust header tube diameter, can have a dramatically negative impact on low-speed performance while showing only marginal gains at the top-end.

Even seemingly unimportant things, such as the size of the exhaust tailpipe, can stifle bottom-end power. I recall one Nissan racer being less than impressed when he took delivery of his car after it had been in the workshop for a full workover. The problem in his eyes was the 3in exhaust reduced down to 2½in just before the rear axle. Surely this had to be hurting maximum power, and didn't all the racers in his class have full 3in systems, some even 4in? Unconvinced by the explanation given a few race meetings later he was back to demand 'a proper turbo exhaust'. Operating on the principle that the customer is always right he was sent to the exhaust shop just down the road. A week later he was back, cap in hand, begging for the original 2½in tailpipe. The reason for his change of heart was that the big tailpipe didn't deliver any power gain, and worse, the bottom-end had 'disappeared' and the mid-range was 'hopeless'.

Related to engine hardware is proper fitting and maintenance of that hardware. The best tube headers in the world can't do much to help performance if flange gaskets are blown, or if flanges have distorted and are leaking. Likewise if the pipes are cracked because of being expected to carry the full load of the turbo and dump pipe. Any exhaust leak prior to the turbo increases turbo lag. Obviously the same applies on the inlet side, so thoroughly check all joints by spraying soapy water onto them with the turbo making boost. Also fix dump valve leaks.

THE IMPORTANCE OF PROPER TUNING

I become rather frustrated with individuals who are willing to shell out thousands on trick hi-tech turbos but are then so broke they can't afford to get the fuelling and spark advance optimised. It makes no sense to run with some cobbled together fuel system which cannot be accurately calibrated, and it is just as crazy to do everything right and then not spend the money for a day or two on the dyno programming the engine management system to get the fuelling and spark right throughout the entire rev range at all engine loads. Remember, turbo lag can be just as much a product of poor tuning as it can be of incorrect turbine size, wrong cam choice, etc. Therefore, if your budget won't stretch to cover dyno time *and* the trick turbo you really want, then opt for a less expensive turbo so as to leave sufficient funds for a full calibration session.

With factory turbo cars, something that will considerably cut dyno time and avoid calibration glitches that could contribute to lag is to retain the stock computer, air-flow meter and injectors to take care of the engine's needs up to the rpm/load area where they are still functioning efficiently. Remember, the factory spends millions mapping the fuel and spark programmes, so don't presume that you can do a better job overall. While it is true factory settings tend to be conservative, with the fuelling a touch rich at higher load levels and the spark a fraction retarded, this really can be to our advantage in providing what our modified engine with big turbo and higher boost really needs. The big proviso here is the intercooler must be up to keeping intake charge temperature down lower at high boost than the stock intercooler at factory boost.

To provide the engine with more intake air flow and fuel, a secondary system is added. As illustrated in *Fig. 8.1*, the secondary system flows straight into the turbo compressor whereas the stock system enters at an angle. The secondary system is equipped with its own servo-operated throttle body which is controlled by a solenoid valve via a pressure switch. Rather than using an air-flow meter, the secondary system is MAP-sensor based with a separate fuel-only computer controlling an extra set of injectors delivering fuel near the mouth of each intake runner.

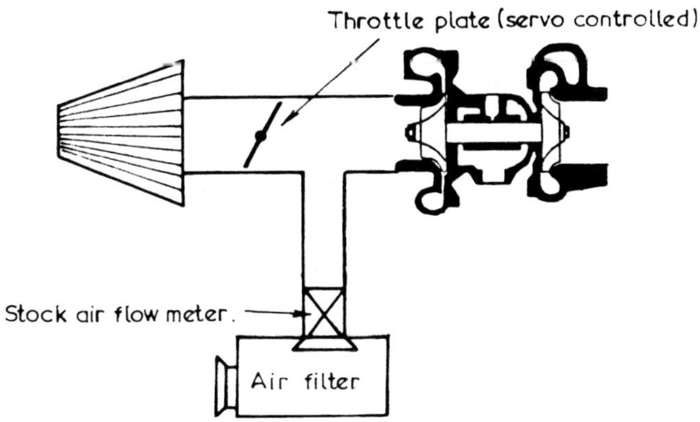

Fig. 8.1. A secondary intake system enables a factory ECU to provide correct fuelling and ignition advance at stock performance levels, while a secondary ECU controlling an extra set of injectors takes care of fuelling at higher air flow levels.

A system of this sort is ideal for a car which serves as both a daily driver and weekend track racer. The factory computer ensures full function of all the car's accessories and provides mild road manners, while the secondary system supplies additional air and fuel at higher power levels. In fact, if competition rules permit exotic fuels, the secondary system can incorporate a separate fuel tank and pump. Then, rather than using some sort of water injection to suppress detonation, the secondary system takes on this role spraying toluol, Avgas or some other high octane fuel into the intake air.

GEARING'S INFLUENCE

We cover the matter of gearing in the final chapter, but while we are on the subject of turbo lag it is good to keep in mind that big gaps between gears and high overall gearing just exacerbate turbo lag problems. Consequently, when planning modifications that will reduce low-speed torque and throttle response we have to be thinking in terms of lowering the overall gearing by means of a numerically higher diff ratio, and possibly also a close ratio gear set.

We have to accept the fact that, with few exceptions, manufacturers gear their cars with a massive overdrive top gear, purely to keep cabin noise levels and fuel consumption down when cruising on the highway. Usually there isn't such a problem with 6 speed gears. Often top gear will be pretty much the same in 5-speed and 6-speed boxes. This means cogs are spaced closer together in the 6-speeder and thereby more capable of covering turbo lag by ensuring that the engine is operating in the 'sweet' area of the power band when exiting the majority of corners.

For example, the stock Subaru Impreza WRX is fairly typical of a 5-speed turbo car. In top gear the engine is spinning at a little over 3200rpm at 75mph and the gap between 3rd/4th is very wide at nearly 30%. For comparison purposes a semi-close-ratio box would have a step in the 22–24% range. In a close-ratio box the steps would be even smaller, with the 2nd/3rd step 26–28% and the 3rd/4th gap 20–23%.

When Subaru developed the high performance WRX Sti with revised cams, lower compression ratio (8:1 vs. 9:1) and bigger turbo, the gearing was drastically lowered (engine revs at 75mph are just over 4000rpm) and the 2nd/3rd, 3rd/4th gear gaps were

Subaru WRX gearing

	WRX			WRX Sti and 22-B Sti			WRX Rally		
	gear ratio	overall	gear step	gear ratio	overall	gear step	gear ratio	overall	gear step
1st	3.45	13.46	+41.5%	3.08	13.68	+33.1%	3.45	14.18	+40.3%
2nd	1.95	7.61	+29.7%	2.06	9.15	+24.8%	2.06	8.47	+29.6%
3rd	1.37	5.34	+29.2%	1.55	6.88	+25.8%	1.45	5.96	+24.8%
4th	0.97	3.78	+23.7%	1.15	5.11	+27.8%	1.09	4.48	+23.9%
5th	0.74	2.89		0.83	3.69		0.83	3.41	
Axle ratio		3.90			4.44			4.11	
mph per 1000rpm in 5th		23.25			18.48			21.08	

closed up to around 25%. Even with the revised gearing the 2-litre Sti didn't have much urge below 4000rpm. Consequently, when the 2.2-litre 22-B Sti appeared, Subaru retained the same gearing.

For some markets Subaru also produced a model called the WRX Rally. To my way of thinking, this version has the ideal 5-speed gearset for a pure road-going WRX. Top gear is fast enough that the engine doesn't drive you crazy at highway cruise, like the Sti does. Also, gear spacing isn't too bad providing you retain the stock cams, high compression (8.5:1 when a front mount intercooler is fitted) and a turbo a shade smaller than the Sti.

As I prepare this material for print, Subaru have indicated that the next WRX (MYO2) will have revised gearing to address the issues raised. The first three gears will all be numerically lower (3.166, 1.882, 1.296) but 4th and 5th will stay the same (0.97, 0.74). However, the differential will change from 3.9:1 to 4.444:1 to provide lower overall gearing fairly close to that in the old WRX Rally (14.07, 8.36, 5.76, 4.31, 3.29). The steps between gears going from 1st to 5th will now be 40.6%, 31.1%, 25.2% and 23.7%, which will get rid of the large 3rd/4th gap.

TURBO TECHNOLOGY TO REDUCE LAG

Back in Chapter 3 we looked at what Porsche did with sequential turbos to overcome the dilemma of turbo lag. Toyota and Mazda went the sequential route with later Supra and RX-7 models, and the Subaru Liberty B4 was sequential from the outset. Also, a few large tuning companies have offered sequential kits, but in general there isn't a lot of enthusiasm for sequential turbocharging. In fact, some tuners derive a fair portion of their income converting RX-7s back to single turbo status after their owners become dissatisfied with

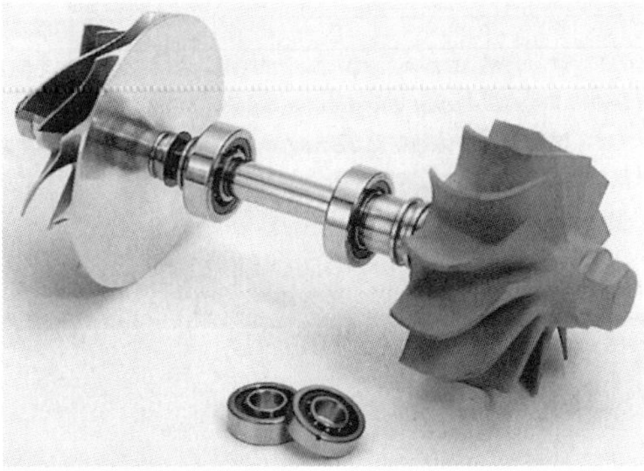

IHI developed fast spool-up ball bearing turbos for Honda's Formula 1 engines and later applied the lessons learned to their regular production units.

A lightweight ceramic turbine wheel aids turbo spool up, but high boost has to be avoided.

system problems in bringing the second turbo into operation. The story is similar with the twin-turbo Supra, with many tuners choosing to ditch the tiny stock turbos and very restrictive exhaust manifolds in favour of an ordinary twin-turbo arrangement.

In general, car makers and turbo manufacturers were looking for less complex and more economical ways of combating turbo lag. In the same chapter we saw that most attention was directed to lightweight ceramic turbine wheels to cut the inertia of the rotating assembly (since then Nissan has also fitted very light carbon composite compressor impellers in Skyline GT-R models) and ball bearings to reduce frictional drag. However, the most exciting development without doubt was variable nozzle technology.

We have already examined the Garrett VNT turbo incorporating multiple vanes. These close when exhaust gas flow is low to provide a small nozzle area so as to maximise turbine wheel acceleration and speed. The vanes then progressively open as exhaust flow increases to minimise exhaust back pressure, control boost pressure, improve fuel economy and increase hp (*Fig. 8.2*). In effect it works as a small A/R turbine at low speeds and like a turbo with a larger A/R turbine housing at high speed. Therefore, turbo lag is quite markedly reduced. In fact, when response times of multi-

Fig. 8.2. The Garrett VNT turbo is a multiple-vane type variable geometry turbine design.

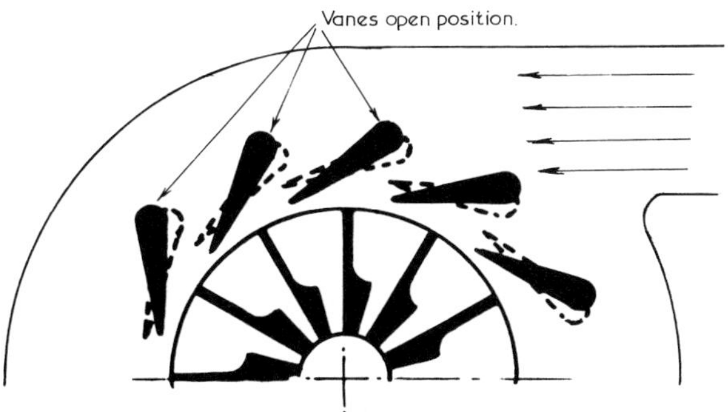

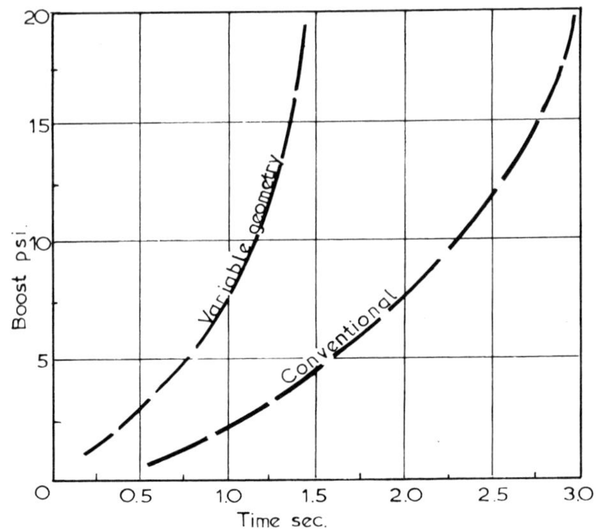

Fig. 8.3. Multi-vane variable geometry turbos build boost much more quickly than conventional units.

vane VNT type turbos are compared with ordinary turbos of similar size it is found that the latter take 70–90% longer to produce the same boost pressure (*Fig. 8.3*).

Garrett is not the only company to manufacture the multi-vane Variable Nozzle Turbine. Mitsubishi and IHI do also, but possibly the Aerodyne Aerocharger is best known among the others. Clearly the multi-vane design is much more complex and expensive to manufacturer, and more moving vanes increase the risk of reliability problems. In some installations, depending on turbo matching and boost restrictions, a wastegate is not required (turbine speed and boost being governed by the vanes opening fully) so there is a possible cost saving in this area.

These factors – cost, complexity and possible reliability issues – have led to the development of other variable nozzle designs. Perhaps the design most likely to find quite wide fitment (because it offers good performance, it is reliable and cheap to manufacture) is the twin-scroll type. Usually two different sized scrolls are used for the primary and secondary. The small primary, in the Mazda RX-7 for example, is open initially for low speed, low gas volume operation. Both scrolls open for high speed, high exhaust flow operation (*Fig. 8.4*). Again the effect on performance is similar to that of a turbo with a small A/R turbine housing at low rpm and large A/R housing for high rpm.

One area for concern with the swinging-flap type twin-scroll design is that when in the closed position the selector flap should present a smooth, gently tapering exhaust flow path rather than an almost flat plate for the gases to slam into. Any design which blocks flow or causes turbulent flow should be avoided as it will hurt performance.

The other twin-scroll type VNT turbo uses a horizontally sliding plate (*Fig. 8.5*) to block flow from the secondary scroll onto the turbine wheel. This arrangement tends to cause more exhaust flow turbulence at the turbine housing entry when the secondary is closed. However, it does have what some manufacturers see as an advantage in that only a single actuator is needed to operate the sliding plate which functions as both a variable nozzle control and a wastegate. The actuator swings the control lever through 90° to

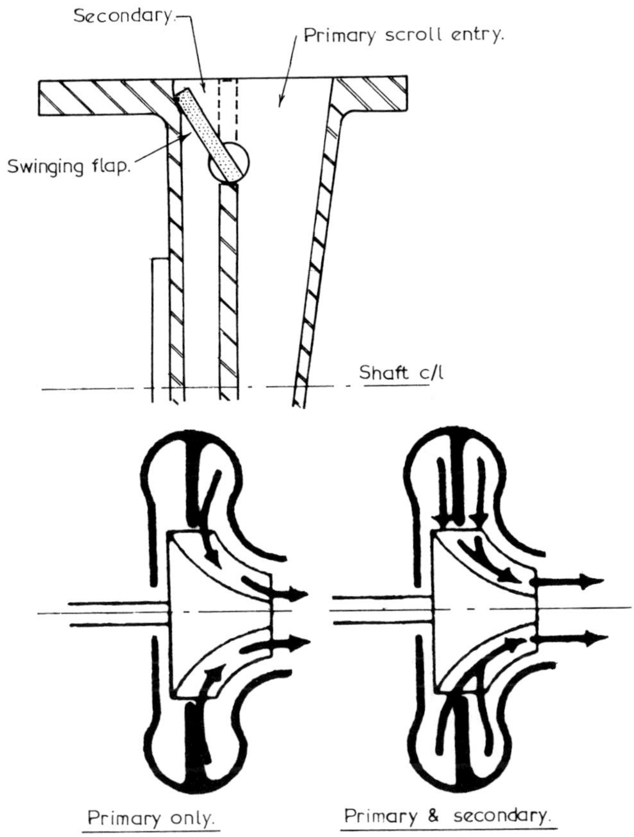

Fig. 8.4. A variable entry turbo may control exhaust flow into the twin scroll turbine by means of a swinging flap.

Fig. 8.5. Alternative variable entry design utilises a sliding plate to control exhaust flow. In some designs the plate functions as a wastegate, moving further to the right to connect the secondary scroll with an exhaust bypass passage.

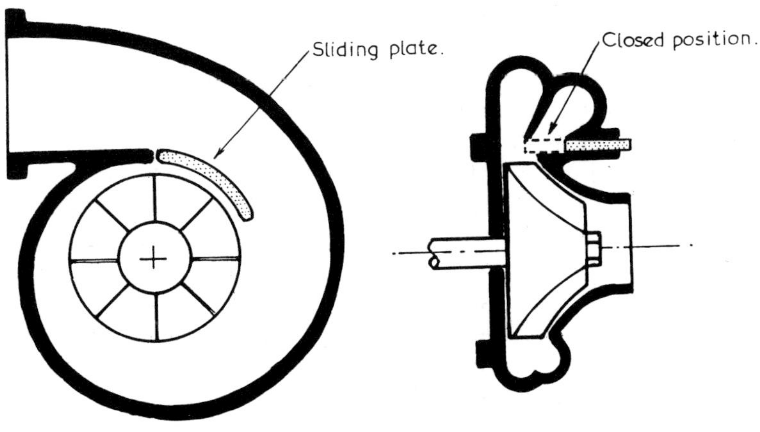

move the sliding plate into the both-scrolls-open position. When boost reaches its limit the actuator shifts the control lever further, pulling the sliding plate away from the edge of the secondary scroll, thereby connecting this scroll with the wastegate passage.

In 1992 Garrett produced another variable nozzle turbine which they called the VAT 25 (Variable Area Turbine). Fitted by Peugeot to their 2-litre T16 model it enabled them to extract 196hp at 5200rpm and 212lb/ft torque between 2600rpm and 4500rpm. On overboost this increased to 216hp and 234lb/ft from 2400rpm to 4500rpm. The design features a single movable vane which controls exhaust gas speed as it approaches the turbine wheel. In the closed position gas velocity increases, which accelerates the turbine wheel faster, similar to a small A/R housing. Then, as exhaust gas volume and speed increases, the vane is moved into the open position to present a larger effective housing area, lowering exhaust back-pressure and improving flow (*Fig. 8.6*). This type is reliable in service, but in the closed-vane mode it does not make best possible use of available exhaust energy. The single vane causes losses because of the wake it produces. Also, gas impingement onto the turbine wheel being less than ideal, from both the aspect of impingement angle and non-uniform velocity distribution, results in more losses.

Perhaps, in endeavouring to overcome these losses, Mitsubishi came up with a double vane design (*Fig. 8.7*) that is almost as efficient as the multi-vane type, but is more reliable and cheaper to make. The two vanes are moved simultaneously and, rather than having a simple open and closed position, they move to provide a continuously variable nozzle area. However, this situation may change as the performance differences between the simple open/closed control and continuously variable opening is quite small. In early tests Mitsubishi found that a conventional turbo of similar size took 70–80% longer to make the same boost and that full boost was not reached until an engine speed 15% higher. On this particular 2-litre engine the vanes remain closed until 1750rpm – at which point the engine was making 25% more power than with the ordinary turbo. Past 1750rpm the vanes gradually open and are fully open at 2500rpm. Commendably, because the internal turbine housing profile has been executed so well, it goes on to make the same high rpm power as a normal turbo, indicating no exhaust flow losses.

Fig. 8.6. The Garrett VAT turbo has a single moving vane.

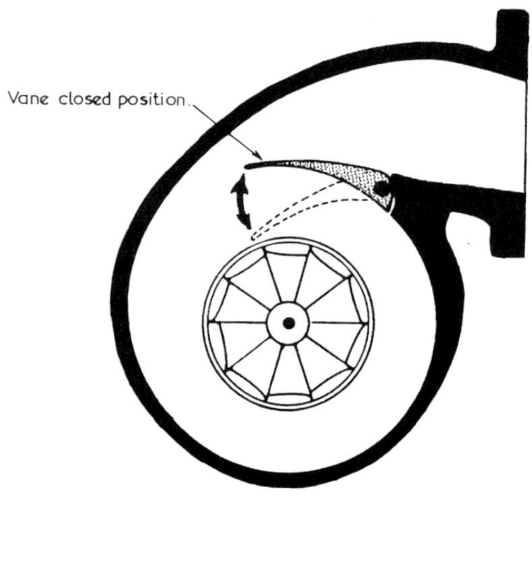

Fig. 8.7. A Mitsubishi double-vane turbine.

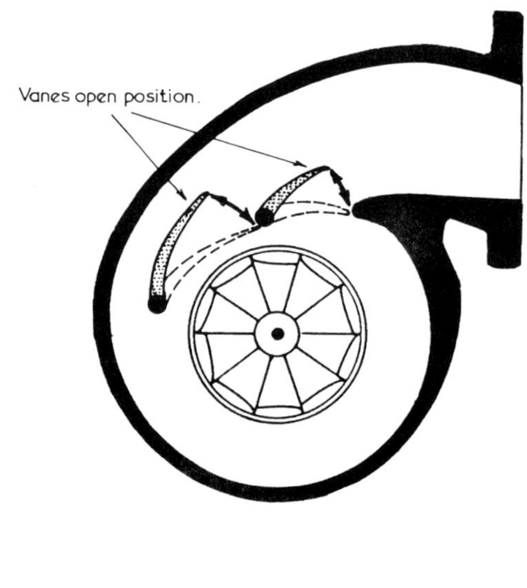

Chapter 9

What you should know about blowers

Many blowers fall into the positive displacement category, but are more usually called Roots type blowers. A positive displacement blower displaces a certain volume of air from the blower inlet into the engine for each revolution it turns. Therefore, if we fitted a 1000cc Roots blower to a 1500cc engine and geared it to spin at 1.5 times engine speed (referred to as 50% overdrive) the blower would in theory displace 1500cc for each revolution of the engine. Because 4-stroke engines only fill half their total number of cylinders with air each revolution, 750cc in this example, the blower is supplying two times the engine's air requirements. Consequently the theoretical boost will be 14.7psi at sea level.

CAM CHARACTERISTICS AFFECT BOOST

In practice, though, it will not be anything like this. At lower engine speeds there is considerably more leakage between rotors and between the rotors and the blower case. Additionally, cam duration and valve overlap have some influence. A stock cam has only a small amount of overlap, so at low rpm only a small volume of the air charge will be lost out through the still open exhaust valve. However, a race cam, with perhaps 100° overlap and the inlet and exhaust valve possibly 0.0250–0.350in open at TDC, will spill a tremendous amount of boost straight out the exhaust at lower engine speeds.

At higher rpm the situation reverses. There is still leakage in the blower itself, but this leakage in comparison with the increased volume of air the blower is actually moving in fact decreases. Likewise during the valve overlap period. Pressure waves generated in the inlet and exhaust tract now restrict charge loss into the exhaust. Less charge loss means pressure in the cylinders and inlet manifold rises, which the boost gauge sees as a higher boost pressure.

In practice what does this mean for our 1500cc engine? Really, cam lobe characteristics will play the leading role in determining boost pressure after the blower drive ratio has been selected. In this example the boost pressure could approach our theoretical 14.7psi at 6000–7000rpm with a stock cam, but it is more likely to be 1–1.5psi less. With a race cam it is unlikely for the boost to exceed 11psi, and it may be as low as 8psi. (A rule of thumb is that boost pressure will decrease by about 4–5% of the absolute pressure for each 10° by which valve overlap is increased.) The top curve in *Fig. 9.1* represents the sort of boost pressure our 1500cc engine would probably see with a sports road cam. For comparison purposes, boost pressures with the same 1000cc blower operating at 15% and 37% overdrive are also included.

Some might jump to the conclusion that a milder cam is better in a blower engine because we get more boost. As I discussed in earlier chapters, boost figures are a rather vague indicator when it comes to the performance potential of an engine. When a blower is driven at high rpm and/or high boost its efficiency falls. It puts a lot more heat into the air. More heat means the boost pressure rises, but rather than power going up it falls because of lower charge density in the cylinders. To give some idea of what I'm getting at here, if we let a steel cylinder fill with air, close the tap and heat the cylinder 100°C above atmospheric temperature, pressure in the cylinder will rise from 0 to 5psi! Even though there was no increase in the amount of air inside, we have a pressure of 5psi simply as a result of more heat. When we run with a mild cam the situation is quite similar, but to better understand what is happening we will look at a single-cylinder engine. With a mild cam the inlet valve closes early, so air being forced out of the blower has nowhere to go. It is trapped between the blower and the back of the inlet valve. Therefore, boost pressure rises and as it does this trapped air tries to reverse its way back through the blower, holding back air that the blower is trying to push into the engine (think of a rising tide holding back the flow of a river where it spills into the ocean). This conflict reduces blower efficiency and pushes up the charge temperature. A hotter charge means lower cylinder charge density and greater risk of detonation, meaning reduced power.

With a race cam the cylinders have more time to swallow output from the blower. This means boost pressure falls because more air from the blower has flowed into the cylinder rather than ending up trapped in the inlet tract against a closed inlet valve. Therefore, at the instant the inlet valve closes, cylinder pressure more closely mirrors boost pressure, whereas with a mild cam cylinder pressure is much lower than boost pressure.

Fig. 9.1. A 1000cc Roots blower overdriven as indicated would provide these boost levels on a 1500cc engine with a sports cam.

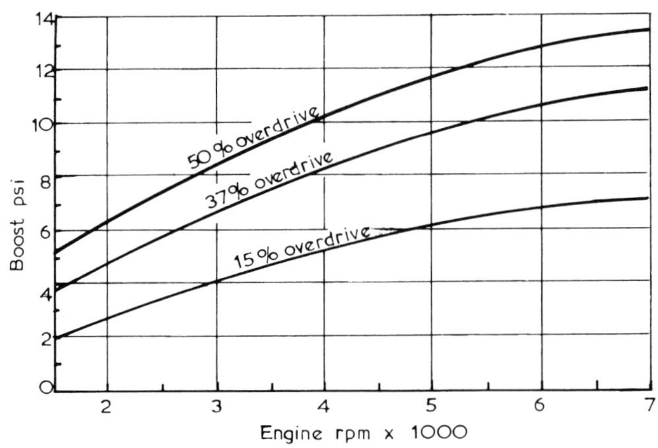

A Toyota SC12 blower is a Roots type limited to 11,000rpm. It displaces 1.2 litres of air each revolution and is fitted to their 162hp 1.6 litre 4-cylinder engine (4A-GZE). Note the electro-magnetic type drive pulley.

Charge loss with a wild cam during the overlap period is a mixed blessing. Obviously there is fuel wastage straight out the exhaust. Also, at low rpm the engine will be down on power because a portion of blower output has gone right out past the exhaust valve without producing any power. And don't forget the blower has consumed power pushing the air into the engine, so we have a lose–lose situation. However, it is not all bad news as at higher engine speeds pressure waves in both the inlet and exhaust systems reduce the through-flow. And some loss out of the exhaust actually becomes a gain in that this action purges any residual exhaust gas out of the cylinder, leaving more volume for a fresh fuel/air charge. Additionally this flow takes with it a lot of heat, particularly from the piston crown and exhaust valve. This lowers the risk of pre-ignition and detonation. Cooler engine components mean less heat transfer to the inlet charge. A cooler charge occupies less space, enabling the blower to force more charge into the cylinders, increasing charge density and hp (*Fig. 9.2*).

Fig. 9.2. At lower rpm large amounts of valve overlap permit inlet charge loss out past the exhaust valve.

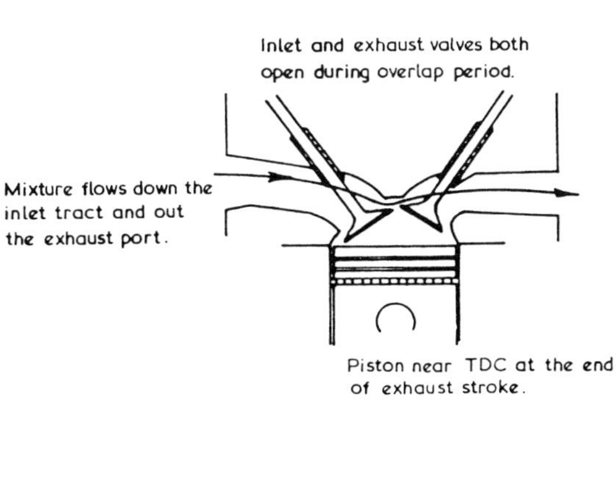

BLOWER SIZE AND EFFICIENCY

Referring back to *Fig. 9.1*, one could get the impression that blower size isn't terribly important as you can simply spin the blower faster and increase the volume of air it pushes into the engine. Such is not the case. All types and sizes of blowers have a band of flow and pressure at which they work most efficiently. Outside that band efficiency falls away, meaning more heat is added to the inlet charge for only a small increase in flow output from the blower.

From *Fig. 9.3* you can see that this large blower is most efficient (59%) when moving something like 48–70lb of air per minute at a pressure ratio of 1.38–1.55. Outside of that range efficiency falls away. Note the amount of heat added to the inlet charge with increasing boost pressure and how much engine power is used to drive the blower at higher air-flow rates.

In Chapter 2 we discussed the advantages of the Lysholm screw blower and the centrifugal blower in both the area of reduced parasitic power drain from the engine and greater operating efficiency producing a much cooler inlet charge. So, while Roots-type blowers are not too bad at 7–10psi boost, they are really out of their depth beyond 14–15psi. However, both the screw and centrifugal type are quite efficient at well over 40psi boost, although as previously discussed the centrifugal blower does not make boost at low rpm like all other types.

Another virtue of both the screw and centrifugal type is that for their air pumping capacity they are usually considerably more compact than the Roots. In fact, very high flow Lysholm blowers, capable of 800hp, can be hidden beneath the inlet manifold in the valley of overhead cam 90° V-style engines! Apart from the drive belt and pulley the blower is not visible.

Fig. 9.3. This Roots blower performance map shows blower efficiency, the amount of heat added to the inlet charge and the power drain to drive the blower.

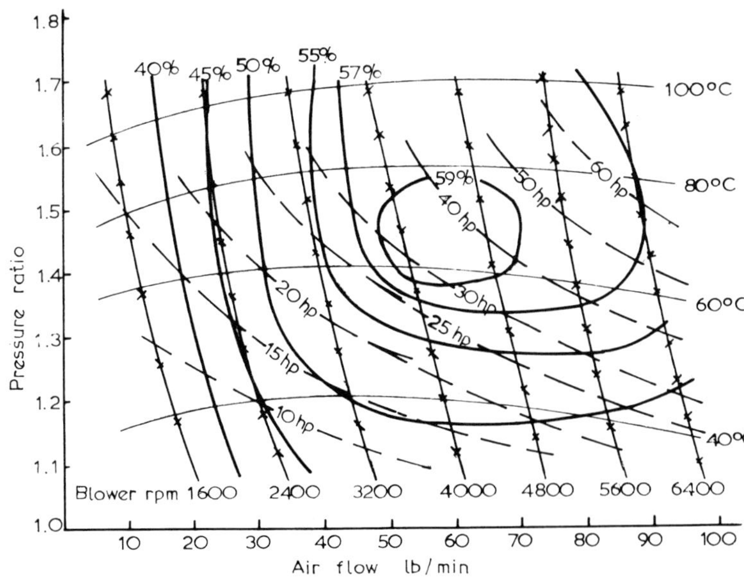

This monster is a GM 16-71 Roots blower.

WHEN TO CHOOSE A BLOWER, OR A TURBO

Fortunately, in the last decade there have been a lot more blowers and blower kits available to choose from, possibly as a result of major auto-makers going the blower route on a few models. That is good, also, from the aspect that it has aided enthusiasts see that there are alternatives to turbocharging – that turbos are not always the best way to go. Many were dragged along the turbo route by all the motoring press hype and possibly also motor sport successes. However, just as race cams are useless in a road car, so likewise must the suitability of turbochargers, and blowers too for that matter, be judged according to the vehicle's intended operating environment.

Personally, I am not a fan of turbos on daily driver road cars and I would not run a turbo on a road car unless I needed high boost (over 12psi), the engine was oversize for the weight of the car (2-litre in a 1800lb car, 3-litre in a 2600lb car) and the gearbox ratios and overall gearing were suitable (see Chapters 8 and 24). In motor sport, turbos are useless in any sort of competition where precise throttle response and low speed torque are required. Therefore they are not for tight race circuits, hill climbs (Pike's Peak type hills are the exception), dirt or tarmac rallies (fast tarmac and/or expert drivers are an exception) and other dirt surface events.

ROOTS BLOWERS

When it comes to blowers, choices have to be made according to the application. For low cost and low boost (up to 12psi) a Roots type is hard to beat. However, they are the most power hungry and the least efficient. Therefore, if used on a road car these factors would have to be weighed up. For example, if the car only travels short distances, fuel consumption may not be a concern, but if you commute 700 miles each week you might want to consider a blower bypass and an electro-magnetic clutch. Likewise, if you don't want the hassle and expense of intercooling then a more efficient, but more expensive, blower may be a wiser choice.

However, the Roots type can be made more efficient by injecting water or fuel into the blower. This closes the gaps between rotors and also between the rotors and the case, which improves pumping efficiency. Boost goes up, too, so a lower drive ratio can be used, cutting parasitic losses a little. The evaporation of this water or fuel takes some heat out of the intake charge and provides an intercooling effect. Methanol, with it's very high latent heat of evaporation, is particularly good in this respect. Remember, though, that when intercooling is employed water or fuel must not be introduced into the air stream before the intercooler.

While carburettors really do not have a lot going for them, if you are stuck with using a carb then you are also probably stuck with running a Roots blower. Remember to always draw through the carburettor, don't ever blow through it. Carburettors work acceptably, just, with screw-type blowers, but if you can afford the high price for this type of blower why would you ever want to impede its performance with a carb? Carburettors don't work on centrifugal blowers and turbos, and in spite of all the pages written on the blow-through vs. draw-through issue my advice is to forget it!

CENTRIFUGAL BLOWERS

Centrifugal blowers, like turbos, are very efficient but, like turbos, they don't produce much boost at low engine speeds, and if they are geared up to give low-speed boost then it is excessive at high rpm. Some view this characteristic of centrifugal compressors as a blessing, others as a curse. To my mind it again gets down to application. Being belt driven these blowers are always more responsive than turbos. Less boost at low rpm cuts traction problems, front-wheel-drive torque steer, transmission wear and damage, etc. Their high efficiency means that in low-boost installations the engine generally does not need to be opened up to lower the compression ratio.

The very nature of centrifugal blowers helps here too. In road cars detonation is more common at lower rpm in high gears, but as there is only minimal boost it is generally easily overcome with a little spark retard or a higher octane fuel. Then at high revs, when the stock cams and inlet system are restricting cylinder filling anyway, the centrifugal type is making good boost to compensate, and to some extent is working a bit like the SAAB variable compression engine with respect to cylinder charge density at the end of the compression stroke.

As they are often considerably less expensive than screw-type blowers they have a place in motor sport too. In most types of competition the engine will be operating at the very top of the power band over a 2500–3000rpm range. Therefore, the centrifugal blower can be geared to provide the necessary boost over that range and a lack of boost at lower rpm should not be a problem in most classes of competition.

LYSHOLM SCREW BLOWERS

This leaves Lysholm screw blowers. Like the centrifugal type they are quite efficient, which means less intercooling is required for the same boost as Roots blowers. This also indicates they are suitable in very high boost applications. Some special designs are

A centrifugal blower is very efficient and, compared to other types of blower, is smaller and lighter.

A Sprintex screw blower.

working up to 45psi and in time it seems 50psi should be within reach. At low rpm they make even more boost than Roots blowers, with boost beginning to register at as little as 1000rpm. Therefore, in conditions demanding the widest possible power range, high compressor efficiency and high boost (13–40psi) the Lysholm blower has no peers.

WHY BLOWERS SUIT ROAD CARS

At this point I should explain why I prefer a blower or a naturally aspirated engine rather than a turbo for road cars. Now, I don't deny that turbo cars have made giant strides in the past decade or so, but in spite of all the figures engineers might quote to me and all the impressive graphs and charts manufacturers might point to when releasing their new models, the simple fact remains that turbo cars lack real urge in day-to-day running. All enthusiasts like the feel of a car that responds with a big surge of power the instant the throttle is depressed, even if the car is in a gear that is too high. Also, they like the very gentle response when the throttle is delicately applied and feathered when traction is at a premium. In such conditions car and driver are as one, with the car seemingly able to respond to a driver's emotional inputs almost as well as to messages from his right foot. It's that level of communication that is missing from road-going turbo cars.

Two things, possibly three, are working against the advances being made in turbocharging. My big bugbear is vehicle weight. Today, consumers want their cars loaded with all the creature comforts of home, and then some. Entertainment systems, fast glass, electric mirrors and antennas, etc., etc. – all add weight. Consequently, mid-size saloons (sedans) that today weigh 1250kg were 1000kg 15 years ago. Full size stuff today is 1650kg, 15 years ago it was 1200kg. The little guys that were 800kg are now 1000kg. That is around a 25% weight increase! All that weight demands bags of low-speed torque to get it moving when accelerating out of corners or overtaking.

Blowers perform very well in motorsport environments such as hill climbing, where instant throttle response is required.

But what has been happening during the time cars were getting more porky? In the same period manufacturers came to realise that power sells and that magazine journalists are impressed by $\frac{1}{4}$-mile acceleration times. The answer was to factory hot-rod their engines. They fitted bigger cams, bigger and more valves, bigger ports, etc. Things which give more top-end hp, but less low-speed grunt. Slap on a turbo and the situation doesn't get better. So, just when cars could do with a good injection of low-speed torque, manufacturers' sales departments forced the engine shop to take it away. Consequently, today many cars are pretty hopeless, even with a 5-speed gearbox on twisty secondary roads.

Which brings us to my third point. To quieten these revvy, highly strung engines and reduce fuel consumption, manufacturers massively overgear their cars, some by about half a gear and others by close to a full gear. Consequently, at cruise these cars feel lethargic. Unless you are travelling in excess of 80mph a slight hill demands a down change and a casual overtake requires the driver to drop back two cogs. Not exactly the stuff of effortless cruising.

WHAT PORSCHE DISCOVERED

However, a blower can change all that. Interestingly Porsche did some comparison testing on a 16-valve 2.5 litre 944 some time back with both a Roots blower and a turbocharger. *Figures 9.4* and *9.5* show two sets of the results. As you can see, the Roots blower made close to maximum boost within about $2\frac{1}{2}$ seconds of the throttle being floored. By comparison the turbo car has virtually no boost registering at this stage. In fact, it takes over 10 seconds for the turbo to make the same boost as the blower.

Things were just as dismal for the turbo in terms of actual acceleration. After five seconds the blown car has accelerated from 1150rpm to 2050rpm while the turbo car has barely increased engine speed. After 10 seconds the blower car has reached over 2800rpm with the turbo car languishing at just 2000rpm.

Fig. 9.4. This Porsche test reveals just how much more quickly a Roots blower makes boost in comparison with a turbo.

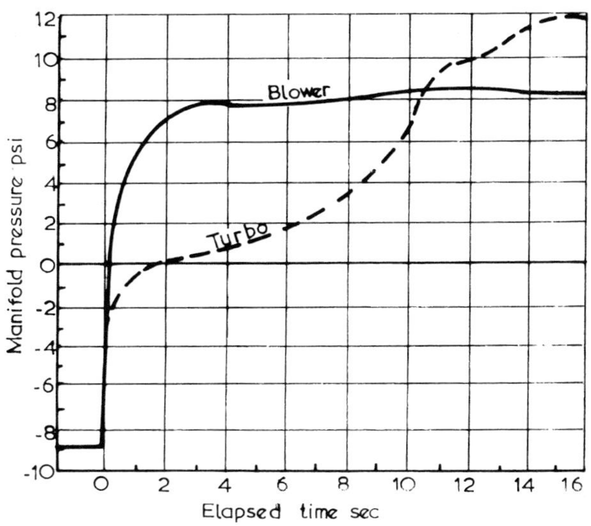

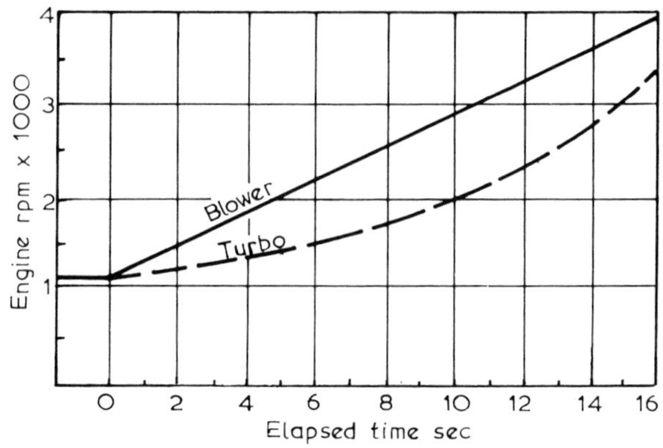

Fig. 9.5. This fourth gear acceleration test again indicates blower superiority in ordinary driving environments.

Porsche reported that at 6000rpm the blower consumed 50hp. However, in the low and middle range it had a torque advantage of 50–70% over the turbo car. Only in outright power and fuel economy was the turbo car superior. Maximum power was up 19% and at maximum speed fuel consumption was down 9%.

For those who believe that flooring the accelerator at 1150rpm in 4th gear is not a valid test as it favours a blower over a turbo, Porsche report that they tested every possible combination of initial speed and gear, but in every test the blower car outperformed the turbo car by 10–40%.

BLOWER LUBRICATION AND DRIVES

In service, blowers are very reliable. Lubrication for drive gears and bearings is usually provided from an internal oil reservoir, although some centrifugal types do have to be connected into the engine's lubrication system in a similar manner to turbochargers.

Possibly the main area for problems is drive belts and mounting brackets. The thing to consider here is just how much power it will take to drive the blower and then engineer the system appropriately. Also, remember that in competition situations these high loads will be sustained for longer periods than in a road car. Therefore, drive belts that gave no trouble for thousands of miles on the road may be unsatisfactory after just a few laps of a race circuit.

Drive belt efficiency is also related to the effectiveness of the mounting brackets and idler pulleys. Obviously these brackets must provide proper mounting of the blower, not stressing or twisting the blower case while correctly aligning the drive pulleys. Additionally, the mounts or idler pulleys must allow for the drive belt to be properly tensioned in the first instance, and also incorporate an adequate mechanism to maintain that level of tension and not slip when the drive system is fully loaded.

BOOST CONTROL AND DUMP VALVES

Boost control of blowers is most usually provided by selecting the appropriate drive ratio. However, in some situations where centrifugal blowers are driven at very high

A big Roots blower with 'bug catcher' on top. Solidly mounted on the inlet manifold and utilising a well engineered drive arrangement, it is reliable in service.

speeds to give more low rpm boost, it may be necessary to limit maximum boost using appropriate vent valves of the type usually employed for turbo engines. Rather than something crude like a couple of radiator caps, it is far better to go with a turbo blow-off valve (if a single dump valve does not bleed sufficient boost, fit additional valves). For best regulation of boost this valve would be controlled from the engine management system. If an intercooler is fitted, the dump valve should be located before the intercooler to improve intercooler efficiency.

In blowers not fitted with a bypass valve it will be necessary to employ a dump valve in any inlet system which has the throttle plate located after the blower. (Personally, I prefer the throttle before the blower. It is simpler and often works better.) Drive to the blower doesn't disconnect when the driver lifts off the accelerator, so the blower continues to force air into the duct interconnecting the inlet manifold and blower. With the throttle open, pressure on both sides of the throttle plate is pretty much equal, but when the throttle slams closed manifold pressure quickly drops below zero to perhaps –9psi on one side of the butterfly plate, while on the blower side it may momentarily rise to 25psi or more. This will bend the throttle shaft, and while it may be possible to force the throttle to open again, the next time the driver lifts off, the throttle may stick open – not a good experience!

BEWARE OF BACKFIRES

Something else that must be considered is the risk of blower damage from a backfire or flash-back in the inlet manifold. Centrifugal blowers are immune to this sort of damage, and usually there is no problem with engines which have electronic fuel injection spraying fuel at the back of the inlet valve. However, if the cam has long duration and big overlap, plus fuel is introduced into the blower or the plenum chamber, there is always some chance of a backfire. When the flame hits all that fuel vapour it explodes. When it does, the rapid rise in pressure has the potential to blow the supercharger apart at its most extreme, but rotor and gear damage is the more likely outcome unless something is done to rapidly vent the pressure.

To rapidly vent that pressure requires a vent of considerable area. How big depends on the volume of air trapped between the blower and inlet valves, but as a guide consider a vent area of 2sq in for each litre of engine displacement as a minimum. Really, two vents located in each end of the plenum chamber are much better than just one located close to the blower. Some racers prefer a spring-loaded pop-off type valve set to vent at about 1.5 times maximum boost pressure. Others, though, prefer to fit a very light gauge metal cover over the vent(s) which ruptures under pressure. This is certainly cheap, the metal often sourced from discarded aluminium drink cans.

As in all areas, prevention is better than cure. Clearly the only way flame can get into the inlet tract is past the inlet valve. Therefore, valve bounce must be avoided and any tendency for an inlet valve to hang in the open position must be properly investigated, and then corrective action taken. Valve bounce is caused by either the driver overspeeding the engine, or weak valve springs. An inlet valve hanging open may be because of a valve gear binding problem, a bent valve stem or insufficient valve stem clearance.

Another problem area is hydraulic lifters. Unfortunately, in some competition classes you don't have any choice in the matter, but if you can get rid of them do so as many blower explosions can be traced back to a single lifter prematurely pumping up and

Waste-spark type ignition has to be treated with respect as engine modifications may open the way for an explosion in the inlet manifold.

holding an inlet valve off its seat. If you are forced to use hydraulic lifters, be sure to observe rev limits, test the lifters often, use an oil which doesn't foam easily and control oil drain-back into the sump, keeping it away from the spinning crank and rods as much as possible. In wet sump systems, keep the oil level well below the crank and rods both by not overfilling the sump and by adequate baffling to prevent surge. In a dry sump system, use a properly designed oil tank with sufficient volume, height and plates to de-aerate the oil.

Waste-spark type ignition systems must also be avoided. This type of ignition shares one coil between two cylinders and eliminates the distributor. As such it fires every cylinder each time the piston approaches TDC. Naturally, during the overlap period on the exhaust stroke the inlet valve could be on its way open when the spark plug fires. With a stock cam and sequential fuel injection this isn't a problem. Except at idle the inlet valve will most likely still be on its seat when the plug fires. Even if it has begun to lift, there won't be any flame travel back into the inlet tract as the injector will not be spraying fuel until well into the induction stroke.

With a modified engine the situation changes. Now, we will most likely be injecting fuel right into the blower to increase its efficiency if the engine isn't intercooled. This then loads the entire inlet system with potentially explosive fuel vapour. Even if we don't inject at the blower, a cam with increased duration will permit a degree of blow-back past the inlet valve under certain conditions. Also, injector duty cycles will have to be extended, which could mean a good deal of fuel vapour will be present in each inlet tract and possibly also the plenum chamber at the time the inlet valve opens and the plug fires on the waste-spark cycle. The potential exists at that time for a backfire.

Chapter 10

Charge air cooling

The more technically correct term 'charge air cooling' has just about been universally replaced by the word 'intercooling', although some will still argue that the word we should be using is 'aftercooling'. Personally, I think charge cooling is better but I've got no problem with any of these expressions, so mostly I stick with intercooling, given its wide acceptance. Whatever word is used, the idea behind the process is to remove as much as possible of any heat added to the inlet charge when some type of supercharging device pressurises inlet air above atmospheric.

As we discussed in Chapter 5, what we are really aiming for is high charge density in the cylinders. Higher charge density means the air is heavy with oxygen. When additional fuel is added to achieve the correct air/fuel ratio the engine's power output goes higher. Hence a 30% rise in charge density will give a 30% power increase.

All forms of supercharging aim to raise charge air density by increasing the pressure of the air entering the cylinders. Unfortunately, pressurising air adds heat to it, and the more you compress it the hotter it gets. Therefore, even if we are using a very efficient centrifugal compressor, as fitted to turbos and centrifugal blowers, pumping 14.7psi boost (double normal air pressure at sea level) we will not double charge density in the cylinders. Without cooling, actual charge density will rise about 53%, indicating that the potential power increase will also be around 53%.

This is where intercooling enters the picture. If we remove a good deal of that added heat, charge density rises, and with it we gain a substantial power increase (there are other benefits which we will cover later). Fortunately we don't have to guess at the likely hp rise, we can get fairly close from a few simple calculations.

CALCULATING AIR DENSITY CHANGE WITH INTERCOOLING

For example, suppose our turbo is pumping 14.7psi boost and we fit an intercooler with 80% efficiency. On a day with an ambient temperature of 25°C, charge air leaving the compressor is found to have a temperature of 118°C, indicating a temperature rise of

This monster air-to-air intercooler out of a truck was replaced under warranty because of a minor crack, and was later purchased for scrap metal value along with several dozen others. Cut down it will provide cores for three or four big car intercoolers.

93°C (118 – 25 = 93). The intercooler will remove 80% of the temperature rise, 74.4°C (93 x 0.8 = 74.4), resulting in a charge air temperature of 43.6°C (93 – 74.4 + 25 = 43.6) leaving the intercooler.

The density change can then be calculated using the formula:

$$Dc = \left\{ \frac{Dt + 273}{It + 273} \right\} - 1$$

where Dc = density change
Dt = compressor discharge temp °C (for °F change formula to Dt + 460)
It = intercooler discharge temp °C (for °F change formula to It + 460)

Therefore:
$$Dc = \left\{ \frac{118 + 273}{43.6 + 273} \right\} - 1$$
$$= \left\{ \frac{391}{316.6} \right\} - 1$$
$$= 0.23$$

This means the intercooler will theoretically increase charge density by 23%, but because of losses the actual power increase will be somewhat less than 23%. Obviously there is some flow resistance owing to turbulence in the intercooler and associated ducting. In a poorly designed system with a number of tight bends and a restrictive intercooler this can easily reduce boost by 20%. Additionally, there is a natural pressure drop owing to the charge air being cooled. However, in a well designed charge cooling system total losses should not exceed 10%. Therefore, if boost pressure is 15psi at the compressor outlet we should see 13.5psi at the inlet manifold. Of course, if the compressor is capable of delivering more boost without losing efficiency some of this loss can be partly compensated for by adjusting the wastegate or blower drive ratio. However, there are definite limits here so don't get greedy, and don't even think about it if the compressor is already operating in a low efficiency area of the compressor map.

A large front-mounted air-to-air intercooler contributes much to power output and engine reliability.

CALCULATING POWER INCREASE WITH INTERCOOLING

By looking at pressure losses we can better gauge just how much power increase will be derived from intercooling. The formula is:

$$HPr = Dc + 1 - \left\{ \frac{Ap + Bc}{Ap + Bm} \right\}$$

where HPr = horsepower rise
Dc = density change from intercooling
Ap = ambient pressure
Bm = manifold boost pressure
Bc = compressor outlet boost pressure

Assuming we are operating at sea level and the intercooler reduces boost from 14.7psi to 13.2psi, the expected power increase will be:

$$HPr = 0.23 + 1 - \left\{ \frac{14.7 + 14.7}{14.7 + 13.2} \right\}$$

$$= 1.23 - \left\{ \frac{29.4}{27.9} \right\}$$

$$= 0.18$$

Therefore, if without the benefit of intercooling our engine produced 306hp, in this example after adding charge air cooling we could expect a power rise of 18% or 55hp. However, before we get carried away with such a large power gain, remember these calculations were based on an intercooler with low internal flow losses, which was sufficiently large to draw off 80% of the heat rise in the compressed charge. Many intercoolers are far less efficient both in terms of flow loss and heat rejection capacity, but before we look at how we can make gains in both of these important areas we really have to check into the various methods of charge air cooling available.

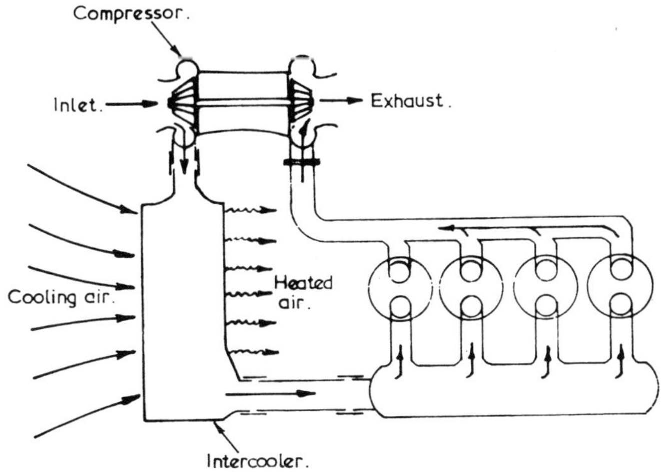

Fig. 10.1. An air-to-air intercooler directly utilises ambient air to cool heated charge air flowing from the compressor.

AIR-TO-AIR, OR WATER-TO-AIR

Figure 10.1 illustrates what is termed an air-to-air intercooler. It is similar to a cooling system radiator through which flows the inlet charge after it leaves the supercharger. It is called an air-to-air intercooler because the charge air flowing through the 'radiator' (more properly called heat exchanger) tubes gives up its heat to the outside air passing across all the fins laid between the tubes.

In a water-to-air intercooler the heat exchanger is encased within what looks like a metal box (*Fig. 10.2*). The heat exchanger looks just like a small, but quite thick, radiator. The usual practice is to circulate cooling water through the tubes with the

Fig. 10.2. A water-to-air intercooler uses water to cool hot air from the compressor.

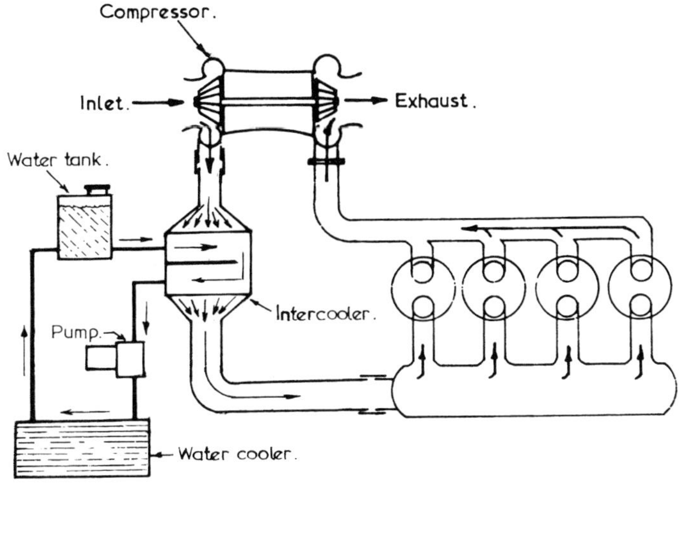

charge air being directed across the fins laid between the tubes. Sometimes this arrangement is reversed. In such cases charge air flows through the tubes and the metal case becomes a water jacket through which cool water is pumped.

In powerboat applications a continuous supply of cool water is available, and when water temperatures are lower than the ambient air temperature intercooler efficiency can exceed 100%. For drag racing and speed trials on salt flats, super-chilled salt water (brine) is pumped through the water jacket, again with efficiencies of well over 100% being possible. For regular automotive use, though, the cooling water will have to be directed to a separate front-mounted radiator to give up the heat it has just extracted from the charge air. With this more usual arrangement charge air cooling is no more efficient than that possible with air-to-air arrangements, but when space limitations or plumbing difficulties indicate that an air-to-air unit would be inefficient, then water-to-air coolers should not be overlooked.

TUBE AND FIN, OR BAR AND PLATE

When we look at the actual design of the heat exchanger there are three main areas of concern. First we have two types of heat exchanger construction. The tube and fin type is fitted to regular production vehicles. Externally it does not look dissimilar to an ordinary cooling system radiator. However, internally it is quite different, for as well as having zig zag finning between the tubes, the same type of fins are located within the tubes to collect charge air heat. The advantages of this type are lower cost from the ease of manufacture using mass production techniques, and that it is often only half the weight of the bar and plate type intercooler. In saloons (sedans) additional weight way out ahead of the front axle line is not to be ignored if roadholding is important. One disadvantage is increased flow resistance owing to the manner in which the tubes join with the end plates, but its main disadvantage is that fins in the tubes are widely spaced and/or smaller, which, while benefiting flow, reduces its heat rejection capabilities (*Fig. 10.3*).

Fig. 10.3. Tube and fin intercooler construction.

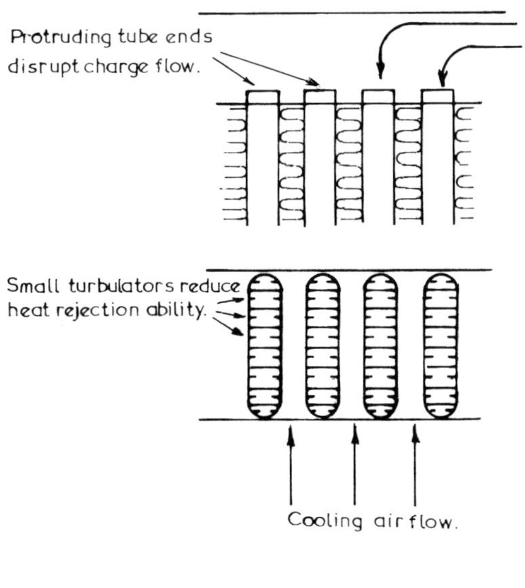

Also, because it is more a mass produced item, the exact size and inlet and outlet positions you want may not be available, which could again compromise charge flow.

The bar and plate design is usually a custom hand-fabricated item. As such it is several times more expensive than the tube and fin type. It can be quite easily distinguished by the square flat face of the tubes compared with the rounded ends, like radiator tubes, of the tube and fin cooler. The main advantages of the bar and plate style is that it can be made to suit your particular installation and it has much greater heat rejection capacity because finning in the tubes is considerably more dense. (Remember, though, in some air/water systems this isn't an advantage. Also, the flat square face of the tubes will restrict charge air flow, whereas the rounded ends of the tube and fin type aid charge flow across the fins.)

OTHER DESIGN CONSIDERATIONS

The second design consideration is that the heat transfer area must be sufficiently large to maximise heat extraction from the inlet charge. The third is that the air charge flow path must be as 'smooth' as possible to minimise flow loss, and it has to provide a spacious flow path of sufficiently large volume to slow charge air velocity right down. The reason being to provide the heat transfer region with the maximum time to draw off heat and transfer it to the cooling medium. In water-to-air charge coolers remember that these zones are frequently reversed, so in all discussions of both concepts you may have to transpose my comments regarding charge flow through the cooler's tubes to charge flow across the fins, depending on how you plan to configure your particular water-to-air system.

Many stock intercoolers are too small to provide adequate air flow and heat rejection capacity. Note the air temperature sensor in the top elbow, and the heavy duty hoses and hose clamps.

For many there isn't a lot of science involved in selecting an intercooler. The steps taken usually follow the sequence of measuring how much space is available, then selecting a new or used intercooler close to the required dimensions with fittings in the most suitable position to simplify duct work. Finally they might compare the power output of the car from which this cooler was sourced with what is expected from their engine.

While this method can work surprisingly well, there's a flaw in that you are trusting that the original manufacturer fitted a performance oriented intercooler to the donor vehicle. If the intercooler is from a true performance car, such as a Porsche 911 or a Nissan Skyline GT-R, then you will know the manufacturer hasn't skimped on cooling capacity or flow capability. However, many car makers are not like that, and because they are building a car down to a price it is accountants rather than the powertrain engineers who call the shots.

Later we will take a look at what can be done to minimise flow losses in all the pipework and fittings from the time the charge air leaves the supercharger until it reaches the inlet manifold. However, while it is correct that all these bits can restrict flow, the fact is the big losses occur in the intercooler itself. Therefore there are much larger gains to be made by directing our efforts primarily to that area.

FLOW vs. HEAT REJECTION

As mentioned earlier, intercooler tubes internally are not like car radiator tubes. Inside there is zig zag finning called by the fancy name 'turbulators'. The idea of this stuff is to agitate the charge air so that as much as possible all the charge air will come into contact with the turbulators or the tube walls and give up its heat. This heat is then conducted to fins between adjoining tubes. From there it is carried away by cooling air flow through the cooler's core.

Naturally all this agitation kills flow, but since zero agitation kills heat transfer we have to strike a happy medium. Every intercooler manufacturer has its own ideas on this. Some use tubes with tightly packed (dense) turbulators, while others prefer the middle ground and still others go for low density turbulators. Personally, I think turbulator density has to be based more on just how long each tube will be and also what

A balance must be struck between flow and heat rejection ability. This core has low flow losses but poor cooling capacity because there are no true turbulators in the tubes.

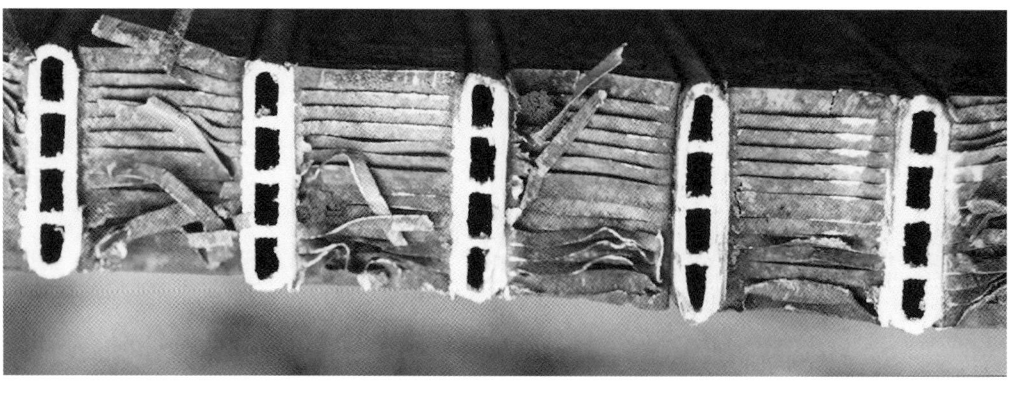

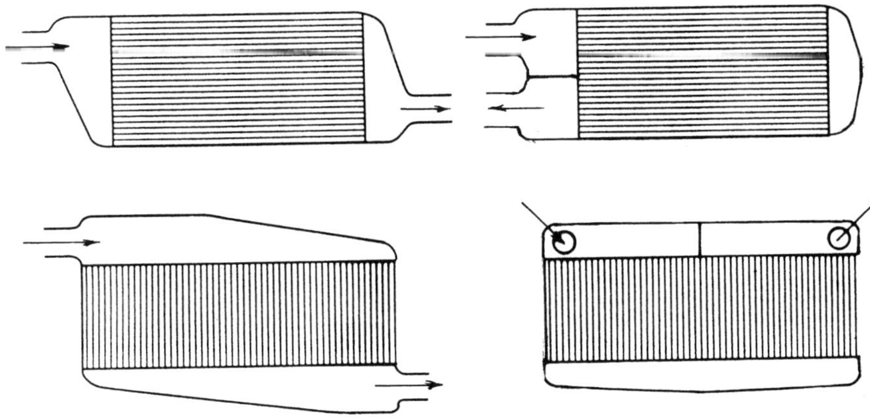

Fig. 10.4. Four air-to-air intercoolers of the same size have quite different flow potential.

the total tube count will be. To some extent these two factors will be determined first by the actual position of the intercooler in the car and then the location of the supercharger and inlet manifold, keeping in mind that we want all ducting to be as simple and free of acute bends as possible.

Taking a look at *Fig. 10.4* you will see what I am getting at. While all four coolers occupy the same space, the total number of tubes and their length varies considerably. Clearly, the two charge coolers at the top have longer tubes and fewer of them. This means flow resistance increases. Then, when a splitter is added to enable the inlet and outlet to be located at the same end, the total number of tubes is effectively halved and their length is doubled, further choking flow.

Therefore, if we have to go with a few long tubes we must recognise that the added tube length increases how much heat is pulled from the charge air, so we should choose low density turbulators to avoid excessive flow losses. On the other hand, if our cooler could be fashioned after the one in the bottom left-hand corner we would have to understand that many short tubes, while reducing flow losses, could be made even more efficient at taking heat out of the inlet charge by using more tightly spaced turbulators. Therefore we could achieve an even higher heat rejection rate without much loss of flow owing to the fact that we have many tubes of relatively short length.

Always we have to strike a balance. There is no point going for massive heat rejection and ruining flow in the process. Conversely, if you opt for flow you can finish up with minimal charge cooling if the space available permits only a small size cooler. When space is no problem, go for many tubes that are relatively short to minimise flow restriction and maximise charge cooling.

TWIN COOLER ARRANGEMENTS

When space is tight and compromises on either flow or cooling are not acceptable, then the problem must be approached from a different direction. Either we can use a pair of air/air coolers, or an air/air cooler and a water/air unit together. When planning for two air/air units, consider all possibilities. Frequently they are positioned in front of either

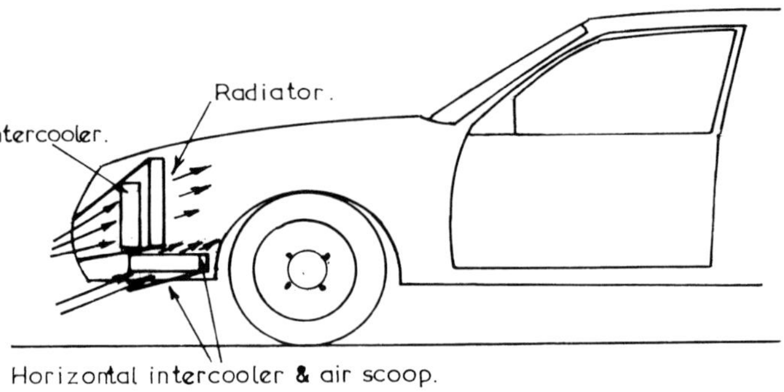

Fig. 10.5. Two intercoolers may be arranged in this manner.

the front or rear tyres, depending on engine location, but there may be other suitable locations. For example, a vertically mounted cooler at the front of the car can be combined with a cooler fixed horizontally under that cooler and water radiator (*Fig. 10.5*). For turbo cars with a small charger cooler mounted above the engine, an additional intercooler can be fitted in front of the radiator. Another possibility for cars with a lot of depth between the front of the car and the water radiator, but a pointy nose which limits intercooler size, is to use the staggered method of mounting (*Fig. 10.6*).

How the two coolers are best connected into the inlet system is debatable. If we are fitting units from production turbo cars, the air flow won't be brilliant, so it is usually best not to connect these in series with all the charge air flowing through one cooler and then the other. Rather, divide the charge to flow into both coolers and then reunite the flow well before the inlet manifold to allow thorough mixing. If one unit isn't cooling as well as the other and that charge air mainly flows into specific cylinders, then those cylinders will detonate.

However, if we have ample tube area it may be possible in a small number of installations to achieve improved charge cooling, without incurring corresponding flow losses, by directing the charge air first through one cooler and then the other. Particularly is this the case when combining air/air and air/water charge coolers.

Fig. 10.6. When intercoolers are staggered, ducting for cooling air flow must be carefully engineered.

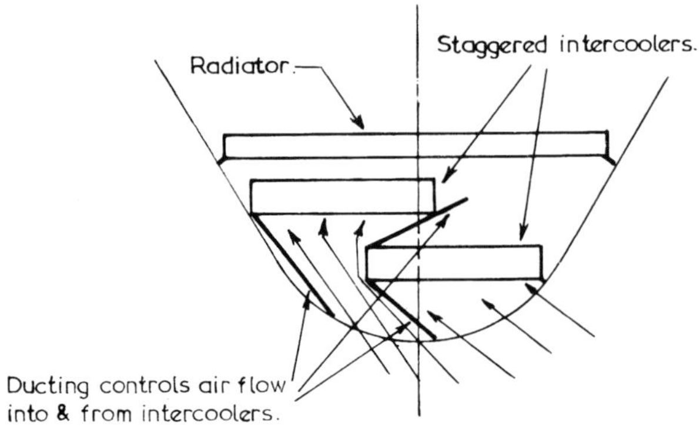

CORE THICKNESS CONSIDERATIONS

At this point some may be wondering why bother staggering two coolers when it would be much simpler to just place them together, one in front of the other? What we have to remember here is that outside air moving through the charge cooler picks up heat as it goes. Therefore, it gets hotter and less able to take up charge heat at the rear of the cooler than at the front. Also, the fins between the tubes, while contributing to heat removal, tend to stifle air movement through the core. In fact, we can almost block flow with a very thick core, and don't forget that often there is a water radiator and possibly a condenser for the air-conditioner in behind the charge cooler.

If you want to check this out sometime in the workshop, set a good fan up in front of a radiator and hold up a streamer behind the radiator, noting how it responds to air flow through the core. Now stack three radiators together and check what little air movement there is, as will be shown by the rather limp streamer. However, hold the streamer above, below or to either side of the radiators and you will see it fly. What has happened is the extra core depth has almost choked off flow, causing the air to stack up in front of the first radiator and forcing it to flow around the edges. We get on to efficient ducting later, but even with suitable ducting and very high air speeds approaching 200mph, air flow through the various coolers can almost stall.

In similar vein, when it comes to deciding on charge cooler core thickness we must be careful not to disregard the fact that cooling air takes up less heat and almost stops flowing with increases in core depth. Therefore, where possible, limit core thickness to 2½in in cars with a 3in or 4in radiator core, or in any road car with an air-conditioner condenser stuck behind the charge cooler. Only where the cooler stands alone should 3½in and thicker cores be used. Even then, remember that the back half of the core may only do one third the amount of charge cooling as the front half. However, core thickness is not an issue

Breakers' yards have many small inefficient intercoolers that nobody wants. However, 'marrying' two together, side by side, with suitably fabricated tanks top and bottom, will double flow and heat rejection efficiency.

with water/air units where the cooler is immersed in water within a water jacket, as water is almost twelve times more efficient than air at removing charge heat. When flow paths are reversed, with water flowing through the tubes and inlet charge flowing across the core, core depths exceeding 6in are best avoided to reduce flow restriction.

WORKING OUT TUBE AREA REQUIREMENTS

Once we have decided on a suitable core thickness we can then work out how much intercooler tube area is required to provide a suitable low resistance flow path for the inlet charge. Based on a tube length of 8–12in, we should aim for a flow area of 8sq in per 100hp in bar and plate coolers with high density turbulators. For tube and fin coolers with low density turbulators we can come down to about 5sq in per 100hp. However, remember that as tube lengths increase flow area must also increase.

Therefore, if we were planning a charge cooler for an air-conditioned road car making 400hp we would require 32sq in of cooling tube end area. Based on a 2½in core, the total width of the cooler at the air entry end, allowing for all of the blank spaces between cooling tubes, will be 28.2in (32 x 2.2 ÷ 2.5 = 28.16). If we had gone for a 3½in thick core, the width would be 20in (32 x 2.2 ÷ 3.5 = 20.1).

For those with access to a flow bench we can be a little more specific. At a depression of 3in water an intercooler should flow around 30cfm per 100hp in a competition vehicle. Down to around 23cfm in high boost environments, where extracting charge heat is more important than air flow, is all right. When we go much lower than that we have to begin questioning whether pushing the boost higher to compensate for flow restrictions is like a dog chasing its tail, and whether we would be better served from a bigger cooler and less boost?

IMPROVING FLOW AND EFFICIENCY

Of course, it is not only the size and number of tubes that determine an intercooler's flow capabilities and thermal efficiency. Other factors also play a very important role. For example, we can improve flow into the tubes by providing each tube with a radiused inlet. Some have experimented with elaborate bellmouths, but these don't give more than a 1% flow increase over something considerably simpler and cheaper. *Figure 10.7* illustrates how it is possible to achieve a 5–7% flow improvement by bonding a half round section between each tube. If you are concerned about weight, then a piece of round pipe can be cut in half and brazed in place rather than a solid strip of aluminium. Once welded in place, carefully blend the tubes into the radius of the half round fillers.

Fig. 10.7. Intercooler flow resistance is decreased when tube entries are radiused.

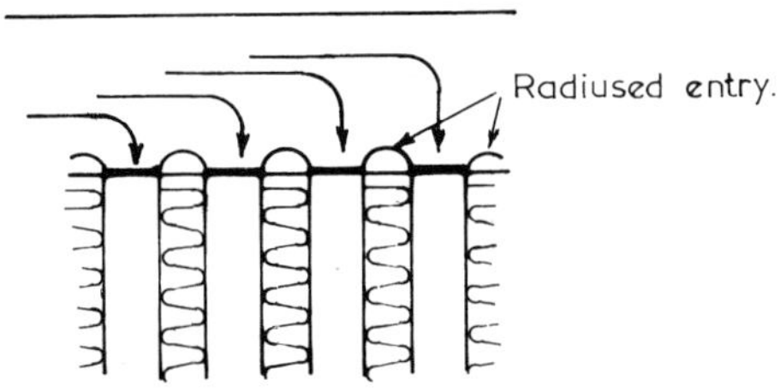

A Nissan Skyline GT-R charge cooler has nicely formed end tanks.

Perhaps the greatest potential for improvement rests in the design of the tanks on each end of the charge cooler. Here the wrong approach can muck up both air flow and cooling efficiency. Always we have to be thinking in terms of equalising as much as possible charge flow down each tube. Keep in mind that flow losses increase dramatically in those tubes flowing more air. Also, because the volume is higher and the flow faster these tubes will draw off less heat.

In some cooler designs the tendency is for those tubes directly in front of the inlet pipe to flow more air (*Fig. 10.8*). The way to make the dead tubes do more work, when the cooler is very tall (crossflow types) or very wide (downflow types), is to allow the in-tube to taper out into the tank and add a pair of deflectors to force flow to both ends of the tank.

Fig. 10.8. Deflectors balance charge flow between tubes, improving intercooler flow and heat rejection efficiency.

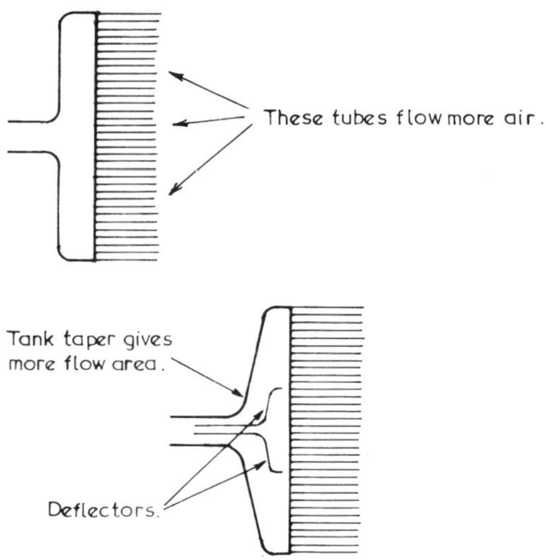

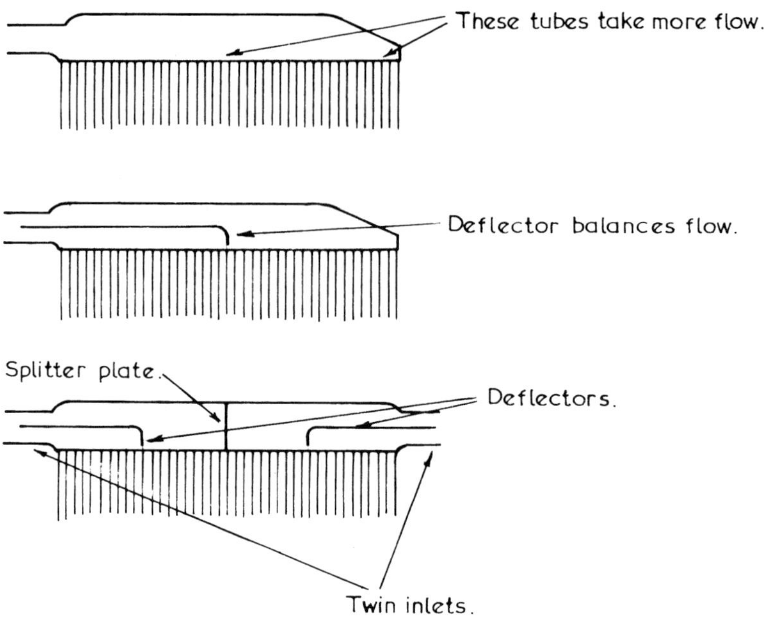

Fig. 10.9. Several methods are available to balance flow and help efficiency.

A common problem I regularly see in wide downflow coolers is that the in-tube enters the tank from the side with no provision to equalise flow between cooling tubes (*Fig. 10.9*). This deficiency can be corrected in a number of ways. If it was a twin-turbo installation on a V-type engine I would add another inlet in the opposite end, put a splitter plate in the middle of the tank to divide the cooler into two separate coolers with a common outlet, and then add a deflector in both inlets to divide the top tank into four sections.

In a single inlet arrangement the top tank can be sectioned into halves or thirds depending on cooler width. Up to about 18in width I would go with a single deflector, but over 20in it is preferable to install two deflectors.

Another thing we must give attention to is the angle at which the in-pipe or the out-pipe enters the end tanks. Never should either pipe join at an acute angle. This really kills flow and creates a whole raft of flow distribution problems (*Fig. 10.10*).

Fig. 10.10. Acute bends should be avoided.

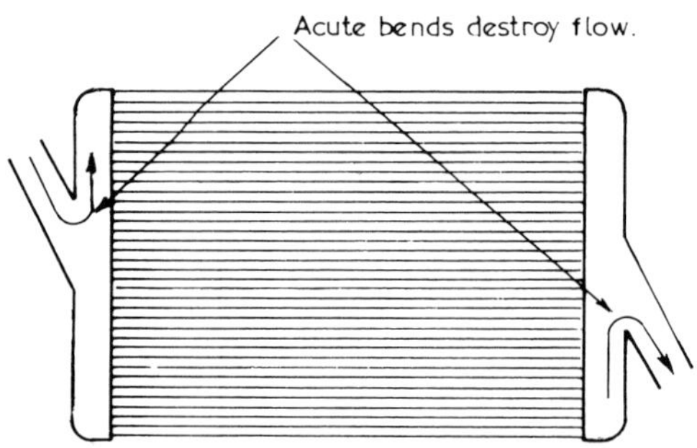

PIPEWORK AND JOINTS

When it comes to ducting to connect the charge cooler into the inlet system it is easy to go overboard with massive tube diameters. You really don't need pipework larger than that indicated in Table 10.1. Some fret over a big intercooler causing turbo cars more lag, without giving any thought to the fact that duct volume also contributes. What we really have to avoid is sharp or deformed bends, and steps when duct sections are joined with flexible hose. Ideally, we want smooth large radius mandrel bends and, where ductwork joins, all sections should butt together without any change in diameter. Where it is necessary to step up or down in diameters of 1/4in or less, get an exhaust shop to stretch the tube using their hydraulic expander. Larger steps are best formed by conical sections with a convergent or divergent angle not exceeding 7°. (If you don't know how to fabricate a cone section, see another of my books *Two Stroke Performance Tuning*, also published by Haynes.)

Table 10.1. Inlet duct diameter requirements

Pipe dia (od)	Horsepower
2in	255
2 1/4in	345
2 1/2in	450
2 3/4in	565
3in	690
3 1/2in	970

Note: To compensate for flow loss caused by bends, decrease horsepower by 5% for each 90° bend. Keep in mind that three 30° bends cause the same flow loss as a 90° bend.

Anytime we join duct sections using flex hose we introduce another possible source for trouble into the system. In low pressure situations no more science is involved than when we connect the cooling system radiator to the engine. All we have to concern ourselves with here is that the joint doesn't leak and allow boost to escape. Some people bring themselves a lot of grief by using a silicone sealant when they should be using a non-slippery non-hardening jointing compound like Permatex No. 3.

In high temperature (over 120°C/248°F) and high boost (over 18psi) environments we have to be a good deal more exacting. Unless replaced frequently, rubber hoses are marginal, and even then they would only be suitable to join two sections of metal duct. A length of moulded rubber hose always spells trouble. Really, this is the place for silicone hoses. Some people rely on the strength of the hose plus double jubilee clips to stop joints from blowing apart, but a tie bar is a much better solution (*Fig. 10.11*).

Fig. 10.11. In high pressure installations a simple tie bar can prevent joints blowing apart.

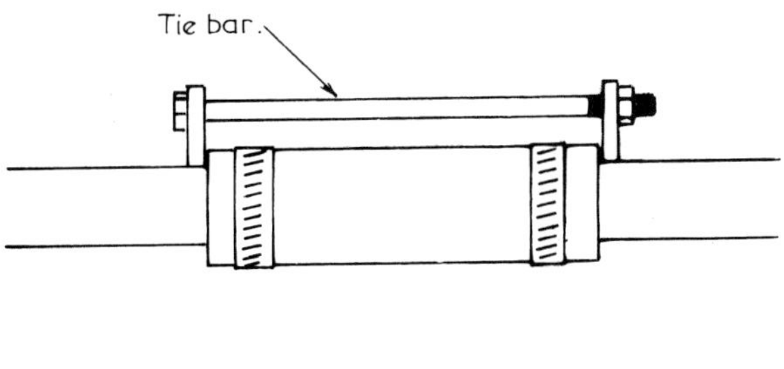

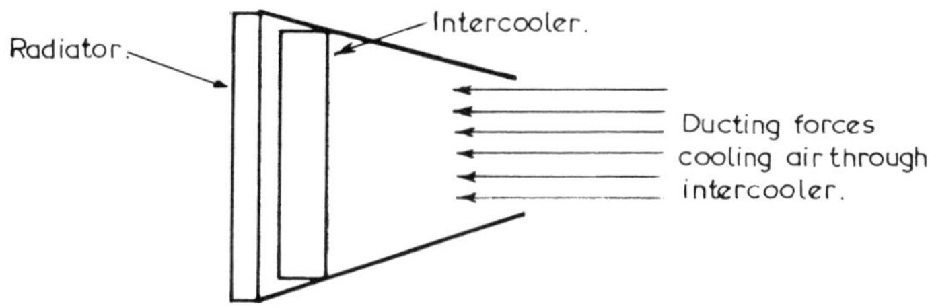

Fig. 10.12. Proper ducting does not allow cooling air to escape around the intercooler.

ENSURE GOOD COOLING AIR FLOW

As mentioned earlier, the intercooler should, preferably, be placed at the front of the car ahead of all other heat exchangers to ensure maximum charge cooling efficiency. Anything we can do to duct a good supply of cooling air to the charge cooler should be done. Never should the cooler stand alone in the air stream, as rather than taking the hard path through the core the air will simply duck around the edges. To avoid this the duct must be formed into the correct shape and sealed where it meets the intercooler (*Fig. 10.12*). In competition cars you need to be a little careful here because should you take a bang in the nose the intercooler and radiator could be wiped out if a very solid rigid duct is used. It is best to use a light duct which will easily collapse, and finish it a couple of inches short of the cooler. Use a long strip of flat rubber glued or tied to the duct and cooler to complete the air path.

Some make a very basic mistake thinking the duct opening must be the same area or larger than the frontal area of the cooler or water radiator. Actually the reverse is true; a duct opening considerably smaller works better as it prevents air, flowing in towards the cooler, simply turning around and spilling out around the edges at the duct opening. To prevent this, reduce the duct opening to 30–60% of the intercooler area. Note, when the intercooler is smaller than the radiator don't just leave the cooler standing free in the duct as air will still want to detour around it. Rather, divide the duct to split a portion of air flow into the cooler, then seal this divider to the charge cooler, thereby forcing all air in that corridor through the charge cooler (*Fig. 10.13*).

Establishing a good flow of air into the charge cooler is only one part of the challenge; you also have to ensure that air can flow through the cooler by providing an environment where the pressure behind the cooler and radiator is much lower than the

Fig. 10.13. With small intercoolers use a divider to ensure cooling air actually flows through the core.

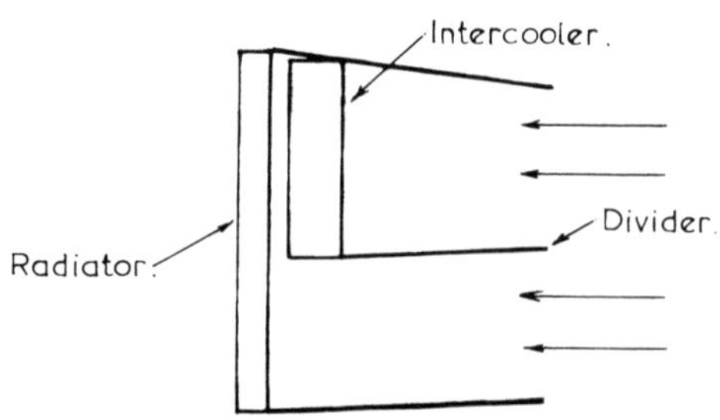

pressure at the front. This means we have to be alert to allow plenty of air exhaust area from the engine bay, plus we have to limit air flow into the engine bay to what is absolutely essential to adequately cool parts which would otherwise cook. Such things as the alternator, battery, fuel lines and injectors, ignition parts, wiring harness, turbo and exhaust manifolds require some flow of cooling air. However, a big bonnet scoop could flood the engine bay with air and if there is a large flow of air under the front of the car also creating pressure, flow through the cooler and radiator could be quite poor.

Often one or two strategically placed small bonnet vents will provide the necessary exhaust path, and if you don't like the idea of cutting into the bonnet you can try raising the rear of the bonnet 4–6mm. Also, don't overlook the use of electric fans to draw air through the charge cooler and radiator.

A simple way to check the pressure differential between the front and rear of the intercooler is by means of a U tube manometer. It sounds fancy, but all that is required is a 30in length of light timber marked off in inches. To that tie a long length of 4mm clear plastic tube formed into the shape of a U. Fill the tube to the 10in mark with water to which food colouring has been added. With the U tube beside the driver, tie one open end in front of the intercooler and the other open end behind the radiator. Now take the car for a high speed run. The fluid level should drop in the branch of the U connected in front of the intercooler, indicating pressure is lower behind the radiator, so cooling air flow is all right.

Air flow into the front of the car can also be checked using a big, powerful dyno blower fan. Set the fan in front of the car with the bonnet closed. With the fan running, spray water or smoke into the air stream and note its direction of flow. Using water, it may not be so easy to identify the exact flow path unless you crouch off to one side of the car while an assistant holds a black panel at the other side. It should be obvious that there is flow into the front of the car, and that it's not turning around and spilling out

Mounting the intercooler directly onto the engine without a large cooling air escape route underneath effectively makes it an 'interheater'. The lack of a heat shield between the head and intercooler, the absence of insulators to prevent heat transfer to the intercooler via the mounting brackets and bolts, plus bent fins restricting air flow all add to the problem.

again. Also, flow shouldn't divide and only go over and under the car. If it does either of the latter then you are bound to have charge cooling and engine cooling problems.

In some production type cars we can reach a point where we have to restrict the size of the charge cooler, either because there isn't sufficient space or because a full width cooler is killing air flow through the radiator, causing engine cooling problems. In the latter situation we should first ensure that we are using a compressor of the correct size, operating at a reasonable boost level, in a high efficiency area of the compressor map, and that the engine cooling system is of sufficient capacity and in good order. Then we have to decide on a solution, either a water spray onto an air-to-air cooler, a combined air-to-air and water-to-air system, or just a water-to-air cooler.

INTERCOOLER WATER SPRAY

A good water spray onto the core of an air/air unit will increase its heat rejection efficiency up to about 8%. That may not sound very much, but as we increase intercooler size it becomes more and more difficult to get big efficiency increases. For example, a cooler that is 70% efficient will require about 25% more core area to achieve the 8% improvement possible with a properly engineered water spray system.

However, to be really effective a water spray arrangement has to dump a lot of water onto the entire area of the charge cooler core. This means a large water tank and correct positioning of spray nozzles. Competition cars regularly have a 30-litre water tank which is sufficient for about 30 miles on a race circuit or rally stage. This raises the question of weight – will the handicap of an additional 30kg of weight, assuming a 30-litre water load, be offset by the performance gain arising from increased charge density and attendant detonation protection?

The Ford Focus WRC rally car uses water spray to increase intercooler efficiency. Note: only the top row of spray nozzles are visible.

A Subaru water-to-air intercooler. Hot air from the compressor enters the lower duct, giving up much of its heat to the fins as it passes across the core to exit through the large duct. Cooling water, circulated by a 2-speed electric pump, flows through the tubes and takes up heat from the fins, entering and discharging via the two small pipes at lower right.

Obviously water loads should be kept to a minimum. This means the spray nozzles and pump must be correctly sized to provide only the volume of water necessary for effective charge cooling. Also, the system must be activated only when the intercooler core temperature or the intake temperature, measured at the inlet manifold, reaches a particular level – perhaps 10–15°C (50–60°F) above our target inlet temperature. Triggering water spray with reference to boost pressure is not very effective as it does not take into account ambient temperature and the additional cooling provided by increases in car speed.

Race teams adjust water flow rates, nozzle position and spray pattern based on readings from perhaps half a dozen temperature probes strategically placed around the cooler core. An alternative is to use a big dyno blower fan and very carefully view the spray pattern. You have to keep out of the air path so you will need very good lighting and perhaps a pair of binoculars to clearly see what is going on. Water flow rates can be adjusted with the car running under load on a wheel dyno or on the track. With a temperature probe into the inlet manifold or cooler outlet, attempt to achieve low charge temperatures with minimal water flow. It makes little sense to double water flow for a temperature drop of a few degrees. It can take a lot of work and expense to take 10kg out of a race car, so always balance out water flow against the added weight you will have to carry. For a race car a pre-race water level check should be a regular item on your check list. However, for a road car it is easy to forget, so you will need a low water level buzzer to remind you to take it easy until you refill with water. Don't rely on a warning light, you may melt a set of pistons before you notice that it is on.

WATER-TO-AIR INTERCOOLER

In comparison with an air/air cooler a water/air arrangement is considerably more complex. However, if space for an air/air cooler isn't available you may have no alternative. Also, if you are able to turn your hands to most tasks and have the ability to seek out cheap parts, a water/air system can be assembled for a fraction of the cost of an equivalent air/air system if you disregard the time component.

As explained earlier, the heat exchanger can be connected as illustrated in *Fig. 10.2* to favour air flow, with perhaps a slight decrease in thermal efficiency. Alternatively, the charge air can be directed through the cooling tubes, as in an ordinary air/air cooler, but instead of air passing over the fins cooling water is used. This reversed system works well with regard to actual charge cooling efficiency. However, flow losses can be quite high unless there is adequate tube area to meet the engine's air flow requirements.

Which system will get the nod depends on the size of the heat exchanger. If it has plenty of tube area (6sq in/100hp with high density turbulators, 3.5sq in/100hp with low density turbulators) and tubes less than 10in long, then the latter system will give much greater cooling efficiency. However, if the tube area is considerably less than the figures stated it would be better to use the tubes for water flow, and this is how most small air/air coolers sourced from breakers' yards will be connected.

At the front of the car you need another radiator, separate from the engine cooling system, to remove heat from the charge cooling water. The size of this heat exchanger will be dictated more by factors other than the temperature of the water exiting the charge cooler. The total volume of water in the system, the pump flow rate, and on-boost time compared to the off-boost time all have a large influence. For example, a road car may be on-boost for say 15 seconds in a long overtake situation but then it may be off-boost or at very low boost for several minutes. As it could take up

A combined inlet manifold and intercooler makes for a very compact package. However, if you decide to undertake something like this yourself be aware of the amount of work that can be involved in equalising air flow into each cylinder.

to 10 seconds (depending on water volumes and the exact flow path) before the front cooler has any heat to dissipate, it really only has to be large enough to handle the heat load during the final five seconds of the on-boost cycle. It can then bring the water back down to ambient during the following several minutes of off-boost operation. As such, the front cooler can be virtually any type of heat exchanger. You will find engine and transmission oil coolers to be favoured, but car heater cores can be used as well as air-conditioner condensers from either car or domestic AC units.

However, in a competition setting the boost cycles could be quite different. Here we are more likely to see a 5–8 second on-boost period followed by only 2–3 second off-boost. Consequently, in all but sprint events like drag racing and hill climbs we would have to run a much larger heat exchanger, with large car AC condensers preferred by those on a tight budget.

In an effort to skimp on the size of the front cooler some put a big reservoir of 5–10 litres in the system. I don't go along with such an arrangement unless there isn't any space for a larger cooler at the front. Really, we want to keep weight out of the car, so if possible use a simple plastic header bottle and cap from a production car and tee it into the system to take care of coolant expansion/contraction. Fix it at a high point in the system to help you bleed all air out. This is most important as any air trapped in any part of the system will greatly degrade cooling performance. Particular attention should be paid to fitting suitable air bleeders at a high point in, or close to, the heat exchanger itself. Even simpler is to connect a vent hose (about 4mm) between the top of the charge cooler and the top of the header tank.

The coolant is moved through the system by an electric pump, and preferably ¾in hose to reduce the load on the pump. All manner of electric pumps have been tried but many are less than successful because of being designed for only intermittent operation. Really, we have to be looking for a pump that will flow 15–20 litres per minute continuously without overheating. The Subaru Liberty RS turbo has an impeller type pump rated at 15 litres per minute. Unfortunately they are not easy to find in breakers' yards. They are at the lower limits of flow capacity and new ones cost an arm and a leg. The best alternative, although they are very noisy for road use unless mounted on a rubber engine or gearbox mount, are pumps made for agricultural crop spraying by the American Flojet company. The 4100-143 model flows just over 20 litres per minute at 13 volts. Larger flow capacity pumps are also available.

In a competition car the pump can be operated at full flow continuously. However, for road use pump life can be extended considerably if operated as a two-speed unit. Some prefer to have the pump switched on only while the engine is operating under boost, but as you have seen already this is a waste of time as most cooling of the water actually takes place after the car returns to a zero boost situation.

Therefore, we want the pump to keep circulating coolant off boost. By placing a 3 ohm resistor (use a high current 5 amp ceramic type or preferably an injector dropping-resistor pack) in the circuit, pump flow drops to around 7 litres per minute and current draw halves to about 3 amps. To make the pump kick in to the full flow mode we have a full voltage circuit operated by a cam on the throttle spindle. When the throttle is depressed, say past 25% open, the cam closes a microswitch which triggers a relay to send fully battery voltage to the pump. If the car is also used for weekend track work, an

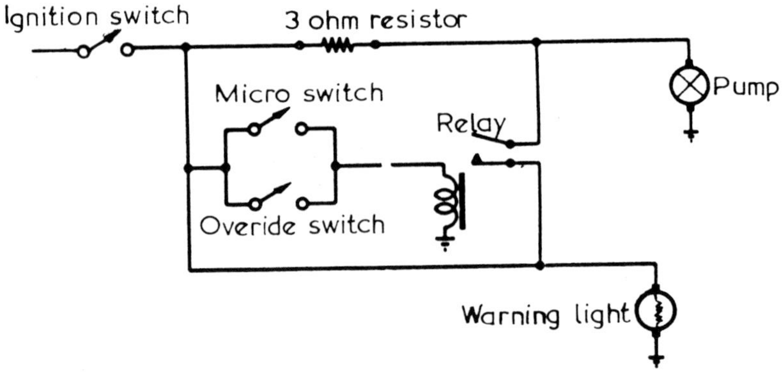

Fig. 10.14. Wiring diagram for two speed pump operation.

override switch and idiot light should also be fitted to give continuous full flow operation (*Fig. 10.14*).

INTERCOOLER MAINTENANCE

Intercoolers, like engine cooling system radiators, are most definitely not a fit and forget proposition. Water/air units demand a lot less maintenance but air/air types require regular attention. If you are fitting a second-hand cooler, thoroughly clean it before use. For starters you don't want rubbish from a previously damaged turbo compressor wheel finding its way into your engine. Less serious, the cooler may have oil sludge in the tubes and turbulators as a result of engine blow-by or leaking supercharger oil seals. This sludge will reduce both air flow and heat transfer.

Externally, the cooler should be cleaned to get rid of mud and small gravel in the fins, and fins should be straightened. All these operations improve charge cooling efficiency, and also engine cooling in installations where the radiator is located behind the intercooler.

Intercoolers require regular maintenance. Corrosion of the fins and rubbish in them restrict cooling air flow and heat rejection. Internally, an oil film also cuts efficiency.

Formula 1 Renault used combined air-to-air and water-to-air-intercooling. Compressed air enters the water-to-air unit at the bottom then passes through the air-to-air unit above. Note how short the tubes are in the air-to-air unit, and the end tank design to ensure minimum flow loss. The DPV variable nozzle inlet developed by Jean Pierre Boudy during his time with Renault is visible at the turbo inlet (see Chapter 3).

This cutaway drawing shows ducting used to force cooling air through intercoolers and radiators. (The radiators are mounted above the ducts and exhaust hot air through louvres in the top of the side pods.)

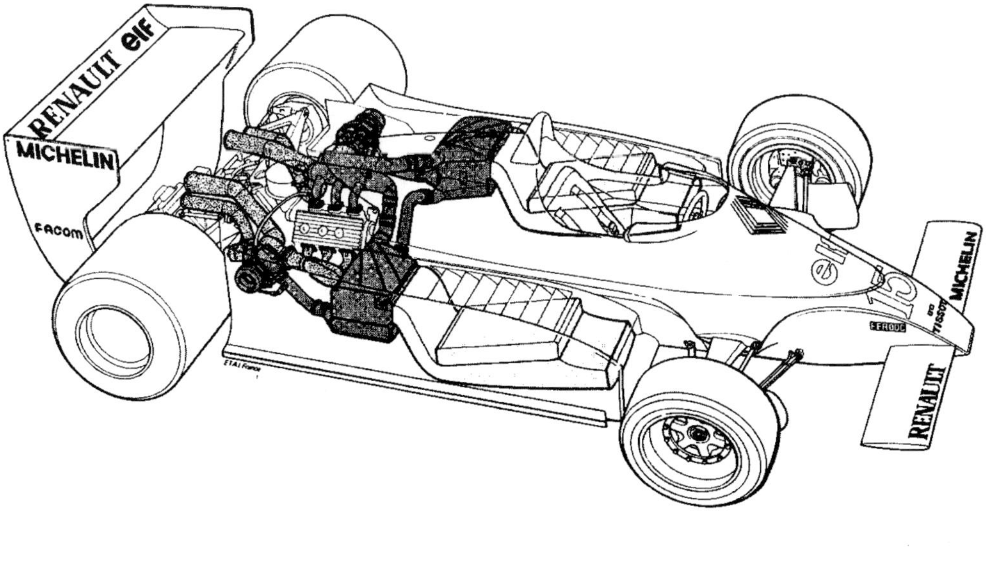

Perhaps the most difficult cleaning exercise involves getting the tyre rubber thrown up off the track out of intercoolers on circuit cars. Really, the only way to go about this is to soak the cooler in a large pan of solvent for several hours. Be sure to cap the inlet and outlet to keep solvent out of the interior. After the soaking, direct compressed air at the back of the cooler to blow most remaining rubber balls forward from between the fins, and finally pick out the few that don't budge. This type of cleaning should be done after each race meeting.

While car intercoolers don't crack as often as those fitted to trucks it is a good idea to subject them to a pressure test reasonably frequently. Radiator repair shops can do this for you, but if you have a competition car you may want to obtain the few pieces of simple equipment required so you can do the test every few races yourself.

OTHER BENEFITS FROM CHARGE COOLING

Apart from the very obvious benefit of increasing charge density, and in turn engine power output, intercooling accomplishes much more. Lower charge temperature greatly reduces the risk of detonation and pre-ignition. Because of this I have seen stock naturally aspirated engines operate at 8psi boost without requiring any change to the ignition map while running on ordinary unleaded fuel. I have also seen factory turbo cars run at double the factory boost (20–24psi) with bigger turbos and intercoolers on 100/130 Avgas on the stock ignition map. In all instances the stock cam was retained, but even so, when you remap the ignition and finish up only ±1° from stock without any real power gain, it highlights just how beneficial charge cooling is in helping the combustion process.

However, it doesn't end there. There is also considerably less heat during the exhaust cycle (a 50°C/122°F drop in charge temperature will produce a 50°C/122°F drop in exhaust temperature). Therefore, the exhaust valve and piston crown in particular run much cooler. This then leads to less heat being added to the fresh inlet charge flooding in on the next inlet cycle. On top of this, any residual exhaust gas remaining in the cylinder is cooler, which also contributes to reduced heating of the fresh charge.

Obviously in the case of turbo engines there is less heat stress on the exhaust manifold, turbine wheel and turbine housing. This may increase their service life, but there is a possible downside. Less heat may reduce turbine speed in turbos with large A/R housings, resulting in slightly longer spool-up times, or lower maximum boost levels because of an inability to drive the compressor at sufficient rpm. This latter situation would only arise if prior to intercooling the turbine housing was so large that there was barely any bypass flow through the wastegate. Obviously it could be easily corrected with a step down in turbine housing A/R.

Chapter 11

Water injection and other alternatives

The first recorded use of water injection was around 100 years ago by an engineer from Hungary named Benki. A decade later, in England, Professor Hopkinson carried out some tests on large industrial engines, and in the 1920s farm tractors were operating with his system. However, it is Sir Harry Ricardo who really studied the effects and benefits of water injection on combustion, and documented his findings for our enlightenment in his book The *High-Speed Internal Combustion Engine*. This was in the early 1930s, and by 1936 Ricardo was granted a number of patents for his 'anti-detonant injections'. Aircraft engineers, eager to establish altitude and speed records, quickly adopted his ideas. Then, as the world went to war for the second time, sophisticated water injection systems were employed on numerous military aircraft to allow leaner fuel mixtures on boost while providing increased power, free of detonation.

We discuss combustion in detail in Chapter 17, but at this point we need to understand that when either the speed at which the fuel burns, or the speed at which cylinder pressures rise, gets out of control the fuel mixture will explode. This is detonation. Mild detonation will destroy head gaskets. Severe detonation can break pistons, piston rings and cylinder walls, and it can damage bearings, rods and cranks. Consequently, it is to be avoided, but the problem is the very steps we take to extract more power or economy out of an engine invariably takes us closer and closer to detonation. In fact, it is on that very narrow divide between rapid controlled combustion and detonation where we find maximum power and best economy (depending on whether the engine has been tuned for power or economy).

Hence, to avoid detonation we have to put the brakes on combustion speed and pressure rise, but it has to be just a gentle tap on the brakes otherwise we will hurt engine power output and fuel economy. One way we can do this is by increasing the knock resistance, the octane rating, of the fuel. However, this course is not all smooth sailing. For example, higher octane fuel for a road car is more expensive, and considering that your engine may only require the higher octane for less than 5% (more usually 2–3%) of the

This early Edelbrock controller permitted some control over water jet flow rates based on engine load and rpm.

time it doesn't make good economic sense. And, there is the problem of finding the high octane stuff when you drive in remote areas. In competition cars, cost may not be a factor if you are permitted to run heavily leaded race fuel or Avgas, but more frequently unleaded race fuel is now being specified. High octane unleaded brews can be ten times the price of Avgas, so to keep costs down and appease the environmental movement sanctioning bodies often specify 98RON unleaded, which is hardly a solution for detonation control.

WATER INJECTION RATES

That leaves us with water injection, plus a couple of alternatives that I will get on to later. First we have to understand that water doesn't burn. In tech speak it is a fire retardant; we have been using it for centuries to extinguish fires. This means that it can put out the fire in your engine too! However, long before the engine refuses to run it will be way down on power. Hence, water injection isn't about simply dumping water into the engine. If power is important to us, water flow and atomisation has to be precisely controlled to provide only the quantity required to stop detonation.

Fortunately for those who manufacture crude water injection systems the window in which the engine will continue to run satisfactorily is quite wide. Accordingly, just as an engine will tolerate air/fuel mixtures as rich as 11:1 down to as lean as 17:1, so likewise with water injection. However, you also know how precise fuel injection transforms an engine previously operating on, by comparison, crude carburettors. There are significant improvements in power particularly in the mid-range, and throttle response is transformed. All this occurs because the engine is being fed the precise quantity of fuel, correctly atomised, that it needs.

In a naturally aspirated engine this varies depending on engine load and rpm, with most engines on full throttle producing best power in the upper rev range at an air/fuel ratio of around 13:1 to 13.2:1. Supercharged engines can be coaxed to make peak power operating a little richer than this, depending on the ambient temperature and intake charge temperature. With low ambients and efficient intercooling it is still usually fairly close to 13:1 but, as air temperatures and intake temperatures rise, the internal cooling and combustion speed suppression produced by richer mixtures, down to about 12.3:1 can deliver more power, at the expense of higher fuel consumption and more pollution.

Relating this to water injection you can see why a system based on a windscreen washer pump and nozzles, even if it is triggered to operate by some sort of pressure switch (oil pressure switches being favoured in some quarters) can at best only prevent detonation. It can do nothing to actually help power output or throttle response. In fact, it most probably is hurting power and throttle response, and possibly also contributing to increased cylinder and ring wear by contaminating the oil lubricating the cylinder walls.

The most basic system that should be considered would at least have a variable speed pump controlled from an electronic module with inputs for engine rpm and boost pressure. Such a kit is about as advanced as a carburettor, in that you will only achieve relatively imprecise control of water flow rates. However, if time is taken to tweak adjustments on the electronic module while you observe power output on the dyno, reasonable results can be obtained.

Always the aim should be to limit water flow as much as possible while keeping the engine out of detonation. To assist you in making such an assessment it will help if you disable the knock sensor so that it is not retarding the spark, cutting the boost or increasing the fuelling. Rather, connect a big warning light to let you know if the engine is detonating as you cut back water flow. A flow meter for both the fuel and water system is helpful here because when fully loaded from peak torque rpm to peak hp rpm, in a worst case scenario best power will be realised when water flow by weight is about 35% of fuel flow. In volume terms this is around 28% depending on the SG of the gasoline. The basic rule is that engines with higher intake temperatures and/or less fuel octane require more water while those with lower intake temps and/or higher fuel octane require less water, with the majority most likely requiring around 10–15% of fuel flow in volume terms.

On your average engine that is just a little low on fuel octane and has intake temps around 70°C (158°F) you should see about a 5% power increase and a 15–20°C (60–70°F) drop in intake temperature when you get water injection right. Naturally, engines way too low on fuel octane and with intake temperatures approaching 100°C (212°F) will record much better results.

SOPHISTICATED WATER INJECTION

That doesn't sound like a big power boost even when the lessened risk of engine damage from detonation is factored in, which is why I find it hard to justify using sophisticated water injection systems costing the same sort of money as a good fuel injection kit. Such

Delco 808 ECU can be reprogrammed to provide precise injection flow rates for minimal cash outlay.

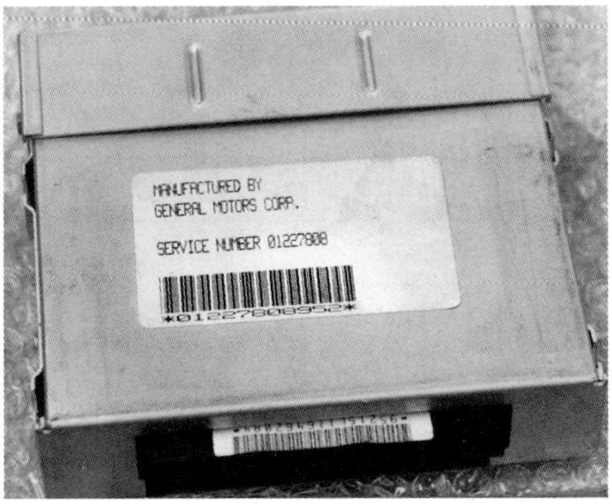

kits have a special water nozzle in each inlet port and a high pressure pump delivering water at anything up to 90psi to maximise water atomisation. To ensure precise water delivery, a programmable computer is included which enables the system to be mapped according to factors such as engine load and rpm, compressor discharge temperature, engine temperature, turbo temperature and boost pressure.

For a competition car restricted to relatively low octane unleaded, or for endurance events where fuel consumption is important and where water isn't counted as part of the fuel load, obviously the choice is to go with the best system regardless of cost. However, for a road car such expense cannot be justified when, as you will see later, you can inject something much more exciting than water and obtain a superior outcome.

Before we get to that, though, there are a few more things that should be mentioned about water injection. When we inject water it can provide a number of benefits, depending on where exactly our supercharged engine has a deficiency. So far we have only mentioned that water helps when we are low on fuel octane. The manner in which it helps here is that during combustion the water is converted to steam. That conversion pulls heat out of the combustion process, slowing the rate of pressure rise in the combustion chamber. Additionally, these steam particles get in between fuel and oxygen molecules which again slows combustion and the pressure rise rate. While this is going on the piston continues moving to TDC squeezing the gases tighter and tighter, but with less heat in the chamber and a slower pressure rise things don't get out of hand to induce detonation. Then as the piston slowly begins to descend a controlled burn continues, but the combustion flame now has to duck around steam molecules plus an increasing number of exhaust gas molecules to get to fuel and oxygen molecules. As such, peak cylinder pressure will be reached at about 15° after TDC. Any later than 18–20° we lose power; earlier than 12–14° we could precipitate detonation.

NOZZLE LOCATION

Clearly, to avoid either of the latter we have to ensure the engine receives not only the correct quantity of atomised water, but also that water must be equally distributed among all cylinders. This is why the expensive systems always feature a separate nozzle for

Fig. 11.1. An injector nozzle located at the intercooler outlet.

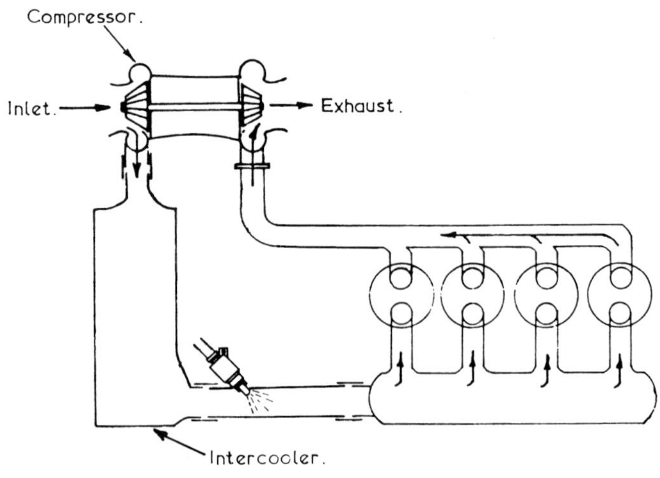

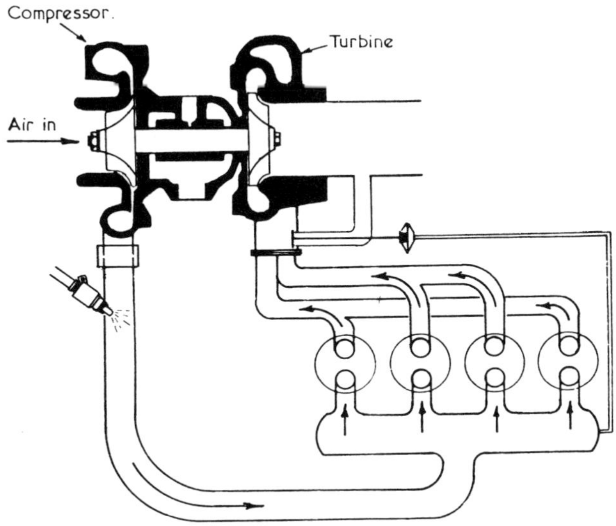

Fig. 11.2. Turbos and centrifugal blowers without intercooling should have the injector nozzle positioned near the compressor outlet.

each cylinder. However, we can get close with a single or double nozzle if we take care where we position it. If we are using an intercooler, the nozzle(s) is(are) best located in the cooler outlet pipe (*Fig. 11.1*). The turbulence here will help to disperse the water fairly uniformly in the charge air. And because there is a fair distance between the cooler and the inlet manifold there is more time for mixing. Additionally, if the charge air is still quite hot when it leaves the cooler this heat will further assist in vaporising the water. In turn, such conversion will draw heat out of the charge air, lowering its temperature by perhaps 20°C (70°F) if the outlet temperature was up around 70°C (160°F).

In non-intercooled systems, nozzle positioning varies with the type of supercharger in use. With turbos and centrifugal blowers, stick the nozzle in the compressor outlet as centrifugal force may sling the water back into liquid, and even though compressor heat will revaporise it, superior atomisation occurs with the nozzle located at the compressor outlet (*Fig. 11.2*).

With most other types of blower, and especially Roots types, water injection into the blower intake is usually superior (*Fig. 11.3*). However, there can be water distribution

Fig. 11.3. Roots and screw blowers usually benefit from locating the injector nozzle so that it sprays directly into the blower inlet.

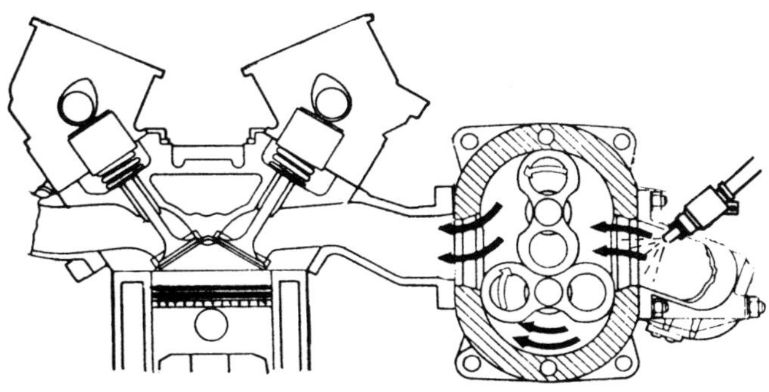

problems when the blower is mounted very close to the inlet ports, and if the amount of water injected is excessive the blower can be damaged.

Earlier, when discussing the Roots blower, it was noted that it isn't terrifically efficient owing to leakage between the rotors and the blower case. However, when we inject water (or fuel) right into the blower this closes the gaps and reduces the leakage. Consequently, boost pressure increases and charge temperatures decrease. A pressure rise of around 20% (e.g. 8psi without water injection to 9.5psi), is typical for a modern, tight clearance Roots blower. In fact, a pressure increase greater than this could indicate we are putting too much water into the blower, risking blower damage. An overload of water closes the gaps between rotors and the case. This, in turn, puts excessive pressure on the rotors and blower case, as well as gears and bearings. A minor overload causes dramatically shortened blower life and if the rotors 'hydraulic' and lock, blower damage could be terminal.

WATER/ALCOHOL MIXTURE

Many advocates of water injection recommend that instead of straight water a 50/50 blend of water and alcohol should be used. Their reasoning is that the alcohol adds some fuel value, and being more volatile it assists atomisation and better dispersal throughout the intake charge.

Alcohol, regardless of what name it is called, should only be added to water injection to prevent freezing.

Always use distilled water for injection, to avoid corrosion, blockages and mineral build-up.

However, I strongly disagree with the addition of alcohol beyond what is actually required to prevent the water freezing, and even then it should only be added when climatic conditions indicate the mercury will fall below 0°C (32°F). As we discuss alcohol fuels in the next chapter, at this time I will make just two brief comments on the matter. First, in many competitions alcohol is illegal and you could be disqualified if it is detected in either your fuel tank or your water tank. Second, alcohol can leave behind white powdery deposits on anything aluminium, but worse it precipitates corrosion of the piston ring lands, the piston rings and the bore wall by washing away lubricant.

USE A FILTER OR DISTILLED WATER

Because of the very fine jets used, often 0.015in or smaller, steps must be taken to prevent blockages and associated engine damage (most likely destruction). Some use ordinary mains water and insert an inline filter in the water injection system. The filtering medium has to be resistant to water damage, so any oil or fuel filter guaranteed as such is suitable. The alternative which I recommend is to keep the entire system clean and use only distilled water. If you keep the distilled water capped and out of sunlight to prevent algal growth, then carefully clean around the water reservoir each time before removing the cap to refill the reservoir, you should not experience any problems with blocked water jets.

ALTERNATIVES TO WATER

Personally, I am not a big fan of water injection. As I have already indicated, many systems are simply too crude; injecting too much water, not properly vaporised, and at the wrong level of engine load. At the other end of the scale, sophisticated arrangements with fuel injection precision are too expensive for what they offer. For considerably less money, and superior results, alternatives to water could be injected. Race cars may have no alternative but to use water injection as competition regulations often outlaw

Fuel injectors spray toluol directly into the blower on this non-intercooled system.

This Racing Fuel 100 is actually green Avgas 100/130. It is an excellent alternative to water but must not be used in CAT-equipped cars.

alternative highly knock-resistant fuels. However, cheaters have found ways around such rules, and those rules do not apply to road cars, so in high boost/high compression engines and non-intercooled installations these alternatives should be seriously considered.

The fuel choices are as follows in order of preference: toluol, 'green' Avgas 100/130 and toluol blended 67%/33%, 'green' Avgas 100/130, methanol. We take a close look at all of these fuels in the next chapter so I will not spend a lot of time on them now except to point out their relative merits as an alternative to water injection.

Toluol has excellent resistance to detonation and as such was the main ingredient in Formula 1 fuel during the turbo era. In similar vein to alcohol fuels, mixtures on the rich side do not hurt power output so relatively crude injection systems can provide quite good outcomes. The really big bonus, though, is that toluol doesn't contain lead and will not damage CAT converters.

High lead green Avgas and toluol blended as indicated has knock resistance similar to straight toluol. It is less expensive, but is only suitable for use in cars not fitted with catalytic converters.

Straight green Avgas is superior to the low lead 'blue' stuff under extreme conditions, but inferior to toluol. In some places it is becoming difficult to obtain, and even then you may have to purchase a 200-litre drum! However, much of the Racing 100 sold in 20 litre drums is actually the old green Avgas 100/130. Have a look at the colour, if it's green you are in business, but you will be paying considerably more per litre than at the airport.

Finally there is methanol. I really do not recommend its use in a road car, but for a weekend track day it is OK providing you are willing to empty the reservoir at the end of the day. Following that, flush the entire auxiliary injection system with petrol to avoid corrosion. Why would you want to go to all this trouble? Some people just love the idea of using methanol; I guess for them it is some sort of nostalgia thing. However, metering is not critical, so quite crude delivery systems are usually not a problem, but do keep in mind that extremely rich mixtures actually promote detonation. Perhaps, though, the major benefit coming from the use of methanol is the resultant cooling of the inlet charge and engine parts because of its very high latent heat of evaporation. Therefore, in installations with inadequate intercooling an injection of methanol at the intercooler outlet will take a lot of heat out of the inlet charge. Where no charge cooler is employed, methanol is injected directly into the blower, but in turbo and centrifugal blower systems injection should be at the compressor outlet.

AUXILIARY INJECTION SYSTEMS

In the most simple auxiliary injection systems, which are only suitable with straight toluol or methanol, a calibrated jet sprays fuel into the inlet stream under high boost or high engine load conditions. This nozzle can be activated by a pressure switch connected to the inlet manifold, or it may be activated via a cam on the throttle spindle closing a microswitch. The complete system incorporates a small tank of 5–10 litres for the fuel, a high-pressure fuel injection pump, fuel nozzle(s), fuel feed line(s) and a fuel return line. To maintain fuel pressure a fuel injection system pressure regulator may be used, or alternatively a 'pill', a calibrated restrictor, may be fitted in the fuel return line.

The next step up in sophistication is to use one or two electronic fuel injectors in place of the calibrated nozzle(s). These injectors can be wired to pulse with the main fuel injectors, but do ensure that the ECU has the capacity to drive additional injectors, and wire the system so that all auxiliary injectors are activated only under high boost/high engine load conditions. Again this is a methanol or toluol only system.

An extreme dual fuel system. The complex primary fuel arrangement utilises two sets of staged 1750cc injectors spraying methanol. When the two-stage nitrous system is activated by the EMS-brand engine management system (first stage at 75% throttle and 3000rpm, second stage at 20psi boost and 4000rpm), the secondary fuel system comes into operation delivering VP racing C-16 (7.2gm/gal lead), supplied from the foreground tank, through the nitrous fogger nozzles.

When precise control of auxiliary injectors is desirable a more sophisticated set-up is called for. Such a system would have an additional injector for each cylinder, either mounted in the manifold outboard from the stock injector, or mounted in the air box lid spraying into the mouth of each inlet tract (see Chapter 13). Additionally, in non-intercooled installations extra injectors may spray into the mouth of non-centrifugal type superchargers. All auxiliary injectors would be pulsed by a programmable electronic management system. Such systems are best mapped to take into account such things as inlet air temperature, engine rpm and load, and boost pressure.

An ECU to serve this purpose doesn't have to cost an arm and a leg. Take a look around at what is available and don't overlook stock type ECUs such as GM Delco '808' units, which are quite cheap in breakers' yards. The main requirement is that it is easily reprogrammable (the Delco is using Kalmaker software), and that you can easily wire it up. There is no sense buying a cheap ECU that is going to cost a fortune to wire, so if your knowledge in this area is limited, purchase a new fuel only unit with clear wiring instructions. Also worth remembering is that some stock ECUs can be modified to run a separate control circuit for such things as water/alternate fuel injection, nitrous oxide, etc. In the USA Jim Wolf Technologies, for example, can supply control modules to 'marry' with certain stock Nissan ECUs.

Whether fuel flow from the stock injectors will have to be reduced at any time when a full set of alternative fuel injectors is installed is a question not easily answered. As a general guide, any engine which is heavily modified from showroom specifications probably will not require any altering of the stock fuel curve. Therefore, an engine that goes from naturally aspirated to supercharged, or from stock turbo to a high boost, modified inlet and exhaust system spec, should not demand any reprogramming of the factory fuel settings. However, if factory mapping set the air/fuel ratio at a very rich 11.8–12:1 at full throttle (some manufacturers do this to cool the pistons and combustion chamber, plus improve CAT efficiency to reduce emissions) and the engine is only lightly modified from factory spec, then it is most likely that the stock injection will have to be leaned off just a fraction after the auxiliary injectors come into operation.

HOW CHEATERS AVOID DETECTION

How, then, do the cheaters avoid detection when competing with anti-knock auxiliary fuel systems? Obviously it doesn't take rocket science to make extra injectors, fuel lines and fuel pump meld into the legal fuel system. And as there is a blending of exhaust odours on the starting grid it is unlikely that even the smell of something as distinctive as methanol is going to be linked to any particular vehicle. When fuel samples are taken, officials naturally enough draw these from the main fuel tank. At times they may ask the competitor if his car has water injection and after taking a quick sniff of the reservoir's contents (it is actually the intercooler spray tank), they move on to the next racer. Possibly all along they have had the actual auxiliary fuel reservoir loaded with an illicit substance in clear view disguised as a dry sump reservoir, an oil catch tank, a fuel swirl pot, a brake booster vacuum tank, etc. Others partition off a section of the fuel tank, and in endurance events they don't even trouble themselves to conceal the extra filler cap. Why should they, twin fillers are frequently installed to expedite gravity refuelling from fuel pails.

Chapter 12

Fuel and fuel boosters

Fuel, to a large extent, is still a great mystery. Most racers and enthusiasts know a little about petrol (gasoline), methanol and nitromethane. They know that an engine will make more power on methanol than on petrol, and they understand that a high octane rating will permit a supercharged engine to make more power without engine wrecking detonation. However, those same individuals would deride anyone who suggested that half a dozen formulations of petrol with the same octane rating could produce an overall variation of 5% in power outputs, or that one formulation could give significantly sharper throttle response.

WHAT FUEL OCTANE RATINGS MEAN

The problem is that over the years we have been unwittingly led to believe that fuel octane ratings are a measure of a fuel's power-producing potential. The thinking was that an engine will make more power on 110 octane petrol than 102 octane, so by inference two formulations measuring 105 octane produce identical outcomes. They do, but only with regard to knock resistance. In areas such as fuel consumption, throttle response and power output there will be measurable differences.

Since there is confusion as to what octane ratings really mean, we should look at the subject before we tackle the tricky field of fuel formulation. As we have already mentioned, most people realise that an engine can make more power and pick-up fuel economy with a higher octane fuel because such fuel at a given boost pressure will allow a rise in the compression ratio and/or more spark advance. However, many don't seem to appreciate that simply changing from 95 octane unleaded to 100/130 Avgas, which has an octane rating of around 105–110, will not necessarily give any power increase and may even cause a slight power loss if the engine does not have knock sensor engine management controls, or was not retuned in any way to take advantage of the octane increase.

VP C-23 racing fuel has a very high Motor Octane Number (MON) of 119. It is leaded (7.2g/gal) and has a specific gravity (SG) of 0.71.

HISTORY OF OCTANE STANDARDS

To better appreciate what octane numbers are all about we have to take a trip back in history to World War I. At that time it was found that aircraft engines would suddenly self-destruct through detonation. An engine might run well on one load of fuel but punch holes in the pistons on the next batch, even though the fuels seemed to be the same, weighed the same and may have come from the same refinery.

The fuel companies tried chemical analysis in an endeavour to achieve parity from one batch of petrol to the next, but despite intensive lab programmes they were not able to identify the batches that were prone to promote engine knock. Because of this problem, special fuel research engines with a variable compression feature were constructed to evaluate and grade fuels. Such a standard heavy-duty, single-cylinder test engine would be warmed up to a standard test temperature, run at standard rpm and load, and then have the compression ratio increased until the fuel being tested just produced engine knock. Its anti-knock rating would then be specified as its Highest Useable Compression Ratio (HUCR).

However, even with every fuel lab supplied with the same type of test engine, and using the same standard test procedure, it was discovered that the same fuel could test out with differing HUCR numbers in different laboratories. Obviously some unvarying standard was needed by which to calibrate the lab test engine. Two pure substances were chosen as reference fuels. The high reference fuel was what we commonly call iso-octane (2-2-4 trimethylpentane), while the low reference fuel was normal heptane (n-heptane).

Now, it was decided that a fuel under test would be run in the variable compression test engine and its HUCR determined. Then a series of runs would be made with various mixtures of iso-octane and n-heptane until a blend was found which produced knock behaviour identical to the fuel being tested. At this point the quality of the test fuel would be rated in relation to the percentage of iso-octane in the reference fuel mixture which gave identical test results. For example, a fuel which behaved the same as 95% iso-octane/5% n-heptane would be called 95 octane petrol. Using this standard test procedure, fuel of consistent anti-knock quality could be refined and supplied for a variety of applications.

RESEARCH AND MOTOR TESTS

Since that time a number of test procedures have come into use to simulate a variety of engine operating conditions. Motor spirit is rated according to the Research and Motor test methods. Both measuring techniques use the same single-cylinder, variable compression test engine but, as indicated in Table 12.1, the Motor method employs a greater engine speed and a higher inlet mixture temperature. Hence the Motor method is a more severe test and generally yields octane numbers 6 to 12 less than the Research numbers, as shown in Table 12.2. This distinction is important as it informs us that the Motor Octane Number (MON) is more relevant in forced induction engines than is the Research Octane Number (RON) as such engines usually operate at higher inlet temperatures. Fuel technicians often state that the RON gives a better indication of a fuel's part throttle knock resistance, while the MON indicates its full throttle knock resistance.

Table 12.1. Comparison of Motor and Research test procedures

	Motor octane test	Research octane test
Inlet air temperature	148.9°C	65.6°C
Engine jacket temperature	100°C	100°C
Engine rpm	900	600

Table 12.2. Octane test comparison

Research octane number	Motor octane number	Pump octane number
92	85	88.5
96	88	92
98	90	94
100	91.5	95.8
105	95	100
110	100	105
113	103	108
115	105	110

The spread between the RON and MON is known as the fuel's sensitivity, and it is quite important to understand what this distinction is exactly. Because intake temperature affects different fuel compounds in various ways it is possible for a refiner to come up with a fuel which has a high RON (or high Pump Octane Number) of, say 98, but by the Motor test that same fuel would rate as, say 86 octane. Hence it would perform as an 86 octane fuel in a typical supercharged engine with higher inlet temperatures. However, on another day the fuel company could use a different blend of fuel compounds, depending on what was in storage or the type of crude oil being 'cracked', and produce a fuel with a RON of 98, but with a MON of 91. This is one reason for the frequent complaint among performance enthusiasts of 'bad fuel'. The RON may legally be up where the company say it is, but owing to the sensitivity of the blending compounds the fuel could be performing like low octane stuff because of the high inlet temperatures in supercharged engines. In the past, when high lead levels were usual, this sensitivity problem seldom surfaced because lead tends to 'cushion' fuel sensitivity. However, with today's unleaded and lead replacement fuels, fuel sensitivity will continue to cause us problems for as long as RON figures are used as the industry standard to rate commercially available fuels.

R + M/2 AND SUPERCHARGE NUMBERS

In the USA, service station pumps carry a Pump Octane Number (PON) rather than the Research number seen on pumps in other parts of the world. This rating, usually expressed as R+M/2, is the average of the RON plus MON divided by two.

The Supercharge test is applied to aircraft fuels which exceed 100 octane numbers, as the other tests obviously become meaningless at anything over 100 octane. The Supercharge Octane Numbers (SON) are really performance numbers obtained by linearly extending the scale beyond 100. In this test the reference fuel is iso-octane plus lead (1 gram of lead added to 4 litres will increase the octane about 6 points). Two tests are involved, the F3 and F4, which explains why aircraft fuels have a dual rating such as 100/130. The first number refers to the F3 test which simulates a supercharged engine running on a chemically correct air/fuel mixture, as when cruising. The F4 number gives an indication of the fuel's performance rating with richened mixture and increased boost, as would be supplied during take-off.

Leaded 100 racing fuel is often straight Avgas 100/130. This is the high lead green variety but, as the drum states, it is not for aviation use. Obviously other hydrocarbons, most likely benzol and/or toluol, which would have lowered the RVP below the aviation standard, have been blended in.

FUEL CHEMISTRY AND OTHER TEST STANDARDS

The anti-knock properties of hydrocarbon fuels are related to their molecular structure. When the hydrogen and carbon molecules are tightly bonded together it takes considerable heat and pressure to break those bonds so as to permit complete combustion to take place. The heat energy required for this process actually comes from the combustion flame itself. Therefore, when tightly bonded molecules are present in the fuel formulation, heat energy which would otherwise have been used to rapidly increase pressure in the cylinder is partly diverted to fracture and rip apart those tightly bonded fuel components. Naturally, this process slows down the rate of pressure rise and combustion speed to an orderly controlled process.

Another factor in all of this is the rate at which fuel vaporises, and at what temperature. If the fuel recombined in the inlet tract or cylinder to the liquid state, it wouldn't burn. On the other hand, if it boiled easily into vapour, controlling combustion speed would be impossible, and there would be huge problems in storing and handling the stuff. Therefore, the challenge facing fuel chemists is to produce a fuel with the right mix of compounds to provide the necessary knock resistance (octane level) and which readily vaporises at specific temperatures.

In the past a lot was made of a fuel's RVP (Reid Vapour Pressure), but relative to fuel performance the numbers are pretty meaningless. Governments regulate the RVP to ensure relatively safe fuel handling (if it boiled too easily, fuel tanks and drums would bulge and eventually explode), reduced emissions during fuel transfer operations and, in the case of aircraft, they don't want planes falling out of the sky through the fuel boiling and causing vapour lock in the fuel lines. The European Standard EN228 calls for car fuel from service stations to have RVP's between 5.08psi and 10.15psi in summer, and between 7.98psi and 13.05psi in winter. Formula 1 fuel regs demand RVP's of from 6.53psi up to 8.7psi. Avgas has to be maintained between 5.5psi and 7psi, and most race fuels fall into the same range.

What gives us much more insight is the distillation curve, as it shows us how rapidly and at what temperatures the fuel turns to a readily combustible vapour. The volume which evaporates up to 70°C (160°F) must be adequate to aid cold starting, but if the proportion is too large fuel will boil in the fuel lines. The amount which vaporises at 100°C (212°F) establishes the engine's warm-up running qualities, and also its acceleration and throttle response at normal engine temperature. The vaporised volume at 140°C (284°F) and 180°C (356°F) should be sufficiently high to minimise dilution of the lube oil, particularly before the engine reaches normal operating temperature.

The European standard for summer blend states that at 70°C (160°F) a minimum volume of 15% must evaporate. By 100°C (212°F) at least 40% has to be boiled off. At 180°C (356°F) 85% of the fuel must have vaporised, and by 215°C (420°F) all must have boiled to vapour. By comparison, Formula 1 fuel, which is supposed to mirror what you can purchase at European service stations (I kid you not!) has at 70°C (160°F) to have lost a minimum of 20% volume to vapour, at 100°C (212°F) 46% has to be boiled off and at 140°C (284°F) the minimum which has boiled off is 75%. By 210°C (410°F) all the fuel must have evaporated, with a 2% residue (tar) allowed to remain.

In America they do things differently. They record at what temperature fuel first evaporates; called the IBP (initial boiling point). Then it is recorded at what temperatures 10%, 50% and 90% of the fuel turns to vapour. Finally they record the end point; the temperature at which all the fuel changes to vapour. Table 12.3 sets out the characteristics of four race fuels for comparison purposes. You will note that Trick Turbo fuel requires 10°C (50°F) more heat than Trick 108 to evaporate 10% of the fuel, indicating that it will not be brilliant when the engine is cold. However, it is going to give good throttle response in spite of the high octane rating as 90% of the fuel is vapour at 113°C (235°F). By comparison, the two VP fuels need a lot more heat initially, but then are completely evaporated at relatively low temperature.

Table 12.3. Race fuel characteristics

		Trick 108	Trick Turbo	VP C-16	VP C-23
Octane	RON	113	119		
	MON	103	109	117	119
	PON	108	114		
Specific gravity		0.73	0.73	0.735	0.71
RVP				1.85	4.54
Distillation	IBP	36°C	34°C		
	10%	58°C	68°C	99.7°C	77.9°C
	50%	94.5°C	96°C	100.6°C	97.6°C
	90%	117°C	113°C	102.1°C	99.7°C
	end point	146°C	144°C	111.8°C	110.1°C

Petrol can be a combination of various amounts of up to about one hundred different components. If the fuel was mainly composed of less volatile anti-knock hydrocarbons, the engine would be hard to start and throttle response at normal engine temperatures would be poor. Also, when the engine was cold some fuel would remain unburned, washing oil off the cylinder walls and diluting oil in the sump. On the other hand, if the

Methyl benzine, Shell's name for toluol, is an aromatic with extremely high knock resistance.

main fuel components were extremely volatile there would be very little resistance to engine knock and fuel boil.

Consequently the composition of race fuel and aircraft fuel, and that for street use, is quite different. Also, street fuel composition changes according to the season. While there are government regulations stipulating how readily the fuel will vaporise in summer and winter (hence the terms 'summer blend' and 'winter blend'), fuel companies actually change the formulation each month. In aircraft the throttle is basically set and only altered for take-off and landing, so throttle response isn't a factor in the blending of that fuel. And because the engine does very little cold running with infrequent start-ups, cylinder lubricant wash and oil dilution isn't an issue. In this regard race fuels are similar but, unlike aircraft, race cars must have excellent throttle response, which calls for a different blend of tightly bonded anti-knock fractions and more easily ignited components.

HOW MOLECULAR STRUCTURE AFFECTS KNOCK RESISTANCE

The paraffins, such as normal heptane and kerosene, are long chains of carbon and hydrogen held together by weak molecular bonds which are easily broken by heat. Iso-octane is a member of the iso-paraffin family. These have branched chain structures that form stronger bonds to better resist detonation. The cyclo-paraffins, or napthenes, also have good anti-detonation properties, with their hydrogen and carbon atoms well bonded in a ring-shaped molecule. The aromatic fuels, such as toluol, also have a ring-shaped structure with very strong bonds, hence their excellent anti-knock characteristics.

As has already been pointed out, the chemical composition determines just how rapidly and completely the fuel will burn, and whether it will be resistant to detonation at high cylinder pressures and temperatures. The fuels required for ease of starting and throttle response have weak molecular bonds, so break up and burn spontaneously (i.e. without being ignited by the combustion flame initiated by the firing of the spark plug) at lower temperatures and pressures than fuels with strongly bonded structures. Some fuel additives, such as the aromatics, make good anti-detonants because they burn slowly and don't oxidise or burn completely until combustion chamber temperature and pressure is very high. Such components therefore inhibit, or slow down, combustion. Hence, a high octane fuel will only increase power in an engine that actually needs a fuel which is chemically stable at high temperature and pressure. Obviously, if the engine does not have a compression ratio, boost pressure and spark advance great enough to produce high combustion pressure and temperature, then the high octane fuel may not burn completely during the early phase of the power stroke, resulting in reduced throttle response and perhaps a power loss.

This is another reason for my fondness for the dual-fuel approach discussed in the previous chapter. With an ordinary, relatively inexpensive street fuel in the main tank you get easy starting, good throttle response and minimal cylinder wash and oil dilution. Then, with either toluol, a toluol/Avgas mix, or straight Avgas in the secondary tank, you achieve excellent detonation resistance with inexpensive, easily obtainable fuels. Also, because these fuels with tightly bonded structures are only being released at higher boost levels, combustion is rapid and complete, ensuring good power output combined with a high level of knock resistance.

AVGAS-BASED RACE FUELS

In the past a lot of race fuel was simply some grade of Avgas relabelled. So 115/145 was sold as 'racing 115', although to be completely accurate some brands of 115 did contain around 20% toluene, not so much to increase its knock resistance but to decrease fuel consumption and to make for easier fuel and oil blending in two-stroke engines. This fuel is no longer available, except to the military and operators of old fighter aircraft. Most leaded 'Racing 100' is, in fact, Avgas 100/130. The old green coloured variety has more lead (4.5–6g/gal) than the more recent blue low-lead type (2–2.7g/gal). Tested by the Research method, the octane rating of both is generally 108–110; by the Motor method it will be 100 minimum, and up to 102. Both attain their high octane by a high percentage of alkylates plus lead. The blue type, though, to get back octane lost from the lower lead levels, also contains a good portion of aromatics, usually toluol.

Avgas 112/160 used to be available in some parts of the USA and is now marketed there as 'Leaded Racing 108', reflecting its Pump Octane Number. It has a Research number in the 112–114 range and a Motor number of 102–104. However, like the old 'Racing 115MB' (for methyl benzine, a Shell trade name for toluol) it also contains a high portion of heavy aromatics which lift its specific gravity (SG) from the 0.69 typical of aircraft fuel to 0.73. This means that there is more energy in every tankful, so the mixture has to be leaned off to maintain the correct air/fuel ratio, but it also allows a race car to go another 5–6% distance, an important factor in endurance events.

SOPHISTICATED RACE FUELS

In time, though, fuel companies recognised the commercial advantages in batching special race fuels which provided not only the detonation resistance possible with Avgas, but also better throttle response and a good power increase. Typically 4–5% at the top of the power curve and even more in the mid-range, (which also serves to broaden the power band) are seen on the dyno. MR#3 from VP is representative of this class of fuel. It is leaded, contains 2½% oxygen, has a MON around 103 and costs $US7.00 per litre!

Motorsport 103 is a street-legal RFG unleaded fuel with a high octane rating (103 by R+M/2 method) and SG of 0.728.

While this was going on, more sanctioning bodies began to ban leaded fuels in the more highly promoted and publicised classes (NASCAR being the exception). The transition has not been too smooth for some racers though. What has caused some confusion is that 'Unleaded Racing 100', which in Europe is rated by the Research method, unlike the old leaded 'Racing 100' which gets its 100 rating with the tougher Motor test. Obviously, a switch to the unleaded 100 without suitable engine modifications and retuning would cause the engine to be destroyed by detonation. In the USA, several unleaded race fuels are available. VP Motorsport 103 is very popular (99MON, 107RON), but the power king in this class is Power-Mist RFG (reformulated gasoline). It contains 6% oxygen and tests at 104MON and 112RON.

By far the most money and resources are devoted to Formula 1 fuel development. F1 must use unleaded with maximum RON of 102 and maximum SG of 0.775. Maximum oxygen content is now 2.7%, and benzene allowable is right down to 1%. Even after the turbo era aromatics were often over 50%, but now have been pegged at a maximum of 35%. Major components in descending order are: toluol, xylene, sopentane, iso-octane, hexene-1, n-butane, 2-methylpentane.

OCTANE BOOSTERS

As you can see, toluol and xylene are the main anti-knock agents in F1 fuel. Unfortunately, straight xylene is difficult to obtain (paint thinners may contain a considerable portion of xylene and/or toluol but as it also contains other compounds its use is fraught with danger), and while such is not the case with toluol there is a question mark over using it as an octane-boosting agent in high-octane unleaded pump fuel. In the past, a reliable method of increasing the octane rating of petrol was to add up to 33% toluol (bigger percentages were not recommended because of cold weather starting difficulties). This percentage would raise the octane rating of leaded fuel (0.2g/litre) about 6 numbers Research and 2.5 numbers Motor, respectively. However, as lead levels have decreased, government regulations permitting, petrol refiners have substituted high-octane hydrocarbons to bring the octane back up, so a reliable octane increase cannot be guaranteed by their addition. A guide as to whether fuels such as toluol, triptane, benzol, xylene, etc. have been added by the fuel company is to check the specific gravity. If the SG is high and there is no government regulation limiting these products to 5% maximum (they are reported to be cancer causing), the petrol probably contains up to 20% of these products either singularly or in combination.

However, if the government limits their use to a maximum of 5%, then with the exception of benzol and triptane all will provide a good octane boost of about 1MON for every 10% added up to a maximum of around 25%. Additionally, in distance racing, by virtue of their high SGs they significantly reduce fuel consumption, and consequently the number of pit stops required.

The situation is similar with another very effective octane booster, MMT (methyl cyclopentandienyl manganese tricarbonyl), which is the base product in the most impressive concentrated octane boosters (Australian-made NF Racing Formula and Nulon Pro Strength). In low octane unleaded (91RON) both will provide a consistent 5RON boost. However, when added to 95RON premium unleaded the results waver between a 2RON and 3.5RON rise, and with super 98RON unleaded it varies between a 1.5RON and 3RON increase.

Concentrated octane boosters vary considerably in potency and should be used with a high degree of caution. NF's top of the line 'Race Formula' will generally provide a 1.5–3RON boost to 98RON unleaded.

Actually, concentrated octane boosters vary considerably in potency. You really cannot take much notice of the octane increase claims. If any of the claims were true, then they must have been obtained when added to the 83RON–86RON stuff which is the staple diet for motor vehicles in some poor parts of the earth. Also, be careful of any change in product which you have previously found to be effective. The word 'reformulated', new packaging, etc. should get the alarm bells ringing. It probably means the product has had to be changed because of government regulations or steep increases in the price of key ingredients, and that it is no longer as effective as previously. 104+ and Super 104+ are a typical example. The old stuff was about the best of the 'bottle boosters'. Then in 1999 it was changed (identified by the eagle logo on the back) and slumped to being a very ordinary performer giving only about half the octane boost (1RON for Super 104+ with 95RON unleaded) as the previous formula.

When using these concentrated additives there are two points you must keep in mind. The first is that American concentrates are usually marketed in US quarts which equal 946cc as opposed to the Imperial quart of 1136.5cc. The US gallon is also different (see Appendix), so if you live outside the USA you will have to carefully work out the blending ratio. The second point is with regard to mixing technique. Don't just pour a can of octane booster into your fuel tank and expect it to blend uniformly. What is preferable is to mix the concentrate with about 2–3 gallons of petrol, give the mixing drum a good shake, and then add this mixture to the untreated fuel.

TESTING, PURCHASING AND STORAGE OF PUMP FUEL

Really, any time you blend a commercial pump petrol you are at the mercy of the fuel companies, so what worked with one batch of fuel may not work with another. Apart from certified racing fuels the only other fuel that is a known quantity and highly regulated is Avgas. Consequently, my advice is don't muck around with fuel blending, it could cost you an engine. If you are in a competition class restricted to 98RON unleaded, test different brands to determine what brand gives the best overall power. Do this, preferably in summer, when fuel turnover is high and when summer blend is being supplied. When you find the 'power king', buy up what you require to do you for the race season. Be sure to store fuel in sealed drums away from sunlight; you don't want vital components evaporating or being destroyed by light (both can happen and high-octane unleaded is more prone to this damage). Also, remember that fuel companies swap fuel between themselves and frequently use a series of common pipe networks, so if your dyno testing of fuel is going to be of any value you really have to be quick to buy fuel from the very same service station where you sourced your winning test sample, before new and possibly different supplies are dumped in the storage tanks.

WHY METHANOL RAISES HORSEPOWER

By now it should be obvious that a fuel's octane rating is not a true measure of its power potential. Its octane rating, its knock resistance, is important but it is not the whole story. When we examine an alcohol fuel such as methanol this becomes even more evident. When we look at Table 12.4 we see that methanol has only a very modest Motor octane of about 89–91 and the high Research numbers are only possible with extremely rich mixtures. The question then is, why does it permit higher boost levels than petrol and provide up to a 20% power increase?

Table 12.4. Fuel characteristics

Fuel	Specific gravity	RON	MON	Fuel/air ratio (lb/lb)	Heat energy (Btu/lb)	Latent heat of evap	Weight (lb/gal)
Acetone	0.79			1:10.5	12,500	225	8
Avgas 100/130:							
'green'	0.69	105–110	100–102	1:12.9			7
'blue'	0.71	105–110	100–102	1:12.7			7
Benzol	0.88	105–110	95–100	1:11.5	17,300	169	8.7
Ethanol	0.79	108–115	90–92	1:6.5	12,500	410	8
Ether(diethyl)	0.71				15,000	153	7
Methanol	0.79	105–115	89–91	1:4.5	9,800	472	8
Nitromethane	1.13			1:2	5,000	258	11.3
Nitropropane	1.05				6,700		10.5
Petrol:							
premium unleaded	0.74	*96	*85–86	1:12	19,000	135	7.4
premium leaded	0.73	96	86	1:12.5	19,000	135	7.3
racing leaded (USA)	0.73	112–114	102–104	1:12.7			7.3
racing unleaded (USA)	0.75	104–106	94–96	1:13.2			7.5
racing unleaded 100	0.75	100	90–92	1:13.0			7.5
Propylene oxide	0.83				14,000	220	8.2
Toluol (methyl benzine)	0.87	120–124	110–112	1:9.8			8.7
Triptane	0.69	110–112	100–102				6.9
Xylene	0.86	117–118	115–116				8.5

*Premium unleaded in some lands has a higher rating (often 98RON, 87MON)

Shell 'A' racing fuel is 96% methanol and 3% acetone.

Methanol has a very high latent heat of vaporisation, i.e. it takes a lot of heat to be converted from liquid into vapour. Petrol has a latent heat of evaporation of 135 Btu/lb, while for methanol it is 472 Btu/lb (One Btu [British Thermal Unit] is the amount of energy required to raise the temperature of 1lb of water 1°F). This heat required for proper atomisation is drawn from the inlet tract, the piston crown, the combustion chamber and the inlet and exhaust valves. An internally cooler engine does not put as much heat into the inlet charge, the result being that mixture density in the cylinders goes up, and hp rises. Also, less initial heat means a slower burn, so the engine may accept more spark lead or more boost, resulting in more hp or improved throttle response.

That's part of the story; the other part relates to the amount of energy available in the fuel burned. Using petrol, the air/fuel ratio for the best power in a supercharged engine will be somewhere between 12.3:1 and 13.1:1, with about 12.7:1 not far from the mark. With methanol, the air/fuel ratio increases to somewhere between 4.5:1 and 5.5:1, with about 5.1:1 being pretty close. One pound of petrol has the energy potential of about 19,000 Btu. In comparison, methanol delivers around 9800 Btu/lb, which means that it produces less than 52% of the heat energy of 1lb of petrol. However, because we are mixing more methanol with each pound of air, we are actually producing more heat energy by burning methanol.

To work out how much more heat energy is produced, we have to divide 12.7 by 5.1, which is 2.49. Next, we multiply 9,800 by 2.49, which gives 24,402. This tells us that, in theory, methanol will produce 28% more heat energy than petrol:

$$\left\{\frac{24{,}402 \times 100}{19{,}000}\right\} - 100 = 28.4\%$$

Right away you can see that an engine burning straight alky will burn more than twice as much fuel. Therefore, the entire fuel system demands a considerable upgrade to flow the required volume of fuel. We cover this in more detail in the following chapter, but for now just be aware that whatever was required in the way of fuel pumps, fuel lines and injectors for petrol will have to be more than doubled for alcohol fuels.

PROBLEMS AND DANGERS OF METHANOL

There are many problems involved in the use of alcohol, some of which will affect you and some of which will affect your vehicle. Since your life and well-being are more important, we will deal with you first.

Methanol is extremely poisonous, and as it is accumulative it can build up and oxidise to form enough formaldehyde to cause blindness and even insanity. It can be absorbed through the skin and lungs, and inhalation of the exhaust gas is also dangerous because vaporised methanol is usually present, and especially so when rich mixtures are being used.

Regarding you car, you need to know that methanol is a very effective paint stripper and it may attack some types of fibreglass resin. It has a scouring effect on tanks and fuel lines so these should be soaked in methanol and then drained so that the residue does not find its way into the injectors when you convert your car from petrol to methanol.

Methanol will absorb huge amounts of water, so it must always be kept in an air-tight container. (In fact, when using lean mixtures, up to 10% water is at times added to the fuel to act as an anti-detonating agent.) Following every race meeting the fuel will have to be drained and the entire fuel system flushed through with petrol. Also, you will need to run the engine to prevent the formation of water-induced rust and oxidation. This can be particularly damaging to aluminium and zinc parts, and usually results in blocked injectors. In fact, it is preferable to use stainless steel where possible for things like fuel rails.

Caution must also be exercised in other areas of the injection system. Petrol tends to lubricate, so flat-slide and barrel-type throttles seldom stick. Methanol does not lubricate

Because alcohol fuels absorb water, either a stainless steel or triple marine grade anodised aluminium fuel rail and methanol injectors are essential.

at all, and there are the added problems of corrosion and possible ice formation, which may allow the throttle to stick, possibly in the wide-open position. Therefore, simple butterfly throttles are preferred.

In colder climates, starting difficulties may be encountered when straight alcohol is being used. Some use other more volatile fuels which they blend with the methanol to help overcome this problem; usually 5% acetone or a maximum of 3% ether. Ether is particularly dangerous because of the possibility of engine damage from detonation, so I do not recommend its use, nor do I recommend the use of starting aerosols which also contain a lot of ether. Personally, I feel the best method is to spray petrol into the inlet manifold if it can be easily got at. Do not attempt to do this while the engine is being cranked over; if it backfired you could be badly burned. The alternative is, before fitting the warm-up spark plugs, spray about 3–5cc of petrol down each spark plug hole.

Many starting problems can be traced to an inadequate ignition or electrical system. Obviously, the system must have enough grunt to spin the engine rapidly enough to fire. Also, the ignition system must be up to firing wet plugs or fuel-loaded spark gaps. Methanol burns much more slowly than petrol, so it is necessary to advance the ignition accordingly. As a guide you can reckon on the engine requiring 15–25% more advance than for petrol, right through the rev range.

NITROMETHANE USE AND BLENDING

Nitromethane is different again, and if you look at the fuel characteristics table (Table 12.4) you will see that it is not even assigned an octane number (because there are too many variables), yet it can easily double the power of an engine burning petrol. This is possible through nitro's very special chemical composition, and also owing to the fact that extremely rich mixtures can be burned satisfactorily. In itself, nitromethane is a very poor fuel, but because it contains approximately 53% oxygen by weight it permits the induction of large quantities of fuel into the engine for conversion to heat energy. Also, because it burns so slowly it continues to apply force to the pistons almost to the bottom of the power stroke.

In the past, nitromethane was burned in the most powerful drag machines in ratios of 80–90% nitro and 10–20% methanol, but today very little methanol is blended in, with most Top Fuel rails running 98% nitro. On the speedway scene and hill climbs, nitro is used (often illegally) in smaller percentages (usually 10–20%) as a power booster. Methanol is the base fuel, and acetone plus other 'doping' agents may also be added (plus about 0.5% castor oil as an odour masking agent).

To deter detonation or other engine damage, it is always necessary to reduce the compression ratio when nitro is used. At all times the air/fuel ratio must be very rich. With an 80–90% blend of nitro it may be as rich as two parts fuel to one part air, or as lean as two parts air to one part fuel. Using 20% nitro this would change to three or four parts air to one part fuel (Table 12.5).

When blending nitro with other fuels, the safest method to avoid error is to mix according to volume. So, for a 20% nitro blend you would add one litre of nitro to four litres of methanol. Some people also blend using a hydrometer calibrated for this purpose. The problem with this method is that the hydrometer will only give a true nitro content percentage reading when the fuel temperature is 20°C (68°F). From Table 12.6

you can see that with a fuel temperature of 38°C (100°F), a fuel with a true nitro content of 80% will read 73% nitro on the hydrometer. Consequently, you should always measure the temperature of the fuels before attempting any mixing.

Table 12.5. Fuel flow increase for nitromethane

% nitro in methanol (by volume)	Flow increase over straight methanol
10	25%
20	50%
30	75%
40	100%
50	125%
60	150%
70	175%
80	200–230%
90	225–260%
98	250–290%

Table 12.6. Nitro hydrometer temperature correction chart

True Nitro (%)	Hydrometer reading at fuel temperature Centigrade/Fahrenheit										
	4°/ 40°	10°/ 50°	16°/ 60°	20°/ 68°	24°/ 75°	27°/ 80°	29°/ 85°	32°/ 90°	38°/ 100°	43°/ 110°	49°/ 120°
100	106	104	102	100	98	97	95	94	92	90	87
90	97	94	92	90	88	87	86	85	83	80	78
80	86	83	82	80	78	77	76	75	73	70	68
70	75	73	71	70	69	68	66	65	63	61	59
60	66	63	61	60	59	58	57	56	54	52	50
50	55	53	51	50	49	48	47	46	44	42	40
40	45	43	41	40	38	37	36	35	33	31	30
30	35	33	31	30	28	27	26	25	23	22	20
20	27	25	22	20	19	18	17	16	15	13	11
10	20	16	13	10	9	9	8	7	5	3	1

Many don't seem to understand that a nitro hydrometer is calibrated to give a true reading only when straight methanol and nitro are being blended. When a methanol/acetone/nitro mix is used, the hydrometer will still give an accurate reading because methanol and acetone share the same specific gravity. However, if any other doping agents were added, such as propylene oxide, the hydrometer reading would be of no value at all.

Igniting high-percentage nitro fuel is always a problem. Top Fuel dragsters run the equivalent of a mini welder delivering 1.2 amps to each spark plug. As higher percentages of nitro are burned, the combustion rate is reduced (unless a fuel ignition accelerator such as propylene oxide is added). Therefore, the spark advance must be increased, with Top Fuel engines typically running a system locked on about 50° advance.

Nitro burns slowly, hence the spectacular flames at night.

NITROMETHANE DANGERS

There is considerable risk to yourself and others associated with the use of nitro. After combustion, relatively large amounts of vaporised nitric acid are exhausted. The higher the nitro dose, the more acid vapour is released. When inhaled, nitric acid vapours cause a muscle reaction, making it impossible to breathe. Therefore, the use of the correct gas mask is essential if the driver is in a position to inhale the exhaust gases. Certainly, mechanics working around the car and those in the starting area will require masks.

Some people have the idea that nitromethane is explosive. It isn't, but like any fuel it can be made shock sensitive. The following are the main causes of nitro becoming dangerous:

* The addition of hydrazine in fuel blending. Hydrazine is banned in many countries because of the danger.
* The use of caustic soda or any other alkaline substance for cleaning fuel tanks or lines.
* The use of 'unpickled' anodised aluminium fuel tanks. After anodising, the tank must be allowed to stand for a few days filled with a solution 90% water and 10% vinegar. This serves to remove any deposits remaining in the tank after anodising.
* The use of excessive fuel pump pressure. Nitro is liable to become unstable when confined and subjected to shock. Even though some Top Fuel dragsters routinely run 500psi fuel pressure, anything over 100psi should be considered dangerous.

When using more than 20% nitro, there is always the danger of a sump fire or explosion because of the large amount of fuel that finds its way past the rings and into the sump. The first signs of such a fire are yellowish flames appearing at any of the breathers. Therefore, it is important to keep an eye on the engine for at least 2–3 minutes after shut-down.

OTHER FUEL COMPOUNDS

All the base fuels – petrol, methanol and nitromethane – may have other compounds added for a variety of reasons.

Propylene oxide (epoxy propane) is used in high-percentage nitro fuel to increase the combustion flame speed. The amount used should never exceed 15% of the amount of nitro in the fuel, because the combustion rate could rapidly increase to the point of severe detonation and engine destruction. For example, if you are using 80% nitro you could use up to 12% propylene oxide.

Propylene oxide can also be used with Avgas or heavily leaded race fuels, and also methanol, up to a maximum of about 5% to give a modest 2–3% power gain. Of much greater benefit, though, is the greatly improved throttle response which makes the engine feel much sharper, as though with less flywheel weight to accelerate. However, caution is in order to get the mixture a little rich and the timing backed off, otherwise expensive engine damage will result.

Once blended with other fuels, propylene oxide is relatively stable, but it can become explosive in its pure state if allowed to come into contact with copper, copper alloys or rust particles. Therefore, it must be stored in aluminium or plastic containers.

Nitropropane is listed in Table 12.4. However, I would recommend its use, with extreme caution, only in sprint type engines operating at modest boost levels. Be mindful of the fact that it is an oxygen bearing fuel and will cause a severe lean-out unless fuelling is reprogrammed. As a starting point, increase the fuelling by three times the percentage of nitropropane added, (i.e. for 10% nitropropane increase fuelling by 30%) and take out some spark lead too as combustion speed with this stuff can quickly get out of hand.

Like propylene oxide, acetone and peric acid, nitropropane is a member of the nitroparaffin family. It can be used with heavily leaded race fuel or Avgas up to a maximum concentration of about 12%, to produce a power gain in the order of 5–6%. It can also be blended with methanol in like percentages to provide a similar power boost as that when added to Avgas. However, if more than about 3% acetone (propanone) is present in the methanol, concentrations over 5–6% should not be used. Be aware that many fuel companies do not supply pure methanol. Often they blend in 2–3% acetone to improve methanol's ability to be mixed with other fuel compounds. Actually, up to 10% acetone will improve the combustion rate of methanol, improve cold weather starting and, in like vein to water, reduce the tendency for lean mixtures to detonate.

ASSESSING OCTANE REQUIREMENTS

Some have stated that an engine requires an octane increase of 2.0 for every 1psi of boost. At best this is a gross oversimplification of a very complex matter, at worst such a 'rule' could easily lead to serious engine damage. I would prefer to state the matter in quite different and non-specific terms.

The first 'truism' is that anytime we significantly increase the intake air temperature, and place the engine under load, we need increased fuel octane to ward off detonation. In large measure this appears to be in full agreement with the idea that we need 2.0 octane numbers increase for every 1psi boost. (Interestingly, are we in need of 2MON or 2RON?) Clearly, every compressor adds heat to the inlet charge, and the heat levels go up with boost increases. However, don't increases in ambient temperature also produce steep increases in intake air temperatures? Unscavenged exhaust gas does the same thing, but actually inside the cylinders. Increased water and oil temperatures also add heat to the inlet charge. The latter can be the product of higher ambient temperature, but

Here the combustion chamber volume is about to be measured so that the compression ratio can be calculated. While the boost/compression ratio relationship is important in assessing octane requirements, other factors such as intercooler efficiency, engine temperature, cam lobe characteristics and cam advance, combustion chamber shape, etc. all have some influence.

it can also be as a result of a very dense, cool inlet charge producing a massive increase in heat energy as a result of all the fuel added to that dense charge to provide the correct air/fuel ratio.

Now I have to make what may appear to be a reversal of my first truism, and it is that anytime we significantly increase charge density, and place the engine under load, we need additional fuel octane to fend off detonation. So, cool dense atmospheric conditions could produce engine knock. In similar vein, adding an intercooler or fitting a more efficient cooler could produce engine knock. As indicated in the sidebar a big drop in humidity could lead to the engine detonating. And, yes, a more efficient exhaust could cause knock. Less unscavenged exhaust gas increases charge density, but less exhaust gas means less corruption of the combustion flame through the mixture. More flame speed can lead to an excessively rapid cylinder pressure rise, out of control combustion – detonation.

As I stated, this is quite complex. It is all very much a delicate balance: a slight increase in charge temperatures or charge density isn't a problem, but a big increase in either will call for more octane. Likewise, a small increase in both charge density and charge temperature will likely require more octane.

HORSEPOWER INCREASES WITH MORE OCTANE

Finally, let's consider in real terms how much additional octane is worth when an engine is retuned to take full advantage of the increased knock resistance. The car is a Subaru Impreza WRX built for club level circuit racing and tarmac rallies. The engine is heavily

Humidity – a natural octane booster

We do not tend to think much about humidity until out clothes begin to stick to us and perspiration drips from us. This indicates that the air is saturated with water vapour. However, not only are there a lower number of oxygen atoms being dragged into the cylinders when the humidity rises, but there is also a much larger number than usual of water molecules interspersed between the oxygen and fuel molecules. The effect is that the space between the two increases, which slows combustion flame speed within the combustion chamber because rather than progressing rapidly from fuel molecule to fuel molecule, the flame front comes up against water molecules that do not contribute to the combustion process. Thus the water suppresses violent combustion – detonation – just as an octane booster does by chemical means. This means that increases in humidity will often compensate for increases in ambient air temperature, which typically demands a 1.0 octane increase for every 13°C (23°F) that the mercury rises. Conversely, a big decrease in humidity and an increase in intake charge temperature or air temperature could see a race engine being destroyed by detonation if octane levels are not increased, or boost pressure reduced and fuel mixture richness increased. A rise in barometric pressure of 1 inch of Hg calls for an octane boost of 1.0 octane, which alternatively can be compensated for by richening the mixture or reducing the spark lead. However, a steep rise in the humidity level decreases the degree of compensation required. As indicated in the accompanying table, we determine humidity levels by taking temperature readings from two thermometers, one with a dry bulb and one with a wet bulb covered by a water-soaked cotton wick. When both thermometers read the same temperature the humidity is 100%. However, if the dry bulb reads 25°C (77°F) and the wet one reads 24°C (75°F), a depression of 1°, then the humidity is 92%.

Percentage humidity

Dry bulb °C	Wet bulb depression °C					
	–1	–3	–5	–8	–12	–18
5	86	58	35			
15	90	71	52	30		
25	92	77	64	45	22	
35	94	80	70	55	37	15

modified, with a lot of attention and money paid to ensure that it will be utterly reliable. The capacity has been taken out to 2.2 litres, not so much for more power, but rather to obtain more torque at the lower end of the power curve and thereby broaden the power band.

The most significant departure from the more conventional approach was to fit two Garrett GT25 ball bearing turbos. Again, this was to minimise turbo lag and fatten up bottom-end and mid-range hp. In this situation it is partly the two small turbos and partly the more efficient exhaust that achieves this objective. It is very difficult to develop a really good exhaust for a single turbo horizontally opposed engine such as in the Subaru and Porsche. If you get the pipe sizes right for one side of the engine, then they will be way off on the other side. The efficient, albeit expensive, solution is to go with twin turbos with proper exhaust sizing on both sides.

Nulon Pro Strength booster provided a minimum 15% power increase over straight 98RON unleaded when turbo boost and ignition timing were increased to take advantage of the extra octane.

In tarmac rally events the car is restricted to 98RON unleaded (the cars have to fill at designated service stations during the course of the events, and no other refuelling is permitted). However, concentrated octane boosters are allowed in certain events. On straight 98RON unleaded the engine produced a maximum of 406.8hp. With Nulon Pro Strength booster at higher than recommended concentration (the CAT is poisoned at this level) power went up to 461.3hp, a 15% increase in peak hp and close to 19% more at the torque peak. For circuit racing and endurance events (6hr and 12hr) the fuel has to be either unleaded 100RON or Avgas 100/130. The naturally aspirated cars mostly choose the more expensive unleaded 100. Because of its high SG, fuel economy improves and pit stops are decreased in endurance events. However, for a turbo car the extra octane is more important, and at any rate by blending in toluol (30% toluol to 70% green Avgas) the SG is brought up to the legal limit of 0.75. With this blended Avgas and toluol, power shot up to 508.2hp.

Does this mean that the high octane brew produced over 100hp more than the super unleaded? No, that extra power was available by virtue of the additional octane (probably around 13–15MON) permitting a much higher boost pressure (up to 9psi). In turn, that much higher boost could be called up because the compressor spec was optimised to provide efficient flow at several levels of boost, and additionally the very large intercooler was still able to maintain intake temperatures at not more than 20–25°C (70–77°F) above ambient on the hottest days.

Chapter 13

Fuel delivery system

The fuel delivery system starts at the fuel tank and ends at the fuel injectors. If any link in that fuel supply chain is deficient in any way then you risk losing your engine in a worst case scenario, and at best the engine will be down on power. Therefore, whenever we plan to add some form of forced induction to a naturally aspirated engine, or plan to upgrade an existing forced induction engine, we must first establish that every part of the fuel system is capable of flowing the quantity of fuel necessary to maintain the correct air/fuel ratio.

In *Fig. 13.1* you can see the various components of a typical competition-type electronic fuel injection system. The basic EFI system can be broken down into three distinct groupings: the air flow and electrical/electronic systems, which are discussed in Chapters 10 and 18 respectively, and the fuel delivery system, comprising the injectors, pressure regulator, fuel pump, fuel filter, fuel lines and fuel rail.

INJECTOR DUTY CYCLE

We cannot just look at an engine's fuel flow requirements simply from the aspect of injector size. Actual fuel flow into the inlet air stream is a product of injector size, injector pulse width and the fuel pressure. Also, because flow peaks when the engine makes maximum hp, we have to factor in the injector duty cycle, which is expressed as the maximum percentage of time the injector is spraying fuel to meet the engine's maximum fuel flow needs. If the injector is open continuously, i.e. 100% of the time, it will overheat and either fail to respond accurately to signals from the ECU, or else it will burn out. In road cars I will not run Bosch injectors longer than an 85% duty cycle, and in competition cars with batch-fire injection (*Fig. 13.2*) I keep the duty cycle below 80%. With ND and Rochester injectors, take 5% off these recommendations, and with Lucas injectors you can safely add 7%. However, before ever running an injector at maximum values, actually test it in an injector test rig and have a look at the spray pattern and flow volumes as weird things can happen (injector fluttering, stalling half

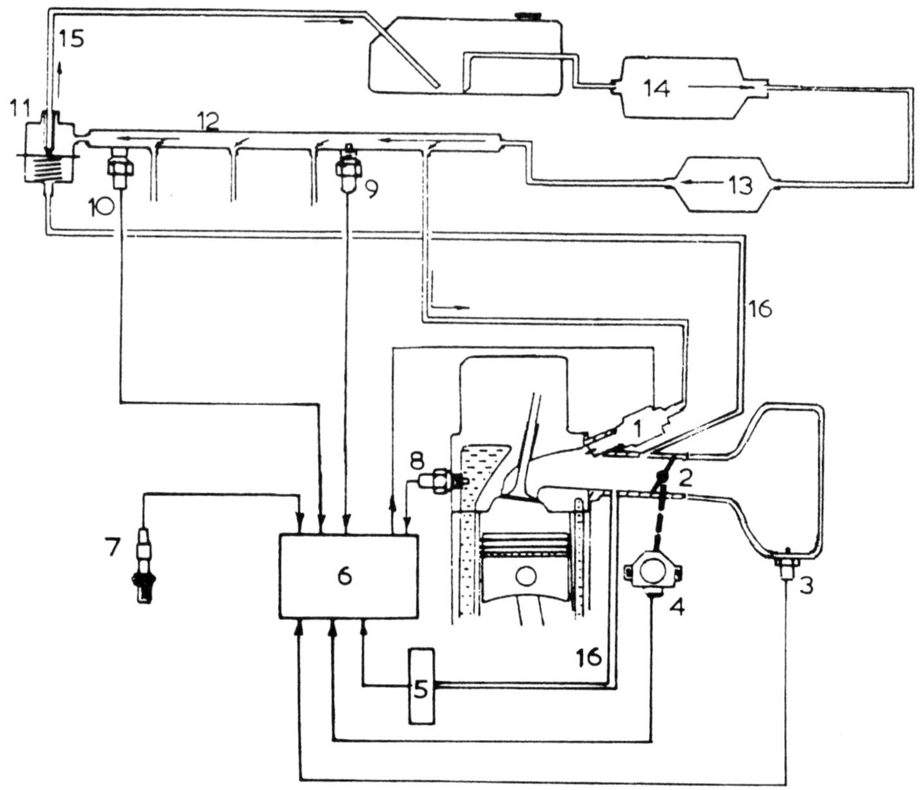

Fig. 13.1. Speed density type electronic fuel injection for a supercharged competition engine. The components illustrated are: 1. injectors; 2. throttle plates; 3. air temperature sensor; 4. throttle position sensor; 5. MAP sensor; 6. ECU; 7. exhaust gas temperature sensor(s); 8. coolant temperature sensor; 9. fuel temperature sensor; 10. fuel pressure sensor; 11. fuel pressure regulator; 12. fuel rail; 13. fuel filter; 14. fuel pump(s); 15. excess fuel return line; 16. manifold vacuum/boost sensor hose.

Fig. 13.2. Batch fire injection sprays all injectors as one every crankshaft revolution.

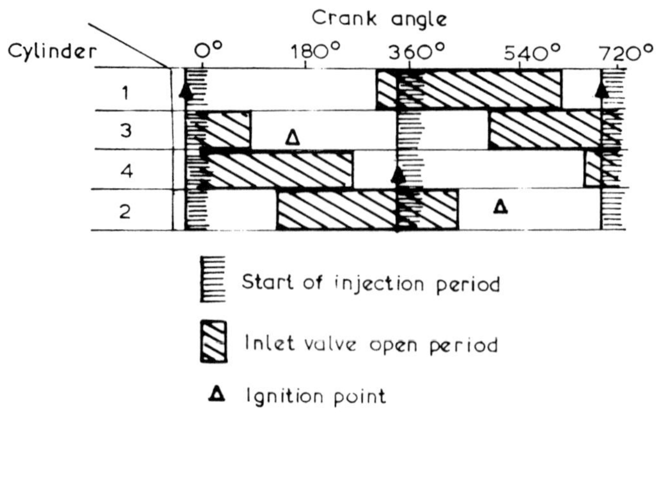

The pulse width and duty cycle relationship

As engine rpm increases, the time available in which to complete the injection of fuel into the intake air stream becomes shorter and shorter. Obviously, an engine running wide-open throttle at 9000rpm has only two-thirds of the amount of time in which to get injection completed compared with an engine running at 6000rpm. However, an injector cannot be held open continuously or it will burn out. In a modified road engine it can be active, or on 'duty', up to 85% of any given time period – this is termed its duty cycle. Clearly its 'rest cycle', where there is no electricity passing through the solenoid winding, accounts for 15% of that fixed time period. Because stock road cars seldom see extended high-rpm, wide-open throttle running, some manufacturers allow up to a 93% duty cycle, but in competition engines, which run at reduced throttle openings for only short periods of time, we should not extend the duty cycle beyond 80%.

This necessary 20% rest cycle impacts on the time that the injector has available in which to pulse and spray fuel into the air stream. We can calculate the maximum permissible pulse width using this formula:

$$\text{Maximum pulse width (milliseconds)} = \frac{M \times 60{,}000}{\text{rpm}}$$

where M = Injector duty cycle.
rpm = Maximum rpm for batch fire, and half maximum rpm for sequential injection.

If during dyno testing we find from our pulse width meter that the injectors are going beyond an 80% duty cycle at any engine speed, we will have to fit larger injectors or increase the fuel pressure. However, remember that the maximum permissible pulse width increases at lower rpm.

Maximum pulse width (milliseconds)

rpm	*70%	80%	85%
4500	18.67	10.67	11.33
5500	15.27	8.73	9.27
6500	12.92	7.83	7.85
7500	11.20	6.40	6.80
8500	9.88	5.65	6.00
9500	8.84	5.05	5.37

*Indicates sequential injection; the 70% duty cycle is suggested as a maximum for best hp. However, the injectors can stay open for an 80 or 85% duty cycle providing 16ms is not exceeded.

Note, however, that in general injectors will not pulse for longer than 16 milliseconds. This usually is not a concern with naturally aspirated batch fire engines, but supercharged batch fire engines, which produce very high torque at low rpm, can run into trouble here, as can all engines running sequential injection.

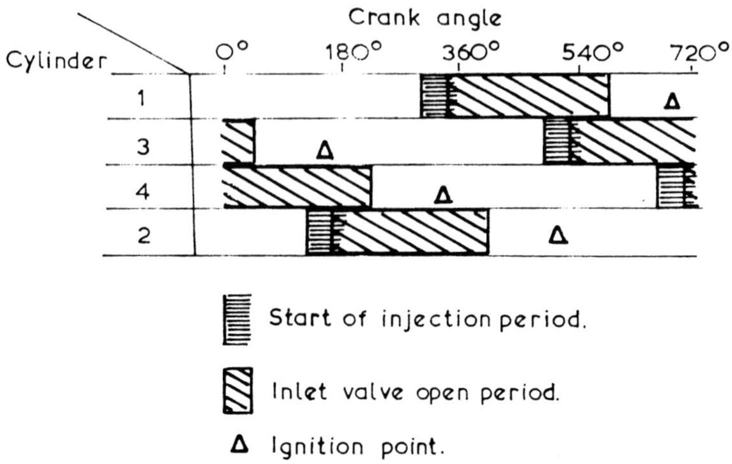

Fig. 13.3. Production car sequential systems typically begin spraying each injector just prior to the inlet valve in that cylinder opening.

open, stalling almost closed, etc.). This means that if the injectors have a factory rated static fuel flow of 390cc/min at 2.7bar (39.2psi) fuel pressure, the maximum quantity of fuel that one injector can flow at this pressure and an 80% duty cycle will be 312cc per minute (390 x 80% = 312).

However, it will often be found during dyno testing that an engine will make best power with correctly phased sequential injection (*Figs. 13.3* and *13.4*) when the duty cycle is 60–70%. A shorter duty cycle of around 40–50%, phased to occur while the inlet valve is open, is often too short for best fuel atomisation and engine cooling. Note here that these remarks apply to sequential injection only; I cannot ever recall seeing any power difference in batch-fire systems at duty cycles ranging from 65% right up to 80%.

Fig. 13.4. Race car sequential systems are often phased to finish spraying just before the inlet valve closes.

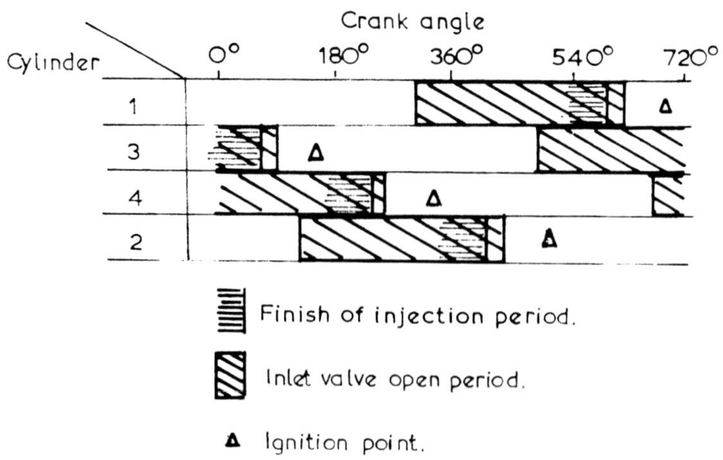

Taking this matter of injector duty cycle a step further, it should be obvious that if we are using a sequential injection system and we are desirous of obtaining maximum possible hp output, we should be looking at running injectors large enough, and with a fuel pressure high enough, to permit dyno testing at a 60% duty cycle. There is not much gain, perhaps no gain at all in running a sequential system in a race engine and using fuel injectors so small, or fuel pressure so low, that you have to go up to a 75–80% duty cycle to get sufficient fuel into the engine at high engine speeds.

CALCULATING INJECTOR FLOW

To determine the approximate amount of fuel flow needed per cylinder for engines burning petrol, I use the following formula:

$$\text{Fuel flow (cc per minute)} = \frac{\text{HP x K}}{\text{C}}$$

where HP = maximum horsepower
C = number of cylinders
K = 5.6 for forced induction engines

Therefore a 280hp engine with four cylinders requires about 392cc of petrol per minute per cylinder to produce that hp.

The next step is to take a look at what size injector/s we require to provide that flow of 392cc/min while keeping the duty cycle down to our required figure, using the formula:

$$\text{Injector static flow} = \frac{\text{TF x 100}}{\text{N x M}}$$

where TF = theoretical flow
N = number of injectors per cylinder
M = injector duty cycle

Assuming that we wish to use sequential injection with one injector per cylinder, and we want to test for best hp at duty cycles down to 60%, the required injector static flow would be:

TF = 392
N = 1
M = 60

$$\text{Injector static flow} = \frac{\text{392 x 100}}{\text{1 x 60}} = \text{653.3cc/min}$$

FUEL PRESSURE AFFECTS INJECTOR FLOW

But before we rush off to purchase four injectors with a static flow of at least 653.3cc/min, we need to consider that such large injectors cost a considerable sum of money, and we can perhaps achieve comparable flow from smaller injectors by upping the fuel pressure. Most of the manufacturers state the static flow of their injectors at a

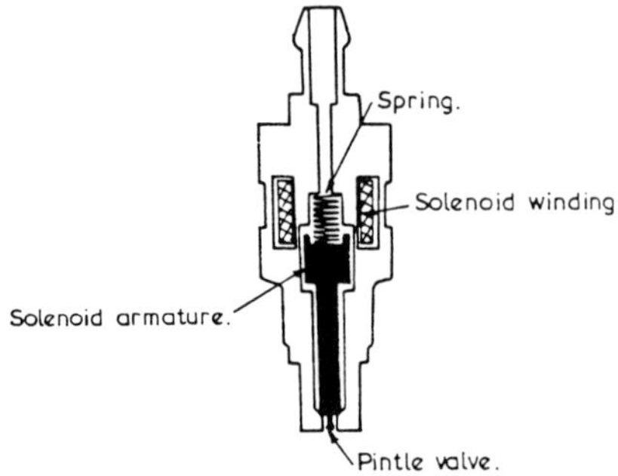

Fig. 13.5. Bosch fuel injector internal construction illustrates why, when fuel pressure is raised, the solenoid is unable to raise the pintle as rapidly.

fuel pressure of either 2.5, 2.7 or 3.0bar. However, we can usually be assured of good flow gains by increasing the fuel pressure up to about 4.0bar. Never push the pressure of stock type injectors past 4.0bar (58psi) otherwise injector life is dramatically shortened, and at 5.0bar injector control could be lost. At any rate I have found that only a few very high flow injectors actually record any flow increase once 4.0bar fuel pressure is exceeded. Smaller injectors often show no flow increase, or not more than a gain of 2–3%, when the fuel pressure is raised from 4.1bar to 4.8bar because the injector takes much longer to open as the pressure increases (*Fig. 13.5*).

At lower fuel pressures (i.e. below 4.0bar) we can estimate the injector static flow with raised fuel pressure using the formula:

$$\text{Revised static flow} = \text{SF} \times \sqrt{\frac{\text{RP}}{\text{OP}}}$$

where SF = injector static flow at manufacturer's standard pressure
RP = revised fuel system pressure
OP = manufacturer's standard pressure

Using this formula we can calculate if the big '351' injectors would be close to what is required. Bosch rate these at 560cc/min at 2.7bar pressure. We will calculate what the flow might increase to at 4.0bar fuel pressure, knowing that these particular injectors will provide good flow gains at higher pressures and so would not be redundant if the engine is developed further at a later stage:

SF = 560
RP = 4.0
OP = 2.7

$$\text{Revised static flow} = 560 \times \sqrt{\frac{4.0}{2.7}} = 681.6\text{cc/min}$$

Obviously the '351' injector should be capable of supplying this engine's fuel needs when operating at a 60% duty cycle. Table 13.1 sets out a list of a wide range of injectors from Bosch, Lucas and ND, and provides a flow estimation chart for various fuel pressures using the above formula. Note, however, that these revised fuel flow rates are estimates only and actual flow may be well down, as is illustrated by the results of testing two injectors at various pressures in Table 13.2.

Table 13.1. Fuel injector flow

Injector		Flow cc/bar	Estimated flow at revised pressure				
			2.5	2.7	3.0	3.4	4.0
Bosch	150208	145/2.7	139	145	153	163	176
	150205	167/2.5	167	174	183	195	211
	150901		180	187	197	210	228
	150203	195/2.7	188	195	206	219	237
	150157	214/2.5	214	223	235	250	271
	150255		226	235	248	264	286
	150013		304	316	333	355	385
	150009	340/2.7	327	340	358	382	414
	150014		360	374	395	420	456
	150803	390/2.7	375	390	411	438	475
	150024		455	472	498	530	575
	150036	480/2.5	480	499	526	560	607
	150351	560/2.7	539	560	590	628	682
	412911	800/2.5	800	832	876	933	1012
Lucas	5207007	147/2.7	141	147	155	165	179
	5207003	164/3.0	150	156	164	175	189
	5207002	188/2.5	188	196	206	220	238
	5208001	188/2.5	188	196	206	220	238
	5207013	201/2.7	193	201	212	226	245
	5207011	218/3.0	199	207	218	232	252
	5208004	237/2.5	237	247	260	276	300
	5208005	237/2.5	237	247	260	276	300
	5206004	260/2.5	260	270	285	303	329
	5207009	299/2.5	299	311	327	349	378
	5207006	323/2.7	311	323	340	362	393
	5207008	368/2.7	354	368	388	413	448
	5208010	422/2.7	406	422	445	474	514
	5207005	503/2.7	484	503	530	564	612
	5207601	503/2.7	484	503	530	564	612
	5207602	540/2.7	520	540	569	606	657
Nippon Denso	195500-2020		536	557	587	625	678
	195500-830		660	686	723	770	835
	N304132500		806	838	883	940	1020

Table 13.2. Effect of fuel pressure on injector flow

	380/3.0 injector		500/3.0 injector	
Fuel pressure	Estimated flow	Measured flow	Estimated flow	Measured flow
2.5	347	353	456	452
2.7	360	364	474	472
3.0	380	389	500	498
3.4	405	401	532	526
4.0	439	423	577	565
4.8	481	429	632	601

FLOW TESTING AND 'RATING' INJECTORS

Really, before running injectors in any forced induction engine tuned to the outer limits, we should have them flow tested at a variety of fuel pressures to ensure that they will give the sort of flow required. Naturally, the injectors should be matched for flow; we don't want more than a maximum 2% flow difference. Ordinary Bosch injectors, for example, have a flow tolerance of ±3% (6% overall) according to the manufacturer, but if you speak to anyone who does injector flow checking they will tell you it can be worse than stated. In a sequential system with individual injector flow trim this is not such a problem, but with a batch-fire system our fuel mixtures end up all over the place.

Another trap to be wary of is the practice in some places of rating injectors as being so many 'horsepower' or so many 'lbs' (pounds). At first glance these ratings appear innocuous, but are they really? To my way of thinking they are quite dangerous, and at best I have seen many people end up with a handful of expensive, useless injectors that are now very second-hand, and as such cannot be returned for a refund or exchange. The problem with the horsepower thing is who rated the injectors. Were they rated mathematically and, if so, was the fact that forced induction engines use more fuel per hp than naturally aspirated engines taken into account, or did the person doing the rating assign the injectors that rating based on outputs he has seen on a dyno? After that we have to question the accuracy of the dyno and the type of fuel used.

In America, injectors are commonly rated in lbs/hr, i.e. the number of pounds of fuel flow per hour, hence the abbreviation of 'lbs'. Sometimes the flow will be further defined as lbs/hr at a certain fuel pressure. This latter practice is much better but still short of what we actually need to know. Injectors flow fuel by volume not by weight, so if rated by weight that figure must then be qualified by also stating the specific gravity of the test fuel. For example, if the injectors were rated at 30lbs/hr using fuel with a 0.72SG they, in fact, flow 4.17 Imperial gallons per hour. However, if we flow those injectors with a high SG unleaded race fuel, typically about 0.78, or methanol with an SG of 0.79, does flow then drop to 3.85 and 3.8 Imperial gallons respectively? Of course it doesn't. The injectors will still flow 4.17gal/hr but, to play another game with such abstract ratings, the same injector could be uprated to 32.5lbs/hr if flowing heavy unleaded race fuel, or to 32.9lbs/hr if flowing methanol, which is even more dense!

Some tuners will argue that rating injectors in lbs makes injector selection easier for them as they just multiply the rating by two and then by the number of injectors. For example, 50lb injectors in a four-cylinder engine, according to their figuring, would be good for 400hp. Rating injectors in this fashion is dangerous; that sort of simple calculation will get you close in naturally aspirated engines, but it is just not on when forced induction is involved. If you are determined to stick with this method, then multiplying the lbs/hr by 1.65, then by the number of injectors will be closer.

LOW FUEL PRESSURE HURTS ATOMISATION

While we are discussing injector flow rates at various system pressures, keep in mind that if you select very large injectors and you wish to keep the injector spray period relatively long, you can lower the fuel pressure to reduce the flow rate. However, the pressure must not be reduced below 2.0bar otherwise fuel atomisation suffers. At this point any possible improvement in atomisation by increasing the injector pulse width will be lost owing to larger fuel droplets issuing from the injector tip when fuel pressure is dropped below 2.0bar.

FUEL PRESSURE REGULATORS

Obviously, to ensure precise fuel flow through the injectors it is necessary to carefully regulate fuel pressure in the system, otherwise at higher fuel pressure the engine would run rich, and at lower pressure it would be lean. Therefore, a fuel pressure regulator is required. There are five main types available to suit a variety of applications: manifold pressure referenced with fixed base pressure, manifold pressure referenced with adjustable base pressure, non-referenced fixed pressure, non-referenced adjustable pressure, and adjustable rising rate manifold pressure referenced.

The advantage of manifold pressure referenced regulators, with either a fixed or adjustable base pressure, is that they increase the dynamic flow range of the fuel injectors. At lower rpm, and when manifold vacuum levels are high, this allows for a

An adjustable fuel pressure regulator referenced to inlet manifold pressure. The fuel pressure gauge allows fuel pressure to be easily checked and adjusted to rapidly alter A/F ratio at drag strip.

longer injector pulse width, with resultant improvements in fuel atomisation. In supercharged applications, because this type of regulator increases fuel pressure as manifold pressure rises, smaller injectors are able to flow at a higher rate to meet the engine's fuel needs.

To help you understand how this comes about, consider the situation when a non-referenced fixed pressure regulator is fitted. A road car at idle or cruise may have an inlet manifold vacuum level of around 15 inches of Hg, or 7.4psi below atmospheric pressure. Obviously, in this situation, with inboard injectors – i.e. injectors located after the throttle plate(s) – the effective fuel pressure at the injector tip will be 7.4psi (or about 0.5bar) higher than the pressure set by the pressure regulator because the high manifold vacuum will actually draw fuel through the injectors during their open cycle. Therefore, if the pressure regulator is fixed at 2.5bar (36.6psi), rather than flowing as if they have a 36.3psi fuel rail pressure, they will flow as if they have 43.7psi fuel pressure (36.3 + 7.4 = 43.7psi). This, in effect, means that the injectors will be flowing about 10% more fuel. We can overcome this by programming the ECU to reduce the injector pulse width. In a road car this is not a good idea as such a move will reduce the intake charge quality because fuel metering becomes erratic when injector spray time is cut down to around 2 milliseconds. (Some Bosch injectors will go a little slower and Lucas injectors will go down to about 0.9 milliseconds, providing the ECU's injector driver will go as slow; many ECUs will not go under 1.5 milliseconds.)

When the engine is supercharged, the situation regarding injector fuel flow reverses when the manifold is experiencing full boost. For example, if maximum boost is 16psi (1.1bar) the injectors will now flow as if they have only 20.3psi (36.3 – 16) fuel pressure. In effect, injector flow will decrease about 25% because manifold pressure will be restricting flow out of the injectors. This problem could be overcome by fitting larger injectors, but then we would have to cut back on injector pulse width at idle and cruise, which would really adversely affect intake charge quality.

To avoid this undesirable situation we can choose a manifold pressure referenced regulator. This diaphragm-operated pressure regulator maintains fuel pressure at a fixed figure, as the injectors see it, relative to either manifold depression or manifold boost. One side of the regulator diaphragm senses fuel pressure, while the other side is connected to the inlet manifold to react to the prevailing vacuum level, or prevailing boost level. A spring is fitted in the engine vacuum side to establish a nominal pre-load of pressure. If this pre-loaded pressure is, for example, 2.5bar and the inlet manifold

Fig. 13.6. A manifold pressure referenced fuel pressure regulator.

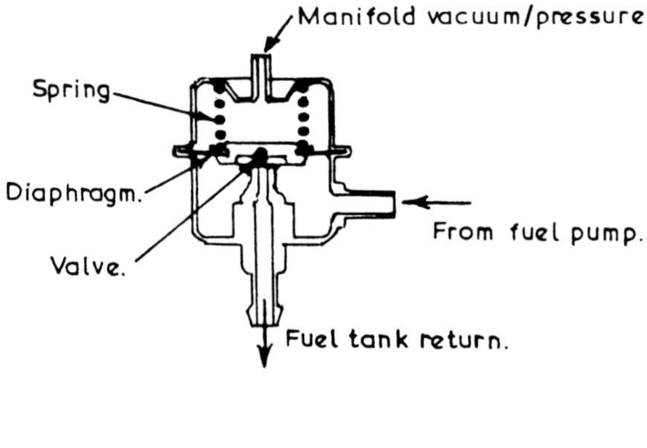

depression at cruise is 15 inches of Hg, this will draw against the spring and reduce its nominal pre-load from 36.3psi down to 28.9psi (36.3 – 7.4), so fuel rail pressure drops to that figure. Then, when manifold boost increases to 16psi, this pressure works in unison with the spring, increasing its nominal pre-load from 36.6psi up to 52.3psi (36.3 + 16), and fuel pressure in the fuel rail will rise to this figure. With this type of regulator, regardless of whether the inlet manifold is experiencing 20 inches of Hg vacuum or 30psi boost, the injectors will flow exactly the same quantity of fuel (*Fig. 13.6*).

In a naturally aspirated competition engine we can choose to ignore such considerations and use either a non-referenced regulator, or more usually an adjustable pressure non-referenced regulator. With the latter option we can very quickly increase or decrease engine fuelling without having to reprogramme the ECU. So, if the track is wet we can back off the pressure a little to lean the mixture and reduce the risk of plug fouling. Conversely, if the humidity takes a dive we might up the pressure to avoid detonation damage.

RISING RATE REGULATOR

The final category of regulator, the adjustable rising rate manifold pressure referenced type, is used in low budget low boost applications (up to about 7psi) when forced

Fig. 13.7. An adjustable rising rate manifold pressure referenced regulator is inserted in the fuel return line after the stock regulator.

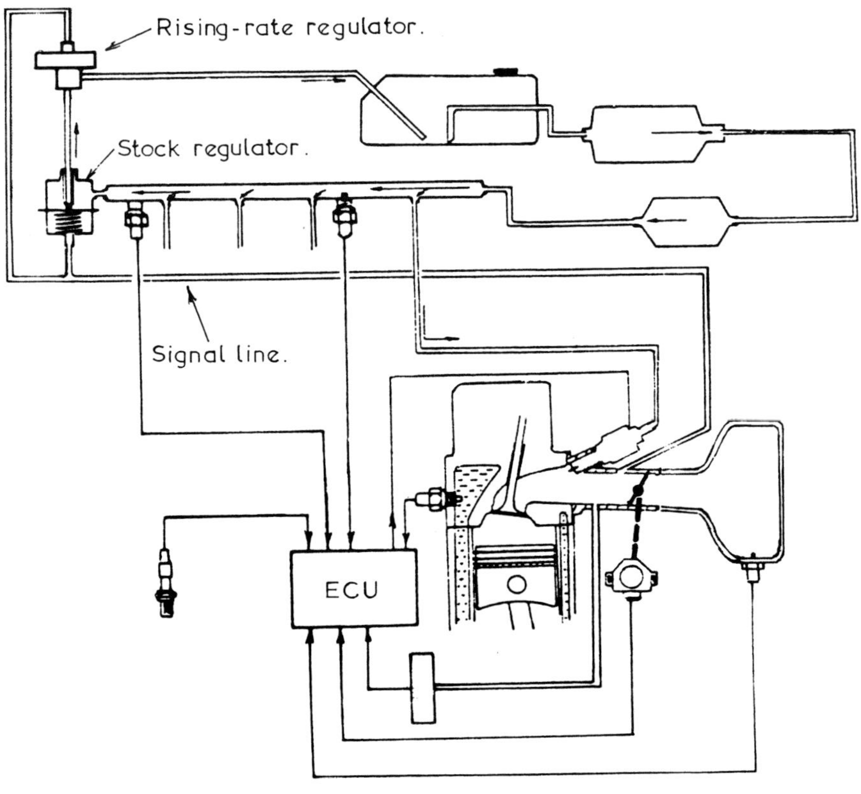

induction is added to an otherwise stock naturally aspirated engine and the client does not want to reprogramme or replace the factory ECU. This type of regulator is added in the fuel return line to override the pressure regulation provided by the stock manifold pressure referenced regulator (*Fig. 13.7*). As such, the stock regulator controls fuel pressures up to inlet manifold pressures of almost 0psi, but as soon as boost builds past 0psi the rising rate regulator takes over, increasing fuel pressure by restricting the quantity of fuel returned to the fuel tank.

The rising rate regulator has two adjustments which are best carried out on a wheel dyno while observing an air/fuel ratio meter. Because many stock cars have a restrictive intake system, it is not at all unusual to record a manifold vacuum of 4–5in Hg (more than –2psi) at wide open throttle. Therefore, when we add some form of supercharging to a car like this, any manifold pressure higher than –4in Hg will in effect be boost pressure, and fuel mixtures will begin to become lean long before manifold pressure climbs above 0psi. To compensate for this situation the rising rate regulator has a spring-loaded centre adjusting screw to set the point at which it takes over from the stock regulator and begins to increase fuel pressure.

Some factory naturally aspirated cars, though, run very rich at wide open throttle, and at a manifold depression below 0psi. Some Fords, for example, are set to run air/fuel ratios down around 12.3:1 even though they make best power closer to 13:1. Consequently, in a situation such as this, when supercharging is added, the point of pressure gain adjustment may be set at a little over 0psi to bring the air/fuel ratio back on track to obtain a power and economy benefit.

The other adjuster is a needle valve which is turned to regulate maximum fuel pressure so that mixture strength is correct at full boost pressure. After making both adjustments we might end up with a fuel pressure/boost pressure curve like that illustrated in *Fig. 13.8*.

Clearly, the rising rate regulator is at best a band-aid solution to providing the quantity of fuel the engine actually requires. Once we adjust the starting and finishing point we basically have no other avenue open to control mixture strength at intermediate boost pressures. As I've already mentioned, there is no guaranteeing substantial flow increase from stock injectors at fuel pressure over about 60psi. Hence massive pressure

Fig. 13.8. Rising fuel pressure provided by rising-rate regulator.

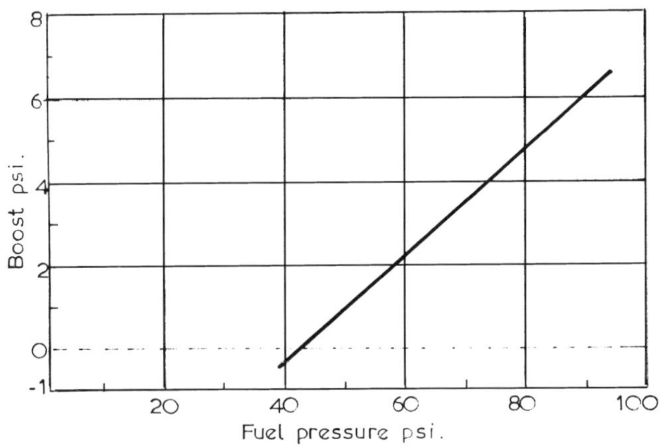

Hidden under the car or in the fuel tank it is easy to forget that the stock fuel pump may not be up to supplying the modified engine's fuel flow/pressure needs. Equally, a forgotten fuel filter could be limiting fuel flow.

increases are necessary to obtain relatively modest flow improvements. Because the injectors are only subjected to these high pressures for a few seconds they will usually continue to function well, but don't ever push Lucas injectors over 90psi or others over 110psi. However, don't just assume that the stock fuel pump will cope. If during the dyno session adjustments to the needle valve on the rising rate regulator do not appear to be yielding the desired increase in mixture richness, do check that the fuel gauge is recording an increase in fuel pressure. If it isn't, then the fuel pump is not up to the task. If fuel pressure is rising steeply, as illustrated, in union with increasing levels of boost, then the injectors are probably not suitable.

FUEL PUMP AND FILTER

Fuel is supplied to the injectors by a high-pressure roller-type electric fuel pump. This pump must be large enough to deliver more fuel than the maximum requirement of the engine so that full fuel pressure is maintained at wide-open throttle and maximum rpm. On leaving the pump the fuel passes through a high-flow fuel filter containing a paper element, which traps anything in the fuel that would otherwise block the injectors. Note that to avoid flow restriction the filter must be replaced regularly; be sure to fit it with the flow direction arrow pointing the correct way. After exiting the filter the fuel continues on through the fuel line up into the fuel rail and to the individual injectors. This all sounds simple enough but, obviously, to ensure sufficient flow of fuel at the fuel rail while maintaining the correct pressure we have to take care to choose the correct fuel system components, and then correctly install them.

First we must calculate the engine's maximum expected fuel flow at wide-open throttle and maximum rpm. We want to choose a pump that can over-supply the engine's

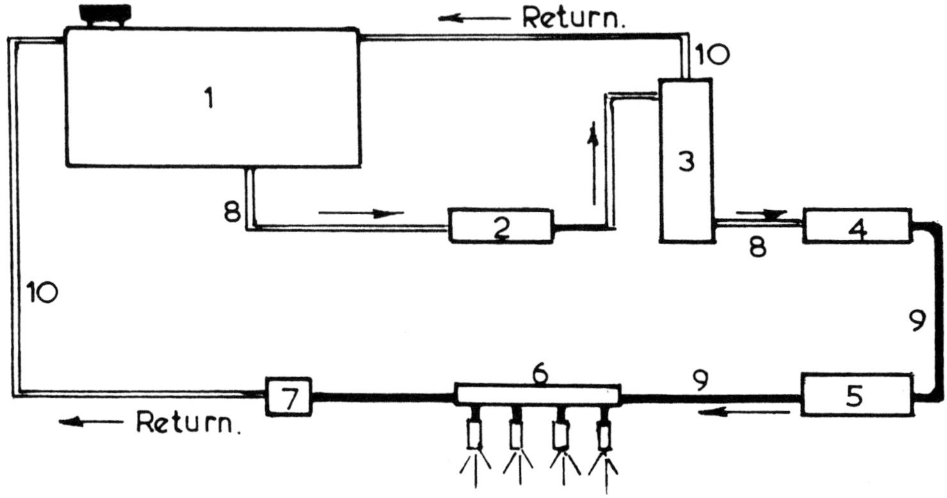

Fig. 13.9. Fuel system plumbing. The components illustrated are: 1. fuel tank; 2. swirl pot feed pump(s); 3. swirl pot; 4. pressure pump; 5. fuel filter; 6. fuel rail and injectors; 7. fuel pressure regulator; 8. suction line(s); 9. high-pressure fuel supply line(s); 10. excess fuel return line.

needs, but we have to be careful not to go too big, otherwise if return fuel is routed back to the swirl pot rather than the fuel tank we would have a lot of fuel flowing in a closed loop gaining more and more heat with each pass through the fuel rail (*Fig. 13.9*). With this in mind I work on a 30% theoretical over-supply using the formula:

Fuel pump flow (cc per minute) = HP x K
where HP = maximum horsepower
K = 7.3 for forced induction engines

Therefore, an engine producing 280hp would require a fuel system capable of flowing 2044cc (approximately 2.0 litres) per minute. Note here that we are not talking about 2.0 litres free flow into an open container, but rather 2.0 litres flowing from the pump at fuel system pressure. When system pressure is regulated higher, pump flow is reduced because the pump has to work harder to push fuel into the fuel line. Typically a system regulated to 3.0–3.5bar fuel pressure will see the free flow figure reduced by 20–30% and at 4.0–4.5bar it will be down 25–35%. Now, if we were to add a supercharger pumping 1.0bar boost, and this is referenced to the pressure regulator, we would see fuel rail pressure rising from 4.0–4.5bar to 5.0–5.5bar. At this pressure pump flow would fall by around 33–45%.

As you can see looking at Table 13.3, flow drops off much more rapidly with stock type pumps than pumps designed for high pressure/high flow applications. This highlights the need for considerable caution in all high-boost applications, and especially so in all applications where a rising rate pressure regulator is installed. In this latter situation, pump flow at maximum boost can easily drop to only one third the flow at the stock fuel pressure of around 3.0bar.

Avoid losing octane

Most racers know that high-octane race fuels, and in particular unleaded fuels, can go 'stale', losing their octane-boosting components to the atmosphere. So they take care to keep all fuel drums, etc., tightly sealed to prevent these additives, often aromatic hydrocarbons, from boiling off their 'light ends'. However, it is easily forgotten that a similar process is occurring in a race car's fuel tank when the fuel is allowed to heat up and vent octane-boost compounds to the atmosphere. Thus, as the race progresses the fuel will suffer a deterioration in octane level, and if the ECU has not been programmed to compensate for fuel temperature changes, the engine will experience both a leaning of the fuel/air mixture and a loss of octane. Clearly, hp will be lost at least, or worse the engine may be damaged. The fuel tank must, therefore, be shielded from heat radiating from the exhaust, brakes, transmission and tyres. Additionally, fuel pump over-capacity must be kept in check otherwise large volumes of fuel heated on its circuit through the engine bay will be dumped back into the tank. Some racers fit a fuel cooler into the fuel return line, but such a cooler must not be installed unless fully protected from stones, track debris, etc., which may puncture it and create a massive fire hazard under the car. Really, a better location for the cooler is within the car, close by the fuel tank, and with a meshed-over duct flowing cooling air across the cooler.

Table 13.3. Fuel pump flow rates

Fuel pressure (bar)	Fuel flow (litres/min)		Flow decrease (%)	
	Pump A	Pump B	Pump A	Pump B
0	2.9	4.2		
3.5	2.0	3.4	–31	–19
5.0	1.76	3.03	–39	–28
7.0	0.78	2.65	–73	–37
8.3		2.06		–51

Pump A – Typical six-cylinder street car Bosch pump
Pump B – Bosch Motorsport pump 0580 152 460

MULTIPLE PUMP ADVANTAGES

To obtain more flow volume we can either install a large fuel pump or we can opt for two or more pumps connected in parallel. Large pumps are expensive, and on a road car all that extra fuel pumping back to the tank when you are cruising will push up the fuel temperature, and drain a lot more power from the electrical system. Usually several smaller pumps are preferable if weight isn't a large concern. With multiple pumps in a road car it is best to wire each pump to a separate relay and bring the extra pumps into operation only when additional flow and pressure are required. This is easily achieved via a Hobbs type pressure switch, connected to the inlet manifold, or by a cam on the throttle spindle closing a microswitch to activate the pump relay(s).

The other advantage of multiple pumps is that you can pick them up cheaply at breakers' yards specialising in European cars. In the early 1980s most of these cars used a Bosch mechanical injection system operating at 4.0–6.0bar pressure. Consequently, these pumps are ideal in supercharged cars as there is less flow decrease at higher pressures (typically similar to the percentage decrease in the Bosch Motorsport pump). Apart from the pumps listed in Table 13.4 you will find Bosch high pressure pumps with flow ratings similar to those indicated for the Mercedes 280 on '80s cars such as the Audi 4000 and 5000 series, Saab turbo, and Volvo 260 series. Later model turbo cars with electronic fuel injection are also a viable source for pumps, but at higher pressures their flow will fall off more rapidly like Pump A in Table 13.3. Japanese cars usually carry Nippon Denso pumps of similar design to those from Bosch, and like the Bosch units they are extremely reliable.

Table 13.4. Bosch fuel pumps

Pump number	Stock fitment	Free flow (litres/min)	Flow @ 3.0bar (litres/min)
0580 464 033[1]	many six-cylinder cars	2.9	2.3
0580 254 942[2]	*Mercedes 280	2.9	2.3
0580 254 975[2]	*Mercedes 380, 450, 500 & 6.9		
	*1981–83 Porsche 911, 924 & turbo	3.3	2.7
0580 254 984[2]	*Porsche 928	3.7	3.1

[1]Barbed inlet and outlet.
[2]Barbed inlet and 12mm thread outlet.
*'80s models with K–Jetronic mechanical injection (some with Lambda sensor).

Fuel delivery has to be taken seriously. Not visible are the two Bosch Motorsport pressure pumps and fuel filter mounted below the fuel tank.

TESTING FUEL SYSTEM FLOW

Some believe that you have to bundle up your entire fuel system and truck if off with the engine into the dyno room to determine if the pump will flow enough fuel at system pressure. Accordingly, the engine is set-up with the car's fuel system laid out in the dyno cell to take into account flow restrictions caused by the fuel filter and fuel lines. Next they disconnect the fuel tank return line and run it into a container to see if at full throttle and maximum rpm there is an excess of fuel at the fuel rail, indicated by fuel flowing from the return line. Apart from the danger of all that fuel in the dyno room, such a test is really a waste of time. You can easily do a quick, safe test in your own garage, which will give a good indication of flow at the fuel rail.

For this test you need to have the entire fuel system installed in your car. The engine does not have to be in place, but you will need to have the fuel rail and injectors connected. The fuel pressure regulator must be adjusted to the pressure at which you intend to operate the system. If it is a supercharged installation with a manifold pressure referenced regulator, you will need to pressurise the regulator to normal boost pressure. This can be done easily by fitting a 'tee' to the regulator; a pressure gauge is attached to one branch of the tee, and a tyre valve to the other to enable a tyre pump to apply the appropriate 'boost' to the regulator. Now disconnect the tank return hose and direct it into a graduated measuring container. Energise the fuel pump to fill the fuel system and get it up to pressure, then empty the measuring container and activate the fuel pump for exactly two minutes. The quantity of fuel collected divided by two is your system's fuel flow per minute. Obviously, the flow must not be less than what you have previously calculated.

The only problem with this test arises if the car's electrics are a bit wimpish. Because the only load is what is being drawn by the fuel pump, the pump may be receiving a voltage boost compared with when the system also has to power the ignition, cooling fan, injectors, transmission oil pumps, etc. You can estimate a reduction in fuel flow of

The swirl pot should be supplied, preferably, by a pump similar to the pressure pump, and be drawing from multiple fuel pick-ups.

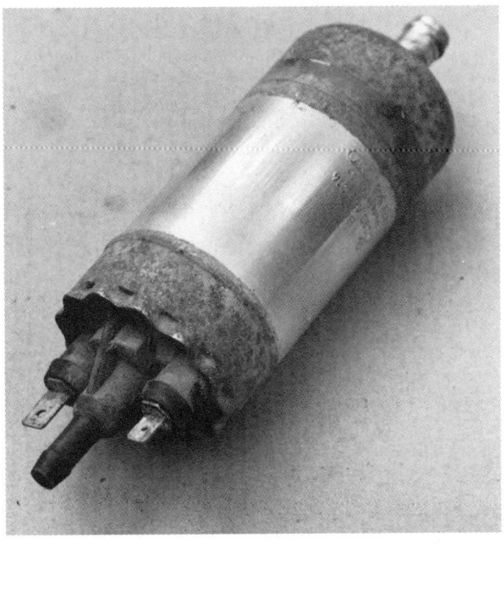

around 12% for every drop of 1 volt at the pump; for example, if voltage at the pump is 10.6 volts rather than 12.6 volts, flow will be down 24%. The way to avoid this unwanted trouble is to ensure that you use good-size wiring, good connections and ample alternator capacity.

GETTING FUEL TO THE PUMP

However, getting fuel to the injectors is only part of the equation – we also have to get fuel to the fuel pump. In a race car with a fuel cell and correctly positioned fuel pick-ups, this is not such a problem, but what if you are running a normal steel fuel tank? As the fuel level falls it may drop below the fuel pick-up when subjected to strong G forces caused by cornering, braking and accelerating. At that time the fuel pump will pump air; fuel pressure will drop and the engine will stumble. Even fuel-injected road cars with well-designed fuel pick-up systems in place in the fuel tank can come up against this sort of trouble when converted for competition use.

The simplest solution is to run a second fuel pump that pumps into a small-capacity swirl pot, which should be large enough to supply the engine's fuel needs for at least half a minute (*Fig. 13.10*); for our 280hp engine, that means about 1 litre. As it is not subjected to pressure it can be fabricated quite easily from 3in tube. Exhaust tubing is acceptable, but if it is to look pretty you might prefer stainless steel or aluminium.

I prefer to use the same type of pump to supply the swirl pot as I would run for the pressure pump, but remember that these pumps do not like to lift fuel, so they should be mounted down low beside the fuel tank. To this pump would be connected either two or four pick-ups depending on the G forces the car was capable of generating, and whether it would run in long distance events where the ability to draw up every last drop of fuel is so important. Alternatively two or four low pressure pumps could be used. Using only two pumps, pick-ups would be positioned in the rear of the tank on both the right side and the left side. With four pumps, pick-ups would be added to both sides in the front of the tank as well. Some, however, choose to use a small pump to feed the swirl pot, and run the fuel

Fig. 13.10. Fuel swirl pot.

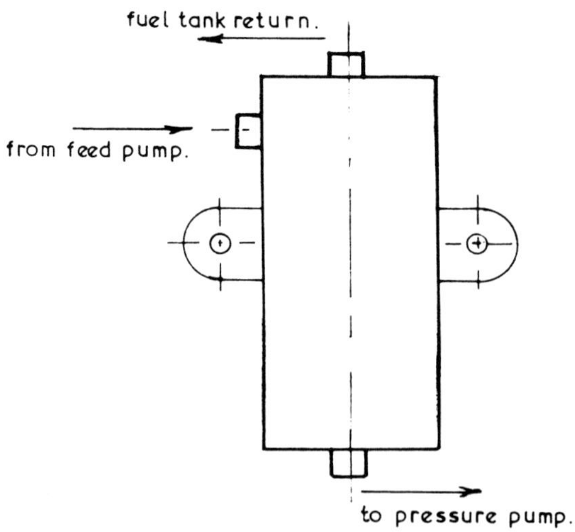

return line back to the swirl pot rather than to the fuel tank. I do not believe that this is such a good idea for two reasons: first, unless the swirl pot is fairly large it can run dry at high engine loads and high rpm, and second, at lower engine loadings the fuel will get hot circulating in a closed loop and the engine could become dangerously lean.

SUCTION LINE, PRESSURE LINE AND FUEL RAIL REQUIREMENTS

Both the suction fuel line from the fuel tank to the pressure pump and the pressure line from the pump to the fuel rail must be of adequate size and with bends of sufficiently generous radii to not unduly restrict fuel flow. Table 13.5 sets out the approximate hp levels that various sizes can support. Note that when stainless steel braided hoses are used these will have an inner diameter smaller than that indicated in the table, but as only small lengths are involved this should not be a concern. Incidentally, these hoses are identified by a dash number representing $^1/_{16}$in, hence a '–6' hose is for connecting to a $^3/_8$in ($^6/_{16}$) hardline tube. On the pressure side, a hose of this size will flow enough fuel for a 500hp engine.

Table 13.5. Fuel line fuel flow

Tube od	Tube id	hp pressure	hp suction
$^1/_4$	.180	165	110
$^5/_{16}$	.243	300	200
$^3/_8$	.319	500	335
$^1/_2$	.444	1000	675

When considering fuel line flow capacity, do not overlook the fuel rail. Compared to the fuel line the fuel rail always appears to be generously oversize, and consequently unlikely to create a flow bottleneck. However, because the fuel is forced to take a sharp 90° turn from the fuel rail into individual injectors, severe flow disruption occurs. The only way you can provide a more generous flow path is to make the fuel rail larger. A fatter fuel rail not only opens up the fuel flow path (this is like being able to swing into a narrow supermarket parking bay from a wide double lane path as compared to turning tightly into the bay off a narrow single lane path), it also considerably increases the volume of fuel stored under pressure and ready to flood down to each injector. When an injector opens there is a slight instantaneous drop in fuel pressure, something like what can occur if someone turns on another tap while you are taking a shower and the plumbing is too small. In a batch-fire system with all injectors flowing simultaneously the pressure drop is even more dramatic. An ordinary pressure gauge will not indicate the drop; it's far too rapid for a heavily damped gauge to detect. However, this pulsing in the fuel rail appears to create flow turbulence that can only be successfully moderated by resorting to larger fuel rails with dual fuel inlets.

ANOTHER LOOK AT A DUAL FUEL ARRANGEMENT

Let's backtrack now to my comments in Chapter 11 concerning the merits of a dual fuel system. At that point we were considering the advantages of such a system in that it allowed us to run cheap ordinary fuel in the main fuel tank and primary fuel system to supply the car's needs at cruise and low boost, and also enjoy the benefit of operating on something such as toluol or Avgas from a secondary system to care for high boost, high engine load situations. Now I want to move that argument to a new plane while keeping these earlier thoughts in view.

From this chapter's beginning you have been able to appreciate that forced induction demands considerable increases in fuel flow. We have looked at big injectors, fuel pumps and fuel lines, and there is the problem of reprogramming the ECU so that all this hardware gives the engine exactly what it needs in the way of fuel flow quantity to make big hp and avoid engine carnage. That all costs money and at the end of the project you may have a box filled with expensive stock parts of little value to anyone. In fact, the entire fuel system, plus the engine management system, could have been made redundant.

Could all or much of that waste have been avoided? Generally, in the case of road cars, I would say yes. With competition cars it could also be yes, but it will depend on the regulations under which you compete and whether you want data logging as most stock ECUs will not support data logging.

As previously explained, the preferred dual fuel arrangement has, in addition to a dedicated fuel tank, fuel pump, fuel pressure regulator and fuel plumbing system, a full set of extra injectors and a simple fuel only ECU programmed for their control.

FUEL INJECTOR POSITIONING

In stock engines we are used to seeing injectors mounted close to the cylinder head, or even in the head, and spraying right at the back of the hot inlet valve. This is fine as it serves to stop the fuel dropping out of suspension and not burning properly at low rpm or when air speed in the inlet tract is slow. Also, inboard injectors improve low-speed and mid-range rpm throttle responsiveness, and in batch-fire systems they stop charge robbing.

However, when we add an extra set of injectors programmed to operate when the engine comes on to boost (in the case of cars previously naturally aspirated) or when higher boost levels are reached (cars with low pressure supercharging from the factory) we can mount these injectors further away from the head. This produces a number of advantages. For example, in cars with an individual throttle plate in each inlet port mounting the injector outboard (i.e. before the throttle plate) ensures that some fuel droplets entering the air stream will smash into the throttle plate and be broken down in size. In all types of inlet systems, fuel droplets will collect some heat on their relatively long journey down the inlet tract. This has two desirable effects: it aids fuel vaporisation and it also pulls heat out of the inlet tract and thereby improves charge density. Additionally, introducing the fuel into the inlet stream a long way from the inlet valve allows extra time for the fuel droplets to more evenly mix with the air as it swirls around and down the inlet tract. Therefore, with improved fuel atomisation and better mixture homogenisation, we gain some power as a result of better combustion.

When a broad power band is desired it is better to have one set of injectors in the stock location and the second set mounted much further out. On the other hand, a maximum effort engine may work better with both injectors mounted outboard of the throttle plate (*Fig. 13.11*). Always the primary injector mounted closest to the head would be staged to fire first, while the secondary injectors mounted further out would spray only at higher rpm and engine loads.

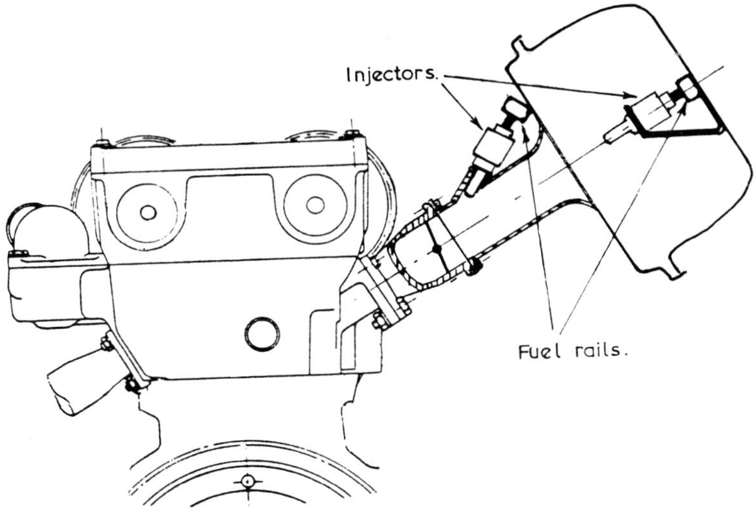

Fig. 13.11. A dual injector arrangement.

STAGING PRIMARY AND SECONDARY INJECTORS

It should be noted that some stock injectors are not suitable for use as secondary injectors; they can only be used to spray at the back of the inlet valve. This type has a very wide spray pattern and, as such, when mounted out along the inlet tract they do a fine job of spraying fuel onto the manifold walls. Hence, what left the injectors as a fine spray mist reverts to a stream of fuel merely trickling into the cylinder. Not exactly what is wanted for good combustion, low emissions and good cylinder and lube oil life!

Some people spend a lot of time figuring out exactly when each injector should fire and exactly what fraction of the engine's total fuel flow each injector should provide at various rpm, boost pressures and load. Others simply run the stock injectors full time on the factory programme and only bring the secondary injectors on line when the engine requires more fuel. This latter arrangement is easy to programme, but it in no way takes full advantage of the system's true power potential. The former staging arrangement is the way to go, but only if you have unlimited amounts of money to throw about.

What will achieve excellent results without a huge outlay is to simply pick an engine speed and load level, and determine that at that point the primary injector's pulse width will be cut by anything from 25–40%. Reduce the fuel by closer to 35–40% in engines quickly making high boost at low rpm, and by 25% if they gain boost more slowly and at higher rpm. The fuel cut from the stock injectors is then provided by the secondaries, and their flow will progressively increase as engine rpm and load continues to rise. Depending on how much octane the engine requires at maximum engine load, at engine revs of approximately 1000–2000rpm below max torque revs right up to the redline, anything from 25–85% of the engine's fuel requirements would be provided by the secondary injectors. Obviously, at less than full load there would be a corresponding reduction in their flow. However, at no time should the duty cycle of any injector exceed 80% (except for Lucas disc valve injectors which are all right up to 90%).

To prevent a stumble at the time when flow in the primaries is reduced and the secondaries begin to spray it will be necessary to provide a spray overlap period. Therefore, as the time for the cut is reached the secondaries must be allowed to pulse twice, and then following that primary flow may be cut by the amount decided upon.

The other point to keep in mind is that you want the secondary fuel pump purging hot fuel out of the fuel rail well before the system begins to spray. In a competition car the pump would be kept working all of the time. In a road car we don't want that as we will simply boil the fuel in our small secondary tank. What works well here is to activate the pump relay via a micro switch which is closed by a cam attached to the throttle plate spindle. Position the cam so that the system comes up to pressure and is purged in advance of the injectors beginning to operate. Also do all you can to protect the fuel rail from heat by wrapping it in the type of heat shield tape used to protect air-conditioner flex hoses from exhaust heat.

INJECTOR RESISTANCE AND ECU MATCHING

If there is no alternative but to run a single fuel system, whether it be with one, two or three sets of injectors, there are a number of factors which must be kept in mind regarding the matching of injectors to the ECU. The first problem is that stock type ECUs generally will only successfully drive a limited number of injectors, and injectors with a certain electrical impedance. Therefore, if the stock ECU is designed to drive small high-resistance injectors (typically 12–16 ohms) then it will baulk at operating big low-impedance injectors (typically 2–5 ohms) or very-low-resistance units (under 1 ohm).

There are ways around this to allow the retention of the stock ECU. For some units the answer is to add a driver box (the electronic equivalent of a relay) to handle the higher current needed to operate the injectors. Nissan take a different approach in the Skyline GT-R. It uses 2.5 ohm 420cc/min injectors in conjunction with a ballast resistor in the wiring harness to stop the injectors drawing too much current from the injector drivers.

Aftermarket ECUs also have their limitations. For example, a unit advertised to drive up to eight injectors may, in fact, be limited to either a maximum of eight high-impedance injectors or four low-impedance squirters.

Chapter 14

The air inlet system

Many enthusiasts, and even some tuners, have got the idea that the basis for improved performance from a forced induction engine is more boost pressure, or a bigger hit of nitrous oxide. However, this is false reasoning. Consider for a moment the CART Champ cars in the USA. The purpose-built 8-cylinder race engines are restricted to a capacity of 2.65 litres and a single turbo pushing a maximum of about 3½psi boost (37in Hg absolute) and yet they produce almost 900hp. In other parts of the earth simple 2-valve push-rod 5.7-litre V8s are making over 1100hp running twin turbos and a maximum of 15psi boost. How are such outputs possible? Obviously, it is not as a result of massive boost pressures, but rather because the inlet and exhaust systems and cylinder heads flow efficiently, consequently encouraging good flow into and out of the cylinders. Therefore, I believe the very first modifications should be directed towards improving the breathing of the engine.

In this chapter we want to look at everything affecting gas flow from the point where the inlet air enters the air intake right the way through to the point where the exhaust port terminates at the face of the cylinder head. So we will be examining, with a critical eye aimed at maximising performance, the whole inlet system plus the cylinder head and the camshaft. As we progress through the entire flow path, keep in clear focus that anything that will impede gas flow in a naturally aspirated engine will present an even greater flow restriction in a forced induction engine because the gas volumes involved are so much bigger. Don't ever think that you can beat flow restrictions simply by upping the boost, and likewise don't ever think that increased boost pressures are the key to improved performance. Just take a look at your garden hose for proof of what I am saying. With the tap (faucet) opened right up and the nozzle on the garden hose closed down pressure in the hose is high, but water flow from the nozzle is dismal, or even non-existent. Opening up the nozzle will drop pressure in the hose right down, but flow from the nozzle improves perhaps twentyfold at least. Getting rid of flow restrictions will have a similar positive effect on any engine, by significantly increasing charge density in the cylinders.

This Subaru Liberty B4 air intake looks efficient, but only actual testing will reveal the truth.

CHECKING AIR FILTER AND AIR BOX FLOW

Many see the air filter as a major impediment to flow, but in a road car, before tossing the filter element and replacing it with something that is supposed to flow better, it is wise to do a little investigation. If the stock element isn't blocked by dirt, insects and other rubbish, changing to a so-called high performance element probably won't do much. Frequently it is found that it is the air box, or other sound muffling devices in the inlet duct that are the real culprits in strangling air flow.

Fig. 14.1. A water manometer can be employed to find where flow restrictions exist.

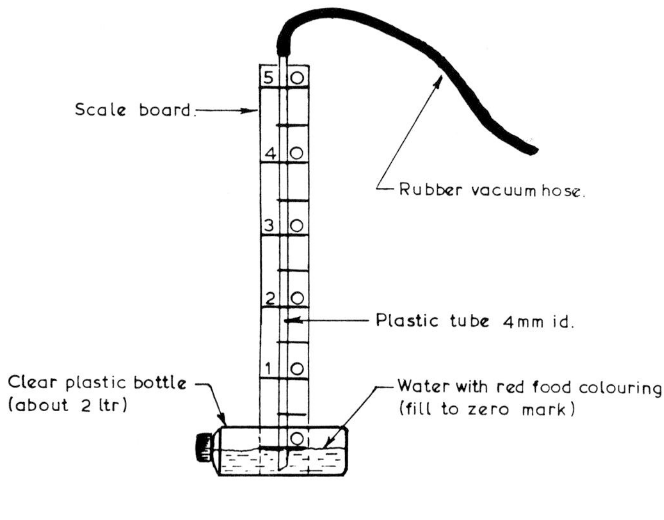

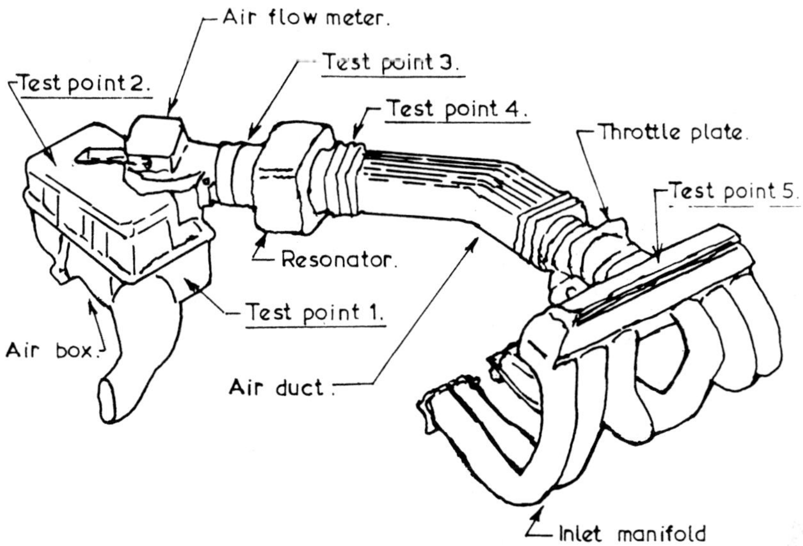

Fig. 14.2. Nitrous oxide assisted engines should be tested, without activating the nitrous, with a water manometer at all five test points to determine what parts of the intake system are causing flow resistance. Turbo and blower engines cannot be checked with a water manometer on the pressure side of the inlet system.
Test point 1 – before the air filter element but after any inlet silencers.
Test point 2 – after the air filter element but before the air-flow meter.
Test point 3 – after the air-flow meter but before the resonator.
Test point 4 – after the resonator but before the throttle plate.
Test point 5 – after the throttle plate.

The easy way to evaluate the inlet system is by means of a flow bench, but an inexpensive and viable alternative is to use a water manometer (*Fig. 14.1*). While this test can be a bit of a fiddle it will help you to discover a good deal about the science of engine breathing, and you will feel challenged to consider flow restrictions possibly existing in the whole inlet system and not just the air filter element itself.

The areas we want to check are shown in *Fig. 14.2*. In naturally aspirated and nitrous assisted engines we can use the manometer to test all the way through to the inlet manifold. However, when a blower or turbocharger is fitted we cannot check out the pressure side of the inlet tract using this method.

Table 14.1 shows the results of tests done on a Subaru Liberty RS turbo while it was running on a wheel dyno. The pressure drop, which can also be referred to as the vacuum or the depression, in the intake system was measured using a 72in water manometer. When you do this test in a car you will be limited to about 42in column height, so past 5000rpm this car would have been off the scale!

When using the water manometer, set your assistant up with it in the back seat. Using full throttle, accelerate from 4000rpm to 7000rpm in 3rd, or preferably 4th, gear up a hill. As you approach each thousand rpm call out to your rear seat passenger. He will mentally record the depression at each call and write the figures down at the end of the run. Extreme accuracy isn't important. For example, if the water level was less than halfway between 20in and 30in, record it as 23in. Again, do three runs at each test point

The innocent looking resonator box located just after the air-flow meter can quite easily cause double the air flow restriction of the stock paper air filter element.

to check for errors. Note that a video camera can be used to record this test, but unless you have another tacho that can be duct-taped close to the manometer you may have problems getting both the manometer and tacho in focus.

Table 14.1. Pressure drop in Subaru intake

rpm	Point 1	Point 2	Point 3	Point 4
4000	12	17	17	20
5000	20	24	26	30
6000	27	32	34	41
7000	36	44	48	60

Point 1 – In the air box before the filter element.
Point 2 – In the air box after the filter element.
Point 3 – In the air duct after the air-flow meter.
Point 4 – In the air duct after the resonator box.

Now, what do these figures mean? Standard air pressure at sea level is 14.7psi which equals 29.92in mercury or 406.9in water. Therefore 1in mercury equals about 0.5psi and 1in water equals about 0.036psi. Clearly, this Subaru has a problem which won't be solved by replacing the air filter element. At 7000rpm there is enough flow resistance caused by the air muffler before the air filter to cause a depression of 36in water. That's a restriction of 1.3psi even before you get to the filter! Again, at 7000rpm the filter restriction is only 8in water (44 – 36 = 8) or 0.29psi. The other major restriction is the little resonator box stuck in the air duct right after the air meter. It is causing a depression of 12in water (60 – 48 = 12) or 0.43psi. In total the system has a pressure drop of 60in before you get to the throttle body (and that's another story). That is almost 2.2psi, so instead of air pressure of 14.7psi to force air into the turbo, this car is working with only 12.5psi. Is that costing horsepower? Well, let's put it this way, by getting rid of that flow restriction it's like getting 2.2psi of free supercharging without adding any heat to the air! In hp terms it means that a 200hp engine when restricted like this is going to be limited to around 175–180hp.

HIGH PERFORMANCE AIR FILTERS

Some will tell you to toss out the intake muffler, the air box and filter and fit a high-flow performance pod-type filter with tapered neck to attach to the air-flow meter. This would be a bad move unless the pod filter is boxed in to prevent hot underbonnet air from entering the intake. The underbonnet air temperature can easily rise by 50ºC (90ºF) above ambient temperature with a turbo engine, and usually about 30ºC (54ºF) for non-turbo engines. You can reckon on a power decrease of at least 1% for every 7ºC (13ºF) the intake air temperature rises. Therefore, if the outside air temperature is 25ºC (77ºF) and the intake air temperature is 75ºC (167ºF) you are giving away 20hp in a 300hp engine. This illustrates why a cold air intake is so important for maximum power. However, at lower engine speeds, particularly during cooler weather, fuel economy may be poorer as a result of inferior fuel atomisation, even with fuel injection.

To give some idea of just how misleading advertised air filter flow figures can be, consider what was found with a 1.3-litre Suzuki Swift Gti modified to produce just over 150hp by using a blower off a 1.6-litre Toyota Corolla. These little engines really get up

A stock air box has been modified with two large inlets drawing cool air from around a headlight. Air enters through a small above bumper grill and is directed by ducting over intercooler to flow behind headlight. The heat shield ensures that no hot air from around the radiator or exhaust manifold can be drawn in.

and go, and looking at the tiny air filter it would seem that the very first modification to obtain an easy 5–7hp gain would be to bin the standard air filter and air box. Table 14.2 shows how wrong appearances can be. When the standard air box was removed and big pod-type filters which flowed about 15% better than stock were fitted, power went up by less than 2% at the top of the power band. Down in the mid-range power was up 1%. All of these figures were obtained with a cold air feed from a blower fan onto the filter. Drawing hot under-bonnet air it is probable these filters would have shown power figures no better than the factory set-up.

Table 14.2. Suzuki 1.3 GTi air filter comparison

Filter type	Air flow (cfm)	Flow increase (%)	hp increase
Stock factory*	195	–	–
Uni-filter*	175	–10	–
K&N*	197	1	1.5hp above 5750rpm
K&N pod**	231	18.5	3.2hp above 5000rpm
Uni-filter pod**	223	14.4	3.0hp above 5400rpm
HKS pod**	226	15.9	3.0hp above 3800rpm

* Filter element enclosed within standard air box.
** Pod-type filter with cold air feed.

One interesting figure was the very poor flow of the Uni-filter element. Its maximum flow when fitted in the standard air box was 10% less than stock, yet the power output was virtually identical to the factory element at all engine speeds. Equally perplexing was the K&N element. It flowed only 1% better than the standard item, but at 5750rpm power was up a full 1%. Also, note the HKS pod which flowed about 2½% less than the K&N pod, yet it produced almost identical maximum power, and from 2700rpm to 4000rpm it produced about 1.6hp more than the K & N.

WHEN A HIGH-FLOW FILTER MAY BE REQUIRED

On the other hand, substantial power increases are possible on some cars when the stock air filter system is scrapped and replaced by a high-flow filter arrangement. However, these power increases are often only realised when other modifications are performed. So, while the factory set-up may be satisfactory, or just marginally deficient, with the engine in standard tune that status can very quickly change as soon as modifications are made to raise the hp.

The R34 Nissan Skyline GT-R is an example of what I am getting at. To keep the bureaucrats happy it has an advertised output of 280PS, but the true output is around 320hp. Adding a free-flow filter system will raise that close to 335hp; about a 12–15hp gain. The big road block to power on this car, though, is the exhaust. If the exhaust is replaced first (using a 4in system, high-flow CATs and muffler), hp will climb from the stock 320hp to 360–365hp. However, when the high-flow filter set-up is then added, the power gain is almost double that achieved when this was fitted to a car with a stock exhaust, simply because the engine's air flow requirement is now so much greater.

This hot wire air-flow meter has a fairly clear air flow path. Removing the mesh screen from both ends may give a 1½–2% power improvement at maximum revs. The main benefit in stock and near stock engines is that removing screens may fool the system to send a marginally lower air flow signal to the ECU, thus shifting stock rich mixture and retarded spark programming closer to what the engine really needs.

A properly-sized air filter will not reduce hp, so don't ever run without one. Unfortunately, some consider that it isn't cool to use an air filter, claiming that they lose power with a filter, but in reality they ditch the filter purely for aesthetics. For those more concerned about under-bonnet beauty than component life I would suggest they have a good look inside engines, turbos and blowers that have been ingesting a steady diet of dust, sand and grit. And, as they do so, it would be wise for them to contemplate just how much sooner everything was ground away, and just how much the engine was down on power because of all the additional leakage past piston rings and compressor blades or rotors.

AIR-FLOW METER FLOW RESTRICTIONS

Immediately after the air filter many cars have some type of air-flow meter. The hot-wire type does not present very much impediment to flow, but the vane type, with its swinging flap, and the Karmen vortex type used on Mitsubishis certainly do. How you deal with each type of flow meter will, to a large extent, be dictated by what you intend to do with the stock engine management system. If you switch to an aftermarket ECU then you will most likely scrap the air-flow meter and go to a speed-density MAP-sensor-based system. However, if you want to stick with the factory ECU, and there are many good reasons to pursue that course if the car is mainly for street use, then you will have to investigate an alternative path to free up air flow because it is often impractical to convert an ECU from being air-flow-based to MAP-sensor-based, and vice versa.

With both hot-wire and vane type systems you can swap to larger air-flow meters and reprogramme the ECU to accurately sense the true air flow. However, with the vane type there will always be some flow resistance. Personally, that doesn't worry me. I am much more interested in reliability and minimal cost in a road vehicle, so I prefer to stick with the ultra-reliable vane type over the less reliable and expensive hot-wire unit. For some, such matters do not enter the equation, so they swap to the more modern hot-wire arrangement.

When a vane type meter is used it is possible to ease flow resistance by reducing spring tension on the swinging flap. Sometimes this is all right (e.g. a large flow meter on a small engine), but caution is in order. First, and this is the least important point, easing the spring tension allows the door to swing open further. The ECU will read this as more air flow and richen up the mixture. Consequently, any flow gains will be well and truly wiped out by excessive richness unless the ECU is reprogrammed. Second, remember the ECU calculates fuel flow based on how far the flap moves. Therefore, with spring tension wound off, the flap may swing wide open long before air flow into the engine peaks. In this scenario the ECU would not sense the increase in flow when air flow does actually peak, so there will not be any increase in injector pulse width, the engine will go lean, detonate and be damaged. Never should the flap swing wide open until air flow into the engine is at its maximum.

DEALING WITH THE KARMEN AIR-FLOW METER

There is no easy way around the Mitsubishi air-flow meter. I have seen crude arrangements that simply bypass air around the air-flow meter, but such a scheme is pretty much hopeless for a road engine because, if the bypass is an effective size, there will be close on zero flow through the air-flow meter at idle and cruise, so the ECU will have little chance of getting the fuelling right.

Because of such difficulties some simply give up and resign themselves to losing a lot of power. When tested with a water manometer a Mitsubishi Lancer turbo was recording a pressure drop at the turbo inlet of 56in water at 6000rpm. With all the inlet ducting and air filter removed, this decreased to 43in water (about 1.6psi) vacuum; showing that 77% of the inlet flow loss was down to the air-flow meter. However, that was for a completely stock engine. When modifications were carried out the losses grew. Simply adding a big exhaust saw the pressure drop grow to almost 50in water, and when a bigger turbo was fitted the water gauge went off the scale. Clearly such inlet flow losses are unacceptable.

The best Karmen vortex meter bypass arrangement, which I call the T-bone system, is not inexpensive but it does endow the car with perfect road manners even in very high

Fig. 14.3. A T-bone system by-passes restrictive air-flow meter.

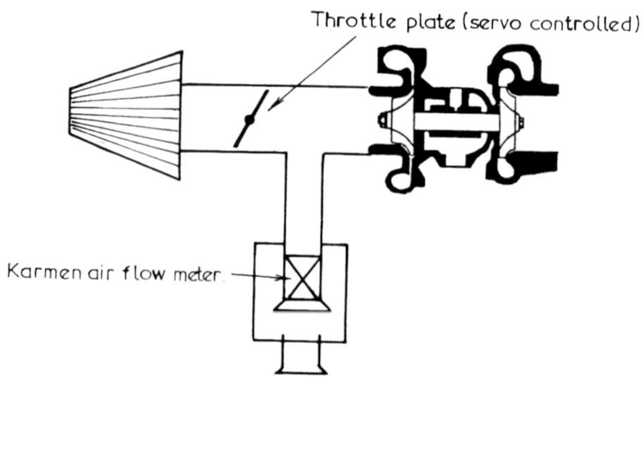

states of tune (380hp from 1.8 litres). As illustrated in *Fig. 14.3*, in normal use at low boost all air flows through the stock air-flow meter and then takes a 90° turn into the turbo. With the servo-controlled throttle plate in the secondary inlet closed, the stock ECU fully controls fuelling and spark on the factory maps. (Note: the car was turboed from the factory so the stock mapping was retained). However, as air flow and boost increases, a pressure switch signals a solenoid to open the secondary throttle plate. Simultaneously a MAP sensor controlled fuel only ECU begins to drive a second set of injectors to provide the engine's increased fuelling requirements. These secondary injectors are part of a dual fuel system, so with an appropriate intercooler upgrade the stock spark maps remain suitable at 20psi boost.

THROTTLE PLATE SIZE AND UPGRADING

The next potential point of flow restriction is the throttle plate(s). In most applications a single throttle body will be used, either fixed before the blower inlet, or in turbo and nitrous applications, attached to the inlet manifold plenum. Table 14.3 sets out the size of throttle plate required. If your stock throttle plate is too small look for a suitable replacement at breakers' yards, keeping in mind that you need to find a throttle body with compatible idle air control and throttle pot. Alternatively, the stock throttle body may have sufficient wall thickness to permit machining to a larger size, thereby alleviating any compatibility worries. Most will bore at least 2mm, and many will allow 5–7mm. Try to go to a size which will allow you to cannibalise a throttle plate off another throttle body. This will save a lot of time and money machining up a larger plate of the correct shape, with the correct angles around the edges.

Table 14.3. Throttle plate requirements

	Maximum Horsepower		
Throttle plate (mm)	Nitrous	*Blower	Turbo
50	170	200	250
56	210	250	310
62	260	310	390
70	330	400	500
75	380	470	600
85	490	600	750
90	550	670	830
100	675	720	1000

*Indicates throttle plate on blower inlet. If downstream of blower, use turbo figures.

Remember, when you opt for a larger throttle plate, that the car will not be so easy to drive around town. With a smaller throttle body, just cracking open the throttle may have given a steady increase in power, but with a big plate a whole lot more air rushes into the engine with just small movements of the accelerator. This may make the car quite jerky and wheelspin on wet roads can be a problem. There are a couple of ways around this problem, apart from resorting to electronic throttle control. We can opt for a multi-plate

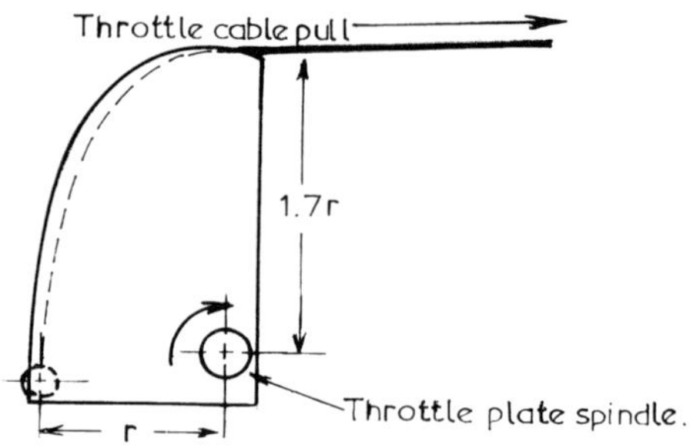

Fig. 14.4. A progressive throttle cam.

progressive throttle body with the small plate(s) opening about 40° before the large plates(s) begin to open, or we can fabricate a progressive throttle cam (*Fig. 14.4*).

Regardless of the size of the throttle body we should be concerned that it is well streamlined. Many are not, so take a look at what can be done to grind out lips and ridges in the flow passage. Also, attempt to streamline the throttle spindle and fixing screws without weakening the spindle.

In maximum effort turbo engines, more precise throttle control and more hp is possible with an individual throttle plate controlling air flow in each cylinder. In this situation, throttle plate size is more a matter of what is needed so that there is absolutely no flow loss as air passes over the throttle plate. In a 4-valve engine this means a butterfly about 1.5 times the inlet valve diameter, or slightly larger, in each inlet tract. However, if low speed performance is also important then the plate would be close to 1.3 times the inlet valve size.

INLET MANIFOLD PLENUM DESIGN

Of much greater importance is the inlet manifold itself. Whether this is a single casting or several bits bolted together the inlet manifold is, in fact, two distinct parts. First there is the plenum – an air chamber from which the individual cylinders draw air. Then there are the runners, which are really an extension of the inlet ports and join the ports to the plenum.

The three things we are interested in with regard to the plenum is its volume, how equally it distributes air to the individual cylinders, and how the openings to the individual runners are formed so as to maximise flow. Some tuners claim that the plenum volume must be carefully monitored, and that volumes of between two and three times the engine size work best. I disagree, having found that anything in the range of 0.8 to 1.5 times engine capacity works well in 3- to 6-cylinder engines, while 8-, 10- and 12-cylinder engines are usually happy with considerably less volume. The big exception is in the case of Roots blowers bolted directly to the inlet manifold without any intercooling. The pulsing flow from the Roots can be quite a problem with 3- and 4-cylinder engines unless plenum volume is in the 2–3 times engine capacity range. Interestingly, a properly sized plenum appears to aid the blower to work more efficiently as it is regularly noted that the intake charge temperature drops when plenum volume is optimised.

Except for the last example noted above, the most important role of the plenum is to equalise air flow to the individual cylinders. Unfortunately, car makers are much more

Multiple butterfly progressive opening throttle body can make low speed throttle control much easier.

The lip in this throttle body should be tapered into the throttle bore. Alternatively, it could be machined out and have a larger throttle plate fitted.

concerned with a neat powertrain package, so instead of air entering the plenum about in the middle it is more usual to find that the air enters at one end. This isn't a problem when the plenum is carefully designed with forced induction in mind, but once we massively increase the air flow large imbalances begin to show up. Usually it will be found that the cylinder closest to the plenum inlet is down on average air flow to the tune of 3–5%, while the cylinder at the blocked off end of the plenum will be enjoying 5–7% more flow. Obviously, this hurts power, but of more concern is the danger of engine damage if the engine is not tuned cylinder by cylinder. Unfortunately, instead of tuning each cylinder, many simply tune the engine as a whole. If the engine is close to stock that is all right, but as we move further away from stock and closer to the engine's limits we have to be more precise. Clearly, the cylinder receiving more air will want more fuel and also less spark advance unless the compression ratio has been lowered to compensate. Such is only possible if the ECU facilitates individual cylinder trim for both fuel and spark. If the engine management system is more basic, then to prevent detonation in the cylinder receiving the extra air the mixture and ignition advance will have to be adjusted to suit that one cylinder. This means all other cylinders will be running rich and retarded, and will be down on power.

The way around such a mess is to modify the plenum to balance the flows as much as possible. There are different ways to tackle this, depending on the layout of the engine and the type of forced induction planned. For a front-wheel-drive transverse engine installation with nitrous it is easy to add another throttle body at the blocked off end of the plenum, and then fit a double filter and air duct system. However, for most other applications we will have to fabricate the plenum like those illustrated in *Fig. 14.5*. The idea is to slow the air as it enters the plenum and to then shape the plenum wall so that it isn't steering the air into the far end cylinder. Deflector ribs can help here, but the main requirement is that the plenum must rapidly grow in size well before the first cylinder and it must also extend well beyond the far end inlet port. If there is insufficient space it is quite acceptable to use a tapering wedge style plenum, but take care not to bring the

Fig. 14.5. A correctly designed plenum helps balance flow into all cylinders.

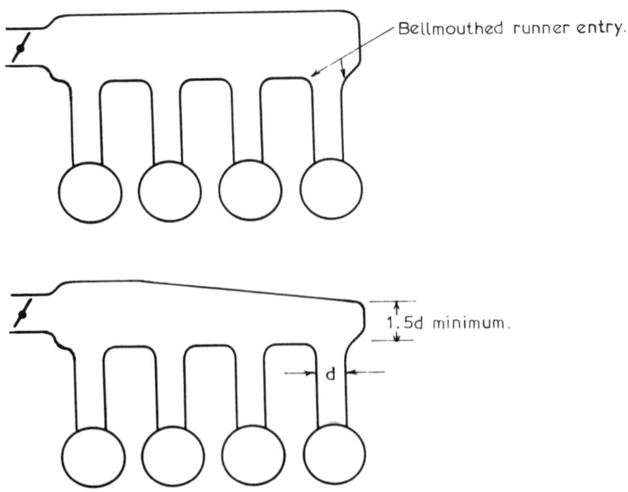

wall in too close over the opening of the end runner. Preferably, the air should not have to make a sharp turn from the plenum into the runners, and in turn the runners should have a nicely formed bellmouth to help flow. Ideally we don't want the plenum wall closer than $1\frac{1}{2}$ times the runner diameter from the bellmouth – twice the diameter is better, if possible. On the Cosworth Ford YB engines, for example, this problem is easily fixed by inserting a 13mm or 16mm spacer plate.

If you are working with the stock manifold it may be hard to form a proper bellmouth entry into each runner as they are often not spaced far enough apart. However, if the runners are more than a ½in apart it can be worthwhile cutting the plenum in half and welding ½in half-round bar stock formed into a circle around each opening. Aluminium epoxy can also be used to form a rolled lip, but be careful that the epoxy's temperature rating is well above the anticipated inlet charge temperature.

INLET RUNNER DESIGN AND SIZE

The runners are a real tuning tool. Both their diameter and length have an influence on the shape of the power curve. The diameter basically fixes the rpm at which the engine makes best hp, so while a big fat runner will let the engine breath freely at high revs, it will be down on power at low rpm. Obviously something to be avoided with turbo engines which are not nitrous assisted. Runner length tends to 'rock' the power curve around the point fixed by the runner diameter. Therefore, long runners will push up bottom-end power, but once past the peak the engine will fall on its face. Short runners do the opposite – they kill the bottom-end and help the engine maintain power way past the peak (*Fig. 14.6*).

Looking at Table 14.4 you can see how this works out in practice. A street turbo engine is going to be quite sluggish with a bigger than stock turbo, so if the stock manifold is shorter than indicated in the table and the runners are larger in diameter this should warn you that the inlet manifold is not going to help you get back lost bottom-end power. Conversely, if the runners are longer and smaller in diameter than shown, then

Fig. 14.6. This graph illustrates how runner diameter and length influences the power curve.

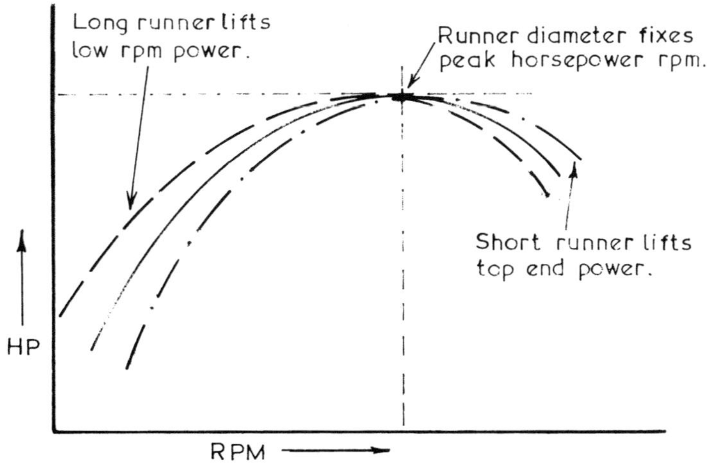

you know that you may need a bigger step up in turbo size to obtain the desired top-end performance with this particular manifold. Roots blowers are different, pumping very well at lower speeds but literally running out of puff at higher rpm. Therefore, to compensate the manifold needs short, large diameter runners. Because of the basic Roots characteristic, a manifold of this size will not hurt bottom-end power and it will help compensate for the blower's lack of efficiency at higher rpm.

Table 14.4. Runner size

	Street/hill climb/Rally		Circuit/Drags	
	2-valve	4-valve	2-valve	4-valve
Nitrous	350–400mm x.82–.9	350–400mm x1.08–1.18	300–330mm x.9–1.0	300–330mm x1.18–1.31
Roots or screw blower	300–330mm x .87–.95	300–330mm x1.13–1.25	200–260mm x.95–1.0	200–260mm x1.25–1.31
Turbo	350–400mm x.76–.8	350–400mm x.99–1.05	260–300mm x.85–.9	260–300mm x 1.12–1.18
Turbo + nitrous or Centrifugal blower	300–330mm x.8–.87	300–330mm x1.05–1.13	200–260mm x.85–.95	200–260mm x 1.15–1.25

Note: Runner length is measured from valve seat to bellmouth face in plenum.
Runner diameter is inlet valve diameter multiplied by the number indicated.

The other way to use the table is to consider the left to right progression of engine specification. This shows that a street engine will always work better with longer, smaller diameter runners than a hill climb or rally engine, while a circuit race engine requires runners just a touch shorter and fatter than either a hill climb or rally engine.

HOW TO MODIFY THE STOCK MANIFOLD

If the standard manifold has runners which are too long and/or too small in the port diameter you may be able to purchase a more suitable performance manifold, but unfortunately these are only available for a few of the more popular engines. Some companies offer a service called extrude honing, where they pump an abrasive mixture through the runners to increase the port diameter. Obviously, this is limited by the casting thickness; go too far and the manifold will fall to pieces from engine vibrations.

The alternative is to cut up the stock manifold and fit appropriately sized runners. Before cutting the manifold, talk to someone well versed in welding aluminium to give you some guidance on where you should cut. If it is a one piece manifold I prefer to make just three cuts; the first just back from the injector mounting bosses, the second just off from the plenum and the third through the plenum itself.

The first and second cuts get rid of a major portion of the runners which will be replaced by 3mm wall aluminium tube, mandrel bent to the required shape. Before

A BMW Formula 1 turbo combustion chamber is a conventional 4-valve design. Note the large separation of exhaust valves. This may have permitted a water jacket or oil gallery to get coolant around the valve seats to prevent a hot spot. Valve sizes are a conservative 35.8mm inlet and 30.2mm exhaust.

Two valve heads benefit most from modifications. Main flow improvement in the combustion chamber came from tapering back the wall opposite the spark plug; by 3mm behind the inlet and by 2mm behind the exhaust valve.

Right: The inlet ports are scribed to match the inlet manifold.

Main flow improvement comes from work in the valve bowl area. Valve seat insert overhanging the valve throat, the sharp lip where the machining meets the rough cast and port 'dingle berries' all upset flow.

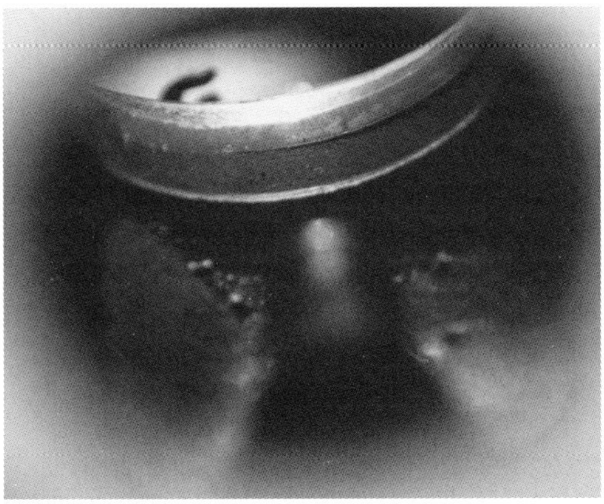

After three angle seats are cut the bowl cleans up like this.

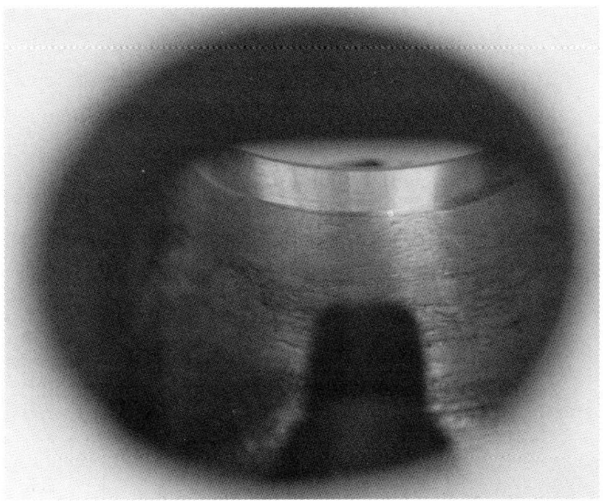

welding these new runners in place grind out the ports on the part of the stock manifold which bolts to the head. After welding, grind or file any welding 'dags' out of the runners. Finally, do whatever you can to add bellmouths in each runner entry.

CYLINDER HEAD MODIFICATION

The single component that separates ordinary engines from great engines is the cylinder head. Mostly we are stuck with making do with whatever the car manufacturer provided, but for a few popular engines there are a vast array of heads available. If our budget is sufficiently large, all alternative heads should be investigated as most will provide superior flow and combustion, and for less money than ported stock heads. When looking at aftermarket heads keep in mind what I have just said about what size manifold runners are required for different applications because the same principles apply when it comes to inlet ports. Also, when looking at air flow charts don't concern yourself with the flow numbers at maximum valve lift. Rather calculate the 70% valve lift figure for your cam and then compare the flow numbers from when the valve starts to lift off the seat up to 70% lift.

It is beyond the scope of this book to go in depth into cylinder head design and modification. However, there are quite basic things you can do to improve flow in the inlet and exhaust ports. The main point to keep in mind is to keep your grinding tools away from the ports. You can clean out the 'dingle berries' but apart from that there isn't much else you should attempt. Really, the part to focus your efforts on is the valve seat proper and the area immediately under the valve seat, called the throat area. Usually there are sharp ridges around where the valve seat meets the port that can be quite easily ground out (*Fig. 14.7*).

VALVE SEATS AND GUIDES

All heads need accurately ground multi-angle valve seats to give good flow and, more important, a perfect gas-tight seal. *Figures 14.8* and *14.9* show in detail what is required. The actual seat should be 45°, 0.050in wide for the inlet and 0.070in for the exhaust. The

Fig. 14.7. Radiusing the lip where the throat cut meets the cast inlet port promotes good inlet flow. If the port is ground right out to the edge of the valve seat, flow is hurt.

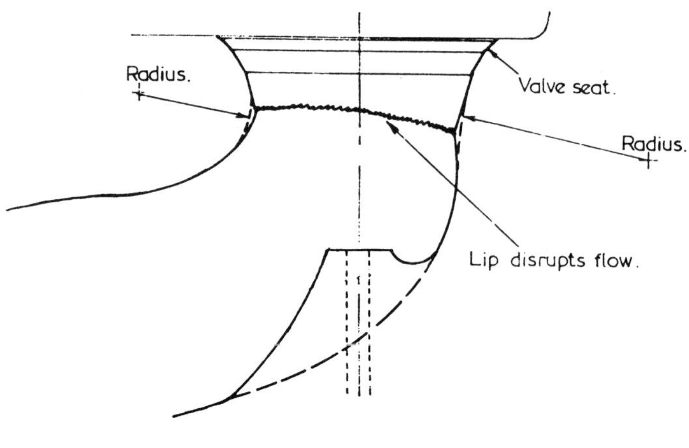

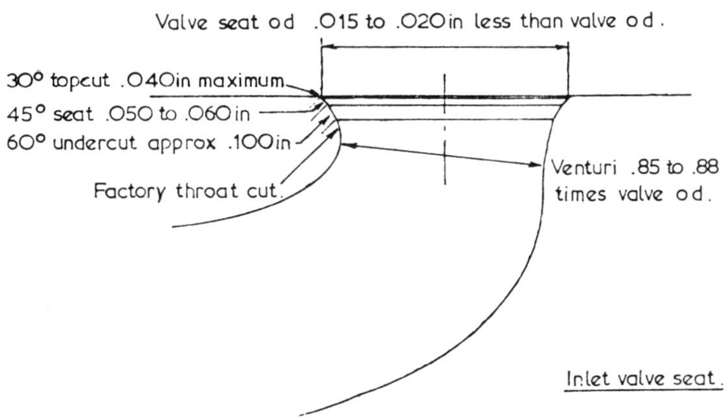

Fig. 14.8. The inlet valve seat is cut 0.050in to 0.060in wide for long life. The outside diameter of the seat should be about 0.020in smaller than the valve head diameter. Because we want the valve as high as possible in the chamber for good flow, the 30° topcut is only as wide as is necessary to blend the 45° seat into the combustion chamber. The 60° undercut will be about 0.100in wide, blending into the factory throat cut.

outside diameter of the seat is 0.015–0.020in less than the inlet valve diameter and up to 0.010in less than the exhaust valve head diameter. The top cut to blend the seat into the combustion chamber is at a 30° angle and just wide enough to blend the 45° seat into the chamber roof. The throat cut is usually made at 60° but this is best determined on the flow bench as some inlets show superior flow with a fourth cut of 70–75° and exhausts can need a single cut of from 55–75°.

Fig. 14.9. The exhaust valve seat is cut wider to get rid of the extra heat. The outside diameter of the 45° cut is almost the same as the valve head diameter because the exhaust valve will grow when hot. The width of the 30° topcut is not critical. In fact, sinking the exhaust valve a little can help reduce mixture draw through and exhaust reversion. If a flow bench is not available, stick to a 60° undercut and radius it into the venturi.

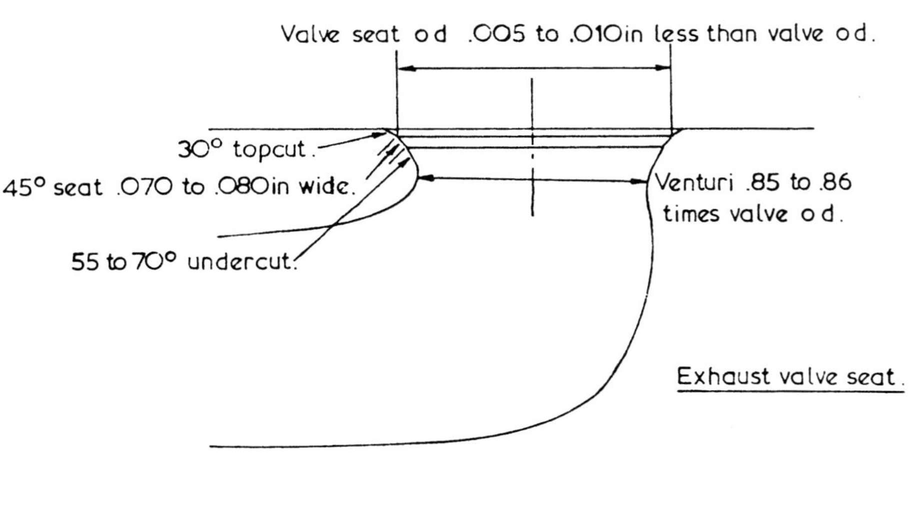

The 3 angle inlet seat and bowl from the combustion chamber side.

Some tuners like to use valve seats much narrower than I have suggested, but this is not wise. Narrow seats flow better than wide seats only at very high lifts; and at low and mid-lift they don't flow as well. On top of that, narrow seats pound out quickly, leading to lost performance and burnt valves. Remember, a valve gets rid of its heat primarily through the valve seat so if the seat is very narrow or becomes rough the valve head overheats, warps and finally burns.

The other heat flow path for the valve is via the valve guide to the engine coolant, so do ensure that the guide and stem are not worn. Primarily the valve guide is responsible for accurately guiding the valve onto the valve seat, to effect a gas seal. When the guide and valve stem wear, this allows the valve to wobble as it seats, wearing the seat into an irregular shape. Again, heat transfer is reduced so the valve head warps and eventually burns. Sometimes the problem can be considerably more serious than a burned valve; it can lead to a major blow-up! What can happen if the valve is very hot and the engine is operating at high rpm, is for the valve head to bend a couple of degrees relative to the valve stem. Each time the valve seats it bends another way, eventually fatiguing the stem. The valve head drops off and proceeds to hole a piston. Usually stock welded two-piece valves are blamed for this type of blow-up, but it will most likely be found that it is worn guides and worn seats which are the real problem.

Narrow valve seats and sloppy valve guides add up to an overheated exhaust valve. Once the hot gases begin eroding a soft spot, this soon happens.

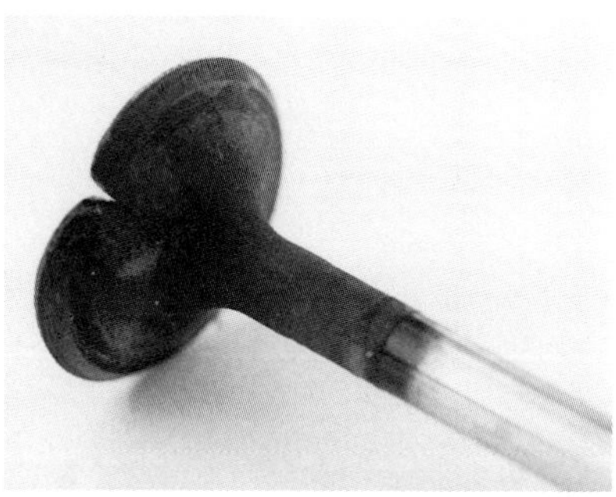

VALVE MATERIAL AND DESIGN

This is not to say that stock valves are always suitable. In competition engines they usually are not, but do keep in mind that the stock valves in some Japanese and European engines are very high quality austenitic stainless steel pieces with hollow sodium-filled stems to improve valve head cooling. If the stock engine was a low hp naturally aspirated unit (under 75hp per litre) and you plan a very mild power upgrade, then the stock valves are probably all right. However, when we start to go past about 100–110hp per litre in such engines we most likely need to go to a better quality valve. Engines starting out with factory supercharging will usually have a better grade of valve material and up to 150hp per litre is safe, but there are notable exceptions like the Impreza WRX which really need the Sti sodium-filled exhaust valve at this sort of power level.

High grade replacement valves have two other virtues in addition to a more heat resistant grade of metal; they are often lighter, which permits a higher rev limit or a softer valve spring, and they are shaped for better air flow. However, the stock valves can also be modified, as shown in *Fig. 14.10* and *Fig. 14.11*, to provide a nice flow

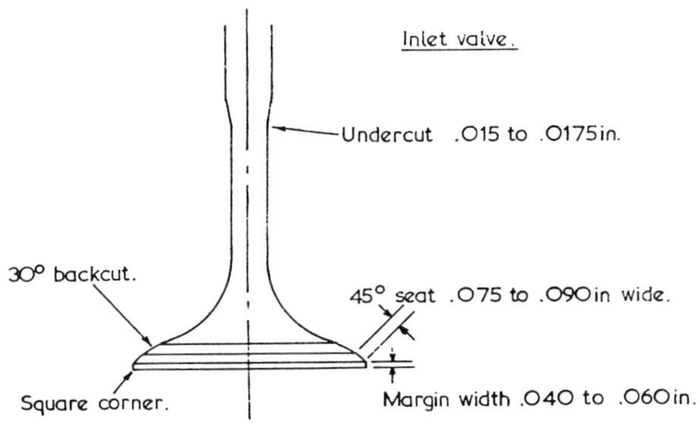

Fig. 14.10. Simple valve modifications can produce as much flow increase as work in the ports. The 30° backcut blends the seat into the underhead radius. Undercutting the stem will help flow but can be dangerous. A square corner between the valve margin and valve face reduces inlet backflow.

Fig. 14.11. A standard exhaust valve can be modified as shown. If high strength austenitic stainless steel valves are used, the stems may be undercut to reduce their diameter 0.035in.

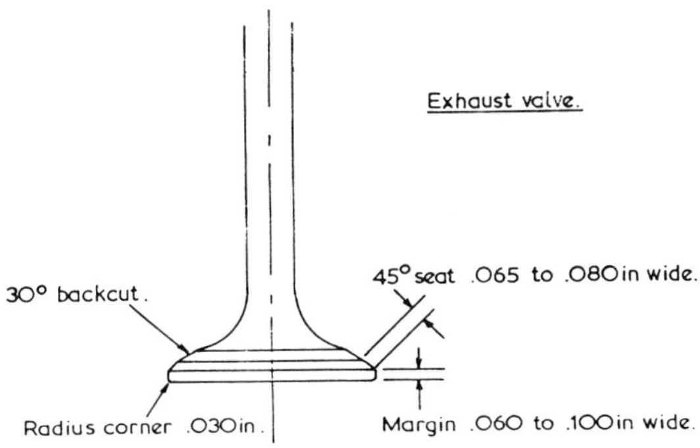

improvement. Inlet valves always flow better when a simple 30° backcut is applied and, depending on the underhead shape, it may be possible to bring this down to a 25°, or even a 20° backcut without weakening the valve. Such a simple modification can provide more than a 10% flow improvement at mid-lift. Another relatively easy modification is to undercut the valve stem by 0.035in, to pick-up another 5% mid-lift flow gain. However, care is needed to avoid valve stem fatigue and eventual cracking. Therefore, the undercut area must be polished and carefully blended into the valve head and also the uncut portion of the valve stem.

Sometimes a 30° backcut will lift exhaust flow by about 5%, but even if it doesn't help flow, it will not hurt it. What really assists is a nice radius applied to the exhaust valve face/margin corner. This will help flow at all valve lifts, and in the mid-lift region it will be up well over 10%. However, unless the exhaust valves are quality one-piece austenitic stainless steel items, do not undercut the valve stems.

CAM SELECTION

In forced induction engines we don't have to be overly concerned with the camshaft and valve train. The very same principles which apply to naturally aspirated engines also hold true in the case of forced induction engines. Although forced induction engines are not terribly fussy and will work quite well with virtually any cam which isn't too wild, each type of engine works best with a cam having specific characteristics. Additionally, in road cars fuel consumption will be an important consideration, so while a blower engine may retain quite good road manners with a long duration cam the engine's appetite for fuel will most likely not be appreciated. Consequently, a softer cam may be chosen even though we have super abundant low speed torque and could well afford to sacrifice some torque to gain more grunt at higher rpm. In general we can say that the better the design of the cylinder head and the more efficient the design of our inlet and exhaust system, then the less cam duration and lift we need to make good power.

Once we move away from the stock cam, turbo engines are the most fussy purely because more cam duration and lift always takes away low speed torque. Therefore, when a broad power range is desirable I would say that in the case of road cars we should be looking for around 10° less duration, and for competition engines it would be closer to 20° less, also maximum valve lift would be around 5% and 15% less, respectively, than for an equivalent naturally aspirated engine (Table 14.5).

HOW CAM LOBE DISPLACEMENT ANGLE AFFECTS PERFORMANCE

The lobe displacement angle (*Fig. 14.12*) plays a much larger role because of the relative pressure differences in the inlet port and the exhaust port of a turbo engine. In a naturally aspirated engine there is only minimal gas pressure in the exhaust port at the time the inlet valve opens during the valve overlap period, and at certain engine rpm there will actually be a vacuum in the exhaust port if the exhaust headers are the correct length. With a turbo, this situation often does not exist because in most instances the exhaust back pressure, owing to flow resistance caused by the turbine wheel and housing, will be higher than the intake boost pressure. Therefore, during the valve overlap period exhaust

Table 14.5. Turbo cam guide

	Road	Rally	Circuit	Drags
1:1 rocker ratio				
2 or 4 valve head				
Duration & lift – inlet	210–225°	220–240°	235–245°	250–260°
	*.320–.350in	*.350–.380in	*.380in	*.390–.400in
– exhaust	200–220°	215–230°	230–240°	245–255°
	*.320–.350in	*.340–.370in	*.340–.370in	*.370–.400in
Lobe displacement	105–110°	104–108°	102–104°	102–105°
Cam advance	3–5°	4–5°	2–4°	2–5°
1.6:1 rocker ratio				
4 valve head				
Duration & lift – inlet	200–215°	210–230°	225–235°	240–252°
	.350–.380in	.370–.420in	.400–.450in	.440–.500in
– exhaust	190–205°	200–220°	220–230°	235–250°
	.350–.380in	.350–.410in	.380–.430in	.430–.480in
Lobe displacement	105–110°	104–108°	102–105°	104–106°
Cam advance	3–5°	4–5°	2–4°	3–5°
1.5:1 rocker ratio				
2 valve head				
Duration & lift – inlet	202–218°	222–244°	232–256°	250–280°
	.420–.450in	.450–.470in	.450–.630in	.500–.700in
– exhaust	200–214°	222–244°	232–256°	250–280°
	.410–.450in	.450–.470in	.450–.630in	.500–.700in
Lobe displacement	108–112°	105–110°	105–108°	105–110°
Cam advance	4–6°	4–6°	4–5°	4–6°
1.7:1 rocker ratio				
2 valve head				
Duration & lift – inlet	198–214°	218–240°	228–252°	250–280°
	.420–.460in	.470–.490in	.470–.630in	.500–.720in
– exhaust	195–212°	218–240°	228–252°	250–280°
	.410–.460in	.470–.490in	.470–.630in	.500–.720in
Lobe displacement	108–112°	105–110°	105–108°	108–112°
Cam advance	4–6°	4–6°	4–5°	4–8°

*Indicates 2-valve engines will require 0.020–0.050in more lift.

Note: Lobe duration measured at 0.050in lift.
Lift measured at zero clearance.

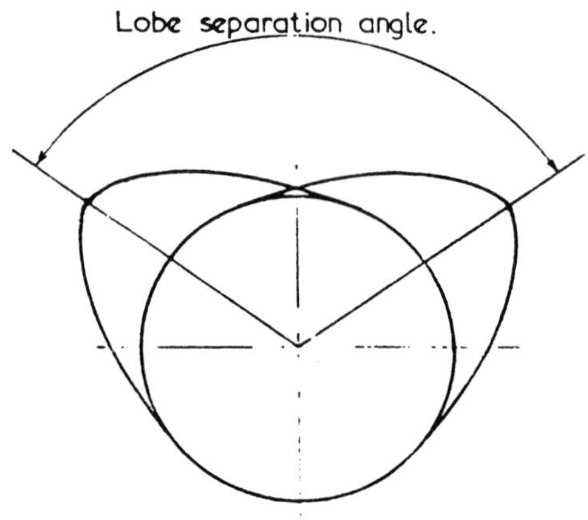

Fig. 14.12. Cam lobe phasing.

gas could flow back into the inlet tract (*Fig. 14.13*). This gas then dilutes the inlet charge, taking up volume which should have been occupied by fresh air/fuel mixture. Also, the hot exhaust gas reduces charge density by adding heat to it, then in severe cases this extra charge heat can set off a chain of events leading to detonation.

This problem is much greater in engines where a small turbine wheel and housing is fitted to improve bottom-end power and reduce turbo lag. Therefore it is more evident in a circuit race engine than a drag engine, but it is worst of all in road engines. There are just three cam related changes we can make to minimise the reverse flow. We can reduce the cam duration and thereby cut the time that the inlet and exhaust valves are both open together. We can redesign the cam lobes to reduce how far the valves are off their seats at top dead centre (TDC), employing a slow opening inlet lobe and a rapid closing exhaust lobe. And, finally, we can move the inlet and exhaust lobes further apart, thereby reducing the valve overlap period. All three options can be employed on competition engines – the engines with less of a problem! Unfortunately there is a limit on just how rapidly we can close the exhaust valves of a street engine. When they hit the seat too fast

Fig. 14.13. Exhaust backflow into the inlet tract.

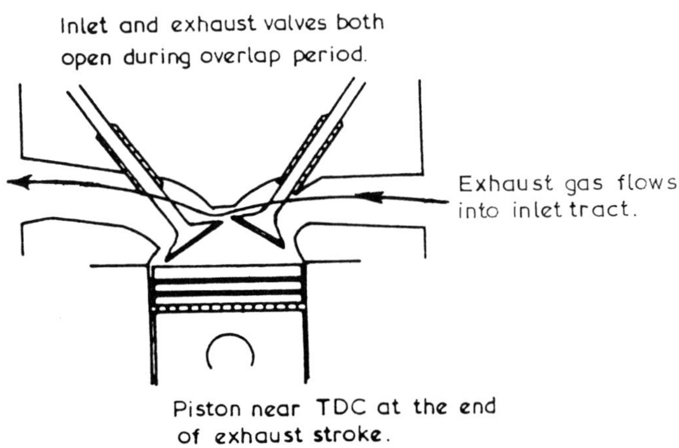

there is a lot of unacceptable noise, as well as undesirable valve and valve seat wear; so ruling out the second option.

Twin-cam engines are at a considerable advantage in that the lobe separation angle is not fixed, and consequently, while running the engine on the dyno, the inlet and exhaust cam can be moved around independently of each other to determine exactly where the engine prefers the cams to be set. (To test the same sort of thing on a single-cam engine we need a selection of cams with identical lobe profiles, but ground with the lobe centrelines at half a dozen different angles.) Therefore, when a big free-flow exhaust system is fitted to an otherwise stock road car we can often get a 5% power increase by moving the cams to reduce the lobe separation angle 4–8°. Many Nissans respond nicely to 3–4° advance on the inlet cam, and 3–6° retard on the exhaust cam, while the YB Ford Cosworth prefers about half that amount of advance and retard. In fact, many successful rally cars and club day circuit racers run with retimed stock cams.

However, keep in mind that we can only successfully run narrower lobe displacement angles (more overlap) when we free up the exhaust and when we are using very mild cams. If we are running cams with lobe characteristics like those recommended in Table 14.5 then we will most likely have to operate with the lobe separation angles suggested. When purchasing cams for a turbo car, always remember small is beautiful. A cam which is too small will not have anything like the hp and rev limiting effects on a turbo engine as it does on a naturally aspirated engine, and if it's a twin-cam engine, increasing the overlap will usually get back some of what you have lost.

CAMMING BLOWER ENGINES

As mentioned earlier, blower engines (particularly those with Roots or screw blowers) are not lacking in low speed torque. Therefore, in theory, it is possible to use very wild cams in street engines. However, blower engines have a problem that is the exact reverse of that in turbo engines; pressure in the inlet port is basically always going to be higher than in the exhaust port, with the consequence that during the overlap period as the piston rises to TDC, fresh air/fuel mixture will flow right out past the open exhaust valve (*Fig. 14.14*).

Fig. 14.14. Inlet charge flows out through the exhaust during the overlap period.

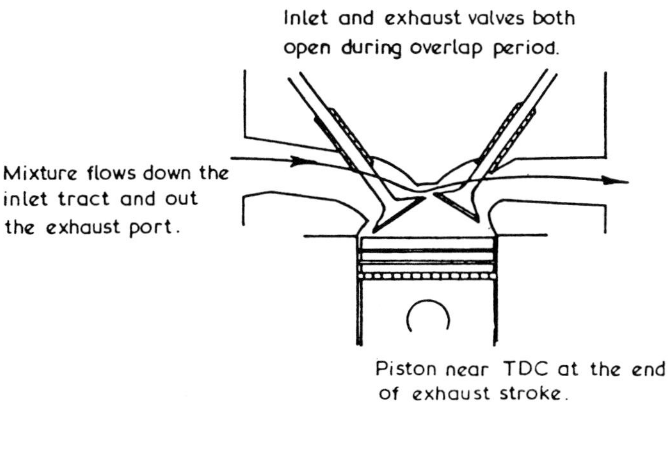

This is particularly good for scavenging exhaust gas out of the cylinder, and it cools the piston, combustion chamber and exhaust valve – things which aid power production and engine life. However, in street engines we don't want the loss of mixture through the exhaust. It adversely affects fuel economy and raises tailpipe emissions – and all that fuel will make the CAT overheat, causing the ceramic core to melt and plug up the exhaust.

In the past it was common for blower engines to use cams with very tight lobe centres of 95–100°, producing massive amounts of valve overlap. That sort of thing is no longer acceptable, or desirable. As already explained, it's bad for street engines, and as we have learned more about producing free-flowing cylinder heads and intake and exhaust systems for race engines, we no longer need massive overlap to scavenge the cylinders of exhaust gas. We now rely more on the exhaust headers and cylinder head to do that, which means we can move the lobe centres apart, reducing the overlap and thereby trapping more of the air/fuel charge in the cylinders to produce power. Cooling the engine components using the fuel mixture is no longer so vital as we have access to better piston and valve materials, and we have learned how to get coolant into critical parts of the cylinder head. Also, truer bores are allowing pistons to shed their heat load more efficiently, consequently most engines work best with cam lobe separation angles in the 112–115° range and cam advance of 4–5°. However, twin-cam engines running stock cams, or cams much milder than suggested in Table 14.6, will usually make more power (but with a loss of fuel economy) by moving the cams to obtain maximum valve lift at 104–108° before and after TDC in order to increase the overlap angle. Also the 5-port 'A' and 'B' series Austin engines in the Mini and MG-B are a notable exception. Because the 5-port design makes it hard to get these engines scavenging efficiently, they tend to work better with lobe centres in the 100–104° range.

Table 14.6. Blower cam guide

	Street	* Sports	Competition
1:1 rocker ratio			
Inlet lobe – duration	215–232°	245–260°	270–285°
– lift	.380–.420in	.410–.460in	.430–.520in
1.5:1 rocker ratio			
Inlet lobe – duration	202–218°	235–245°	260–280°
– lift	.430–.450in	.510–.530in	.600–.700in
1.7:1 rocker ratio			
Inlet lobe – duration	198–214°	230–240°	255–275°
– lift	.450–.470in	.540–.560in	.670–.850in

*Indicates street cars used for club circuit racing, hill climbs and rallying.

Note: Lobe duration and overlap measured at 0.050in lift.
Lift measured at zero clearance.

Another point to note with blower engines also relates to the design of the cylinder heads and the relative flows of the inlet and exhaust ports. Cylinder heads with 2-valve hemi combustion chambers and those with 4-valve pentroof chambers have relatively large exhaust valves, typically 85–90% the size of the inlets, and very high flow exhaust ports. Therefore, most of these engines will have an exhaust lobe similar to the inlet. However, some competition engines at times benefit from about 4–5° less exhaust duration and/or up to 5% less lift than the inlet lobe. Conversely, the cylinder heads on most American V8s have relatively small exhaust valves and poor flowing exhaust ports, consequently they work best with exhaust durations 5–15° longer than the inlet.

CAMS FOR NITROUS ENGINES

Nitrous oxide cams have much in common with blower cams. We don't want the nitrous flowing out of the exhaust port during the overlap period, so that means we have to run wide lobe centre angles. Also, because exhaust gas volumes are way up, we generally have to add duration to the exhaust lobe. Perhaps the biggest dilemma we face when choosing suitable camming for a nitrous engine is deciding which way it is going to be biased. For a pure competition engine that question doesn't arise; we simply select a cam to work with nitrous. Road engines, though, are more of a problem. If we cam the engine to work best with nitrous, when running naturally aspirated, performance can be a bit lethargic. On the other hand, if we cam it for best performance when naturally aspirated we lose out when we hit the gas. When we take a middle of the road decision we lose out both ways.

Personally, I feel that the decision regarding road cars has to be based on answers to questions such as: Is the car a daily driver? How highly is the engine tuned (i.e. does it have a peaky, top-end power band)? Is the engine small relative to the weight of the car? Is the engine a twin-cam unit with belt drive cams? Will nitrous only be used at track days, hill climbs and drags? Generally, I would say that a yes answer to some of these questions would steer me in the direction of selecting a cam for naturally aspirated operation. What would really clinch that decision would be if the engine had two belt driven cams, because on track days using adjustable pulleys it would be a simple matter to swing the cams from their naturally aspirated 105–110° lobe centre position, to something like 115–118° centres.

With a blower engine, any low speed torque we lose from moving the lobe centres apart is more than compensated for by the blower. However, with a nitrous engine we only get back that lost torque when we are on gas. Modern, peaky 4-valve engines are worse in this regard because they usually work best on nitrous with lobe separation angles 2–4° wider than the 112–114° separation that older engines with two in-line valves prefer. Also, whereas these older style engines generally work best with 5–8° cam advance, which helps pick-up some low speed torque, the modern 4-valvers don't seem to like much cam advance on nitrous. Therefore, when referring to Table 14.7 keep in mind that 4-valve engines, and a few 2-valve hemis also, are usually best on the road with street cams set at 107–110° lobe centres and sports cams set at 105–108°; all with 3–5° advance.

Table 14.7. Nitrous cam guide

	Street	*Sports	Competition
1:1 rocker ratio			
Inlet lobe – duration	215–232°	245–250°	270–285°
– lift	.300–.420in	.410–460in	.430–.520in
†Exhaust lobe duration	+4°	+6–8°	+6–15°
1.5:1 rocker ratio			
Inlet lobe – duration	202–218°	235–245°	260–280°
– lift	.430–.450in	.510–.530in	.600–.700in
†•Exhaust lobe duration	+10–12°	+12–14°	+14–16°
1.7:1 rocker ratio			
Inlet lobe – duration	198–214°	230–240°	255–275°
– lift	.450–.470in	.540–.560in	.670–.850in
†Exhaust lobe duration	+10–12°	+12–14°	+14–16°

*Indicates street cars used for club hill climbs or drags.
†Indicates the number of degrees by which exhaust duration should be increased over the inlet.
• Indicates cars with 2-valve hemi or 4-valve heads should use the same exhaust lobe duration increases as those listed for 1:1 rocker ratio engines.

Note: Lobe duration and overlap measured at 0.050in lift.
Lift measured at zero clearance.

If we are using nitrous oxide in combination with some other form of forced induction, the situation regarding camming changes. On turbo engines with nitrous assist, always select the cam to suit the turbo as the gas is only being used to help overcome turbo lag. If you want more top-end power, use a larger exhaust housing and let the nitrous fill in for any lack of low rpm power, rather than trying to get more peak power with extra cam duration. The situation is similar with centrifugal blowers; select the blower and blower cam to give the top-end you want, then let the gas fill in for any lack of low speed grunt. With Roots blowers it is usually best to cam for the nitrous. The blower will provide the necessary mid-range urge, and when it starts to run out of puff let the nitrous and the cam provide the required peak power.

Chapter 15

Nitrous oxide injection

An alternative way to increase air flow – actually oxygen flow because it is the oxygen content which produces combustion – is by means of nitrous oxide injection. Atmospheric air contains 23.3% oxygen by weight; N_2O is 56% higher than that at 36.4% oxygen. Consequently, nitrous oxide can be viewed as a chemical means of supercharging the engine. In fact, no other method of supercharging, with the possible exception of Roots and screw type blowers, can provide the instant torque increase and the low speed rush of power like even the most basic nitrous system.

WHY NITROUS CAN DAMAGE ENGINES

This, however, has probably led to nitrous being its own worst enemy. Because the power rush is so addictive, and because the bottom of the range nitrous systems appear on the surface to be so simple to install and tune, many enthusiasts have been caught out and have very quickly turned expensive engines into worthless piles of scrap metal! It must always be remembered, when using nitrous, that there is no room for complacency. Never be fooled by its benign appearance, and be sure to afford it the same respect, and devote the same attention to detail and tuning, as you would to a similar power upgrade using a blower or turbo.

In reality N_2O is only slightly more dangerous than air, but when it is injected into an engine we have to provide a corresponding increase in fuel flow, and to counteract the much higher cylinder pressures which result from the combustion of that extra fuel we have to retard the ignition timing. The real essence of a reliable nitrous installation, in fact, revolves around three things: the correct nitrous to fuel ratio, equal distribution of nitrous and fuel to every cylinder and the appropriate amount of ignition retard. Of course, there are also mechanical considerations, but in this area the requirements are similar to what is necessary in any supercharged engine and will be dealt with in later chapters.

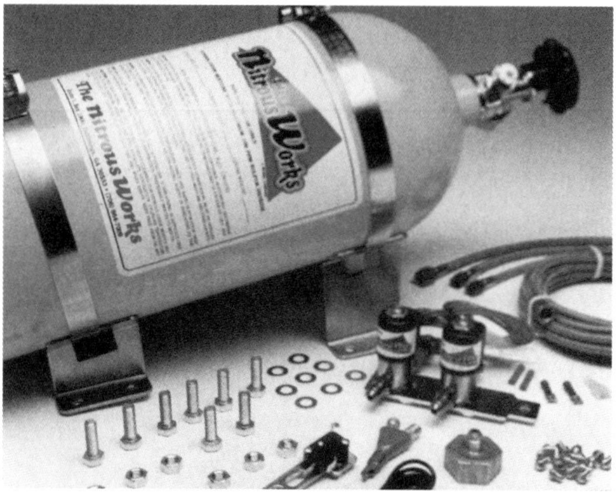

A wet nitrous kit from Nitrous Works includes all the basic hardware, including a big selection of nitrous and fuel jets.

WET AND DRY SYSTEM CHOICES

Back in the era of carburettors, nitrous systems were selected on the basis of how much power you wanted. However, when electronic fuel injection became the norm the choices doubled. As such, an early consideration is whether we will run with a 'wet' or a 'dry' system. Originally all systems were wet. That is, a fogger nozzle (or nozzles) containing both a fuel and a nitrous calibration jet was used to dump a mixture of atomised fuel and N_2O into the inlet tract. A dry system uses just a nitrous metering nozzle (or nozzles) to discharge the gas into the inlet tract, while relying on a modified fuel map in the ECU to increase the injector pulse width, or drive extra injectors to provide for the engine's increased fuel requirements. (Some dry systems rely on increasing fuel flow from the injectors by raising the fuel pressure. Such crude arrangements are impossible to tune. In some parts of the rev range the engine will be way down on power owing to mixture richness, but in other places it will be dangerously lean. Keep right away from this type of dry system.)

The dry system sounds good and some people have been able to get good results, but the reality is that it is difficult and expensive to map the ECU for increased fuelling, and fitting larger or additional injectors adds to the cost. The other problem is that most enthusiasts need to be able to quickly alter how much extra power they can reap from their nitrous system. Some start out greedy and, after cooking a couple of tyres, recognise that a smaller hit may be better. Conversely, the conservatives among us begin with a small dose, get addicted to the power rush and then want more. A dry system doesn't permit much leeway in this regard but with a wet system it is simple to increase or decrease the amount of fuel and/or gas injected using a few inexpensive metering jets.

The single area where the dry system has it all over the simpler wet arrangement is with regard to safety. Nitrous oxide by itself doesn't really present a problem, but a cylinder of compressed nitrous carries with it some risk because of the pressure of the contents. However, if you use your head and don't do anything to compromise cylinder integrity then you will be quite safe. That means never weld anything to the cylinder,

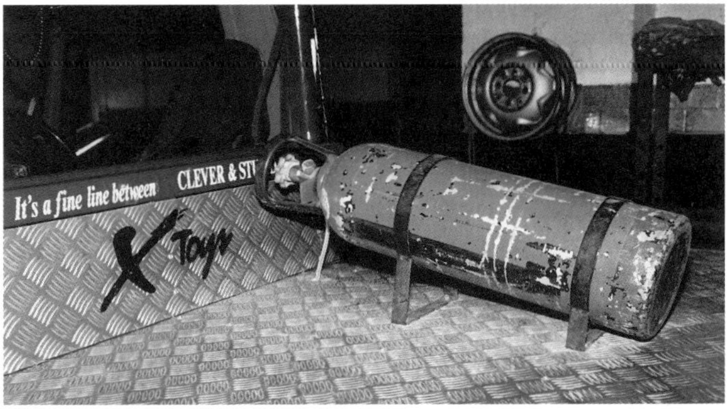

A nitrous bottle exposed to the elements on a Suzuki 4WD. The sign reads 'It's a fine line between CLEVER & STUPID'.

don't grind or scratch its surface, don't drill it and take care not to drop it. While there is very little risk with N_2O by itself, once it is mixed with petrol the situation changes. Just as a petrol-vapour-and-air mixture in a sealed enclosure becomes a bomb when ignited, so likewise with a nitrous-and-petrol mix. Consequently, in a wet system with the petrol and N_2O most likely being blended together in ducting before the inlet manifold there is quite some volume of a potentially explosive mixture present which could be ignited by a backfire. While explosions are rare, the risk does exist. In a dry system mixing of the fuel and nitrous does not take place until very near the inlet valve, so the volume of potentially explosive vapour is much smaller, reducing a backfire to a subdued 'pop'. However, that minor benefit does not sway me and I would always choose a wet arrangement.

NITROUS CYLINDER MOUNTING

As stated earlier, a basic nitrous system is a pretty simple arrangement. At the start of the system is a cylinder of compressed gas. The cylinder is usually aluminium and it will contain, depending on its size, anything from 5lb to 15lb of gas. To ensure that liquid nitrous rather than gas is flowing from the system it is necessary to correctly mount the bottle as illustrated in *Fig. 15.1.* Unless an inverted bottle arrangement is chosen, the bottle will have a siphon tube extending to one side at the bottom. Normal practice is for

Fig. 15.1. Correct cylinder mounting to ensure liquid nitrous flows to the engine.

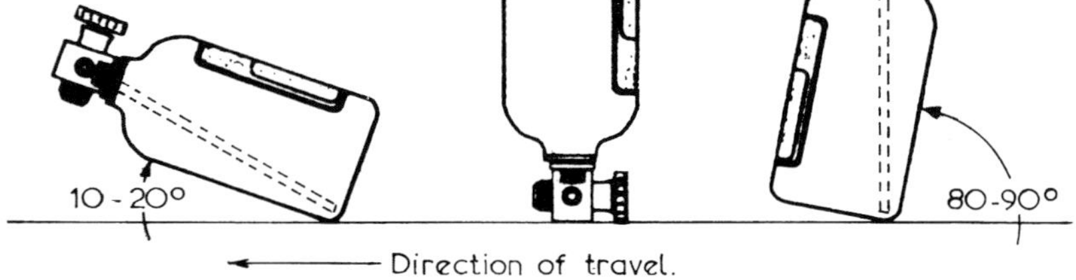

the pick-up to be located directly in line with the valve body outlet, but at any rate the cylinder or kit manufacturer should tell you the location of the pick-up. It is preferable for the bottle to be mounted at about 10–20° from horizontal with the valve towards the front of the car. This position takes up a lot more space than upright or inverted mounting, but it allows for a greater volume of liquid to be used before the siphon tube begins to pick-up nitrous in the less desirable gaseous form.

For a road car the cylinder would usually be located in the rear of the car away from direct sunlight. If, because of the design of the vehicle this isn't possible, or if you live in a hot climate which may see bottle temperature exceed 40°C (104°F), then it would be wise to cover the bottle with an insulating jacket and a silver thermal blanket. These steps are necessary to keep cylinder pressure below 1200psi. At approximately that pressure the safety pressure-relief disc ruptures, releasing expensive gas from the bottle. Even if not a legal requirement it is wise for those living in a hot climate to connect a tube to the pressure-relief side of the valve. This tube should pass through the floor of the car to ensure, in the event of the safety disc blowing, that released gas passes to the atmosphere rather than filling the vehicle with a fog of nitrous.

For competition use, first consult the rule book before deciding on a location for the nitrous bottle. Usually you will want the bottle in a location that will help you with weight distribution. Also, to keep the length of the N_2O line as short as possible in order to reduce weight and line flow losses, as well as ensuring a hard hit, the bottle will often be positioned as low as possible alongside the driver.

SUPPLY LINE ROUTING

After exiting the valve at the top of the bottle, the nitrous flows through a special line with braided steel covering. Even though it may look similar to the stuff used to convey fuel, water or oil in race cars it is quite different and should never be replaced by ordinary braided line or fittings which cannot take the high pressure and extremely low temperatures involved in the delivery of nitrous oxide.

A pair of Oz Noz solenoids control fuel and nitrous flow to a single fogger nozzle.

There are just a few things to remember when routing the supply line. Basically, you want to keep it away from rubbing against anything as the outer braiding is a very efficient file capable of very quickly wearing a hole through anything, especially aluminium or light gauge steel. Also, you want to keep the line cool so that the N_2O stays liquid until it erupts from the fogger nozzle. Therefore, keep it away from the exhaust or any other hot area. If the line passes by the exhaust, install a reflector shield and put an insulator jacket over the line such as is used on water heater pipes and refrigerator or air-conditioner cold pipes.

NITROUS SOLENOID AND JET

The nitrous solenoid is the next component in the nitrous supply route. Its function is to start and stop nitrous flow. As such it is quite different from the fuel solenoid (used only in wet systems) as it has to operate at very high pressure and very low temperature. Consequently it requires a powerful coil to create a magnetic field sufficiently strong to rapidly pull a valve open against 800psi pressure, and the rubber seals must resist chipping or shattering at temperatures close to freezing. In view of this, always follow the manufacturer's recommendations regarding the size of electrical cables required and don't use anything but genuine seals. Also keep in mind the total current draw of the entire system. That may include several solenoids, a bottle heater and an additional fuel pump. All up, the load may exceed 30 amps. If the electrical system isn't up to providing the additional power, the voltage will drop. This could slow the fuel pump, causing a leanout serious enough to damage the engine. On the other hand, the low voltage could bring the ignition system to its knees, resulting in a difficult to trace misfire.

Next in line are the calibration jets used to regulate the amount of nitrous flowing from the fogger nozzle. These jets are usually brass inserts that can easily be changed to increase or decrease the flow of nitrous, and also fuel in the case of wet flow systems.

DISCHARGE NOZZLE SELECTION

Following this we get to what I consider to be the most important part of the nitrous system, the discharge nozzle. A variety of nozzles are available. The simplest EFI arrangement consists of a plate which is sandwiched between the inlet manifold and throttle body. Stretching across the plate are two thin brass tubes (we are talking wet systems here) positioned so that the discharge holes in the nitrous tube cause the high pressure/high velocity nitrous to blast into the fuel flowing at 35–70psi from orifices in the fuel tube. Other arrangements follow a similar pattern, always with the nitrous being used to help in breaking down the fuel into, it is hoped, droplets of a uniform and easily combustible size.

I say 'it is hoped' because these schemes are far from perfect. Something we must never forget is that fuel must be atomised to burn efficiently. No matter how much oxygen and no matter how much fuel we can cram into an engine, we could be way down on power unless we manage to properly break the fuel down into the smallest particles and then thoroughly blend it with the correct portion of oxygen so as to ensure complete and rapid combustion. What makes this proposition all the more difficult to actually accomplish in a nitrous engine is that the inlet system plus the air and the fuel are all being super-chilled by the nitrous. On the one hand this is desirable because it

This engine runs a two stage system of fogger nozzles to help traction and transmission life.

increases the engine's volumetric efficiency and it cools the combustion chamber, exhaust valve and piston crown. (The former opens the way for more hp, and the latter decreases the possibility for detonation.) On the other hand, all that cooling going on just makes it so much more difficult to vaporise the fuel. For precisely this reason, in earlier times manufacturers routinely added heat to the inlet manifold and arranged the air cleaner snorkel so that it drew in air pre-heated by the exhaust manifold.

I'm not suggesting that we revisit that sort of 'engineering' in our quest for maximum combustion efficiency. However, I do believe that if we are going to spend money on a nitrous system, we should not baulk at paying out whatever is necessary for the best possible fogger nozzle(s).

NOZZLE NUMBERS AND LOCATION

Having decided on the type of nozzle to use, the next decision revolves around the number of nozzles required and where they should be located in the inlet tract. Ideally we want a discharge nozzle well down each inlet runner close to the stock injector. This will ensure that every cylinder receives an equal amount of nitrous and fuel and that no cylinder will receive too much fuel in relation to the amount of nitrous injected (or, worse, no cylinder will receive an excess of nitrous relative to the quantity of fuel injected). With no distribution problems (no rich or lean cylinders) or no variation in the quality of the fuel/air mix (no cylinder with poorly atomised fuel not homogenised with the air) we can run closer to the ideal air/fuel ratio and closer to the ideal spark advance to produce more power with less risk of engine damage than a similar engine consuming considerably more nitrous and fuel.

Such a system is the way to go if money isn't a problem and you are after the least possible risk of engine damage even at very high power levels. Mostly we are after a much smaller power increase, closer to 15–20% of the engine's peak power without nitrous. If we are talking about a 150hp naturally aspirated engine, that means a 22–30hp increase at all engine speeds. In the case of a 200hp turbo engine, we are looking at a

Nitrous oxide assist in the 3200 to 4200rpm range transforms the B4 Liberty by bringing the secondary turbo up to boost more rapidly.

30–40hp rise, even at low engine rpm. That probably means that the engine will be making twice as much power or more at lower rpm! In such a situation a single fogger nozzle is quite adequate. If the engine has a dual intake system then you should install a fogger in both inlet tracts.

The place to best position the nozzle(s) will be dictated by the actual inlet system. In a blower or turbo installation, position the fogger at the intercooler outlet. This allows for the fuel and nitrous to fairly equally disperse throughout a long length of inlet tract. For naturally aspirated engines the positioning will to some extent depend on the length of the inlet duct and on the type of air-flow meter. Speed density and Alpha-N systems don't have an air-flow meter, so it is possible to inject the nitrous just down stream of the air box. Vane type (flap) air-flow meters aren't much affected by nitrous, so the fogger can be positioned in the flexible trunking soon after the air-flow meter. Hot wire and hot film meters are affected; the nitrous will cool the air-flow meter, richening the mixture issuing from the injectors, so the nozzle must be mounted further down the flexible duct. To avoid tuning problems I reckon about 300mm is correct, but sometimes it must be less because the duct has a more suitable mounting surface closer to the flow meter.

FUEL DELIVERY SYSTEM

The single black spot in any nitrous system often has its origins in the fuel delivery system. Always remember that nitrous does not damage engines. Most engine damage comes about because there isn't sufficient fuel injected. That can be because the fuel metering jet is too small, but it can also be the result of restricted fuel lines (lines too small, deformed bends blocking flow, rubber lines swollen internally or breaking up and blocking flow, blocked fuel filter, etc.) or a fuel pump not stout enough for the job (this could also be a low voltage or poor earth problem). While insufficient fuel is at the root of many engine problems, sometimes the culprit can be excessive ignition advance or excessive rpm (perhaps owing to wheelspin or a slipping clutch).

Really, with a fuel injected vehicle we should never have a fuel flow problem in the majority of low hp nitrous upgrades as the stock fuel pump will most likely meet the engine's needs. However, it is always wise to consider what was discussed in Chapter 13 rather than to assume the stock pump and line are all right. If after doing the calculations you find the stock pump and line will flow the extra fuel required then it is very easy to route the fuel from the fuel rail. Simply tee off from the flexible hose connecting the pressure side of the fuel pressure regulator to the fuel rail. Alternatively, and this is what I prefer, connect the fuel line directly to the fuel rail and then tee off this line to the pressure regulator. This eliminates a 90° bend and lets the fuel flow directly from the fuel rail to the fuel solenoid. It's only a minor thing but gets rid of just another possible flow restriction.

If our calculations reveal that the stock pump is inadequate, we may be able to follow the same course as above, with the exception being we install a bigger fuel pump or another pump running in parallel with the stock pump. Running a second pump is the better option. It is often cheaper as small pumps are easy to find at breakers' yards, but it also means there will not be the fuel recirculation problems discussed in Chapter 13 as the second pump need only be activated when the nitrous system is armed.

In all out systems we have to take another route. Probably large bore fuel lines will be required along with an upgrade to high output pumps. Most likely this system will not be connected to the EFI fuel system, but rather will be a dedicated fuel supply serving only the nitrous system.

Regardless of the fuel system employed, the fuel must first flow to the fuel solenoid before passing along through the calibration jet(s) and then finally the discharge nozzle(s). Externally the fuel solenoid appears pretty much the same as the nitrous solenoid. However, internally it is quite different. First it only has to work against fuel pressure of 35–70psi as opposed to the 800psi in the nitrous line, so the opening and closing coil doesn't need to be so powerful. Secondly, seals in the fuel solenoid do not have to withstand low temperatures, but they do have to hold together and not swell or

VP racing C-23 is specially formulated for nitrous race engines.

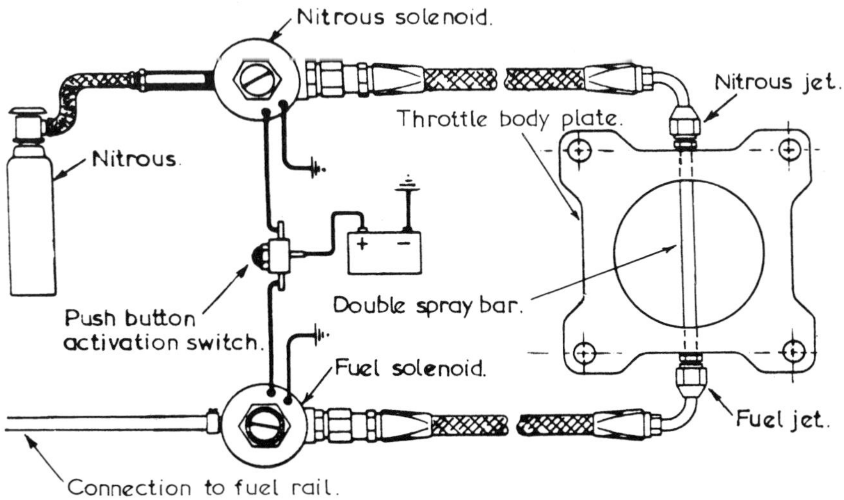

Fig. 15.2. A wet nitrous system with simple push button activation connecting to stock EFI fuel rail.

deteriorate rapidly in the presence of various fuel compounds. Do keep in mind that for alcohol and nitro fuels you need a different type of solenoid. It has a different seal material, usually teflon rather than viton, and also the capacity to flow a greater volume of fuel.

NITROUS ACTIVATION ARRANGEMENTS

Looking at *Fig. 15.2* you can see how all the basic hardware making up the full nitrous system comes together. As we have already discussed there are two systems involved, the nitrous and fuel systems, but it is the control system (what is usually referred to as the activation system) which enables the driver to control when and how the nitrous hits. Early activation arrangements were both crude and dangerous. Typically there was an 'arming switch', a simple toggle switch which in the on position made battery voltage available to a second 'firing switch' mounted on the dash or steering wheel. This spring-loaded second switch, like a buzzer switch, only activated the nitrous and fuel solenoids when depressed. You can imagine what a thrilling ride it would have been keeping the car under control while keeping the firing switch depressed! In similar vein you can guess how many engines were grenaded when drivers inadvertently kept the switch depressed as they stomped on the clutch to grab another gear.

The next evolution of the activation system replaced the push button firing switch with a wide open throttle microswitch. This was better as the driver could concentrate more on keeping the car under control, and any loss of control could be easily corrected by quickly lifting off the throttle. Engines lived longer, too, because lifting for a gear change also shut down the nitrous. There were still problems, though, as forgetful drivers would leave the arming switch on when the car was parked, and anyone accidentally depressing the throttle would fill the inlet system with petrol and nitrous. If the car was restarted in the next few minutes a backfire would soon have the engine bay and bonnet redecorated with bits of shrapnel! Clearly there is a need for the system to be

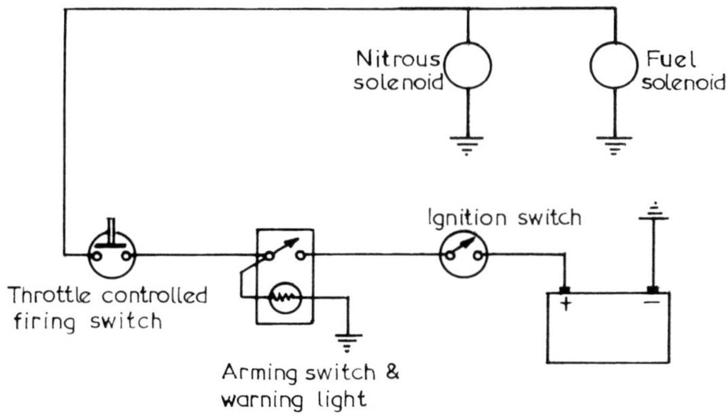

Fig. 15.3. Most basic activation systems should include a throttle-controlled firing switch.

wired in series with the ignition switch (*Fig. 15.3*), and if you think you may still get inadvertent activation you can wire it through the oil pressure switch as well.

While such a control system is inexpensive and simple, it leaves much to be desired as you get only full throttle nitrous power regardless of what gear you are in and regardless of the degree of traction available. The way around this is to use one of the many electronic controllers available. It is true that these add more complexity and cost but the cost can fairly quickly be recouped from the amount of nitrous saved, and reduced tyre and transmission wear can also be factored in.

The most inexpensive controllers usually contain only a rpm activated on and off switch. You adjust the control knobs to set the low rpm point where the N_2O kicks in and the high rpm point where the nitrous shuts off. Being wired through a full throttle switch ensures the gas only hits at full throttle between those preset rev points. This, of course, doesn't get around traction problems in 1st and 2nd gear but it does reduce the risk of detonation when nitrous hits hard at low rpm, and it prevents an expensive over-rev. To get around the lack of traction in 1st and 2nd gear you have a number of options. The simple solution is to mount the firing switch on the gear stick or steering wheel in a position where you can easily flick it on as soon as you are out of 1st or 2nd gear. A more tedious arrangement is to use a second micro-switch which is closed by the gear shift mechanism only in the appropriate higher gears. Finally, there is a time delay switch which can be adjusted to close the electrical system after a number of seconds and so delay the nitrous until the car is in the appropriate gear. This scheme, however, is best suited to drag racing.

In turbo cars, if you want to get the most distance out of a bottle of gas you can also use a pressure switch alone or in series with the low rpm/high rpm controller. This will allow you to run a big exhaust turbine for maximum top-end hp, while the nitrous fills in for any lack of grunt when off boost. The other benefit of this arrangement is that you can greatly cut the risk of detonation by shutting down the N_2O as the boost builds past a certain pressure. Also, in the lower gears the rpm controller will shut down the gas to protect the engine from an over-rev because, when it hits, engine rpm soar extremely quickly.

The next level of controllers from Jacobs Electronics, NOS and Schnitz Motorsports in the USA are much more specialised and costly. These units have varying degrees of sophistication but the general idea is that they are set to provide progressive nitrous flow according to engine rpm and the gear selected. Nitrous flow can be adjusted from 0–100% and the timer option can delay when injection begins and regulate at what rate the flow increases. They may also incorporate an ignition retarder with the degree of retard electronically adjusted to match rpm and nitrous flow percentage. For those who grew up with carburettors, such electronic controls may be a bit of a mystery but the principle is not a lot different from the ECU pulsing the fuel injectors to vary fuel flow according to the engine's requirements. The main difference is that the controller is adjusted by the driver to rapidly pulse a second set of nitrous and fuel solenoids to limit the power going to the tyres so as to achieve the best power/traction compromise and the lowest elapsed time down the drag strip.

With so much emphasis on sophisticated controllers it is quite easy to be made to believe that they are the only route to solving traction problems or transmission breakages. However, not to be overlooked is what you can do to adjust just how hard or soft the nitrous hits by means of altering the length and/or internal bore size of the section of nitrous delivery line between the nitrous solenoid and the calibration jet. A short standard bore line has a smaller volume, so when the nitrous is activated it fills quickly and delivers a hard hit. On the other hand a long big bore line fills more slowly when the nitrous solenoid is activated. This softens the hit, so the power comes on a little more gently.

ESTABLISHING THE NITROUS TO FUEL RATIO

Near the beginning of this chapter it was mentioned that an area of major concern with nitrous oxide injection is actually getting the nitrous to fuel ratio correct. One thing that works against us in this quest is that the pressure in the nitrous cylinder varies according to the bottle temperature. If the bottle is exposed to sunlight or exhaust heat or freezing cold air, the nitrous pressure changes. Clearly, when the pressure rises the nitrous calibration jet will flow more N_2O and when the pressure falls it will flow less. In turn, this will affect just how rich or lean the mixture is. Table 15.1 sets out the approximate

The Schnitz progressive controller provides a big range of control options.

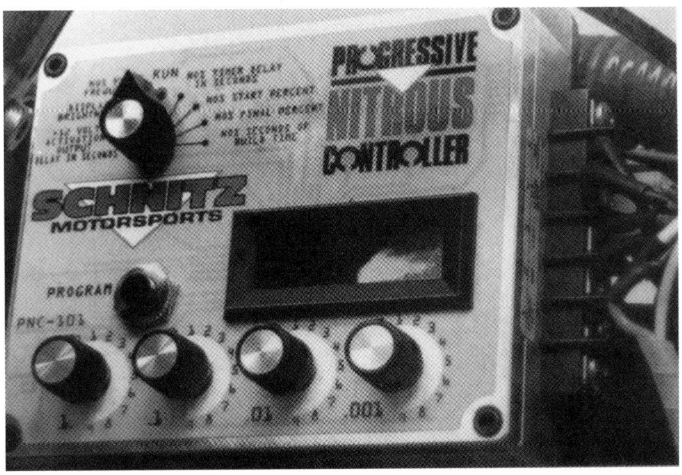

pressures which result from varying bottle temperatures. Some people attempt to tune around this by altering the fuel and nitrous jet sizes to compensate. Unfortunately, this is not by any means straightforward as at very high bottle temperatures the liquid nitrous may be partly gaseous. So, rather than nitrous flow increasing it may actually decrease! Long time racers know from experience and well kept records what size fuel and nitrous jets are required, but it is far safer and considerably less bothersome to install a fully automatic thermostatically controlled bottle heater jacket to maintain bottle temperature at 24°C (75°F). Such a temperature maintains bottle pressure at a steady 800psi.

Table 15.1. Nitrous temperature and pressure

Bottle temperature		Nitrous pressure	
°C	°F	Bar	psi
–18	0	19.7	285
–9	15	24.8	360
0	32	31.7	460
7	45	38.3	555
16	60	46.5	675
24	75	55.2	800
32	90	67.6	980
41	105	81.7	1185

With the nitrous flow rate thus stabilised it is then relatively easy to inject the correct quantity of fuel. Nitrous kit manufacturers will usually give a base-line jetting combination. Invariably with a carburettored engine this will be a safe fuel-rich mixture with about 5lb of nitrous injected for each 1lb of petrol, i.e. 5:1. This is about equivalent to an 8:1 air/fuel ratio. On the surface that appears to be way too rich for the engine to even run but, in fact, on N_2O engines prefer to run rich in much the same manner as alcohol engines. That isn't meant to imply that 5:1 is the preferred N_2O/fuel ratio for best hp. Depending on the design of the combustion chamber, inlet manifold and the compression ratio, best power will be found at a nitrous to fuel ratio of around 6:1, equivalent to an air/fuel ratio of 9.5:1. Some engines can run leaner, but because of manifold distribution problems it is risky. This means that for each 10hp of nitrous flow we will have to inject an additional 53cc of petrol per minute at 5:1, or 44cc of petrol per minute at 6:1.

Tuning a port fuel injected engine on a dyno using an air/fuel ratio meter we can be a lot more exacting. By observing the dyno readout in comparison to the A/F ratio we can establish what the engine actually prefers. Also, because a good portion of the engine's fuel requirements are being met by an EFI system squirting precise quantities into each cylinder there is less of a fuel distribution problem. This is even allowing for the fact that when the nitrous switches in, the extra fuel may only be supplied from a single fogger nozzle. Consequently, with a single fogger we can run good and safe with an overall A/F ratio of around 12.0–12.2:1. With a fogger in each port we may be safe and gain some power a little leaner, but once we get lower than 12.5–12.6:1 we are in dangerous territory.

It has been suggested that nitrous engines should be tuned to a particular brake specific fuel consumption (bsfc) number. The idea has no merit. You cannot dictate to an

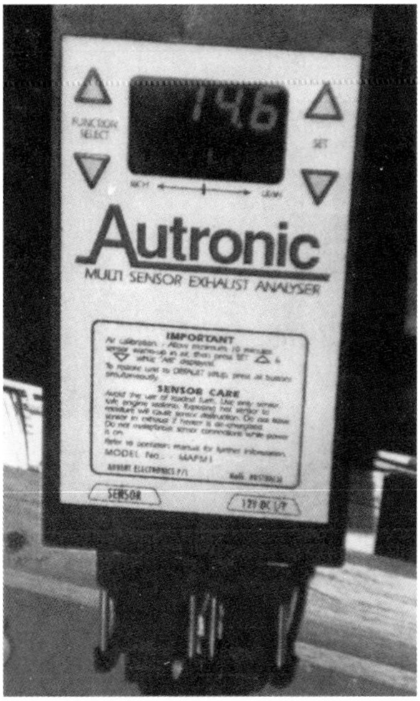

Setting up using an A/F ratio meter while running on a dyno enables fuel injected engines to use nitrous with little risk of engine damage.

engine just what bsfc it will operate at. You have to give it the fuel it requires to make the best power. Very efficient engines with well-designed combustion chambers, a properly homogenised and atomised air/fuel mixture, an efficient exhaust that scavenges exhaust gases and which has low frictional losses will make more power on less fuel than an engine with deficiencies in any of these areas. On top of that, the bsfc changes throughout the rev range. At maximum torque rpm the figure is at its lowest, and at peak engine speed it can quite easily increase by 30–50%.

The bsfc is a useful figure in the dyno room because it provides the tuner with a meaningful yardstick to measure engine efficiency, and particularly the air/fuel mixture quality and combustion quality. If an engine is making 200hp at a particular engine speed and is consuming 80lb of petrol per hour to make that power, then it is operating at a very good bsfc of 0.4 (80 ÷ 200 = 0.4). On the other hand, a less efficient engine may use 110lb/hr, a bsfc of 0.55, to produce 200hp and if we cut back the fuel the hp will fall.

Really, what we have to do when we install a N_2O system is run the engine with the suggested jetting. If all the spark plugs are black and sooty then we know there is an excess of fuel, so we would go down one fuel jet size at a time to get rid of the soot on the plugs. When they get the fuelling about right, some then begin to increase the nitrous jet size. This is dangerous. Before you begin to inject more nitrous you should already be too rich on the fuel jet.

IGNITION TIMING AND SPARK PLUGS

The ignition timing will also have to be retarded just as is required for any of the more usual forms of supercharging. How much retard is required is difficult to predict. There are so many variables. All engines are different, but basically you can expect that the spark will have to be retarded more when bigger doses of nitrous are used, and when it is activated at lower rpm even more retarding will be necessary. The other concern is that for best performance the engine really requires varying degrees of retard at different rpm. Consequently, an engine requiring 10° less advance at 2000rpm may require only 5° retard at peak hp revs. Clearly, a simple system which pulled out 10° advance when the N_2O switch was activated would be robbing us of power further up the rev range. Therefore, when we are chasing maximum performance we really need to be thinking of using an ECU with an additional timing map for when operating on nitrous. Alternatively, we could fit a simple ignition controller allowing varying degrees of spark retard at different rpm.

We discuss spark plugs in Chapter 17, so for now I will very briefly make a couple of points. In high power nitrous upgrades stock type projecting nose spark plugs will most likely give trouble. Often the very long earth electrode extending deep into the combustion chamber will overheat and burn away. More serious is the situation where the earth electrode becomes so hot that it works like a glow plug in a diesel and pre-ignites the air/fuel mixture. This can destroy engines. The solution is to use conventional gap style spark plugs, while in maximum effort nitrous engines retracted gap racing plugs will likely be necessary.

COMPRESSION RATIO AND PISTON REQUIREMENTS

With regard to the compression ratio there is generally no need to reduce it in N_2O engines, providing the fuelling and spark advance is in order. Some all out drag engines are an exception. For example, the big 650cu in Pro Street engines typically run a 17:1 compression ratio, but on nitrous this is reduced to around 15:1 to permit a much larger hit of N_2O.

Piston requirements are no different from other supercharged engines. High silicon content hypereutectic pistons are to be avoided. They are just too brittle and will not hold up with mild detonation. Pressure cast solid skirt pistons are OK, but for big hp you really need forged pistons or pistons machined from solid bar stock. Ring placement then becomes an issue. This is discussed in Chapter 19.

Apart from spark timing adjustments, the main change necessary when using nitrous doses amounting to more than a 20%–25% power increase, is the need for strong pistons with wide ring lands.

Chapter 16

The exhaust system

The primary role of the exhaust system is to rid the cylinders of exhaust gas and encourage the inflow of the inlet charge to completely fill the cylinders. Any exhaust gas remaining behind takes up cylinder volume which should have been occupied by potentially power producing air/fuel charge. However, as explained in earlier chapters, this isn't the only consideration. Exhaust gas has considerable heat content. That heat is transferred to the inlet charge, and apart from the consequent increased risk of detonation the inlet charge expands considerably and in so doing charge density is reduced. Then, during the combustion phase, stale pockets of exhaust gas can block flame propagation to some of the air/fuel mixture causing further power losses, as well as increasing exhaust emissions of unburned hydrocarbons.

THE EXHAUST AFFECTS TURBO PERFORMANCE

These considerations apply to all types of engines whether they be naturally aspirated or employ some form of forced induction. However, in the case of turbo engines there is an additional concern, namely the free flow of exhaust gases to initially rapidly accelerate the turbine wheel and then keep it spinning at high speed in order to quickly build and maintain boost. Clearly, energy expended overcoming flow restriction is no longer available to power the exhaust turbine. Because of this aspect, which applies uniquely to the exhaust of turbo engines, every effort must be made to ensure that the exhaust faithfully transfers the highest level of exhaust energy possible, over a wide band of engine operating conditions, to the exhaust turbine. Any deficiency here will result in increased turbo lag, or possibly a high speed power loss because of a smaller than ideal turbine housing having been employed purely in an attempt to claw back lost low speed boost. Typically, the exhaust turbine has been regarded as the engine, or driving force, of the turbocharger. I believe this is not totally correct. In reality the exhaust turbine in unison with the exhaust system is the engine of the turbo, and any improvements we can make in this area will profoundly influence not only the shape of the power curve but also the 'feel' of the engine – how it responds to driver inputs.

Consequently, when dealing with factory turbo cars, it is in this area that we can most easily make massive performance gains. Not only that, you also stand to reap more hp for the money spent than on any other area of the engine. And all this is possible without compromising engine reliability. However, there is a downside. If the plumbing isn't done properly there can be problems with the exhaust cracking with monotonous frequency, and there may be exhaust resonance problems characterised by incessant drumming at certain rpm; most often at constant motorway speeds. Even when expertly executed there is the issue of accessibility around the exhaust side of the engine. Even

How not to do it. The power output of the Toyota Supra (IJZ-GTE) is incredible considering these tiny exhaust manifolds impeding exhaust flow....

... The turbines are tiny and equally restrictive, as is the shared dump pipe joining both turbos. Note how poorly the turbine inlets match the exhaust manifolds.

factory installations can range from tortuous to impossible, and invariably re-engineering the exhaust, particularly the manifold section prior to the turbo, just exacerbates the problem of almost zero space for hands and spanners to work in. Don't let these negatives put you off because there are impressive hp gains to be made working downstream of the turbo. In fact, for most road-going turbo cars I prefer to keep right away from the pre-turbo part of the exhaust. Of course, there are exceptions. Some cars have incredibly horrible power-robbing plumbing which cannot be ignored. The twin-turbo Toyota Supra is one such beast which demands a better exhaust manifold. Also, if the client is after a big power increase the stock manifold usually has to be discarded regardless of whether the vehicle will be operated on the road or in competition.

MODIFYING THE STOCK MANIFOLD

Retaining the stock exhaust manifold does not mean that we have to put up with factory production tolerances and casting practices. While we have to take care not to weaken the stock item in any way, there is much we can do to improve its gas flow properties, and it is at times possible to alter the factory manifold to compensate for design weaknesses which may have left it prone to cracking in specific areas.

The first thing we want to do is make sure the internal surfaces of the manifold are smooth. Keep in mind that a smooth interior not only aids gas flow, it also discourages the build-up of carbon deposits which in themselves restrict flow. Therefore, grind out rough areas where you see casting slag and casting sand imbedded in manifold runners. Where core box halves have joined there are sharp edges which have to be smoothed. Take a look at how well the manifold matches the exhaust ports. Invariably you will find that at least in some runners their edges intrude into the exhaust gas flow path. When doing a match-up in this area, also check whether the manifold gasket needs to be trimmed.

At the very least, an exhaust manifold should look and flow like this half of the stock manifold off a Nissan Skyline GT-R. Note the large step up in size between the manifold and turbine inlet.

The crossover pipe in single turbo installations on flat and Vee engines always impedes exhaust flow and limits turbo performance.

Don't restrict your matching efforts to the cylinder head port face alone. Some manifolds are two or three piece bolted items. At each joint do all you can to streamline flow, and don't overlook the manifold to turbo joint.

Often there will be areas you cannot reach with a grinder. This leaves you with two options. Either you can fix what you can and forget the rest, or you can send the manifold off to be extrude honed. For those not familiar with the practice, extrude honing involves pumping a thick grinding 'mud' back and forth through a manifold (inlet manifolds and cylinder heads can also be 'ported' using this method). Extrude honing will do a great job of smoothing the hard-to-reach areas of the manifold, but there are some concerns. If employed over-enthusiastically it can weaken the part in critical areas, with heat or vibration cracking being the likely outcome. Even when this isn't the case, greatly increasing the internal size of the runners should be generally viewed as a negative. The reason is that high velocity flow does a better job of accelerating the turbine wheel. When we increase the runner diameter we lose velocity, particularly at lower engine rpm when we really want turbine driving forces to be as high as possible. Naturally, any time we remove material from inside the manifold we end up with thinner manifold walls. This has an impact, too, as any reduction in wall thickness increases exhaust gas heat losses. Heat is energy, and its loss here will cut the force available to accelerate and drive the turbine. That escaping heat, in turn, can cut performance in other areas. By adding heat to the underbonnet environment we can wind up with engine cooling problems, and we can be putting heat into the inlet air as it flows through inlet ducting and the inlet manifold. To avoid all this we have to find an extrude hone operator who will follow our specific instructions to, as far as possible, not increase port sizes while cleaning up the numerous bends in the manifold.

 Having given the internals of the factory manifold a work over, we should turn our

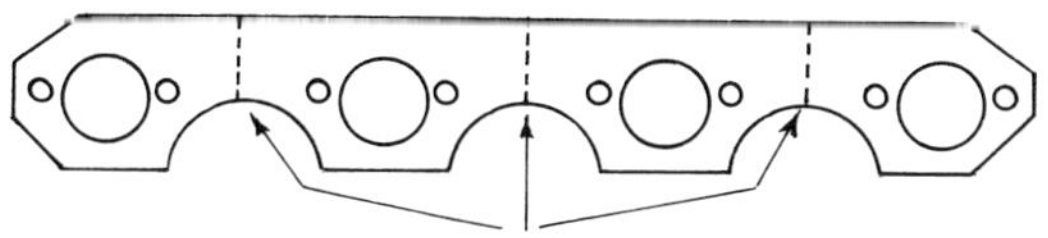

Fig. 16.1. Making expansion cuts in a linked manifold can help prevent manifold cracking and stud shearing.

attention to the outside. What we are aiming to do here is get rid of stress raisers on the surface and ensure perfectly flat mating faces at all joints. By grinding off sharp edges left behind after the casting process we are giving the manifold a much better chance of not cracking as it undergoes the extreme range of temperature change typical of turbo engines. As a side issue while we are on the subject, I have noticed that some manifolds are cast with all of the individual port runners linked together where they bolt to the cylinder head. This isn't a good idea. On the surface it may appear to give strength to the manifold, but it is usually cast in that fashion to make machining easier and faster. Unfortunately, that bridging at the manifold face can stress the manifold as it heats and contracts, or it can fracture the manifold studs. All that it required to rectify the problem is to use a hacksaw to make a cut through the web between each runner (*Fig. 16.1*). This simple modification drastically reduces the amount of expansion and contraction the manifold is subjected to at the manifold face. As well as taking stress out of the manifold this also minimises the possibility for leaking or blown manifold gaskets.

GASKET SELECTION

All gaskets in the area between the cylinder head and turbo dump pipe perform two very important, though frequently overlooked, roles. For some the gasket simply has to prevent exhaust leaks. Clearly, that is a vital service as any leakage of exhaust gas upstream of the turbo spells a loss of energy to power the turbine wheel. On top of that, each gasket can work as a heat barrier or conductor, depending on its composition. To help gaskets do their job we must ensure that every joint has been machined or lapped so that it is perfectly flat, and we have to select the correct gasket material.

In some joints it doesn't matter if the gasket is not a heat barrier. Therefore, we are free to use a material such as copper, providing it delivers effective gas-tight joints. An example would be between the exhaust manifold and the turbine inlet. Both parts are operating at similar temperatures, and heat flow from one to the other is not an issue. The joint between the cylinder head and exhaust manifold is different. Here, we don't want to lose exhaust heat energy back to the cylinder head, and we certainly don't wish to work the cooling system harder, needlessly getting rid of heat which we have allowed to transfer from the manifold back to the head. The type of gasket that is best here will both seal and insulate to some degree.

A metal reinforced gasket with fibre composition material on both faces is a good insulator but may not survive very well in this environment. The alternative, which doesn't insulate so well, consists of a fibre thermal barrier sandwiched between two light metal facings, usually made of soft steel and/or copper. We would use a similar type of gasket on the turbo exit side, where we want to avoid heat loss from the turbo to the dump pipe. And, of course, we don't want the dump pipe becoming any hotter than that which is unavoidable.

If we are using an external wastegate we would also use a thermal barrier style gasket between it and the exhaust manifold, the reason being that we want to insulate the wastegate diaphragm from as much heat as possible. However, on the exit side of the wastegate there may be a non-insulating gasket fitted if the wastegate has a totally separate tailpipe, as this would allow for heat transfer away from the wastegate.

FABRICATING A TUBE TYPE MANIFOLD

If we are striving for maximum hp we have to dump the stock exhaust manifold and fabricate a multi-branch system employing tuned length pipes. Also, in situations where we are adding a turbo to what was a naturally aspirated engine we would most likely follow a similar route. As well as figuring out a suitable design we have to decide on what material to use. To some extent there is a close relationship between the two, so it is wise to not separate them in the planning process.

When the aim is to locate the turbo as close as possible to the cylinder head we have the choice of two materials. The simplest material to use is steam pipe and its associated bends which are available in a variety of angles and diameters. This stuff is strong and will support a turbo without problems in much the same manner as a cast manifold. Because of its thick walls it retains heat well and, with careful welding and appropriate grinding of internal welds, it flows very well too. The downside is that, like cast manifolds, it is heavy.

The alternative material is stainless steel tubing. It is suitable for both short and long exhaust tubes. Being smooth internally it flows very well, provided all bends are formed in

A tube header ensures minimum restriction to gases exiting the cylinders and makes more flow energy available to drive the turbine. Note how smoothly the wastegate pipe blends into the collector.

a mandrel bender. Other types of bender seriously deform the inner radius of bends and dramatically reduce the available flow area. Welding sections to form bends should be avoided if possible. Every weld reduces flow because the filler material always bleeds through to the inside of the pipe, and usually you can't get in there to grind it out. Also, every weld adds a degree of stress to the manifold and exposes it to cracking in service. Even though it has thin walls, stainless tube retains heat well because it is a poor conductor. That property can be further enhanced by applying a ceramic coating to its inner surface.

When fabricating a manifold do not skimp on the thickness of plates at the cylinder head and turbo mounting faces. I consider 10mm to be the minimum, and don't overlook surface grinding these when fabrication has been completed.

Another area for attention with header-type manifolds is adequate support for the turbo. If you have chosen a short tube stainless manifold, and the turbo is a small lightweight unit, the manifold may support the turbo without much additional assistance. However, in all other installations a substantial support arrangement is required if you wish to eliminate exhaust and turbo reliability hassles. Remember, the turbo cannot be mounted solidly to the engine. To avoid exhaust and turbo cracking it must be free to move as the exhaust expands and contracts. Hanging the turbo from a spherical bearing in the top of a tripod which is solidly fixed to the engine is a common method of getting around the problem.

CALCULATING MANIFOLD DIMENSIONS

When it comes to the actual physical dimensions of a turbo engine exhaust manifold there are numerous schools of thought. Personally, I hate anything which adds weight, complexity or unreliability, so I will gladly trade a little performance to achieve that outcome. Often, though, the sport's rule makers have a vastly different agenda, and to win back throttle response and performance which their rules have pretty well destroyed means that sometimes a lot more complexity than I want has to be accepted.

If you are free to use whatever turbo you want, and there are no boost limits or inlet restrictions, then I would choose a pipe length of about 20in, but 4in longer or shorter will not make a lot of difference. The inside diameter of the pipe will be dictated by the size of the engine and whether we want a flexible power band or a top-end biased power band. Table 16.1 sets out what I would consider to be a basic starting point. Note, however, that if the car has a relatively large or relatively small engine for its weight this would indicate that pipe sizes slightly larger or smaller would be in order.

Further development of the manifold would be undertaken only after the engine was fully sorted, with the car running in its specialised environment. If more funds were made available, there would be further investigation, and most probably this would be first undertaken on the engine dyno. Length would be progressively added to the manifold, and its effect on the power curve noted. If gains were made, a new manifold would then be fabricated for actual in-car testing, and so the process would continue.

When planning an exhaust manifold it is always wise to start out with small diameter tubes. These keep the gas velocity high and so make for a more responsive turbo that will spool up to boost more quickly. Only in situations where all out power is more important would the larger sizes indicated in the table be selected. Also, whereas it is considered normal practice for all of the exhaust tubes to come together as one at the turbo, don't be afraid to try a different pipe joining arrangement. This applies in particular when space around the engine is at a premium, so in a 4-cylinder installation,

Table 16.1. Basic turbo exhaust manifold

	Engine configuration	Exhaust tube diameter
4-cyl single turbo	1300cc	1.25–1.5in
	1600cc	1.25–1.625in
	2000cc	1.375–1.75in
	2200cc	1.375–1.75in
	2400cc	1.5–1.875in
6-cyl single turbo	2000cc	1.125–1.5in
	2300cc	1.25–1.5in
	2500cc	1.25–1.625in
6-cyl twin turbo	2600cc	1.25–1.625in
	3000cc	1.375–1.75in
	3500cc	1.5–1.75in
	4000cc	1.625–1.75in
8-cyl twin turbo	4000cc	1.375–1.62in
	5000cc	1.5–1.625in
	5700cc	1.5–1.75in
	7000cc	1.75–2.125in
	8200cc	1.875–2.375in

Note: Use smaller diameter tubes for widest power band and larger size where top-end power is more important and sluggish throttle response is acceptable.

The minimum requirement for a single turbo straight-6 engine is to run a split pulse turbine and divided exhaust manifold, otherwise exhaust gas will backflow into other cylinders, upsetting combustion and fuel efficiency.

for example, it can be worth working with a 4-2 pipe system connecting to a split pulse turbine housing. In such an arrangement it is preferable to keep the primary pipes at least 12in long – although closer to 16in is usually better. Two primaries from cylinder 1 and 4 would come together into one secondary, while cylinders 2 and 3 would come together into the other secondary. This serves to separate the exhaust pulses to give the turbine more drive. In a 4-cylinder installation the secondary pipe length could be up to 24in. Its length won't have much influence on the power curve, but its diameter will. Usually $^1/_4$in larger than the primary will be pretty close if we are looking for a wide power band and good throttle response. For a single turbo 6-cylinder arrangement we have three pipes coming together into a secondary. In this situation the secondary is usually $^3/_8$in larger than the primaries, although at times secondaries $^1/_2$in larger will give better hp without a significant loss of throttle response.

SPECIALISED DESIGNS

In a specialised environment the situation is quite different. In this case the engine and turbo would be constructed according to what was permitted by the rules. Several exhaust designs would be tested on the dyno to determine which configuration was getting close to what was needed in the way of power and throttle feel. Having arrived at something that could prove suitable, a new manifold would be fabricated to neatly package the engine and turbo within the space available in the car. Testing in its competition environment would then follow, and the exhaust would be tweaked to achieve better throttle response, a better power range or more hp, depending on what was deemed to be the biggest need.

To give a small indication of just how far removed from the norm such specialised exhausts are, take a look at the dimensions of exhausts on CART Champ open wheelers running on American speedways. As you can see in Table 16.2, these exhausts appear to consist of very long, large diameter tubes. However, remember these engines were tuned to work over only a very narrow rev range and, because of the high speed nature of the turns, excellent throttle response was not a priority. Also, because turbo specs, boost and fuel loads were tightly regulated, the exhaust was a major component in attempts to win a power advantage.

Table 16.2. CART Champ exhaust manifolds

Engine	Manifold specification
Buick pushrod V-6 3.4 litre 805hp at 8200rpm	6 primary pipes 28in x 2in dia 2 secondary pipes 19in x 2$^1/_4$in dia
Chev pushrod V-8 3.4 litre 720hp at 8600rpm	8 primary pipes 14in x 1$^3/_4$in dia 2 secondary pipes 36in x 2$^1/_2$in dia
Cosworth DFX V-8 2.65 litre 820hp at 10,700rpm	8 primary pipes 24in x 2in dia 2 secondary pipes 29in x 2$^1/_2$in dia
Drake DT-160 V-8 2.65 litre 795hp at 9200rpm	8 primary pipes 27in x 1$^3/_4$in dia 2 secondary pipes 30in x 2$^1/_2$in dia

Note: Exhaust dimensions relate to engines operating at boost pressures of approximately 13psi (pushrod) and 8.4psi (twin-cam) and not today's low pressure engines.

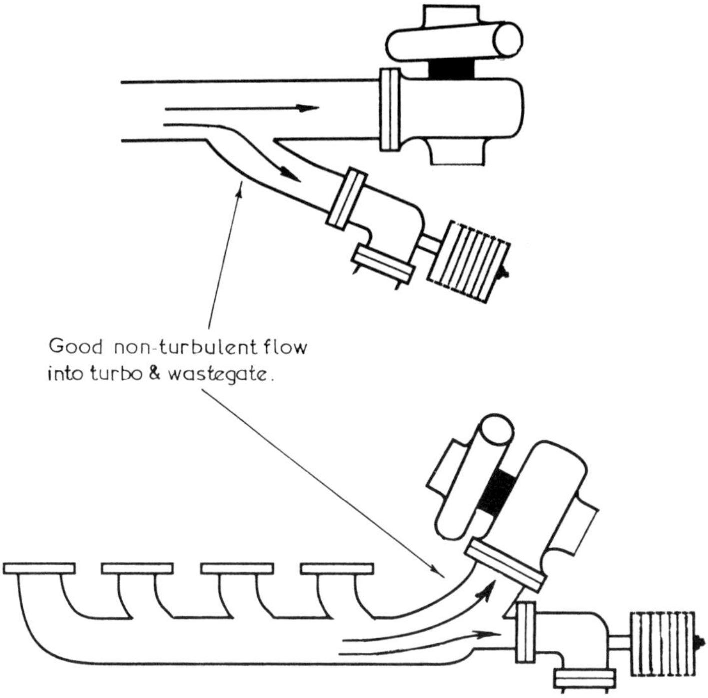

Fig. 16.2. An external wastegate connection should allow good flow into the wastegate and not disrupt flow into the turbine.

CONNECTING AN EXTERNAL WASTEGATE

If we have chosen to use an external wastegate we have to give a good deal of thought about how we can best tee off from our exhaust manifold, without disrupting exhaust flow, and thereby reduce drive to the turbo. At the same time the connection should maximise flow through the wastegate in order to keep exhaust back-pressure as close as possible to, or preferably lower than, the boost pressure. The best place to locate the wastegate is after all the exhaust tubes have come together at the turbo, or even on the turbine inlet itself. The wastegate should not tee off at an abrupt angle. If it does, then exhaust flow into the wastegate will be poor and there will be a lot of turbulent flow in the main exhaust. Don't forget this is right at the mouth of the turbine where we really want the flow to be as undisturbed as possible to achieve maximum drive.

One of the problems is that too many simply view the wastegate as a wastegate – some sort of exit through which we dump useless exhaust gas. The correct perspective is that the wastegate is a vital component that when correctly fitted will help our turbo get up to boost more quickly and help us gain more hp. For it to function in that manner we have to think about how we can streamline exhaust flow from the main system into the wastegate and beyond. *Fig. 16.2* illustrates the sort of flow path for which we should be striving. Keep in mind that an external wastegate does not have to be fitted in close to the exhaust manifold or turbo. In fact, mounting it further away enables you to keep it cooler so that the diaphragm lives longer, and you will most likely achieve better, less turbulent exhaust flows.

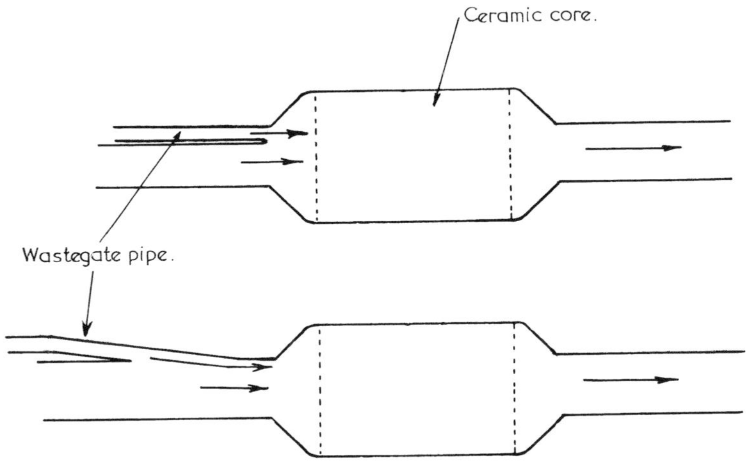

Fig. 16.3. Wastegate pipe connections back into the main exhaust to minimise flow turbulence.

EXTERNAL WASTEGATE DUMP PIPE

Where the wastegate exhaust goes after it exits will, to a large extent, be dictated by various laws. Preferably, the wastegate pipe should remain separate and not rejoin the main exhaust. However, road laws and some competition rules require that wastegate flows reconnect to the main pipe before the catalytic converter. When this rule applies we have to ensure that the two pipes are brought together as far away from the turbo as possible. Whenever the two flows rejoin there is some turbulence, so we don't want that turbulence reaching back up the dump pipe and hindering flow out of the turbo. Therefore, we have to connect the wastegate pipe to the main pipe at a shallow angle or, preferably, straight into the CAT itself if there is no law against CAT modifications (*Fig. 16.3*).

In some lands the law stipulates that for road cars the CAT cannot be moved from its stock location. Where that rule is enforced we have to comply even if it means that our wastegate pipe will increase turbulence at the turbo exit. However, where such laws do not exist we should relocate the catalytic converter further away from the turbo. This will allow the external wastegate pipe to rejoin without upsetting gas flow out of the turbo. In fact, even for installations not using an external wastegate we should, if possible, move the CAT further away from the turbo. The CAT possibly won't work so well, but it will cause less power loss if it is placed where the exhaust gas is cooling down. Also, there is much less risk of the ceramic honeycomb core melting and blocking exhaust flow.

INTERNAL WASTEGATE DUMP PIPE

When we have an internal wastegate in the turbo it is considerably more difficult to achieve non-turbulent flow out of the turbo. In Chapter 7 we looked at what we could do in the turbo proper to ease the problem. Now we have to rework the turbo dump pipe so that we prevent, if possible, flow from the internal wastegate port cutting across and impeding flow out of the exducer. In some cases, owing to turbine housing and

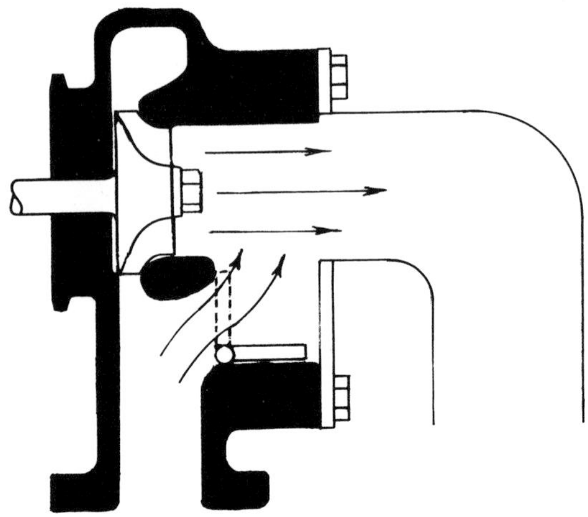

Fig. 16.4. Exhaust flows from both the internal wastegate and turbine are disrupted by poor dump pipe design.

wastegate passage design, there isn't much we can do. However, if the exit of the wastegate port runs parallel with the exducer bore, there is much we can achieve using some quite simple dump pipe designs.

Fig. 16.4 illustrates the problem with many factory installations. As you can see, on exiting the wastegate gases are forced to make two acute turns to flow into the dump pipe. This drastically cuts how much flow the wastegate can handle. This, in turn, increases exhaust back pressure, reducing exhaust flow out of the cylinders. On top of that, exhaust gas exiting the wastegate port is forced to cut across flow from the turbine. The turbulence which results impairs flow from the turbine, reducing its ability to accelerate rapidly and requiring that a smaller housing be used to achieve acceptable throttle response times. The smaller housing and disrupted flow, in turn, create more back pressure, reducing cylinder scavenging and the ability to refill with fresh air/fuel mixture.

There are several ways around this problem. Different tuners have their pet ideas of what is best. Personally, I think it is more a matter of horses for courses. Do what is most practical for a particular installation. While there are some hp differences I feel they are over emphasised. The twin pipe arrangement shown in *Fig. 16.5* usually yields the best power curve, but it is more complex and expensive, so is probably best reserved for pure competition vehicles or those used in weekend motorsports. The pipe within a pipe setup is a good compromise. It is simple to fabricate but carries a weight penalty as the inner pipe has to be around 12in long to be more effective than the third arrangement. This features what I call a tongue within the dump pipe to keep the flow paths separated for at least 6in. Further is better, but I feel the main thing we want to accomplish here, in addition to dividing the two flows, is to allow the main turbine dump tract to taper out to the wall of the dump pipe. By permitting the gases issuing from the exducer bore to progressively expand, turbulence in the turbine exit is reduced. To allow wastegate flows to rejoin as calmly as possible, a series of 6mm perforations are drilled in the tongue close to where it is welded to the dump pipe wall. I have tried leaving a gap between the tongue and dump pipe wall as a wastegate flow tract, but in some installations I found the tongue cracking up, possibly because of exhaust harmonics.

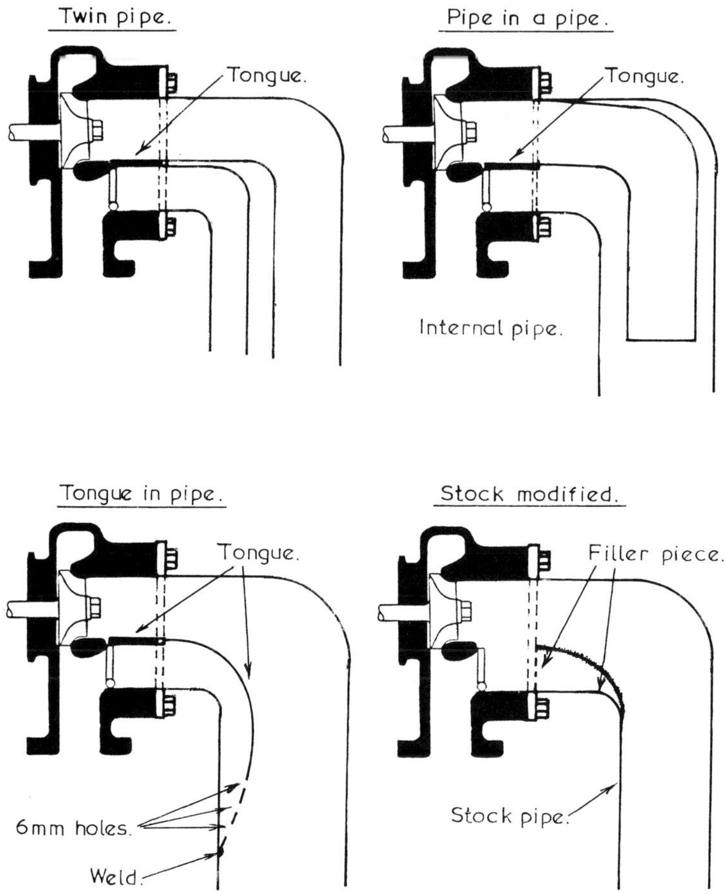

Fig. 16.5. Four methods of helping exhaust flow and performance of internal wastegate turbos.

The fourth arrangement is better than the stock situation but really is quite inferior to the above three. The main problem is the lack of a divider tongue and the massive step up in exhaust flow area causing quite turbulent flow. Some, rather than cutting a section out of the stock pipe and welding in a taper piece as shown, simply add a large bore dump pipe to enable wastegate flow to travel straight on out of the turbo. This suffers the same turbulence problem, but it is stronger as the cutting and welding to insert the taper piece invariably weakens the dump pipe, giving rise to cracking.

DETERMINING EXHAUST PIPE SIZE

How big the diameter of the exhaust pipe should be has been a matter of debate for a long time. Some tuners maintain that they achieve excellent results using a pipe of similar size or up to 15% larger than the exducer bore. Others would never consider anything smaller than 3in for engines up to 2 litres, 4in up to 3 litres, and twin 3–4in systems on anything larger. Really, I do not believe that you can look at these things in isolation. For a competition car operating where ground clearance isn't an issue, use the pipe size that provides the best power curve and throttle response. For other environments we have to be more pragmatic. Not only is there the matter of maintaining ground clearance, we also have to keep in focus that big pipes, and in particular big diameter silencers (mufflers), carry a massive weight penalty. Consequently we have to ask whether the performance gain will be overshadowed by the weight increase? Also, don't forget that extra weight increases fuel consumption, so if you travel hundreds of miles each week the increase in operating costs could be telling.

With these factors in mind Table 16.3 sets out pipe sizes for turbo cars operating on the road. If at all possible I prefer to maintain the pipe size indicated to the back axle, then if there are clearance problems the tailpipe size could step down by 1/4in without any hp penalty. In fact, in many instances there may be a slight low rpm performance gain, and as there is often a noise reduction a smaller less restrictive silencer (muffler) may be all that is required to pass noise tests.

Table 16.3. Turbo exhaust size

Engine size (litres)	Exhaust size (in)
1.3–1.6	2.25–2.50
2.0–2.3	2.75–3.00
2.5–3.0	3.5 or 2 x 2.5
3.5–4.0	4.0 or 2 x 2.75
5.0–5.7	2 x 3.0

Note: The above pipe sizes apply to road engines developing 120–150hp per litre. Competition engines developing 200–250hp per litre will use pipe sizes 0.5–1.0in larger.

The dump pipe connected to this big external wastegate does not rejoin the exhaust system.

Competition turbo engines generally develop closer to 200–250hp per litre. Usually, but not always, they will show a useful performance increase with larger exhaust sizes. There is no substitute for actual testing. I've seen fairly ordinary 2-litre circuit race engines way down on performance with anything smaller than a 4in exhaust. On the other hand I've seen other 2-litre race engines make excellent power on 3in pipes while being hopeless on 3½in pipes (incidentally on a 4in pipe some picked up power and were showing a fairly similar power curve as on the 3in pipe).

Remember, when dealing with turbo exhausts the greatest restriction is caused by the turbine wheel and housing, so anything we can do which will enable us to run a bigger housing without introducing massive turbo lag is worth more than can ever be achieved elsewhere in the exhaust tract. The next most restrictive area is the manifolding between the cylinder head and the turbo, so improvements here will provide us with good gains too. This is not meant to imply that what happens after the turbo isn't important, just that you need to be alert to the greatly overstated gains which are claimed for modifications down-stream of the turbo.

UNEXPECTED PROBLEMS

Do keep in mind that replacing the exhaust may cause other tuning issues to arise. A common problem can be that you are no longer able to control boost pressure at the desired level! What is happening in this situation is that engine breathing increases so dramatically that the stock internal wastegate is too tiny to bleed off sufficient exhaust flow to keep boost at the required pressure. The Nissan Skyline GT-R, for example, operates at about 0.8bar (11.6psi) with the factory air box system and exhaust in place. However, when the stock items are replaced it is not at all unusual to see the boost jump up to 1.1bar–1.2bar (16–17.4psi). This, is turn, gives rise to claims that the new exhaust and inlet system increased power by 70–80hp. In truth they added less than half that amount, the rest being because of the increase in boost. So what is the problem? Well, the car may now require more expensive, or difficult to locate in some lands, higher octane fuel in order to avoid engine detonation. Fortunately the Skyline GT-R has a reasonably efficient intercooler which is able to keep inlet temperatures in hand even in fairly warm weather. Some other cars are not so well endowed, the Subaru WRX for example. So, to avoid engine damage in summer, the choices are: use high octane fuel, fit a bigger front-mounted intercooler or don't work the engine so hard.

BLOWER AND NITROUS ENGINE EXHAUST PRINCIPLES

Blower engines and nitrous engines are totally different from turbo engines. They share the basic principles applying to naturally aspirated engines and, unlike turbo engines, they generally exhibit quite low pressure in the exhaust tract; often less than 1psi. Consequently the exhaust responds more readily to conventional exhaust tuning techniques using either free flow or acoustic wave tuning principles, or a combination of both. Here, rather than just relying on the upward motion of the piston on the exhaust stroke to force exhaust gases out of the cylinder, we want (if at all possible) to make use of wave movement throughout the exhaust headers to create a vacuum which will actually pull exhaust gas out of the cylinder. This action has many beneficial outcomes which ultimately translate into a more powerful and efficient engine.

First, this vacuum (or pull) on the exhaust gas improves exhaust flow out of the cylinder. Consequently, even an engine with restrictive exhaust valves and/or poorly designed exhaust ports stands a better chance of ridding the cylinder of exhaust gas in the short time available at high engine rpm. This, in turn, means less contamination of the inlet charge as it flows into the cylinder on the inlet stroke. Further, with less exhaust gas left unscavenged in the cylinder there is more cylinder volume available to be filled with potentially power producing fuel/air mixture. Also, you will recall from earlier chapters that less unscavenged gas means less heat is added to the inlet charge, so it expands less on entering the hot cylinders, and thereby helps us achieve a higher charge density before the inlet valve closes on the compression stroke. Also, reduced charge temperature means there is less risk of pre-ignition or detonation, and during the combustion phase the mixture burns more uniformly early on in the power stroke. This increases the push on the crankshaft, which we read on the dyno as more hp, or feel through the seat of our pants as better acceleration.

When our engine has very good exhaust porting and suitable cam timing, proper exhaust tuning provides an additional service during the valve overlap period (i.e. when both the inlet and exhaust valve are open). If, when the inlet valve begins to open, the exhaust header is exerting a vacuum on the cylinder this will begin to draw in the inlet charge even though the piston is still rising at the completion of the exhaust stroke. In effect this extends the time available for the cylinder to completely fill with fuel/air mixture, leading to a higher power output. Consequently we can attain the hp we desire with less cam timing, which gives us a more flexible engine with a broad power band.

REDUCING PUMPING LOSSES

Not to be forgotten when we discuss the exhaust system is the issue of pumping losses. Most people don't have too much trouble accepting that it takes power to drive a blower to force air into the cylinders, but they have difficulty understanding that power is consumed pushing exhaust gas out of the cylinders. It is true that, as soon as the exhaust valve opens, exhaust gases flow into the exhaust owing to the pressure in the cylinder being considerably higher than in the exhaust tract. However, don't forget that we opened the exhaust valve early, wasting energy that could have been used to provide more push on the crankshaft, in part to have more time in which to rid the cylinder of combustion gases. The prime reason for this action, though, is to reduce pumping losses.

We have to remember that as soon as the crank swings past bottom dead centre (BDC) and the piston starts its way up on the exhaust stroke it has to literally ram the exhaust gas out past the exhaust valve and through the exhaust system. This action consumes power, power provided by another cylinder on its power stroke, so there is less power measured at the flywheel and less power to accelerate and drive the vehicle.

There are two routes available to cut these pumping losses. The lazy way is by opening the exhaust valve very early, providing more time for cylinder pressure to 'blow down' as the piston travels to BDC, thereby leaving a smaller volume of exhaust gas for the piston to have to push out on the next stroke. Even though opening the exhaust valve very early costs us some power and economy at cruise, at higher rpm and wide open throttle the losses if the piston had to ram out a larger volume of exhaust gas would often be much greater, so in effect this action provides a net gain in the upper half of the power curve.

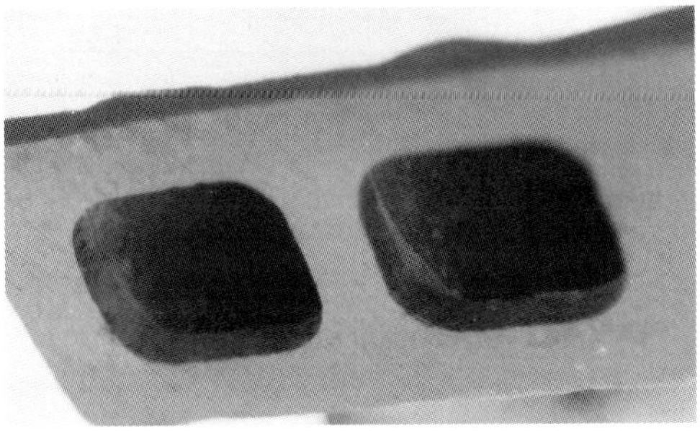

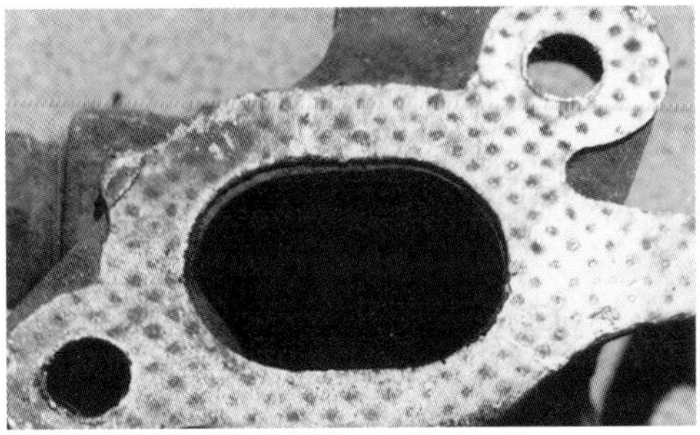

Left: Header tubes overlapping the exhaust flange impede flow and increase pumping losses.

Right: Grinding the manifold ports to match the gasket ensures that no 'step' exists to restrict flow and increase pumping losses.

The other route, which is the focus of our discussion, is to ensure that the entire exhaust flow path is developed to be free of obstacles to flow and also actually incorporate elements to encourage flow. As was pointed out in Chapter 14, this begins with correct modification of the combustion chamber, the exhaust valve and seat, and the exhaust port. From that point on it is the exhaust headers and the rest of the exhaust system, including the CAT and silencer, where we direct our efforts to minimise pumping losses. When correctly done this will unleash a good deal more power. Also, with road engines we have the potential to reduce fuel consumption at cruise from reduced pumping losses and from being able to successfully ignite a leaner fuel mixture. (Exhaust gas separates the air and fuel molecules, so to ensure good combustion more fuel has to be injected.) Also, there is the possibility of employing slightly shorter cam timing than would otherwise be practical. So, with exhaust valve opening delayed a few degrees we extend the power stroke a few degrees and waste less energy out of the tailpipe when operating at small throttle openings.

EXHAUST SIZE CONSIDERATIONS

We reduce obstacles to flow by ensuring that the flow path is sufficiently large to move the gas volume in the time available, and by eliminating sharp bends and steps. On the surface that doesn't sound as though one would require a degree in rocket science to come up with something suitable, but there are other factors which seriously cloud the issue as to what design is best. The first problem to confront is that engines do not operate at a fixed rpm point and a fixed throttle opening. Rather, at times, the throttle will be wide open with the engine barely breaking into the lower reaches of the power band, while at other times the converse would be true. Consequently flow volume is never fixed, and what is a suitably sized flow path at one particular combination of rpm and throttle will be way too large at another point and too small at yet another.

Obviously a compromise has to be struck, but again the answers are in no way clear cut. For instance, with a nitrous engine: Are you going to size the exhaust for when the engine is on gas, or when it is running naturally aspirated? With a road car or a pure competition vehicle the answer is to build the exhaust to favour the car's intended use.

Therefore, for the road car the exhaust would be biased towards normal road use, and for the drag car the exhaust would be sized for a full dose of nitrous. The same idea applies to turbo engines employing nitrous injection to help the turbo spool up.

With blower engines I would nearly always build the exhaust to help the engine when operating under boost. Generally, such engines have good low speed torque and a small exhaust will not help them much to gain any more, but it will certainly choke the engine and hold it back from reaching its potential at full throttle and higher rpm. The time I would opt for an exhaust to help bottom-end grunt is when you have a small engine in a heavy car, especially if the gearbox has big gaps between gears and a massively overdriven top gear.

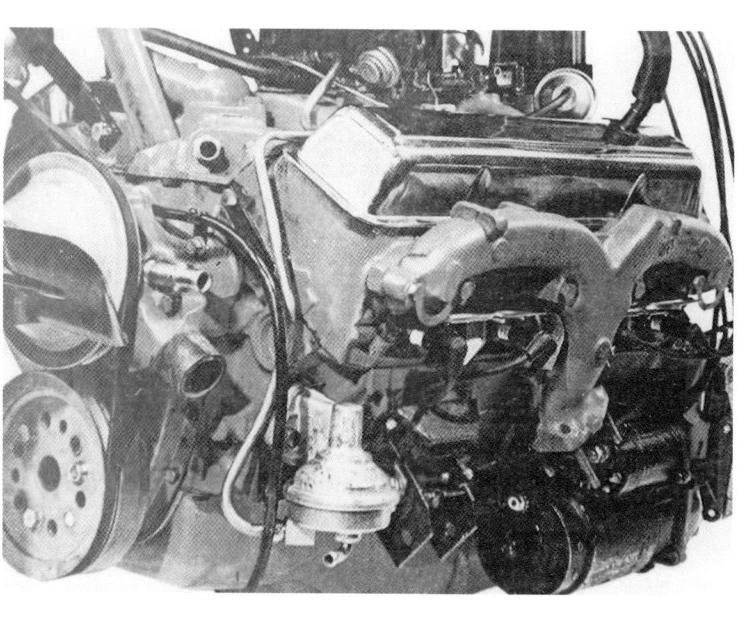

Above: When mild cams are employed 'split' type stock manifolds provide good exhaust flow on straight-6 engines.

Left: The centre dump Chev 'rams horns' exhaust manifolds work well with mild cams.

EMPLOYING EXHAUST PRESSURE WAVES

How we encourage flow is by means of wave resonance tuning. This means connecting the cylinders in a particular order using exhaust tubing of a specific length to take advantage of exhaust pressure waves to pull exhaust gas out of the cylinder and then, during the valve overlap period, suck in the intake charge to maximise cylinder filling. With small cams this isn't so important, but as we get more aggressive with our camming so the gains to be made when we get it right, or the losses we make when we ignore it, grow exponentially.

Consequently, when short duration cams with very little valve overlap (i.e. what I rate as street cams in Chapter 14, Table 14.6) are employed, it may be quite feasible to stick with the stock cast iron or tube type exhaust manifold. What we are looking for in this instance is that the header joins the correct cylinders together and that the exhaust 'branches' are sufficiently long to reduce exhaust contamination from one cylinder to another. This means that when we consider the firing order of the engine we join the exhaust primary branches so that the exhaust gas from one cylinder will not pressurise another cylinder.

EXHAUST HEADER DESIGN

Taking a typical 4-cylinder engine with a 1-3-4-2 firing order, it can be seen that at the end of its exhaust stroke the No. 1 cylinder will tend to draw in exhaust stroke gas being expelled under high pressure from the No. 3 cylinder if a log type cast-iron manifold which connects all four exhaust ports together is used (*Fig. 16.6*). Naturally No. 3 will draw in gas from No. 4, No. 4 will draw in gas from No. 2 and so on. To overcome this problem a tube type header or a cast-iron manifold with branches up around 10in to 12in long is necessary for what I have rated as street cams.

The exact design of the header may see the branches connected 4 into 1 or 4 into 2 into 1 (also called tri-y). In 4-cylinder engines operating at less than 8000rpm I get better results, in the form of a much broader power band, with the 4-2-1 style (*Fig. 16.7*). When top-end power is more important the 4-1 header would be chosen, but this design requires considerably more dyno time to get pipe diameters and lengths right.

Fig. 16.6 When an exhaust manifold connects all the exhaust ports of a 4-cylinder engine together, the gases exiting one cylinder can contaminate another cylinder, diluting the air/fuel mixture.

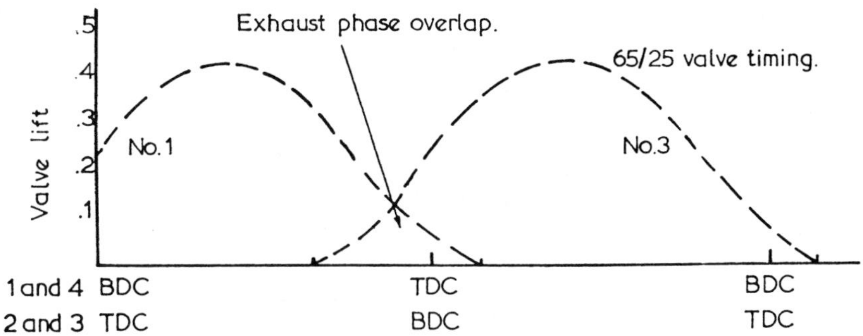

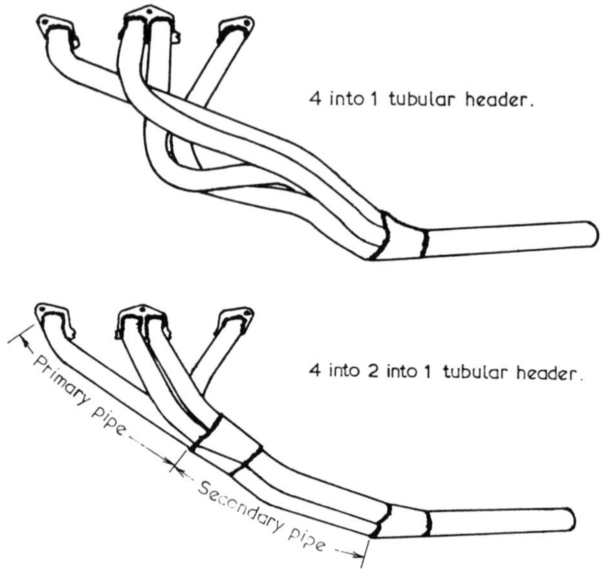

Fig. 16.7. 4-cylinder engines may use either 4-1 or 4-2-1 style tube headers.

For V8s with 'street' cams I always favour cast-iron manifolds like the centre dump Chev 'ram's horn' style or the 4-2-1 type tube headers as this gives considerably more working room around the engine. With bigger cams the 4-1 type always works better. The main thing not to be overlooked with V8s is that the exhaust pipes from the left and

Fig. 16.8. V8 tri-y headers and balance pipe.

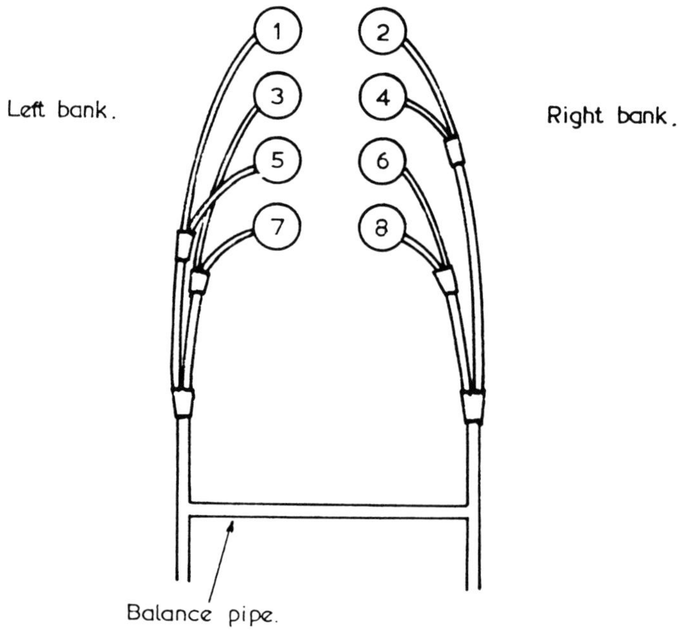

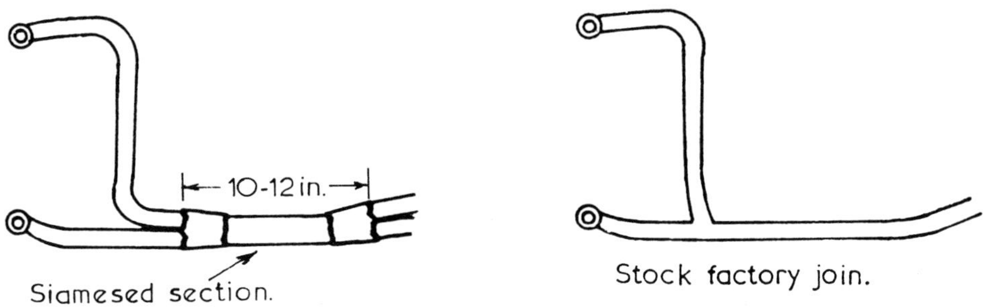

Fig. 16.9. When V6 and V8 exhaust pipes come together, every effort should be made to make the 'Y' connection between the two engine pipes as free flowing as possible.

right banks should be joined together as one pipe over a length of approximately 10in to 12in. Alongside the gearbox is about the earliest you can do this. If you find that it is simply impossible to get the two pipes running side by side in order to be joined as one, then you should use a balance pipe to connect between them (*Fig. 16.8*). This should be of the same diameter as the main exhaust pipes. Unfortunately, a balance pipe doesn't give anything like the power increase and it usually only works in the lower half of the rev range. On the other hand, a 10–12in sleeve will often give double the hp increase possible with a balance pipe, and this is over the entire power band.

Stock 6-cylinder exhaust manifolds fall into two categories; very good and plain horrible. Fortunately, very few straight-6s use a cast-iron manifold which connects all exhaust ports together and dumps through a single outlet. Such a manifold creates a lot of exhaust back pressure and, because of overlapping exhaust open periods, each cylinder will draw in exhaust gas from two other cylinders during its exhaust cycle. V-6 engines, of course, are not a problem in this respect and most modern straight-6s use a 'split' type cast manifold which joins cylinders 1-2-3 to one exhaust pipe and cylinders 4-5-6 to a second pipe.

Probably the biggest problem I see with V-6 exhausts is the manner in which the pipes from both banks are brought together as one. Frequently one pipe will almost form a right-angle tee with the other pipe. This type of connection totally destroys exhaust flow and creates massive back pressure. *Fig. 16.9* illustrates how the two pipes should be brought together to form a single exhaust. However, as with V-8s, most sixes show a power increase when the two pipes join over a length of 10–12in up beside the gearbox and then split again to form two separate exhausts.

DETERMINING HEADER DIMENSIONS

As previously mentioned, when cam timing and valve overlap periods increase, so the need to be more specific with exhaust header dimensions becomes a necessity. Part of the reason for this is that at certain rpm the cam will be causing a number of undesirable things to happen, and we can use good exhaust design to reverse, or at least minimise, those problems. Also, we can call on the exhaust to work in harmony with the cam, particularly at higher engine speeds, to maximise both the scavenging of exhaust gas from the cylinders and their being refilled with high density air/fuel charge in the very short time frame open to us.

This means that not only do we have to connect the exhaust side of the cylinders in the correct order, but also we have to select exhaust tubes of the appropriate diameter and length. Obviously, the best place to do this type of experimentation is on an engine dyno. However, it is possible to achieve reasonable results if we understand the principles of exhaust tuning and then make choices in accordance with the basic rules.

The first thing we must appreciate is that exhaust gas is expelled from the cylinder at a velocity of between 200ft and 300ft per second, but pulses or pressure waves are moving through that gas at around 1500ft to 1700ft per second. By understanding the behaviour of these waves we can use them to improve cylinder scavenging, and to improve cylinder filling with air/fuel mixture. What happens is that, as the initial charge of burnt gas bursts from the cylinder into the exhaust, it creates a wave of positive pressure that travels at the speed of sound through the gas to the end of the pipe. As it surges into the atmosphere the positive wave dissipates and produces a negative pressure wave (suction wave), which returns along the exhaust pipe into the cylinder. Because its pressure is much lower than cylinder pressure it arrives at the exhaust valve with a certain amount of evacuation power. The art of exhaust tuning is to ensure that this suction wave arrives back at the cylinder during the valve overlap period and at engine rpm when its effect will bring greatest benefit to the shape of the power curve.

Header pipe diameter fixes the velocity of exhaust gas through the pipe. A large diameter pipe, relative to the size of the cylinder, will reduce gas velocity. Because engines typically produce peak torque at a mean gas velocity of around 250ft/sec the header pipe diameter, in turn, influences at what rpm peak torque will occur. Thus a large pipe size moves the rpm at which peak torque is produced further up the rev range.

Changing the length of the pipes tends to 'rock' the power curve of the engine around the point of maximum torque. Adding length to the pipes will increase low speed and mid-range power, with a reduction in power at maximum rpm. Shorter pipes give an increase in high-speed power at the expense of a reduction in the mid-range. However, there will be little change in the peak torque number or the engine speed at which it occurs (*Fig. 16.10*).

Fig. 16.10. This graph illustrates how header tube diameter and length influence the shape of the power curve.

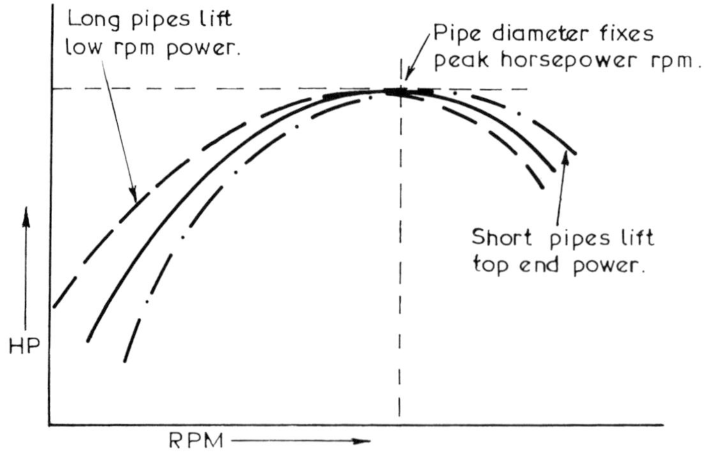

With these factors in mind, a good starting point – whether you are selecting an off-the-shelf header or you are setting out to design your own – can be found by referring to Table 16.4. For a 4-cylinder road engine or a competition engine operating at less than 8000rpm I would always choose a 4-2-1 interference header style. Keep in mind that the length of the secondary pipe also includes the length of the first collector. All street engines, regardless of the number of cylinders or the type of header, would use the smallest pipe diameters listed. Engines intended for hill climb or ski boat events would generally use an intermediate pipe diameter. In most instances only maximum output engines would benefit from headers using tubes of the largest size.

Table 16.4. Header pipe sizes for blown engines with sports and competition cams

Cylinder Size (cc)	4-2-1 interference header		4-1 header 3-1 header
	Primary pipe (in)	Secondary pipe (in)	Primary pipe (in)
250	1.375–1.5 dia x 26–30 long	1.625–1.75 dia x 8–12 long	1.625–2.0 dia x 29–38 long
325	1.5–1.625 dia	1.75–1.875 dia	1.625–2.0 dia
400	1.625–1.75 dia	1.875–2.125 dia	1.75–2.125 dia
500	1.75–2.125 dia	2.0–2.25 dia	1.875–2.125 dia
600	1.875–2.125 dia	2.0–2.5 dia	*1.75–2.0 dia
700	1.875–2.125 dia	2.0–2.5 dia	*1.875–2.125 dia
800			1.875–2.25 dia
900			2.0–2.375 dia
1000			2.125–2.5 dia

*4-cylinder engines would use pipe sizes 0.125–0.25in larger.

Note: All pipe lengths are as shown for 250cc cylinder and include the length from the exhaust valve to the end of the exhaust port. The secondary pipe length also includes the length of the first collector. All pipe diameters are outside diameters.

As mentioned previously, engines with street cams rely more on exhaust flow characteristics than pulse tuning. Therefore, pipe diameters which provide adequate flow are chosen, while tube length should be such that there should not be any back flow of exhaust gas from one cylinder to another. As you can see from Table 16.5, this leaves header choice quite open. However, for street engines I would tend to choose headers of the smallest diameter listed when using centrifugal blowers, particularly if the engine is small relative to vehicle weight.

Table 16.5. Header pipe sizes for blown engines with street cams

Cylinder size (cc)	4-2-1 interference header		4-1 header 3-1 header
	Primary pipe (in)	Secondary pipe (in)	Primary pipe (in)
250	1.25–1.375 dia x 12–18 long	1.5–1.625 dia x 15–24 long	1.25–1.375 dia x 15–30 long
325	1.375–1.5 dia	1.625–1.75 dia	1.375–1.5 dia
400	1.5–1.625 dia	1.75–1.875 dia	1.5–1.625 dia
500	1.625–1.75 dia	1.875–2.0 dia	1.625–1.75 dia
600	1.75–1.875 dia	1.875–2.125 dia	+*1.625–1.75 dia
700	1.875–2.0 dia	2.0–2.125 dia	*1.75–1.875 dia
800			1.75–1.875 dia
900			1.875–2.0 dia
1000			1.875–2.0 dia

*4-cylinder engines would use pipe sizes 0.125 larger.
+V6 engines would use pipe sizes 0.125 larger.

Note: All pipe lengths are as shown for 250cc cylinder. The secondary pipe length also includes the length of the first collector. All pipe diameters are outside diameters.

THE COLLECTOR'S INFLUENCE

Regardless of the style of tube header used, in most instances the individual exhaust tubes would be joined together into a single pipe by means of some sort of collector. The main effect of this action on engine performance is to increase torque output at engine speeds below the torque peak. Once past the maximum torque rpm the collector no longer influences the power curve.

Generally, we want to boost engine torque whenever possible. However, in some circumstances excessive torque at lower rpm may be a hindrance, actually reducing overall performance. Consequently, in some classes of drag racing the pipes from individual cylinders are not brought together with a collector, as any increase in mid-range torque just makes traction problems more difficult. You may have witnessed the same sort of thing with sprint cars on dirt speedways. When there are no noise rules, and when traction is limited, you will see these cars racing without collectors in order to get rid of unwanted torque. Sprint cars aren't blown, but the same principle applies.

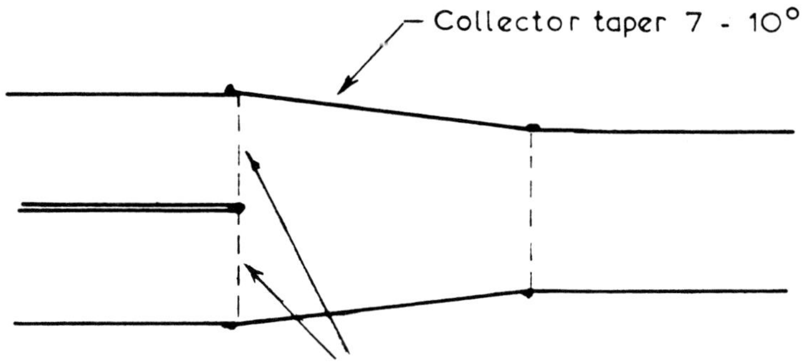

Fig. 16.11. Baffle-type collector.

BAFFLE COLLECTOR DESIGN

Fig. 16.11 illustrates an ordinary collector that I personally prefer to call a 'baffle' collector. With this arrangement the primary tubes terminate abruptly into an open tapering chamber that feeds into the tailpipe. When the primaries terminate abruptly like this a pressure wave returns up the primary. Now, if a negative pressure wave (suction wave) arrives back at the exhaust port during the time when the exhaust valve is next open, it will work in harmony with any vacuum created by exhaust gas momentum to aid in efficient evacuation of that cylinder of exhaust gas, also pulling in fresh air/fuel charge. Also, depending on primary pipe length, at certain engine rpm a positive pressure wave will arrive just as the exhaust valve is about to close to reduce the amount of fuel/air charge that 'spills over' (called over-scavenging) into the exhaust during the valve overlap period owing to the effect of exhaust gas momentum.

That is the positive aspect of what this type of exhaust collector will accomplish. The negative side is that at certain rpm a positive pressure wave will arrive back at the exhaust port during the early phase of the exhaust valve opening period and actually impede exhaust flow out of the cylinder. Likewise, in other parts of the rev range, a negative pressure wave will arrive just as the exhaust valve is closing and draw even more air/fuel mixture out into the exhaust. Obviously, both scenarios lead to reduced hp and increased fuel consumption.

Consequently, this type of collector is better suited to road engines and competition engines using milder race cams with a duration of up to about 260° measured at 0.050in lobe lift, and lobe centres in the range of 105–110°. In this kind of application the baffle collector works very well.

Some competition headers using a baffle collector have a 'spear' welded to the centre of the baffle where all the primaries terminate. This is supposed to reduce flow turbulence, but I believe that devices like this are really just a marketing gimmick. The main thing to watch with any collector is the angle at which it tapers. In general terms, I have found the best angle to be 7° or 8° (14° or 16° included angle). However, some engines will work best with a steeper taper of 9°, or even up to 10°. Very shallow tapers of less than 7° do not seem to work too well, probably because of excessive collector volume allowing massive flow turbulence.

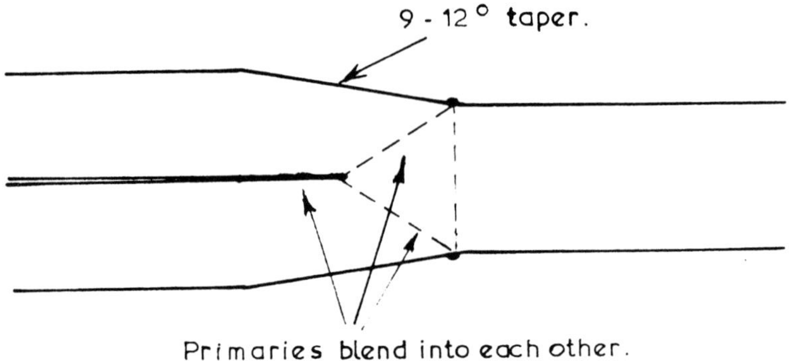

Fig. 16.12. Merge-type collector.

MERGE COLLECTOR DESIGN

The collector that is better suited to maximum-effort race engines with long-duration, big-overlap cams is known as a 'merge' collector (*Fig. 16.12*). With this type the primary pipes do not terminate abruptly, but 'merge' or blend into the tailpipe. This has the effect of reducing the intensity of reflected pressure pulse waves in the primaries, which in effect tunes out of the header the negative aspects of pulse tuning mentioned previously. Of course, using this type of collector means that the headers lose out on the positive aspects of pulse tuning. However, as we are talking about maximum-effort engines, operating over a very narrow rev range, this gives the exhaust tuner more scope to build the headers to scavenge the cylinders effectively, relying on exhaust gas momentum or inertia alone.

With the merge design, the primaries must generally come together fairly abruptly to give a collector taper of about 10°, and I would always begin testing at this angle. Some very peaky high-rpm engines respond well to even steeper angles of around 12°, and a number also work best with slightly shallower tapers around $8\frac{1}{2}$° to 9°.

HEADER FABRICATION AND FINISH

When it comes to header design and fabrication I regularly see many errors, and very basic errors at that. Enthusiasts waste hours polishing the inlet tract to a mirror finish, then throw on a header with welding 'dags' hanging into each tube where they are welded to the flange. A few minutes with a grinder or round file would rid the header of such impediments to good gas flow.

Another factor that many do not seem to understand is that the header is an extension of the exhaust port, so any abrupt change in direction as gas flows from the port into the header is to be avoided. Many headers leave much to be desired in this area. Often tubes bend through a full 90° right at the flange, and in the very worst examples I have seen, pipes cut off at around 50°, then welded directly to the flange to avoid the time and expense of actually bending the tube (*Fig. 16.13*).

To compensate for sloppy manufacturing and to ease fitting problems, grossly oversized flange holes are common, possibly allowing the header to overlap the exhaust

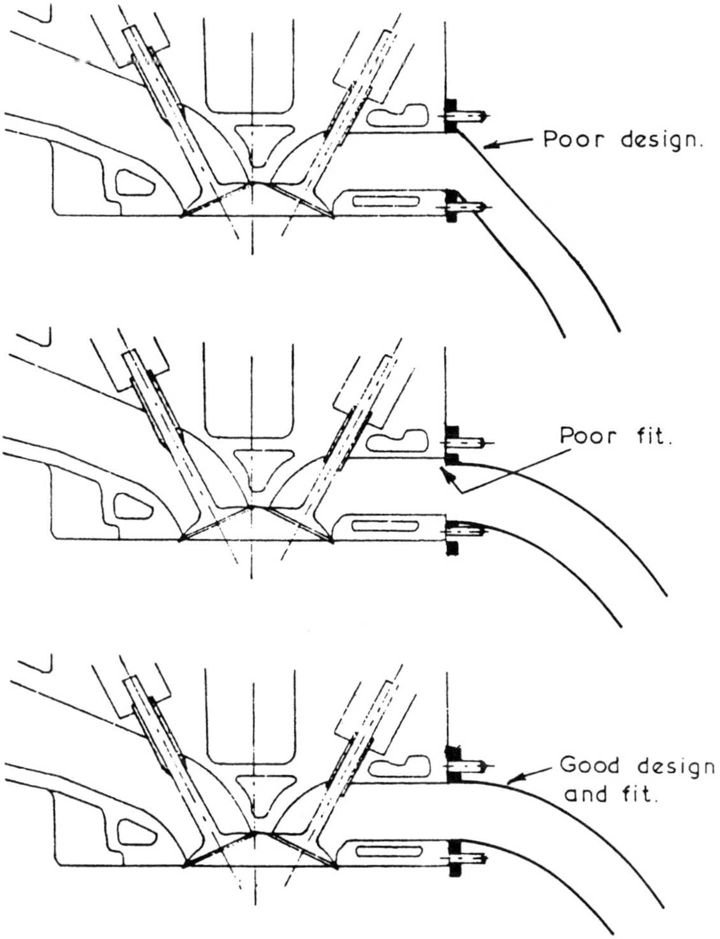

Fig. 16.13. The header is an extension of the exhaust port.

port, thus restricting gas flow. This can be rectified by pushing sleeves into the oversize holes, then checking for overlap using the manifold gasket as a template. It should be noted, though, that with turbo headers this often isn't a good idea. With these we have to allow for massive expansion and contraction so if the header is a neat fit on the studs they are quite likely to be sheared right off.

Another problem area that has arisen since mandrel pipe-benders have become more commonly available to general exhaust workshops is a real lack of design planning in the placement and severity of bends in the header pipes, and indeed in the entire exhaust system. Years ago, when most pipe-benders noticeably reduced the tube diameter at bends, a real effort was made to limit the number of bends and their severity. Just a quick glance at the exhaust told you that any sort of bend was bad news for gas flow. However, when mandrel benders came along the resulting bends looked so much better that many forgot that bends do impede gas flow. Also, be aware that few so-called mandrel benders produce a perfect constant-diameter bend; many, in fact, reduce the pipe size by about 1mm.

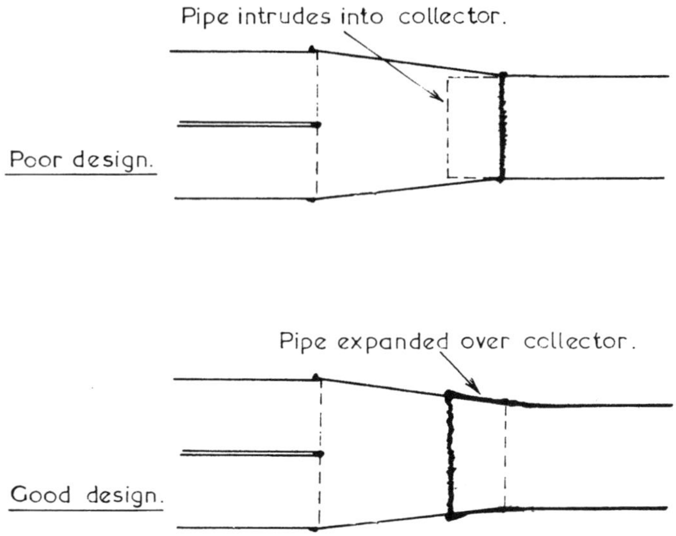

Fig. 16.14. Collector connecting methods must not impede gas flow.

In some turbo applications, small tube headers are called for, but there is confusion over what to do when the exhaust port is larger than the header id. Obviously, the lip formed when a 1.6in diameter port meets with a 1.375in id header tube will cause massive flow turbulence, but this is exactly what some small tube headers are like. Some try grinding out the flange to get rid of such a lip, but this does not really help very much. In a situation like this I prefer to flare the tubes gently over a length of at least 1in, and preferably 1½in. However, if the pipes have to bend very close to the exhaust port this may not be practical, so it may be necessary to come off the ports in larger diameter tube for, say 5–10in, then step down to the smaller tube size. However, again flare the small tube up to the size of the larger over a length of 1–1½in to avoid a lip.

The final thing we have to consider is that the method employed to join the collector(s) to the next section of exhaust tube does not impede gas flow (*Fig. 16.14*). Many mass produced headers are very bad in this respect. I have seen many where, to expedite welding operations, the exhaust tube protruded up to half an inch into the collector exit. At the very least the collector exit should butt with the next section of tube. However, for strength, and to avoid any possibility of flow restriction, it is preferable for the exhaust tube to be expanded and actually lap over the tail of the collector. Of course, the opposite applies at the entry of the collector.

TAILPIPE DIMENSIONS AND FABRICATION

After the exhaust header we have either just a tailpipe in the case of some competition vehicles, or the rest of the exhaust system in the case of cars required to meet emissions standards and/or noise rules. If the exhaust is open, the tailpipe length will usually be just slightly shorter than the overall header tube length when maximum mid-range torque is being sought. Do keep in mind that the tailpipe length also includes the length of the collector. Less length has the effect of reducing low speed torque, but it may bring some increase in peak hp or allow the engine to over-rev well past maximum power rpm.

Therefore, some high speed engines will have a tailpipe which terminates at the end of the collector.

Tailpipe and exhaust system diameter is best determined during dyno testing. However, a good starting point can be found by referring to Table 16.6. When a catalytic converter has to be used the pipe size indicated may have to be increased. This applies particularly to road cars where the CAT is required to be located in the stock position. The reason is that a CAT that is actually doing its job adds considerable heat to the exhaust gas, so to keep flow velocity and losses down after the CAT we have to go to a larger pipe size.

Table 16.6. Tailpipe and exhaust system size for blown and nitrous engines

hp	pipe od (in)
80–120	1.875
110–140	2.0
130–150	2.125
140–185	2.25
180–220	2.5
210–265	2.75
250–320	3.0
280–360	3.5
400–500	4.0
480–630	4.5
580–750	5.0

Note: When two exhausts are fitted, divide engine hp by two.

Naturally, the entire system must have gentle bends as well as adequate bore size to minimise flow restriction. Unfortunately, with road cars it can prove quite difficult to find sufficient space to fit a big bore exhaust while still maintaining ground clearance. When this is a problem, remember that it is quite acceptable to ovalise the exhaust over

Ordinary exhaust pipe benders visibly deform the pipe (below). With a mandrel bender it isn't so obvious (above), but the white background shows that it still deforms the inside radius, and over about twice the length!

much of its length. You can even 'flatten' it by a further 0.25in over a 6in length, if necessary, to fit closer to crossmembers, etc. Also, because exhaust gases cool and contract as they get towards the rear of the car, it is in order to reduce the pipe diameter after the rear axle. This won't have much effect on power output and it may help reduce noise, thereby allowing the use of a less restrictive silencer (muffler).

SILENCER DESIGN AND SELECTION

The type and number of mufflers fitted will depend on what the law requires and on how quiet you want your car. My preference is for a quiet mellow tone which doesn't give me a headache. If fitted as far to the rear of the car as possible, a well-designed silencer will not reduce power by more than 3–5%. Fitted closer to the engine, where exhaust gases are very hot, they cause more flow restriction and power loss.

There are many designs around, but the types most suited for performance applications are of either straight-through or reverse flow (also called 'S' flow) design. Straight-though types don't have to be loud; in fact, many quiet high quality silencers are of this design. I prefer this type, but on non-turbo cars two silencers or a silencer and resonator may be necessary to prevent exhaust 'popping' on over run. The quietest silencer designs often have an open resonating chamber in the middle for this reason.

Reverse flow silencers are quite different and generally do not contain any sound absorbing material. This cuts their weight considerably but, unfortunately, poorly designed units tend to be noisy, and the noisy ones often make less power. The main benefit of the reverse flow design is that it is less prone to popping when you lift off the throttle and coast. Consequently, on non-turbo cars often only one silencer, or a silencer and small resonator, will be all that is required to produce an acceptable exhaust tone.

CAT PROBLEMS AND MODIFICATION

When it comes to CAT converters, the advice is to have a look at what is available, check the rules to see where it must be located and, if there is no specific placement rule,

If the core of the CAT looks like this on the inlet side, it is probably all right. A more reliable method of checking for CAT blockage is to measure the back-pressure before the CAT when it is brand new, and then at regular intervals thereafter. Do the check at full throttle in 3rd gear holding the car on the brakes at 5000rpm.

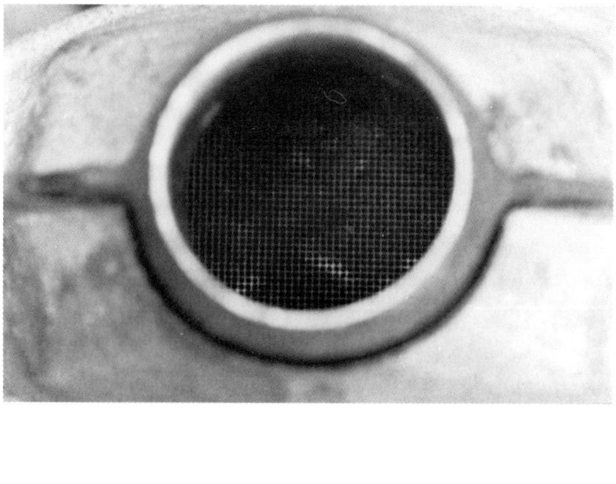

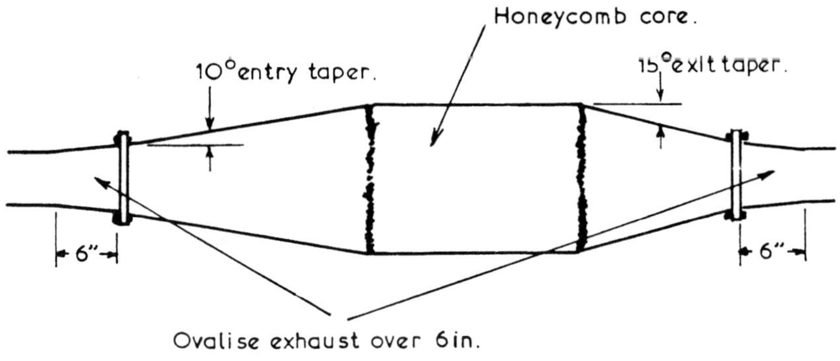

Fig. 16.15. CAT modifications improve flow.

move it around while testing on the dyno. Note that in high-performance and competition environments CAT flow performance must regularly be checked as many do not hold up well to extended wide-open throttle running. A collapsed CAT can easily rob a 300hp engine of in excess of 20hp.

However, it is not just the honeycomb innards of the CAT that cause an exhaust flow problem. Even when the honeycomb is in good condition, a poorly designed CAT will impede gas flow owing to turbulence caused by the exhaust diverging at anything from 35° to 60° into the honeycomb, then converging again into the tailpipe at a similarly steep angle. To overcome this problem the CAT must have a gentle entry and exit taper. On the inlet side a taper of 10–12° allows the exhaust gas to expand gradually to the full size of the honeycomb with minimal turbulence. Similarly, on the exit a taper of 12–15° forces the gases to converge into the tailpipe without unduly disturbing the exhaust flow. When modified in this manner ordinary street CATs will double their flow rate (*Fig. 16.15*).

Chapter 17

Combustion and the ignition system

When engine performance is discussed, the main thread of the conversation invariably involves such things as the size of the engine, cylinder head flow potential, camshaft timing and lift, etc. And, in the case of forced induction engines, we could add the turbo

In spite of what the manufacturers of aftermarket ignition components would have us believe, combustion involves a good deal more than the ignition system. The reality is that these parts only start the fire.

or blower boost. While all these aspects of performance are relevant, they are actually only one half of the performance equation. The other half is equally, if not more, important for it focuses on releasing every drop of energy from the fuel/air charge which has been drawn into the engine. It is relatively easy to come to an understanding and give numbers to the things to which the performance industry tends to divert our attention. However, the process of igniting the fuel/air mixture and ultimately totally consuming it by means of the process of combustion is not so easy to quantify; it is considerably more abstract. Possibly this is why this most important facet of performance is so seldom discussed, nor elevated to its rightful position of importance.

Most of the people involved in engine development appear to understand the importance of filling the cylinders with fuel/air charge, but unless we take proper steps to release the energy locked up in that mixture we have pretty well wasted our effort in getting it into the cylinders in the first place. Consequently, if we are to have our engine operating at its true potential we have to ensure that the fuel charge is ignited at the correct time with a spark of sufficient intensity to initiate combustion. Then we have to use a number of 'tools' to keep the flame, which started at the spark plug, moving through the entire combustion space at the correct speed. If flame travel is slow, cylinder pressure will rise too slowly to give the crankshaft a good sustained shove. On the other hand, if combustion is too rapid, combustion can run away and become violent, and like a nuclear reactor meltdown, eventually destroy the engine.

FACTORS INFLUENCING THE COMBUSTION PROCESS

From earlier chapters you will have gathered that there is a lot more involved with combustion than the ignition system. In reality, the ignition system's true function is to initiate combustion. To that end, it must be robust enough to get the flame started. The engine management system controls when the spark plug fires to start that flame off, but there are a whole host of factors which influence the speed at which the flame then proceeds through the whole combustion space. When we come to understand these factors we are in a much better position to extract the maximum potential from our engine.

The major factors are mixture quality, mixture movement (or turbulence) within the combustion space, and the design of the combustion space. All appear to be relatively simple concepts, but actually each is extremely complex owing to the fact that internal combustion engines operate over a broad range of rpm and load conditions. Therefore, what is optimal under one set of conditions will be less than optimal at other rpm and throttle open angles.

MIXTURE QUALITY AND TURBULENCE

Mixture quality involves filling the cylinder with fuel particles of the correct size, evenly dispersed throughout the air. During the preparatory stage prior to combustion we have to break the liquid fuel down to a vapour. That means that outside of the cylinder we have to choose the correct type of fuel and fuel injector, and we need to inject that fuel into the airstream at the optimal location and angle. Following that we have to rely on the physical shape of the inlet port and valve, and their angle to the cylinder to impart a swirling or tumbling motion to this fairly poorly blended mixture of fuel and air as it enters the cylinder (*Fig. 17.1*).

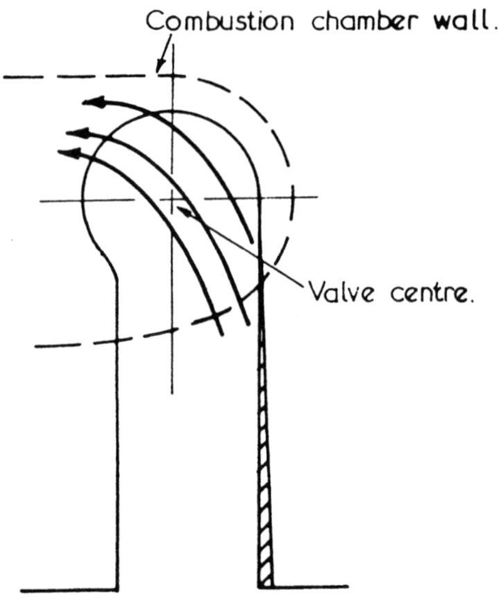

Fig. 17.1. An inlet port offset from the valve centre promotes swirl. Moving the port further to the right will increase swirl.

Once in the cylinder we want this action to continue. The shape of the combustion chamber and piston crown then become a major controlling influence. Four-valve combustion chambers and flat-top or dish-top pistons are superior in this regard as they do not impede the tumbling or swirling action of the mixture. All of us have observed the swirls and eddies in a flooded river, and have no doubt marvelled at how peaceful the flow can be in one area while right alongside there is great turbulence. A similar action is in progress in many 2-valve combustion chambers, and also in those engines with high-top pistons. Even very minor redesigns of shallow valve cutouts in the crown of flat-top pistons fitted in 4-valve engines has resulted in power increases in the range of 2–3%. Therefore, the impact of the shape of the combustion space on mixture motion and combustion flame propagation should not be underestimated, as the possible gains are much greater in engines with less favourable designs.

The purpose of this mixture motion is often said to be to homogenise the mixture – to have an even distribution of fuel molecules and air throughout the cylinder. However, this agitation also serves to further break down the fuel into smaller more readily combustible particles. An equally important aspect, and at times an even more important consideration, is the fact that this motion tends to even out the temperature of the mixture. Cool pockets tend to burn slowly and so contribute nothing to power production. (This isn't totally correct. In some turbo engines and under certain operating conditions these cool pockets of mixture may continue burning late into the exhaust stroke and thus help turbine spool up.) The real danger, though, comes from hot pockets. These may cause the main body of mixture to self-ignite before the spark plug fires. This is what is termed pre-ignition. Another more common scenario is for a hot pocket to self-ignite after the spark plug fires. This flame front and the one initiated by the spark plug collide and set off violent combustion of the remaining fuel charge. This is called detonation.

THE SHAPE OF THE COMBUSTION SPACE

We have already touched on how the shape of the combustion space influences mixture motion prior to ignition. Following ignition this area continues to exert its influence. Its design can help the combustion flame spread through the moving fuel mixture at the desired speed. Also, in the area furthest away from the spark plug the combustion chamber will be pulling some heat out of the mixture, so stalling ignition in these areas where the pressure is rapidly rising until they are reached by the spark-plug-initiated flame.

In both respects the 4-valve chamber is superior. First off, the spark plug is in the middle of the combustion space. Hence, the combustion flame does not have to travel further to get to one end of the combustion chamber. An added bonus with this design is that the deepest area of the combustion space, and therefore the area where there is the greatest concentration of fuel/air mixture, is across the centre of the combustion chamber. The outer areas further away from the spark plug are shallower and in contact with a relatively large surface area. Consequently, the mixture here is less inclined to bring on detonation (*Fig. 17.2*).

Actually, the radial 4-valve chamber is even better in this respect. As the combustion chamber is shaped like a saucer turned upside down, rather than like a pentroof as in the traditional 4-valver, the entire outer perimeter of the chamber is shallow. Unfortunately, arranging the valves radially makes for a more complex valve actuation arrangement, so outside of pure competition engines the design has not gained much acceptance.

Fig. 17.2. A traditional 4-valve arrangement works well, not only owing to high air flow potential, but also because the combustion chamber shape promotes good mixture burn.

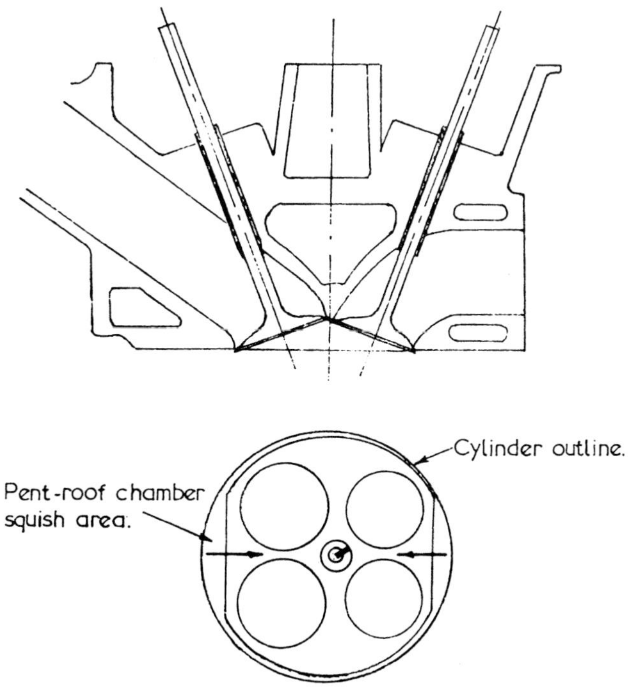

OTHER ASPECTS OF MIXTURE MOTION

In naturally aspirated engines, coming to grips with the complexities of mixture motion and combustion is extremely difficult, and when we throw forced induction into the mix there are many more aspects that we have to consider. For example, in a naturally aspirated engine these matters are always a compromise. Therefore, a road engine intended to provide acceptable cruise performance coupled with good fuel economy and low exhaust emissions will favour a design which provides a high level of mixture motion and rapid combustion at very small throttle open angles, most likely in the 2500–4500rpm rev range. However, at full throttle and high revs the situation changes. The mixture motion will now be excessive, limiting the air flow potential of the engine, and it may even throw some fuel into the liquid state in extreme cases owing to centrifugal action. Then, during the combustion phase, even though the engine is not detonating, flame speed may nonetheless be so high as to limit hp output and induce engine harshness.

In high performance road and competition engines we would eliminate these problems by changing the porting and/or the shape of the combustion chamber to reduce swirl or tumble. Then, to control flame speed, we would perhaps reduce the combustion chamber squish area, open up the clearance between the piston crown and the combustion chamber wall and change the shape of the piston crown, depending on what gave the most favourable outcome. Naturally, these changes would all contribute to a road engine being less efficient at cruise.

When we add forced induction, the same limitations arise, with turbo road engines being the most extreme. Basically we have to decide at what throttle openings and rpm

A BMW Formula 1 turbo piston has minimal squish to help stabilise combustion. Note the piston ring placement to limit heat stress on rings and to maximise ring land strength.

we want the engine to work best, and then plan our modifications accordingly. Personally, I feel there is much to be gained in 4-valve road engines by designing the inlet system around a two inlet port per cylinder arrangement. Then, at cruise, a butterfly valve can be used to block off one inlet port, thereby increasing both air speed in the other port and swirl into the cylinder. As engine speed increased and the throttle was opened wider, this valve would open, reducing air speed and swirl.

IGNITION ADVANCE

Apart from the above factors, which play such a major role in how well the engine performs, there are other influences at work which combine to dictate how late into the compression stroke we actually initiate the fire in the cylinder. Most engines produce max hp when we start ignition in order to reach peak cylinder pressure at around 12–14° after top dead centre (TDC). To arrive at such a situation we have to get the fire started some time before TDC. This is called the ignition advance angle. At lower engine speeds this may be 10° before TDC, but as the revs rise this may increase to more than 20° advance in a forced induction engine (*Fig. 17.3*), and almost double that in a naturally aspirated engine.

These changing ignition advance angles are necessary to give the air/fuel mixture the correct amount of time to burn properly. Clearly, at 7000rpm the combustion flame must have progressed the same amount through the combustion space as it would have at 3000rpm. The only way we can make up for this time deficit is to start the combustion flame off earlier.

Apart from the time factor, there are several other reasons why the ignition advance angle must be varied. The main contributory factors, apart from engine design, are mixture density, the air/fuel ratio and fuel type.

Mixture density is, of course, also influenced by the atmospheric pressure, the boost pressure, intercooler efficiency, compression ratio, engine rpm and how wide the throttle

Fig. 17.3. A full load wide open throttle ignition advance curve.

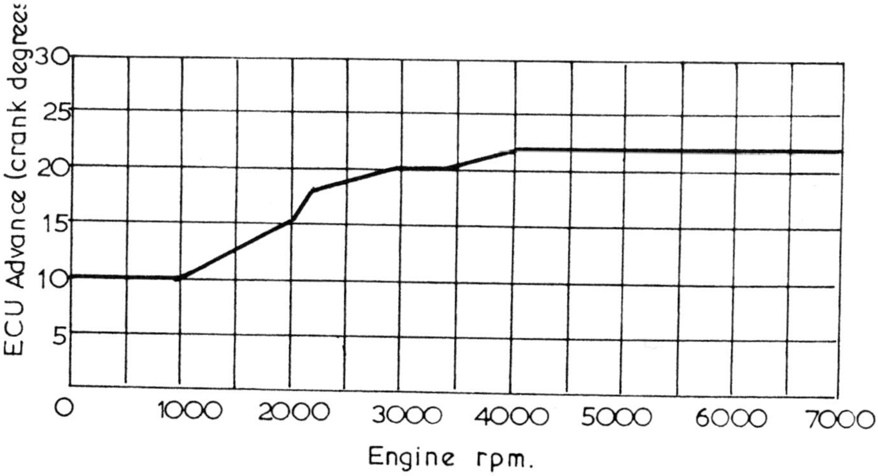

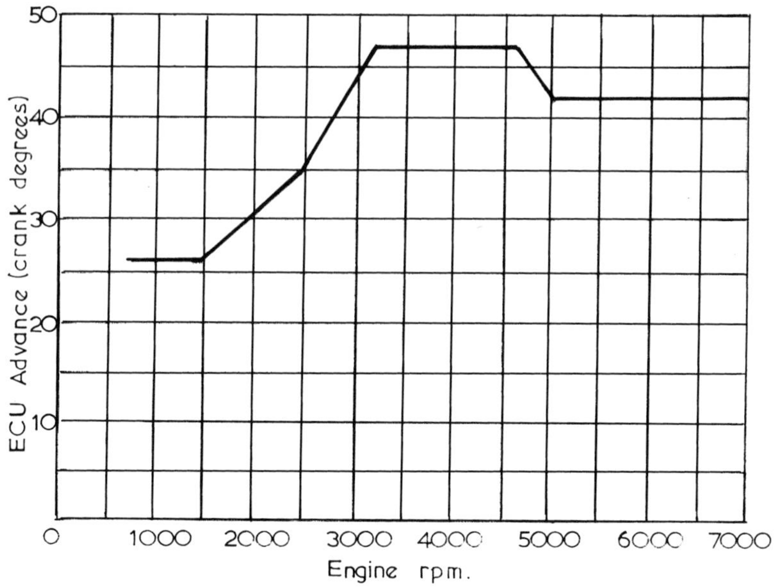

Fig. 17.4. A light load cruise ignition advance curve.

is open. When the throttle is wide open and we are operating at maximum boost with an efficient intercooler, then we are packing more air and fuel into the cylinders. Conversely, when the car is just cruising with the throttle barely cracked open, and boost is at zero, the cylinders do not become crammed with air/fuel charge. Therefore, even when compressed as the piston rises to TDC, the oxygen and fuel molecules are separated slightly, which in turn slows the rate at which the combustion flame travels across the combustion chamber. To compensate for this slower burn rate we have to start the fire much earlier before TDC (*Fig. 17.4*).

Both very lean and rich mixtures burn slowly and require more spark advance. On the other hand, an air/fuel mixture close to full power lean (around 13:1) burns the fastest and requires less advance.

When an efficient exhaust is fitted, the amount of exhaust gas left unscavenged in the cylinder, or reverse flowing into it, is reduced. This leaves more room in the cylinder to pack in air/fuel charge, so mixture density increases and spark lead must be reduced. The other effect of upgrading the exhaust system or turbo housing is that there are fewer exhaust gas molecules to get in between the oxygen and fuel molecules. Consequently, flame speed through the combustion space is faster, which again calls for less advance.

The fuel compounds used also affect combustion speed. Petrol burns fairly rapidly, so requires less advance than other fuel types. Alcohol fuels burn more slowly and nitro is slower still, which means that these fuels need more spark advance. However, when a flame accelerator compound is added to either alcohol or nitro, spark lead must be reduced.

Obviously, in a modified engine, some (or perhaps all) of the above factors will have been changed from the stock engine. Therefore, the amount of spark advance needed under various engine operating conditions will be quite different from that programmed into the engine management system by the factory. Therefore, during dyno testing an

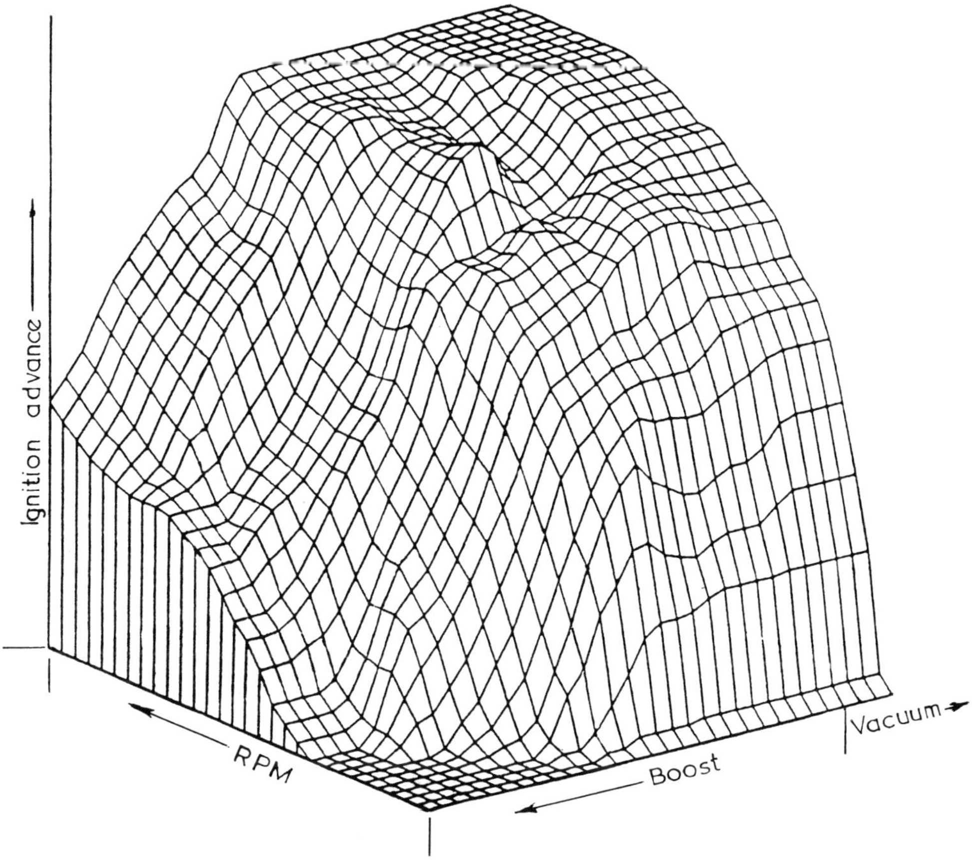

Fig. 17.5. A computerised spark timing map for a supercharged engine.

ignition map such as that illustrated in *Fig. 17.5* is built up and programmed, usually via a laptop computer, into the ECU memory. This process must be carried out at all engine speeds, boost pressures and throttle plate open angles. Also, compensation factors to adjust the base map for inlet charge temperature, engine coolant temperature and exhaust turbine temperature would be added.

POINTS-TYPE IGNITION

Before the days of electronics, ordinary ignition systems relied on a points-type distributor to time and distribute the spark to the cylinders in the correct firing order. Also, there was at least one coil, sometimes two coils in 8- and 12-cylinder engines, to provide a high-voltage spark (*Fig. 17.6*).

The distributor has two switches, the contact-breaker points and the rotor, and also an advance mechanism to vary the time at which the points open and close.

The switching of the primary (low-tension) circuit is accomplished by the contact breaker points, which are opened and closed by the distributor cam. When the points are closed, electric current flows through the coil primary winding, then through the points, to earth. The current in the low-tension winding produces a magnetic field that surrounds

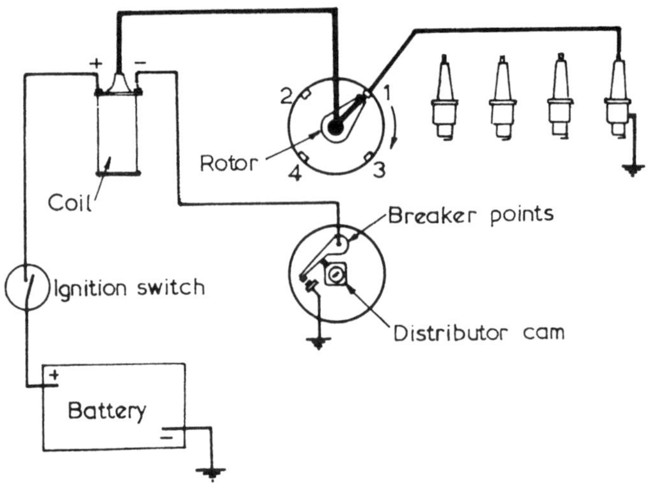

Fig. 17.6. A points-type negative earth ignition system.

the secondary or high-tension winding. As soon as the points open, current flow through the primary stops and the magnetic field collapses, causing a current to be induced in the secondary winding. This creates a high-voltage current (up to 25,000 volts) capable of jumping across the spark plug gap, to fire the fuel mixture.

The high-voltage current flows from the coil to the centre of the distributor cap and through the carbon brush to the rotor button. The turning rotor then directs the current back through the distributor cap via the distributor points to the individual cylinders.

At normal engine speeds and low-compression pressures, this conventional system works very reliably, with just periodic maintenance required to replace the points. However, when an engine is modified the ignition system has to work much more efficiently under adverse conditions.

For example, the very first problem we see with points-type ignition is that as engine rpm increase the time that the points are closed between each plug firing is decreased. This means that the coil has less time to build up a full magnetic field between each plug firing, so the ignition energy available to fire the plugs is reduced and a high-speed misfire occurs.

HIGH ENERGY ELECTRONIC IGNITION

The common electronic transistorised ignition is similar to the contact breaker system in that it is also an inductive storage-type ignition system. However, rather than using mechanical contact points to open and close the low-voltage circuit it relies instead on magnetic pulses and electronic circuitry to perform this function (*Fig. 17.7*).

A number of different designs are in use, but the basic mode of operation is similar. The distributor shaft turns a pulse generator rotor inside a stationary permanent magnet. Functioning like a mini-alternator, this induces a signal in the pick-up coil located within the distributor. This electric signal flows through the electronic control module (also called the ignition module, amplifier or igniter) causing the electronic module to switch 'on' or 'off'. When the circuit is switched on, current flows through the primary

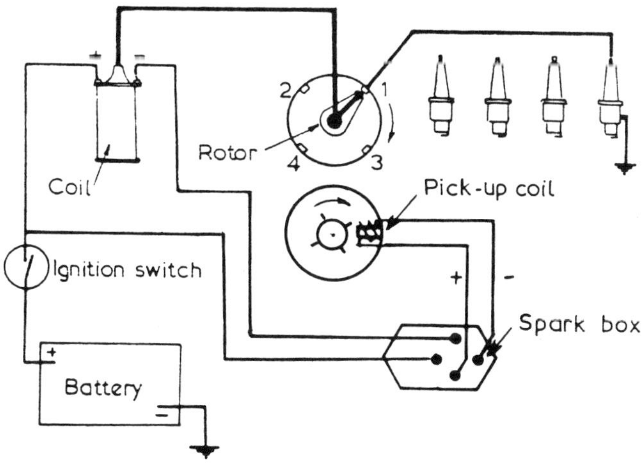

Fig. 17.7. Transistor HEI ignition.

windings of the ignition coil. When it is switched off, the magnetic field collapses, inducing a high voltage in the coil's secondary winding. As in the points-type system, the high-voltage current then flows to the centre of the distributor cap, through the carbon brush and rotor button out to the individual spark plugs. Because the transistor ignition system does not contain tungsten contact points that pit and burn, a substantially higher primary circuit current and voltage can be used to saturate the coil fully, even in the short time available at higher engine speed. Thereby a high-intensity, or high-energy, spark is provided, which has led some manufacturers to call this 'high-energy ignition', or HEI.

Whereas a points-type system can supply about 18,000 sparks per minute (which corresponds to 9000rpm for a 4-cylinder engine, but only 6000rpm for a 6-cylinder and 4500rpm for a V8), a stock HEI ignition raises this figure to just over 20,000 strong sparks per minute, and a good competition HEI system should increase this to 30,000. This latter figure means that a top-of-the-line race transistor ignition system should enable a good ignition burn with a single coil at engine speeds of 15,000rpm, 10,000rpm and 7500rpm in 4-cylinder, 6-cylinder and 8-cylinder engines respectively.

MULTI-COIL AND CAPACITOR DISCHARGE IGNITION

However, what if we require the engine to run at higher speeds, or what if we are burning a fuel that requires a very high intensity spark of long duration to effect good combustion? Depending on race regulations, we could opt for an HEI multi-coil system, or capacitor discharge (CD) ignition, with either a single ignition coil or multiple coils. With a multi-coil system, more time is available to saturate each coil fully because usually individual coils may be firing only one or two spark plugs, although in some V8 and V12 applications each coil may be firing three or four.

When race rules bar the use of multi-coil systems or we wish to avoid such complexity, we can choose capacitor discharge ignition. Unlike the previously discussed inductive storage ignition systems, CD ignition does not store ignition energy in the ignition coil, but rather in a capacitor. Even the best transistor ignition amplifier and coil

This model Subaru has multi-coil ignition with all four coils combined as a single unit. Such systems improve timing accuracy and ensure a good spark even at high engine speeds. Note the intercooler being 'iced down' prior to another attack on the hill.

combination begin to run out of ignition at about 7500rpm on a V8. At higher speeds a single-coil HEI system runs out of time for the amplifier to 'soak' the coil, discharge all the energy from it to fire the plug, then fully saturate the coil in readiness for the next plug firing. Thus above 7500rpm, in single-coil V8 applications, race HEI systems yield less and less secondary spark voltage and energy, the end result being a more and more feeble spark and reduced hp at best, or a misfire at worst.

Some tuners, however, believe that CD ignition is inherently superior to transistorised HEI, and it is therefore the logical choice for a competition engine, even if the engine is running at speeds below that at which a transistor HEI system is running out of ignition energy. The reasoning is that a bigger, longer-duration spark must produce a superior burn, and therefore it has to be good for power. However, my testing indicates that an excess of ignition energy does not mean that an engine will produce more power. Only when an ignition system is inadequate will a switch to a better system allow hp to increase.

CD ignitions in single-coil V8 applications should not begin to run out of ignition energy to speeds in excess of 10,000rpm because it takes considerably less time to charge and discharge a capacitor than it does to charge a coil. In CD systems, current in the primary circuit powers a mini-oscillator/transformer, which in turn charges a capacitor at about 400 volts. The distributor has either a magnetic or LED (light-emitting diode) trigger system. In this latter arrangement a tiny infra-red light beam passes across a gap between an LED and a photo transistor. A slotted distributor rotor interrupts the light beam to produce the triggering signal, which passes to the control module for amplification. This amplified signal then breaks the primary circuit, at which time the capacitor almost instantly dumps its stored energy in the coil's primary winding. The coil, acting as a pulse transformer, steps up the 400 volts or so delivered from the capacitor to something in the order of 30,000–40,000 volts, which is directed to individual spark plugs via the rotor button, distributor cap and plug leads in extremely quick time (*Fig. 17.8*).

Rise time for a conventional ignition is 75–125 microseconds, but CD systems have a rise time of only 20 microseconds. This enables a CD ignition to fire badly fouled or wet plugs, which means that an engine that has been 'loaded up' during a spin and stalled will quickly fire again. With other systems the rise time may be so slow that a wet plug will bleed off voltage across the fouled insulator, and there will be insufficient ignition energy left to heat the plug gap and get the engine running. A CD ignition, however, delivers full energy to the spark plug, because the voltage is produced so quickly that it jumps the plug gap before the spark energy has time to bleed off.

MULTI-SPARK CD IGNITION

A disadvantage of ordinary CD ignition is that it produces a spark of very short duration. What happens is that there is ample ignition energy to bridge the spark plug electrodes, but because the spark is of such short duration, turbulence within the combustion chamber blows out the spark before it gets sufficient heat into the fuel/air mixture to initiate a good combustion flame. This problem led to the development of multi-spark CD systems, which typically produce multiple sparks of up to 20° of crank rotation. In turn the introduction of multi-spark CD ignition has led to a lot of misunderstanding about CD systems, and this has opened up the floodgates for advertising propaganda to add to the confusion.

First, let us consider the multi-spark aspect. Some advertising seems to indicate that the ignition will supply a sequence of, say, six sparks per plug firing, regardless of engine rpm. This is just not so, except at very low engine speeds when the ignition can

Fig. 17.8. A CD-type ignition system.

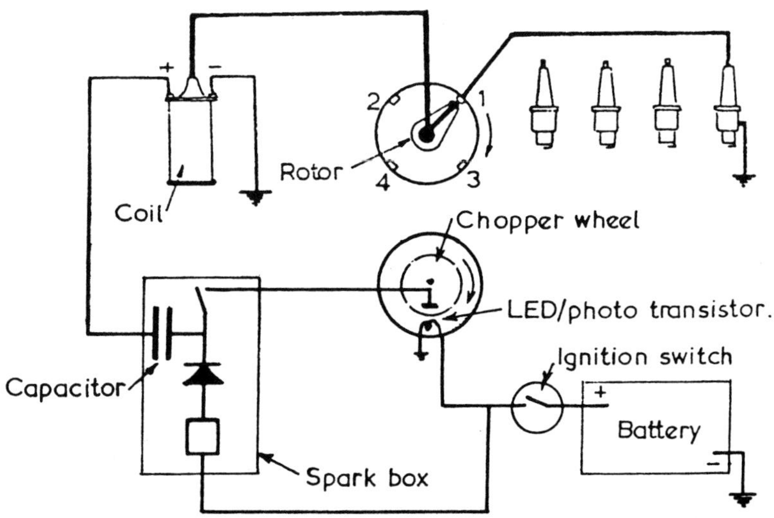

get in six quick sparks in the relatively long time it takes the crank to rotate through 20°. Therefore, in CD systems fitted to V8 engines, by about 3500rpm the system will have reverted to just one fat, long-duration spark. Obviously, in a race engine we want to know how much spark energy is available at high rpm per spark, not per spark sequence. Whether the system gives three or six sparks per firing sequence at idle does not matter too much; what we need to consider is how much ignition energy is available at 8,000rpm. One system that produces, say, six sparks per firing sequence at idle may give a relatively feeble, by CD standards, 85 millijoule (mj) stored energy level at 8000rpm. By comparison, another system, which only produces a three-spark sequence per plug firing at idle, may be banging out 140mj of stored energy at 9000rpm.

To further complicate matters, manufacturers often do not specify the energy value per spark. Instead, the multiple sequence of sparks is frequently added together to give a

If you look carefully at the distributor in this big block Mopar you will see the word 'ICE'. This company, in Melbourne, rebuilds the Bosch HEI distributor (used by Ford and General Motors on their Australian models during the 1980s) to fit a large variety of engines. Because it was designed to get a fire started across huge plug gaps of up to 0.065in in a combustion chamber filled with poorly atomised fuel from a carburettor, and contaminated by big quantities of exhaust gas intended to keep emissions down, the Bosch HEI ignition had to produce an intense, very long duration spark. When compared with CD ignition, current output isn't fantastic (60 milliamps from HEI, 120mA–150mA from CD, 5mA–6mA from twin points distributor), however, spark duration is something else. Most CD ignitions don't exceed 0.1–0.15 millisecond, which is even shorter than a twin points system (0.2 millisecond at 7500rpm), but the Bosch HEI at 8000rpm has a spark duration of 0.4 millisecond (19.2° of crankshaft rotation). So long, in fact, that the module cuts it short to keep the whole spark length within the combustion cycle, and give the coil time to be recharged with energy that will be used to fire the next spark plug.

big millijoule energy figure that could lead you to believe that you are reading the spec sheet for a welder rather than that of an ignition system. When a stored-energy-per-spark-sequence figure is quoted, find out from the manufacturer at what engine speed, on how many cylinders and at how many sparks per sequence they obtain that figure. It is the stored energy level per spark at high rpm that really counts.

Even then we run up against a problem in that energy losses vary from one ignition system to another. One CD ignition may have a stored energy value for a single spark of 140mj, but only 50% of that energy is actually available to fire the plug. Another CD system may have a stored energy figure of only 100mj, but it is very efficient and delivers 75%, or 75mj of ignition energy, at the plug gap. Clearly, there is more involved than merely looking at raw numbers, as these two systems could actually be quite closely matched in real-world performance. Note that by comparison race HEI and points-type systems will be hard pressed to deliver better than 50% of their peak stored energy, about 95mj and 28mj respectively, at the spark plug.

IGNITION VOLTAGE REQUIREMENTS

At this point we should consider how an engine's secondary voltage requirements enter the ignition system equation. Some people believe that we should be looking at systems that will deliver, say, 50,000–60,000 volts in a race engine. This is wrong thinking, because at anything in excess of 40kV – regardless of whether the system is a distributor or distributorless type – you will see sparks jumping to earth from plug wires and boots. Every time this sort of thing happens there is no fire at the plug gap, so hp falls off.

Really, the engine's voltage requirements depend mainly on the width of the spark plug gap, and the cylinder pressure at the time the spark plug fires. At the bottom end of the scale an engine running 0.028in spark plug gaps and a cylinder pressure equivalent to a compression ratio of 11:1 will require about 20kV to get a spark across the plug electrodes. A 0.040in gap and a 12.5:1 compression ratio will push the voltage requirement to about 27kV, while a 0.060in gap and a 15:1 compression will see this requirement rise to around 35kV.

The effect of this secondary voltage requirement in drawing off stored ignition energy is substantial. In the above three examples, with voltage requirements of 20kV, 27kV and 35kV, the stored energy given up just to get a spark started will be in the order of approximately 10mj, 18mj and 31mj respectively.

Obviously, from these figures you can see that if the ignition system is capable of delivering, say, 20mj across the spark plug gap, we would have to run with narrow plug gaps otherwise we would get no spark at all in an engine with a voltage requirement of 35kV. If the voltage requirement was 27kV we would have enough energy to get a spark, but there would only be 2mj of energy left over to keep the spark gap ionised and get the combustion flame started (20mj – 18mj = 2mj); in reality there probably would not be enough heat produced in the plug gap to initiate combustion. However, if we bring the voltage requirement down to 20kV we should have enough ignition energy to bridge the spark plug electrodes in the first instance, and have 10mj left over to pour heat into the gap and get the fuel/air mixture burning (20mj – 10mj = 10mj).

From the above we should understand that the more secondary ignition energy we have left after all of the losses in the ignition system, the better are our chances of starting and sustaining combustion at high engine speeds. Also, we must take note of the fact that as the voltage requirements increase because of big plug gaps and high compression pressures, we must switch to an ignition system with more available spark energy to ensure that we have enough heat energy in the plug gap to initiate and sustain combustion. Keep in mind that we also need more spark energy to burn exotic fuels such as methanol and nitro efficiently. It is much more difficult to get these fuels to burn initially, then to keep them burning for complete combustion.

The converse must also be taken into account, i.e. when race regulations or our budget preclude the use of high-energy ignition systems, we must do something about reducing our engine's ignition energy requirements. The place to start is to reduce the spark plug gap in line with what is discussed later in this chapter. I know a lot of noise has been made about the modest power gains possible with wider plug gaps, but remember that much of this is based on what tuners are discovering when working with the small-block Chev V8 and other American V8s. However, most other race engines do not use high-dome pistons, which block up the combustion chamber and retard flame travel through the fuel/air mixture, so it is not so important to have such a big flame right at the start of the combustion process.

CHOOSING AN IGNITION

While we are on this point, don't automatically assume that the factory ignition isn't up to the task. The aftermarket ignition manufacturers might want you to believe the stock ignition is junk, but this usually is not the case. Modern ignitions have to operate reliably without misfires, otherwise the manufacturers fail government emissions standards. Such testing is considerably tougher than the environment in which many performance and competition cars operate as the ignition has to ignite very lean mixtures, contaminated by high levels of exhaust gas (usually purposely introduced to keep nitrogen oxide emissions low) while firing across extremely wide spark plug gaps. The main area where stock systems have a problem is at very high rpm. However, you can reliably assume that as you will invariably be operating on considerably richer mixtures and with smaller spark plug gaps, the power differences between the stock system and high performance ignitions will be negligible up to the rev limit set by the car manufacturer. In fact, I've seen many aftermarket systems give marginally less power and a few gave a lot less at such rpm. When the revs rose way beyond the factory rev limit, though, or when a rich methanol mixture had to be ignited, then some aftermarket ignitions significantly outperformed the stock unit.

Another important factor in choosing an ignition system is its reliability. Most modern stock systems are extremely good in this regard, but there are a few which regularly fail owing to some basic design problem. The same is true of aftermarket high performance ignitions. Therefore, ask the manufacturer about their standard 'burn-in' and testing procedures, and about the ease and cost of repair. The ignition amplifier (also called the spark box or black box) contains numerous electronic components, which, if they are going to fail, will do so in the first few hours of operation. With this in mind we should be looking for a unit from a manufacturer where every spark box coming from the factory is given a 4–6 hour burn-in and testing with normal ignition loads before it is packaged for sale. Obviously, an amplifier subjected to this type of testing may cost a good deal more than if the manufacturer were to pull one box off the production line out of every hundred, or a random unit every couple of weeks, for a burn-in and test.

The other concern is amplifier repairability. Some manufacturers, after mounting all of the electronic components, fill the spark box with either a rubbery or a hard epoxy compound to keep out air and moisture. However, when the amplifier breaks down all this filler material makes it virtually impossible to replace failed components, so the entire ignition box has to be scrapped. The alternative sealing method used by some other manufacturers involves dipping the mounted amplifier components, but the case is not filled with any epoxy. When any parts break down the sealer can be penetrated, the defective part replaced, then the assembled components are dipped once more to reseal the electronics.

HELPING THE IGNITION LIVE

However, having said all that, most high-energy ignition troubles are caused by the user, not the manufacturer. The basics to keep in mind are that these systems will not tolerate excessive heat, vibration or moisture, and they quickly fail from high voltage build-up within the system. Therefore, before fitting the ignition into the car, read the installation instructions carefully. If you are not clear on some point, talk to the manufacturer. On the other hand, if the instructions appear to make sense after a quick read through, re-read them; no instructions are that easy to follow, and you may have misread something!

Be sure to mount the spark amplifier and the coil away from the headers and turbo, and shield them from heat being radiated or blown off these. If it is practical, direct a small cool air duct over the spark box. Whenever components mounted on a heat sink are replaced,

Multi-coil distributorless ignition eliminates problems with timing accuracy and distributor cap flashover, and it extends the rpm potential of transistorised HEI ignition. However, a waste spark system like this has to be treated with caution. Adding any type of fuel some distance from the inlet valve could result in an explosion. Therefore, a wet nitrous system or fuel injection into the blower may not be feasible.

take care to apply the appropriate heat sink grease when the new components are being fitted. To control vibration use the mounting rubbers provided by the manufacturer, and if the rubber pads are through-bolted do not squash them tighter than specified.

Do not blast any part of the ignition system with a pressure washer. If any sealer or insulator is less than perfect, a tiny fraction of moisture could see a high-voltage flash-over and subsequent system failure when the engine is started – even after several days' drying time. Keep in mind, also, that detergent and degreaser residues are corrosive and conduct electricity.

To avoid power overload in any part of the system, never disconnect any wires with the engine running. In similar vein, never crank the engine over with any of the spark plug leads or the coil lead disconnected, without first isolating the ignition system. With many systems the engine can be cranked while the leads are disconnected providing that the low-voltage positive (+) ignition wire is disconnected at the coil, but check this with the manufacturer.

Battery chargers and welders have also taken their fair toll in destroying ignition components. When using a battery charger with the battery still fitted in the car, disconnect both battery cables before hooking up the charger. When welding on the car, disconnect both battery cables and disconnect the main harness going to the spark box as well as any earth straps connected to it. Also, put the welder earth clamp in a place where it can make a good ground connection close to where you want to weld.

IMPROVING SPARK TIMING ACCURACY

With the switch to electronic advance control we no longer need a distributor with bob weights, etc., to effect ignition advance. Consequently, we now have the option of discarding the distributor and going to a distributorless system. This move offers many advantages, but the most significant is improved spark timing accuracy. Distributors are generally driven by the camshaft, or else by a jackshaft driven by the timing chain. Either way, harmonics in the valve train play havoc with the motion of the distributor and the problem worsens as valve spring pressures and engine rpm increase. On top of this, some engines have the cam drive at one end of the camshaft and the distributor drive at the opposite end.

This does not present a problem at ordinary engine speeds, cam lobe lifts and valve spring pressures, but in a competition engine the story changes dramatically. Engine speeds rise spectacularly, cam lobe lifts rise and the camshaft diameter may decrease to achieve such a big lift, and valve spring pressure may double or treble. What happens now – particularly in blown push rod V8 engines – is that the cam tends to screw up, with the distributor drive end of the shaft lagging behind the other end where the timing sprocket is attached. However, the distributor end of the cam does not trail behind for ever; depending on which valves the cam is pushing open, the camshaft will wind up so far, then instead of the distributor end trailing, it will unwind and spring forward. This means that the triggering signal intercepted by the ECU to reference TDC, and thereby determine the spark firing angle, is unreliable and inaccurate. In reality this means that what should have been a 32° before-TDC firing angle at 8000rpm may become a 35° angle on cylinder No. 1, a 29° angle on No. 2, 33° on No. 3 and 32° on No. 4. Obviously, because in this example No. 1 is firing 3° advanced owing to camshaft 'spring', we have to retard the timing of all of the other cylinders to avoid detonation in the No. 1 hole. This costs us power and it may cost us in another way. Those cylinders that are now retarded may 'afterburn' because of late ignition. In turbo engines this could reduce turbo life and limit our boost pressure. In other forced induction engines the temperature of exhaust valves and valve seats could tip over the edge and introduce reliability problems.

Distributors driven from the camshaft by a helical gear experience timing error problems in a different manner. Unless the distributor shaft is shimmed to keep end play to a minimum – 0.003in is preferred – the distributor gear will ride up and down on the cam gear and upset timing accuracy. Likewise, a camshaft 'walking' backwards and forwards in the block owing to excessive end-float also contributes to erratic ignition timing. Cam end-float should be 0.003in, controlled either by a thrust plate or, where one is fitted by the factory, a timing cover incorporating a thrust button and appropriate reinforcement, to keep cam walk to a minimum. However, even when both distributor shaft end-float and cam walk are kept to the specified minimum, a measure of ignition timing error still occurs.

When we get rid of the distributor we also rid the engine of another potential trouble spot, namely the distributor cap. With small spark plug gaps and low secondary ignition voltages, ignition flashover, or crossfiring, within the cap is rare. However, the situation changes as plug gaps are opened up and ignition voltages increased. With 4-cylinder and 6-cylinder engines it is possible, by going to a large cap diameter, to put more space, and hence a bigger air gap, between the terminal posts. However, with 8-cylinder engines it can be difficult to find sufficient space to fit such a large distributor cap, and the problem is compounded when we go to 10-cylinder and 12-cylinder engines, as the cap diameter grows ever larger.

If we have enough ECU capacity, accurate distributorless ignition opens up the prospect for ignition with individual cylinder trim, which means that we adjust the spark firing angle to suit the characteristics of each cylinder. In this way each cylinder can be tuned as if it was an individual engine. For example, because of engine breathing characteristics, less than perfect coolant flow through the block and head or fuel or nitrous distribution problems, one or more cylinders may run hotter or cooler than others. With ignition trim we can get more power out of the cool-running cylinders by giving them more spark advance. Also, we may get more power, or at least better reliability, when we cut back on the advance angle for the cylinders that are running hot.

When the distributor is discarded we have to use another means of informing the ECU when each piston reaches TDC. Instead of using a pulse trigger within the distributor, this function is transferred to a pulse trigger that is usually attached to the front of the crankshaft. The alternative is to have 'firing' pins embedded in the circumference of the flywheel. Regardless of their position, as each firing pin moves past the stationary magnetic pick-up a pulse is relayed to the ECU. The ECU in turn calculates the appropriate advance angle and switches the ignition primary (low voltage) circuit on and off.

Probably the main things to watch with these crank trigger-type systems are that the firing pins are precisely located, that they are of the correct diameter and that the air gap to the magnetic pick-up is correctly adjusted. That the timing pins must be precisely located to ensure accurate spark timing is quite clear, but the need to use pins of the correct diameter is less obvious. The manufacturer of the system should state the minimum pin diameter, but in any case it is a good idea to use a diameter not much smaller than the size of the pick-up. Therefore, if the pick-up is about 6mm in diameter, use firing pins with a 6mm head. If you use 3mm pins the pick-up may not 'read' them accurately, leading to all kinds of problems. Likewise, the air gap where the firing pin passes the pick-up must be adjusted to that recommended by the manufacturer, usually 0.020–0.040in, otherwise you will have trouble getting the engine to fire up.

DETERMINING OPTIMUM IGNITION TIMING

Regardless of the type of ignition system used, it is imperative to get the ignition timing spot on for your engine. Many enthusiasts have a tendency to over-advance the timing in an effort to pick-up every last fraction of performance, but my advice is to use the least amount of advance conducive with peak performance. Years ago the advice I regularly heard was to bump up the timing until the engine rattled, then back off the advance 2°. This sort of tuning may get you close with a low-performance close-to-stock road engine, but it is not good advice for a forced induction engine, particularly a competition engine. You do not hear a race engine rattle, and if you do, the damage is already done.

The place to get the timing sorted is on the dyno. Keep adding spark lead while watching the read-out, and when adding additional advance brings no power increase, or even perhaps a drop in hp, back off the timing, as you have gone too far. If the timing is adjusted using a transient 'acceleration' dyno test rather than a 'steady state' test, the timing could be way too far advanced, depending on the acceleration rate programmed into the dyno. This occurs because the load is not held for sufficient time to get a lot of heat into the combustion chambers and pistons. Therefore, the engine will take more advance before it detonates or the power falls off than it would in the real world. Acceleration tests are useful for initial quick setting up, but keep the rate down to 100rpm/sec, and back up the results with steady state dyno pulls.

Something to keep in mind with turbo engines is the effect spark timing can have on turbo performance. When the spark is slightly retarded, exhaust gas temperature increases. If not overdone this can improve turbo spool up and increase boost pressure. The reverse is also true. When we give the turbo engine excessive advance, not only do we hurt the hp because of what is happening inside the engine, we also lose because the turbo is now not working as efficiently.

If you run very close to the maximum total advance that the engine will tolerate without going into high-speed detonation, you must be very careful to compensate for changes in the atmospheric conditions. A 7–10% increase in the relative air density (RAD) could result in a fuel/air mixture lean enough to produce a set of melted pistons, if you do not richen the mixture accordingly. Also, a decrease in the relative humidity could see the engine detonating unless the spark advance, or the amount of cam advance, is reduced.

ABNORMAL COMBUSTION

Detonation and pre-ignition are both engine wreckers. Pre-ignition is self-ignition of the fuel caused by a hot spot within the combustion chamber, or because of the fuel becoming unstable because of excessive pressure or heat. Detonation is a violent burning of the fuel (almost an explosion) caused by colliding flame fronts after the spark plug has fired. When engine damage results from either condition, the culprit can usually be identified by an examination of the pistons and spark plugs.

Careful tuners regularly use an illuminated magnifier to check all spark plugs for the tell-tale signs of detonation.

Pre-ignition damage is caused by extreme combustion temperatures melting the top of the piston and possibly also the ring lands. If a hole is present in the piston crown, it will appear to have been burned through with a welding torch, and the metal around the hole will have a melted appearance. Spark plugs exposed to sustained pre-ignition quite often have the centre electrode melted away, and in extreme cases the insulator nose may also be fused.

Usually, pre-ignition can be traced to combustion chamber or exhaust valve deposits becoming incandescent, but it may also be because of blocked water jackets creating a hot spot, or a glowing spark plug with a heat range too hot for the engine. Although less likely, it can also be caused by an overheated piston, perhaps because of inadequate lubrication, improper clearance or a broken ring.

A piston damaged by detonation will show signs of pitting on the crown; in extreme examples the piston crown may be holed. The hole will appear to have been punched through, with radial cracks and a depressed area around it. A spark plug subjected to detonation will usually show signs of cracking at the insulator nose.

When detonation occurs, a portion of the fuel/air charge will begin to burn spontaneously owing to excessive heat and pressure after normal ignition takes place. The two flame fronts ultimately collide and the resulting explosion hammers the engine's internal components. Detonation is usually attributed to excessive spark advance and/or lean fuel/air mixtures, but in supercharged and turbocharged engines excessive boost and high intake charge temperatures most likely lead to this condition.

Piston erosion almost directly under the spark plug or around the edges is a sure indicator of detonation or pre-ignition.

THE ROLE OF THE KNOCK SENSOR

Many modern engines, both forced induction and naturally aspirated, have a knock sensor right from the manufacturer. In these cars the sensor alerts the ECU of any abnormal combustion condition. The ECU then takes action, depending on how it has been programmed, to bring combustion back to normal. This may be achieved by retarding the spark, adding more fuel or lowering boost. In practice this usually works reasonably effectively with stock and lightly modified engines. However, as we get into higher levels of tune, the ECU most likely will not be able to bring combustion back on track owing to programming limitations. Consequently, the engine may be damaged if we leave it to the knock sensor and ECU to control combustion.

Another problem when we modify an engine is that the knock sensor may respond to 'phantom noise', assuming it to be abnormal combustion and setting about reducing hp to 'protect the engine'. In reality those vibrations detected as knock could be owing to increased piston or bearing clearances, wilder cams and increased valve train harshness, or even the fact that the engine is now spinning at more rpm. Some stock engines also generate phantom signals from such sources and manufacturers get around the problem by filtering out signals from the knock sensor above a certain number of rpm, or over a certain rev range. Too bad if the engine actually detonates during those periods!

Consequently I prefer to disconnect the knock sensor from the ECU and instead have it activate a warning light and a buzzer to alert me to knock, either real or phantom. Immediately I can lift off the accelerator to protect the engine and then set about lowering the charge temperature, adjusting the boost and fuelling and perhaps the ignition advance. If the knock light still came on I would then keep running the engine in that condition for a few seconds. Following that, the plugs would be pulled and carefully examined, and likewise for the tops of the pistons using a borescope poked down the spark plug holes. If there was no evidence of detonation, the inference would be that engine harmonics or rough combustion at those rpm and throttle opening were setting the knock sensor off.

SPARK PLUG HEAT RANGE

Owing to increased combustion temperatures in a modified engine, consideration must be given to finding a spark plug with the correct heat range. A hot plug transfers combustion heat slowly and is used to avoid fouling in engines with relatively low combustion chamber or cylinder head temperatures, i.e. in a relatively low-horsepower engine being driven at cruise with the throttle barely open. A cold plug, on the other hand, transfers heat rapidly from the firing end and is used to avoid overheating where temperatures are high, as in a racing engine held for long periods at wide open throttle.

The length of the insulator nose and electrode alloy composition are the primary factors in establishing the heat rating of a particular plug. Hot plugs have long insulator noses, and therefore a long heat transfer path, while cold plugs have shorter nose lengths to transfer heat more rapidly from the insulator tip to the water jacket via the metal spark plug body (*Fig. 17.9*).

Motors in sports tune will probably require a standard plug or one not more than a couple of steps colder than standard. Semi-race engines cause us more problems, as two plug types may be needed to cope with city driving and high-speed running. A plug two or three grades colder than standard should be tried for everyday street use and short full-throttle bursts; it must also be able to resist low-speed fouling and pre-ignition at a constant 80mph. The second plug type would be used under competition conditions or sustained high-speed driving.

Race engines also require two plug heat ranges, one for engine warm-up and one for racing. If there is sufficient space in the combustion chamber, projected-nose or regular-gap plugs are best used for warm-up. Otherwise use the hottest retracted-gap racing plug available. Whenever warm-up plugs are fitted, tag the tacho or steering wheel as the engine could be destroyed if, in the heat of the moment, your memory lets you down and you race with hot plugs fitted.

After you have decided on a plug of the correct heat range, you must test to ensure

Fig. 17.9. The heat rating of a spark plug is dependent on the length of the insulator nose and the electrode material composition.

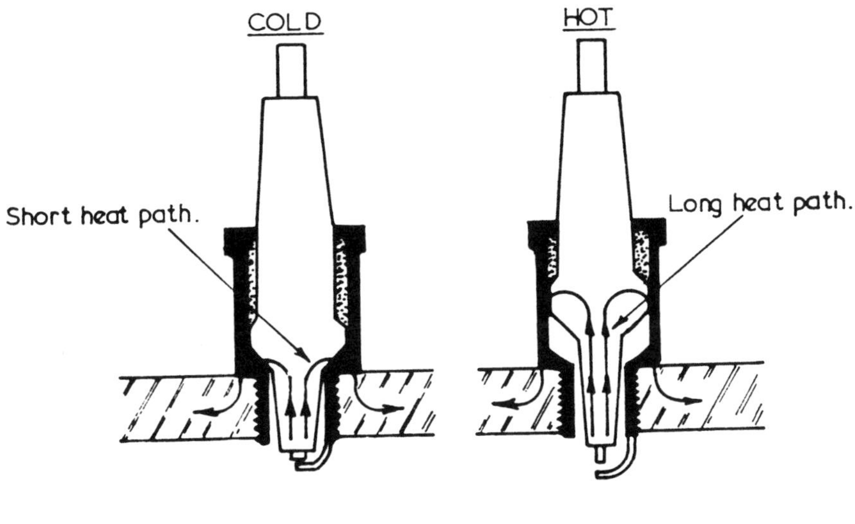

Table 17.1. Spark plug readings for correct heat range and other conditions

Spark plug condition	Indications
Normal – correct heat range.	Insulator nose white or very light tan to rust brown. Little or no cement boil where centre electrode protrudes through insulator nose. Electrodes not discoloured or eroded.
Too cold – use hotter plug.	Insulator nose dark grey or black. Steel plug shell end covered with dry, black soot deposit that will rub off easily.
Too hot – use colder plug.	Insulator nose chalky white or may have satin sheen. Excessive cement boil where centre electrode protrudes through insulator nose. Cement may be milk white or meringue-like. Centre electrode may be 'blue' and rounded off at edges. Earth electrode may be badly eroded or have molten appearance.
Pre-ignition – use colder plugs and remove combustion chamber deposits.	Insulator nose blistered. Centre electrode and side electrode burned or melted away.
Detonation – retard ignition and richen mixture.	Fractured insulator nose in sustained or extreme cases. Insulator nose covered in tiny pepper specks or even tiny beads of aluminium leaving the piston. Excessive cement boil where centre electrode protrudes through insulator nose. Specks on plug shell end.
Insulator glazing – replace with plugs of same heat range. If condition re-occurs fit plugs one grade colder.	Shiny yellow, green or tan deposits on insulator nose, particularly close to centre electrode.
Ash fouled – clean or replace with plugs of the same heat range.	Thick yellow, white or light brown deposit on insulator, centre and side electrode.

that your choice is the right one. Providing that the engine is in good condition and the air/fuel mixture ratio is correct, reading the nose of the plug will indicate if it has the correct heat range. To avoid engine damage, it is always advisable to begin testing with plugs that are too cold, or else test the machine at moderate load and speed, then check the plugs before you engage in any full-power running.

For the plug reading to be accurate it will be necessary to run the engine at full throttle and maximum speed on the track (or road), then cut the engine dead. If you allow the engine to keep running as you bring the vehicle to a stop, the plug reading will be useless. We are not just interested in the colour of the insulator nose. Many fuels will not colour a plug, and it takes miles for some types to colour the insulator. Therefore, all of the plug firing end exposed to the combustion flame must be examined and read. The signs to look for are indicated in Table 17.1.

Of course, for those involved in road racing or rallying, the spark plug heat range must be tailored to each race environment. Courses with long straights or high-speed banked turns usually require a colder plug than those with short straights and many esses and hairpin corners. In drag racing when the nitrous dose is changed or when the fuel blend is changed, the plug range will have to be changed accordingly. Anything we do here to increase engine power invariably means that a colder plug grade is needed.

Once you have determined the correct plug heat range, do not swap over to another brand with an 'equivalent' heat range. Heat range conversion charts should be used as a guide only when you swap from one plug brand to another, as different plug manufacturers use different methods of determining the heat range of their plugs. If you cross-referenced the conversion charts from all the plug manufacturers, you would find that they disagree with each other because of different test procedures.

Even spark plugs with the same number from a particular manufacturer can have a wide heat range tolerance. This is why some mechanics are very particular about the brand of plug that they use in their engines. They claim that one particular brand gives a more consistent heat range, so they do not have to test dozens of plugs with the same number to find a set that has the same heat range.

SPARK PLUG GAP STYLE AND ELECTRODE MATERIALS

As well as the heat range, the gap style of the plug and the composition of the electrode materials must be considered. In forced induction applications we have to select a plug with a very thick earth electrode made of a material that will very quickly conduct heat away from it into the plug shell. A thin electrode made of a poor conductor is a recipe for disaster. Either the electrode will burn off and cause a misfire, or else it will hold so much heat that it will glow like a glow plug and pre-ignite the mixture, possibly destroying the engine.

When burning methanol we also have to keep away from platinum electrode materials. The problem is that, in combination with methanol, platinum acts as a catalyst and may cause pre-ignition. Some racers have been blissfully ignorant of this and not suffered any problems, but others have not been so fortunate.

The best plug to use in most engines, where there is sufficient physical space between the tip of the plug and the valves and piston crown, is the projecting nose, or projecting core, type (*Fig. 17.10*) with copper implants in both electrodes to increase heat transfer rates. It has a very wide heat range to resist both fouling and pre-ignition.

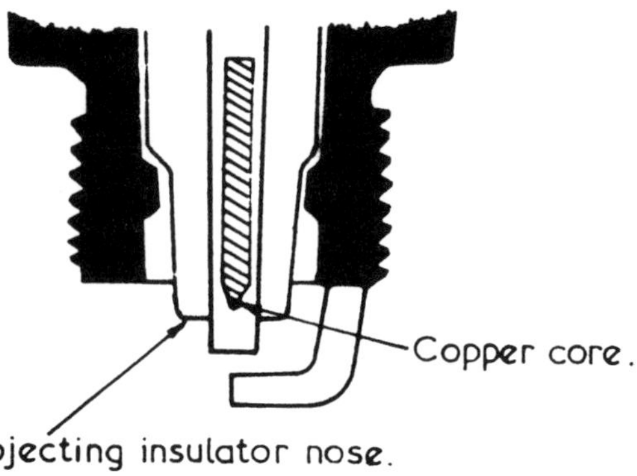

Fig. 17.10. Copper implants in the spark plug electrode(s) increase heat transfer rates, giving both projecting nose and regular gap designs a wider heat range.

At high speeds the long insulator nose is cooled by the incoming fuel charge to increase its cold heat range, while at low speeds the long tip runs hotter to prevent fouling. Projecting nose plugs are not recommended for highly supercharged engines or those using more than 20% nitro fuel.

If you change from a regular-gap or retracted-gap plug (*Fig. 17.11*) to a projecting nose type it may be necessary to retard the advance slightly. The projecting nose starts the ignition flame burning physically deeper in the combustion chamber and this reduces the length of flame travel within the chamber. Often, projecting nose plugs will raise the power output of the engine right through the power range because of improved combustion.

The conventional-gap plug is next preferred after the projecting nose type, and can usually be obtained in slightly colder grades than projecting nose plugs. This style has a

Fig. 17.11. Common spark plug gap styles.

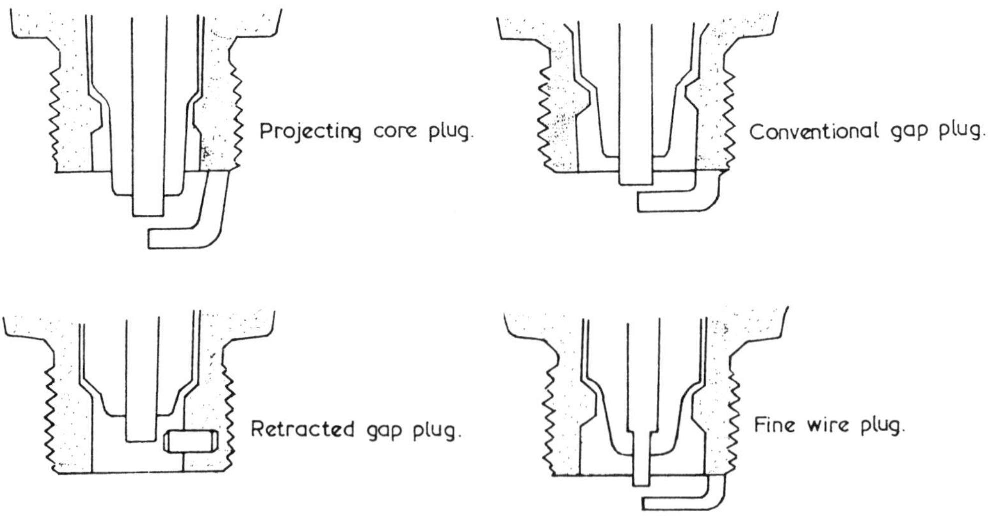

wider heat range than a retracted gap plug and provides superior ignition flame propagation.

The retracted-gap racing plug is available in the coldest grades and is necessary in highly modified supercharged or high-percentage nitro-fuelled racing engines. It may also be used in racing engines where there is insufficient clearance between the spark plug and valves or piston to use either a projecting nose or regular-gap design. A plug of this type should be used only when absolutely necessary. It has very little resistance to fouling and generates a poor ignition flame front.

The fine wire plug (e.g. Champion Gold Palladium) has some use in semi- and full-race engines. It was originally developed for racing two-stroke engines, but can be used in four-stroke applications requiring a wide heat range (similar to projecting nose plugs) in the colder grades. These plugs are expensive, owing to the use of a fine semi-precious metal electrode, but they perform very well in high-performance and racing engines requiring a cold plug relatively resistant to low-speed fouling. They are not recommended for highly supercharged or high-percentage nitro-burning engines. In addition to the regular gap style illustrated a few grades are also available in the projecting core style.

Whenever a cylinder head with larger valves, modified combustion chamber or relocated spark plug is used, or when pistons with a non-standard crown are fitted, always check that the spark plug is of the correct reach and gap style to avoid mechanical damage to the engine. Also, with the head removed, check that the plug reach (i.e. the length from the plug seat to the end of the thread) is neither too long nor too short. A plug that does not extend the full threaded length of the plug boss in the head will reduce performance by masking the ignition flame and invite hot spots from carbon build-up in the unused thread. A plug that is too long (i.e. has exposed threads in the combustion chamber) can cause damage and stripping of the plug boss threads in the head as carbon build-up on the exposed plug threads makes plug removal very difficult. Also the exposed threads can become a hot spot to initiate pre-ignition. In some instances a change to a plug with a different reach may be in order, but in most cases the use of a single solid copper gasket on each plug will ensure the proper depth fit. The range of Champion gaskets is listed in Table 17.2.

Table 17.2. Champion spark plug gaskets

Plug diameter (mm)	Gasket thickness (in)	Part No.
10	0.045	Y-674
	0.055/.045	Y-678
12	0.057	P-674
	0.070/.052	P-678
	0.095	P-677
14	0.057	N-675
	0.135	N-677
	0.070/.052	N-678
	0.080	N-673X1
18	0.080	A-675
	0.065/.050	A-676
	0.075/.055	A-678

SPARK PLUG GAP

The width of the spark plug gap for best performance depends primarily on the compression pressure of the fuel/air mixture, the engine speed and the coil saturation time. Increasing the first two factors and decreasing the latter calls for a decrease in plug gap. It is safe to say that all modified engines using factory ignition require a plug gap narrower than stock when operating at rpm higher than the factory rev limit.

Manufacturers stipulate a relatively wide gap as this improves the engine idle and low-speed performance on very lean air/fuel mixtures. The ignition spark from a wide gap is larger than that from a narrower gap, so a larger initial combustion flame is generated, assisting low-rpm running. However, at high engine speeds and combustion pressures the coil is not able to supply electrical energy of sufficient intensity to jump the spark gap, so the gap must be reduced to avoid a high-speed misfire, or the possible breakdown of the high-voltage system insulation.

From experience I would recommend that no high-performance engine using a conventional high-energy electronic ignition should use a plug gap exceeding 0.035–0.040in. Semi- and full-race engines may need a gap as small as 0.028–0.032in with conventional plugs, and down to 0.025in when using retracted-gap racing plugs. Race engines running methanol will require gaps 0.005–0.010in smaller than these figures because it is much more difficult to get methanol burning initially, then keep it burning for complete combustion.

Engines with a capacitor discharge ignition can pick-up some power with wider plug gaps as this type of ignition is relatively insensitive to high engine rpm. In this instance the plug gap should be set to the ignition manufacturer's specifications for short-distance events and ordinary road use. However, those involved in long-distance racing and rallying should not use a gap exceeding 0.045in, as any weakness developed in the CD unit during a race could lead to total ignition failure. A closer plug gap reduces the load on the 'spark box', minimising the chance of any weakness developing, and if some trouble does arise, the system may have just enough reserve to finish the event.

COIL POLARITY

Always ensure that your coil polarity is correct before you try experimenting to determine what plug gap gives the best performance. The equivalent of 40% coil energy is lost when the polarity is reversed. The spark should always jump from the centre electrode of the plug to the side electrode as this considerably decreases the voltage required for ignition. Because the centre electrode is at a much higher temperature than the side electrode, less voltage is needed to produce a spark, as electrons will leave the hotter surface at a lower voltage. This is why it can be difficult to get a cold motor started; because the centre electrode is cold, a much higher voltage is required to produce a spark.

The coil polarity can be checked by looking at the coil's low-voltage connections (electricity supply wire from ignition switch to positive; module/amplifier wire to negative), or by using an oscilloscope. The oscilloscope's trace pattern will be upside down if the polarity is not correct. A dished or eroded spark plug side electrode also indicates wrong polarity; the dish is caused by metal leaving the electrode each time a spark jumps across to the centre electrode.

Filing the earth electrode back to the middle of the centre electrode and tapering the corners has two effects. Reducing the length of electrode exposed to combustion heat makes the plug run cooler by around half a grade. More important, because there is less earth electrode in the way to block its progress, the spark initiated flame kernel moves out through the entire fuel/air charge more uniformly, thus assuring more complete combustion.

SPARK PLUG MAINTENANCE

A road vehicle with electronic ignition should have the plugs filed, gapped and tested every 6000 miles, and a race machine after each event. Retracted-gap plugs cannot be filed, and fine wire electrode plugs should not be filed. Projecting nose and conventional-gap plugs should have the earth electrode bent back far enough to permit filing of the sparking surfaces shown in *Fig. 17.12*. A point file can be used to file a flat surface with sharp edges on both the centre and side electrodes. This lowers the voltage requirement to fire the plug, not only because electricity prefers to jump across sharp edges, but also because the electrical conductivity of the electrodes is improved. The heat and pressure of combustion tends to oxidise and break up the electrode firing surfaces, increasing the electrical resistance. Filing removes this 'dead' metal and exposes new highly conductive material.

Spark plugs should never be cleaned with a wire brush as metallic deposits will impregnate the insulator and short out the plug. I also do not recommend cleaning with a

Fig. 17.12. Spark plug electrode filing improves performance and extends useful life of projecting nose and regular gap plugs.

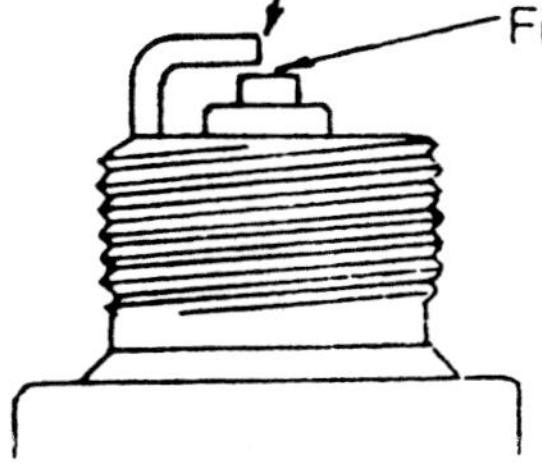

plug cleaner, as some abrasive material always seems to become wedged between the insulator and the plug shell. If this cannot be probed out with a scriber, it will drop out in the engine and possibly cause some damage. If you choose to have your plugs abrasive blasted, be sure to remove all abrasive grit from inside the plug nose and from the threads.

Personally, I prefer to leave plugs uncleaned. If they are fouled by fuel or oil I clean them with a toothbrush and ether, or some non-oily solvent. If the insulation is breaking down owing to leaded fuel deposits, I throw the plug away.

SPARK PLUG LEADS

The spark plug leads provide the electrical connection between the distributor cap and spark plugs. Most stock road cars use radio suppression cables with a carbon-impregnated rayon cord to conduct this high-voltage current, which will ultimately jump across the plug gap and initiate combustion. With age, the electrical resistance of this cable increases, so less and less energy is available to provide a big hot spark. Eventually one or more of the plugs will not fire at all. Therefore, carbon suppression cables should be discarded and replaced by high-quality induction-type suppression leads. These leads have a metal core to conduct full electrical energy to the plugs, and they have a metal induction spiral wound within the cable to provide both noise and induction suppression. Remember that inadequate suppression could seriously upset the proper functioning of the ECU, and other electronic or radio-type devices, so do not use ordinary copper core or stainless steel core cable.

The other good reason for using induction spiral leads is to minimise the incidence of induction crossfire. This is a real problem with today's high-energy ignitions. Basically, owing to magnetic induction, whenever high-voltage wires are grouped together, or run in parallel for any distance together, one with current flowing through it can induce a corresponding voltage in a neighbouring wire. In fact, this is the way in which your ignition coil works, but it is not what we want going on between plug wires! Induction crossfire in its mildest form causes the engine to run roughly, but in situations of high engine load or high rpm such as are typical under competition conditions, it can result in serious engine damage.

To help you better understand the seriousness of the problem, let us take a look at a typical blown V8, as these engines most frequently experience a crossfire problem. A crossfire that will destroy cylinder walls, pistons, bearings, cranks, heads or head gaskets is most likely to occur between consecutive firing cylinders when these are located alongside each other in the engine. Taking the most common V8 firing order of 1-8-4-3-6-5-7-2, we see that No. 7 cylinder is one likely to suffer crossfire damage, because when No. 5 fires at say 25° before TDC, No. 7 is trailing only 90° behind, or 115°BTDC, and it is almost full with fuel/air mixture. At this stage No. 7 piston is moving up on the compression stroke and, depending on the cam profile, the inlet valve is closed or at least about to close. It is in this setting that No. 5 plug wire has a high-voltage current surging through it. If the resulting magnetic pulse induces current flow in the No. 7 plug wire, that plug also fires, but as it is 90° early violent combustion occurs. Depending on how often this happens, and under what sort of engine load conditions, engine damage can range from a blown head gasket through to a wrecked block or crankshaft.

The way to avoid this sort of problem is to use induction spiral suppression leads, but this is not the total solution. The plug wires should also be separated, preferably by about an inch – $^{1}/_{2}$in is the minimum – and in V8 engines consecutive firing cylinders, usually No. 5 and No. 7, should have their respective plug leads crossed once. Crossing the problem wires once inverts the magnetic field in the second wire and cuts out any induction. Never tape plug wires together or run them through metal clips or other metal fittings. Not only can this lead to induction crossfiring, but can also open the way for misfiring.

For the plugs to fire, the plug wires have to prevent the coil's high-voltage current from tracking to earth. With today's high-energy ignition systems, and tuners insisting on running race engines with big plug gaps, plug wires need all the help you can give them. If you do not keep plug leads separated at least half an inch, and if you do not keep them a similar distance away from the engine and other metal parts, you could run into a misfire problem. With most race engines pushing out about 80–140hp per cylinder, that means you lose 80–140hp each time the spark plug fails to light.

The average race engine needs at least 20,000 volts, and anything up to 35,000 volts if the plug gaps are very wide, to get a spark across the spark plug. When you compare the amount of insulation that you have available to prevent all that electrical pressure flashing across to earth with what the power companies use on their overhead power lines, you begin to appreciate what you are up against. To work in this sort of environment you must also use only top-quality silicone cable and boots. I recommend 8mm silicone wire for the majority of road applications and race engines that will not see voltage much over 20kV. This means plug gaps down around 0.028in and 11–12psi boost. When we open the gaps up to 0.040in and push the boost past 15psi, the voltage to fire the plug rises to about 27kV. At this sort of voltage and higher we have to be careful to choose only the best grades of ignition wire. Some people think that 9mm or even 10mm wire is the solution. However, 8mm Moroso spiral core wire has a dielectric insulation strength of well over 50kV, which is superior to anything else I know of, and it combines the highest levels of electro-magnetic suppression with the least electrical resistance to ensure maximum ignition energy at the spark plug.

DISTRIBUTOR CAP AND ROTOR BUTTON

Besides the plug leads, other potential high-voltage leakage or flashover areas are the distributor cap, rotor button and coil tower. These parts must be free of accumulated dust, moisture, cracks and carbon tracks to ensure that the full voltage is reaching the plugs. Therefore, a periodic check must be made to ensure that none of these conditions exists.

Unknowingly, some tuners bring trouble upon themselves by removing carbon from the distributor cap terminals and the rotor button contact. Actually carbon in these areas assists electrical conduction. The real problem is that quite often the knife or screwdriver being used to scrape the carbon from the terminals slips and scratches the glaze inside the cap. This drastically reduces the dielectric strength and can lead to a carbon track forming during wet conditions. Often a piece of abrasive paper is used to polish the rotor button contact. This cuts straight through the insulating glaze and results in voltage leaking to earth.

When high-voltage ignition systems are used, particular attention must be paid to keeping all the insulating mediums clean, otherwise less voltage will be available to fire the plugs than when the standard ignition system was fitted. Actually, when a high-output ignition is fitted, only top-quality distributor caps and rotors, made of alkyd material (being resistant to carbon tracking even when cracked) should be used.

As mentioned earlier, another thing we have to worry about with high-energy ignition is the problem of crossfiring within the distributor cap. This is usually not a worry with 4-cylinder and 5-cylinder engines, but it begins to surface with 6-cylinder engines and becomes a real difficulty once we get to eight cylinders or more because the high-voltage posts are just so close to one another. The first move is to swap to a larger-diameter cap containing well-designed anti-flashover ribs; this will put more space – a larger air gap – between posts.

However, the location of the rotor button when the ignition fires, and the length of the rotor button's brass tip, also have a bearing on internal flashover. We tend to think of an ideal situation with the rotor button being dead centre with the distributor cap post at the time the ignition fires, but it is not like that at all, which is why there is such a long brass tip on the rotor button to help place it somewhere close to the distributor cap when the ignition fires. This is not a problem when ignition voltage is low, but as we begin to increase electrical pressure in the system to shoot a spark across a plug gap in a high-pressure environment in the combustion chamber, the situation changes.

Now, the easiest path to earth – and remember that electricity always takes the easiest path – may not be across to the distributor post, to which the rotor button is closest, and down the plug wire. Instead, that wide brass tip on the rotor button may be closer – not physically, but from the aspect that the electrical resistance is less – to a distributor cap post either directly in front of, or else directly behind the cylinder that is supposed to be getting ignition energy. When this situation exists the electricity jumps from the tip of the rotor across to a post that may be half an inch or more away, then travels down the plug wire to fire a plug in that empty cylinder. Because it is not that cylinder's turn to fire, there is less pressure in it, so it is easier for the electricity to get to earth via that route, even though it has to jump a much bigger gap to get across to the distributor terminal for that cylinder.

The brass contact on this rotor button has been shortened at both ends to help prevent crossfiring within the distributor cap.

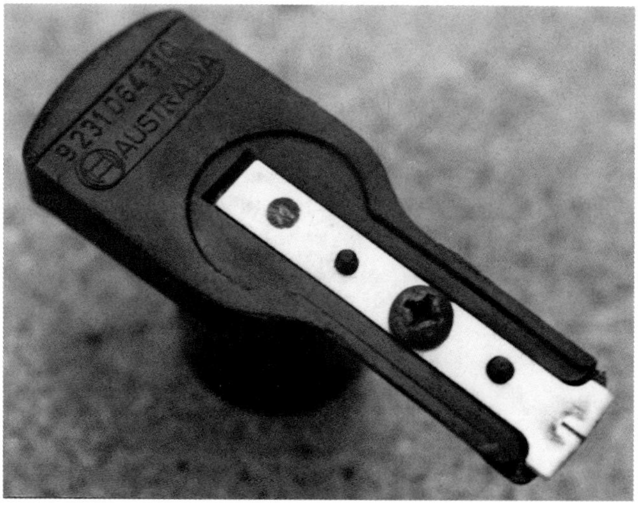

There are a number of ways to minimise this sort of flashover. If it is possible, the easiest way is to cut back on the length of the rotor button's brass tip. To check this, remove the distributor cap and turn the engine over to 10° before TDC, then accurately scribe a line on the brass tip to correspond with the centre of the plug wire post. Move the crank back to 35°BTDC and again scribe the rotor button to correspond with the distributor cap post. The brass tip extending beyond the boundary of those two scribed lines can now be cut off, thus widening the air gap to the adjacent plug wire terminal in the distributor cap.

When doing this check you may be surprised to find that the rotor button is way off lining up with the plug wire post. Here there are two options: relocate the rotor or relocate the distributor cap. If you choose to relocate the rotor, have a new locating slot cut into the distributor shaft. Some enthusiasts knock out the locating tab of the rotor, then either epoxy or pin the rotor to the distributor shaft, but this is not always successful.

Another way to reduce flashover within the distributor cap is to vent it of ionised air. What is needed is a 10mm hole between each plug wire terminal around the circumference of the distributor cap, and at about the same height in the cap as the rotor.

This brings us to the final solution when we are forced to run a distributor-type ignition system, and that is to narrow the spark plug gaps. When we reduce the plug gap we reduce the electrical resistance to earth, so the high-voltage ignition pulse will choose to take this path of less resistance and fire the plug as it is supposed to, rather than crossfire to another cylinder within the distributor cap.

PRIMARY VOLTAGE INFLUENCES IGNITION PERFORMANCE

Before we finish with the subject of ignition systems, we need to give some consideration to how the power being consumed by today's very-high-energy ignitions can affect the performance of other electrical components on the vehicle. Also, we should look at the idea that pushing up the primary-side voltage will increase the energy available to fire the plugs.

The electrical energy consumed by high-intensity electronic or capacitor discharge ignitions is not insignificant. Added to this, racers have over the years added more and more electricity-consuming devices to their race cars. With electronic fuel injection, electric fuel pumps, electric cooling fans, etc. having become a standard part of the modern race car, loads on the electrical system are dramatically greater than they were previously. Because some racers have not given proper attention to this aspect of race car engineering, they find that their cars are down on power at best, or they may experience reliability problems, with the car stopping out on the track for apparently no reason.

For example, CD ignition can have a current draw two or three times higher than a transistorised HEI system – up around 8 amps to 10 amps. However, while the average power draw of an HEI system is relatively low, the instantaneous amperage draw, when the module first switches the primary circuit to charge the coil, is very high. It is for this reason that a transistor HEI must be wired directly to the ignition switch using heavy 10 gauge wire, and nothing else should be connected in that circuit. Now, when the current begins to surge into the coil, that high instantaneous power drain can, if the electrical system has not been designed properly, cause it to malfunction – lose the plot for an instant. However, it can occasionally cause the ECU to shut down completely – then the engine stops. Incidentally, I have seen the same sort of thing happen when the electric cooling fan cuts in and sends a surge through the electronics.

Also, the idea that increasing the system voltage from, say, 12 volts to 16 volts to achieve a bigger output from CD ignition, is a fable. All the CD systems I know give a constant output regardless of whether the feed voltage is 10 volts or 16 volts. If you see increased ignition output with CD systems when the supply voltage is increased, you can be certain that you have a problem in the primary feed circuit. Either the cable is too small or too long, or you have 'hot' joints with cables poorly crimped to connectors, or connectors not making perfect contact and cutting supply voltage at the amplifier to less than 10 volts.

Transistor HEI systems, however, are very different in this respect. Any reduction in supply voltage will cause a corresponding reduction in coil output. Conversely, any increase in system supply voltage will provide an increase in secondary coil energy. As a minimum we should be expecting the ignition supply voltage at the coil positive (+) terminal to be the same as the battery voltage. Therefore, if the battery measures out at 12.6 volts, we should see the same voltage at the positive coil terminal with the engine running and all of the car's electrical equipment turned on. If the voltage is less than the battery voltage, you are losing out on maximum spark energy, so increase the size of the ignition circuit wiring and ensure that nothing else except the ignition is connected in this circuit. Note that you must take care when carrying out this test as there could be 350 volts at the coil negative (–) terminal.

Many transistor HEI systems will accept higher supply voltages and thereby provide more secondary spark energy to fire the plugs at higher engine speeds, or else fire bigger plug gaps or exotic fuel. For example, a V8 system that works at 7500rpm on a six-cell battery (up to 12.6 volts) will work up to about 8200rpm with a seven-cell battery (up to 14.7 volts), and up around 8800rpm when connected to an eight-cell battery (up to 16.8 volts). Remember to check with the ignition manufacturer before you increase the input voltage, otherwise you may overload some part of the system and 'blow' the electronics.

When purchasing a higher-voltage battery, obtain one with extra terminals that allow you to tap the correct voltage supply for various electrical items on the car. Therefore, while the ignition may work adequately at 16.8 volts, or even up to 18 volts, with the exception of the starter motor nothing else will survive for any time at this level. The fuel injection pump also pumps more fuel and at higher pressure at higher voltage, but most become unreliable at more than 14 volts. The remainder of the electronics, however, should not be operated at more than 13 volts.

Chapter 18

The engine management system

As the name suggests the engine management system (EMS) serves as the manager or overseer, controlling other systems which directly impact on engine performance and reliability. Its two basic areas of management are the fuel injection system and the ignition system. And, in many forced induction engines we could add boost control and maybe nitrous control. In some vehicles it may have oversight of the gear change mechanism, security system and exhaust emission control systems. To exercise such stewardship the electronic control unit (ECU), commonly called the computer, has to make 'decisions' based on inputs which it receives from numerous sensors. However, of all the information which is gathered by the ECU, by far the most important inputs relate to how much air is flowing into the engine and the rpm at which the engine is operating (*Fig. 18.1*).

For example, if air flow into the engine is very low and stable and the rpm are about mid-range and stable, then the ECU would figure that the car is cruising with the throttle almost closed. Consequently, the fuel injectors would be opened for only a relatively short time to provide a lean 'economy' air/fuel ratio. Also, because air density in the cylinders was quite low, additional spark advance would be provided to ensure that the low density mixture was completely burned in the time available.

What, though, if air flow was very high and stable, with engine revs in the mid-range and steady? Obviously, the engine is heavily loaded with a wide throttle opening. Air density in the cylinders is also high, so the ECU would have been programmed to open the injectors for a larger period to supply a slightly rich 'power' air/fuel ratio. Also, because air density in the cylinders is very high, but engine revs are only mid-range, combustion flame speed will be rapid and the risk of detonation will be high. Hence spark advance will be reduced.

From this you will appreciate that before the ECU can determine the appropriate spark advance angle and injector pulse width it must first be provided with information about what the engine is actually doing, and in particular precisely how much air is flowing into the cylinders at that particular moment. There are a number of methods that we can use to determine air flow into the engine: we can measure it using an air-flow

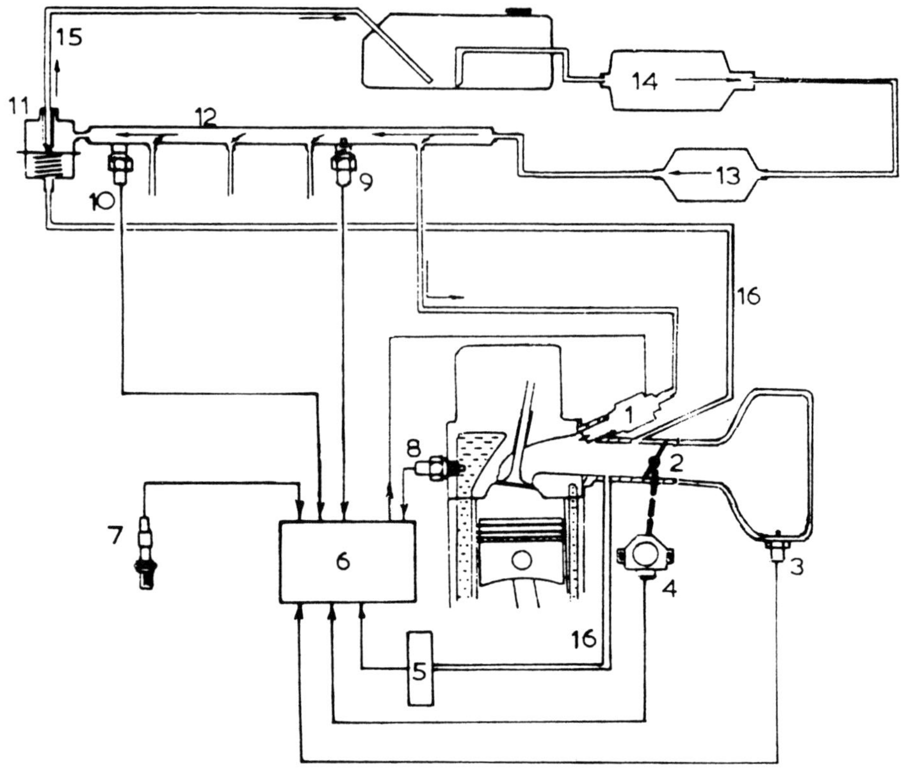

Fig. 18.1. Controlling electronic fuel injection is just one part of the engine management system's responsibility. The components illustrated in this speed density type system for a competition turbo engine are: 1. injectors; 2. throttle plates; 3. air temperature sensor; 4. throttle position sensor; 5. MAP sensor; 6. ECU; 7. exhaust gas temperature sensor(s); 8. coolant temperature sensor; 9. fuel temperature sensor; 10. fuel pressure sensor; 11. fuel pressure regulator; 12. fuel rail(s); 13. fuel filter(s); 14. fuel pump(s); 15. excess fuel return line; 16. manifold vacuum/boost sensor hose.

meter; we can calculate it based on engine rpm and manifold pressure levels; or we can calculate air flow from the engine rpm and the throttle plate open angle. As each system has its advantages and disadvantages we will look at each in turn.

MASS AIR FLOW SYSTEM

The first arrangement using an air-flow meter is what we call a 'mass air flow' system. The earliest air-flow meter, often called a vane air-flow meter, is a simple unit that passes the air flowing into the engine through a passage blocked by a spring-loaded flap. As the air flows through the meter it 'blows' the flap open to a certain angular position, depending on the force of the spring and the amount of air flowing in the passage. Connected to the flap pivot is a variable resistor – a potentiometer – which changes the ECU's input voltage according to how wide the flap has been swung open.

Remember that air density changes with the ambient temperature, as well as other factors, so air-flow input in itself is not adequate for accurate fuel metering and ignition timing. So the air-flow meter also houses an air temperature sensor, and this signal, along with the air-flow signal, is used by the ECU to calculate mass air flow. The air-flow meter also has a bypass passage with a mixture adjusting screw. By varying the amount of air that is permitted to bypass the sensor flap, the air-flow meter signal can be changed to alter the air/fuel ratio at idle.

The other common air-flow meter, originally called a 'mass air flow' (MAF) sensor, or 'hot wire meter', is now usually referred to as a MAF sensor. It is basically a tube in the inlet duct through which the air flows to enter the engine. Stretched across this tube is a very

This vane air-flow meter, still attached to part of the air box, is characteristic of the shape of air-flow meters of this type. A large slotted adjusting screw, located under the plastic cap (upper right), is used to regulate air flow in a bypass passage to control idle mixture strength.

This Nissan hot wire type air-flow meter has a large slotted adjusting screw (top left beside terminals) that permits the voltage signal going back to the ECU to be altered, thereby changing the A/F ratio and spark timing. From the factory, cars typically run slightly rich and retarded. Tweaking this screw while running on a chassis dyno allows this to be corrected.

fine platinum wire, which is heated by an electric current to maintain a constant temperature above ambient. An air temperature sensor in the meter signals the ambient temperature to the ECU so that it sends out the correct amount of power to keep the wire at the required temperature. However, air passing into the engine flows over the wire and cools it in direct proportion to mass air flow. Consequently the ECU must increase the current flow in the wire to keep the temperature constant, and it then reads this signal as mass air flow.

A third type of air-flow meter, found in Mitsubishi vehicles, is the Karmen vortex flow sensor. It uses an ultrasonic sender and receiver to measure the rate at which vortices pass through a control passage. The signal from the sensor is measured by the ECU, and after being corrected for inlet air temperature and barometric pressure, the mass air flow is calculated and the injector pulse width and spark advance angle are determined.

Many manufacturers use a mass air flow type system as it provides very accurate engine management when air flow into the engine varies over a very wide range. Also, because the air-flow meter is the major source for the ECU in determining the injector pulse width and ignition timing, this type of system can compensate for slight differences in engine specifications and long-term engine wear. However, it does have its drawbacks. Car manufacturers originally looked at other systems because air-flow meters are costly. From the aspect of performance, though, the major negatives are air-flow restriction, particularly of the flap-type meter, and inaccurate metering owing to reverse pulsing in the inlet tract when long-duration, high-overlap cams are employed.

SPEED DENSITY SYSTEM

To get around the high cost factor of air-flow meters, manufacturers developed the speed density system. In this system there isn't any air-flow meter to provide information to the ECU about air flow into the engine. Rather the ECU uses manifold pressure and engine rpm as the central keys to establishing the engine's fuel and spark requirements. The ECU also receives input from the air charge temperature sensor and the barometric pressure sensor (*Fig. 18.2*).

Fig. 18.2. Inputs for a speed density type system.

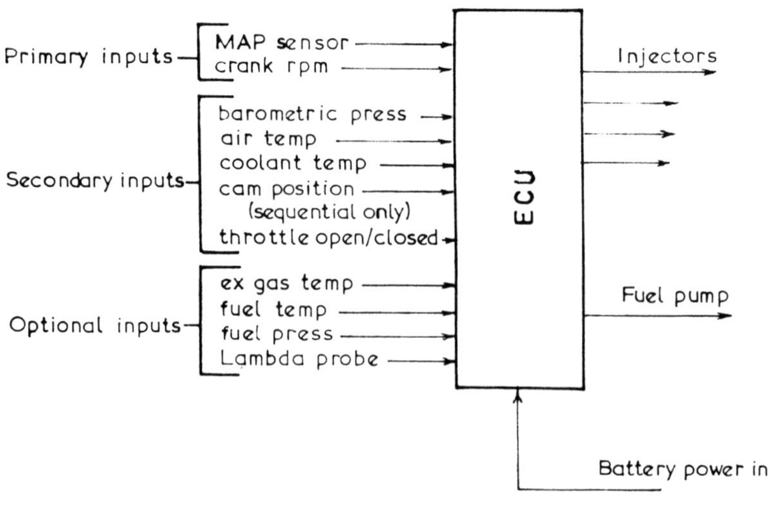

The MAP sensor is connected to the inlet manifold by a hose. It converts that engine load (boost/vacuum) signal to an electrical signal which the ECU uses to adjust fuelling and spark timing.

The advantage with this arrangement is that there is no air-flow meter to cause any flow restriction, and the manifold absolute pressure (MAP) sensor will respond to both positive and negative manifold pressures. Hence the system is well suited to turbo and supercharged applications. However, wild camshafts that cause strong reversion pulses in the inlet tract can give rise to metering difficulties with the speed density system. In more sophisticated ECUs (the GM Delco is one example), the effect of this rapid pulsing can to some extent be filtered out by calibrating the MAP sensor filter factor up to a full 100%, but this only works on cams up to about 230° duration with 108° lobe centres. Above that we often need to cross reference the MAP sensor with a throttle position sensor. In effect the system becomes a hybrid using inputs common to both speed density and Alpha-N systems. This arrangement is also recommended when a throttle plate per inlet port is used.

ALPHA-N SYSTEM

The third type, the Alpha-N system, was originally developed for competition engines. The primary ECU inputs for determining inlet air flow come from the engine speed sensor and the throttle position sensor (*Fig. 18.3*).

The advantage with this system is again that there is no air-flow meter involved, so there are no obstructions to air flow. Also, with no manifold pressure sensor being required, the Alpha-N arrangement is immune to metering difficulties caused by wild camshaft timing and massive reversion pulses. However, this does not mean that it is free from metering problems. With this system, accurate throttle position input is important because the assumed air flow into the engine is based on the exact throttle plate open angle, and the engine speed. This is not a worry with competition engines, but with road cars the first few degrees of throttle opening are most important for ease of driving. However, it is in these first few degrees of throttle movement, where big changes in inlet air flow are occurring, that potentiometers have their greatest resolution

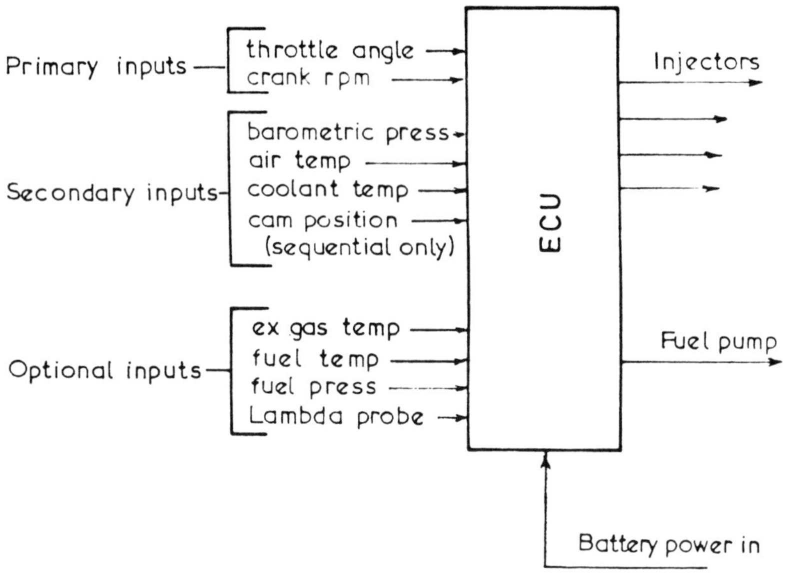

Fig. 18.3. Inputs for Alpha-N type system.

problems. Employing a logarithm type pot rather than a linear pot helps, but does not fully provide the very fine resolution that is really required. This will not stop the engine from running, but it will take away the low-speed and cruise smoothness that we have come to expect from modern cars. Rather, the engine may feel as though it is not up to temperature and has Weber carbs fitted with overly big chokes. Consequently, I advise against employing this system on turbo cars, as the hybrid system previously described is often far superior.

The throttle position sensor informs the ECU how far the throttle is open.

SECONDARY INPUTS

Apart from inputs which relate to air flow into the engine, the ECU will also be receiving input from at least one other source, and perhaps several. When the engine is cold, more fuel must be delivered to ensure that it will run and not stumble on acceleration. Therefore, according to input from the coolant temperature sensor, the ECU will increase the injector pulse width until the engine comes up to normal operating temperature, and on road cars, emissions laws permitting, it may increase spark advance too.

Because today's race engines are being tuned to finer and finer levels in search of more power, the dividing line between engine survival and destruction from detonation is narrowing. Hence, in the higher levels of competition it is no longer possible to run a 'safe' rich fuel/air mixture, but rather we are running as close as possible, just a hair's breadth away from detonation, with a full-power lean mixture. This means that more inputs are necessary to keep the ECU fully informed as to just what is going on, not only within but also around the engine.

For example, there are fewer fuel molecules in 1cc of hot petrol than in the same volume of cold fuel. So the true fuel/air ratio will become leaner and perhaps send the engine into detonation if the ECU does not adjust the injector pulse width as the fuel temperature increases. Conversely, without compensation, a decrease in fuel temperature will increase mixture richness. This will reduce engine hp slightly, and in distance race engines where we may be chilling the fuel to a low temperature prior to dumping it into the car's fuel tank in an effort to stretch the distance between fuel stops, we will not be gaining the full mileage advantage from the cold fuel. Therefore, when the ECU has input from a fuel temperature sensor it will adjust the pulse width to compensate for the temperature of the fuel being injected.

An air temperature sensor signal may be used by the ECU to not only adjust fuelling and spark but also to activate intercooler water spray and/or intercooler fan, water injection, adjust water-to-air intercooler pump speed, etc.

Another factor which the ECU needs to know about is reduced fuel pressure. Obviously, when the fuel pressure drops, less fuel will flow though an injector in a given time than when the pressure is normal. This is not normally a problem because fuel injection systems are designed to provide a reliable and steady fuel pressure. However, if we are running high fuel pressure, or are running distance events, a fuel pressure input to the ECU could save the engine from a meltdown, or in endurance events could help us reach the finish.

For example, if we have mapped our engine on the dyno with 60psi fuel pressure, and battery voltage drops off during the race, we could lose the engine because pump speed and fuel pressure will fall, resulting in a lean-out. In distance races the same sort of thing can happen, but a more usual problem is a fuel filter blockage knocking pressure at the fuel rail, or a pump losing speed because of a bearing seize-up and not being able to maintain system pressure.

Another source of input, the exhaust oxygen sensor, or 'Lambda probe', was originally developed for road cars expected to meet government emissions requirements. This sensor is fitted in the exhaust header collector, and is a metal-coated ceramic probe that functions like a weak battery, producing about 100 millivolts when oxygen content in the exhaust gas is high, as when the fuel/air mixture is lean. When the oxygen content falls, indicating a rich mixture, the probe's voltage output to the ECU increases. With this input the ECU can check on itself and when necessary make programme adjustments.

With time, the wide band type O_2 sensor has found a home on many race engines. During dyno mapping sessions an exhaust oxygen input can save a lot of time, and it can help avoid an engine meltdown. For example, you can key in a safe air/fuel ratio of 12.7:1 and quickly build a map at all load sites at that mixture richness. Also, in a race

The single wire Lambda probe is a narrow range sensor primarily designed to maintain cruise fuelling at an A/F ratio close to 14.7:1. Wide range sensors (the four wire AC Delco is the cheapest) are used in high performance applications and for engine tuning/mapping.

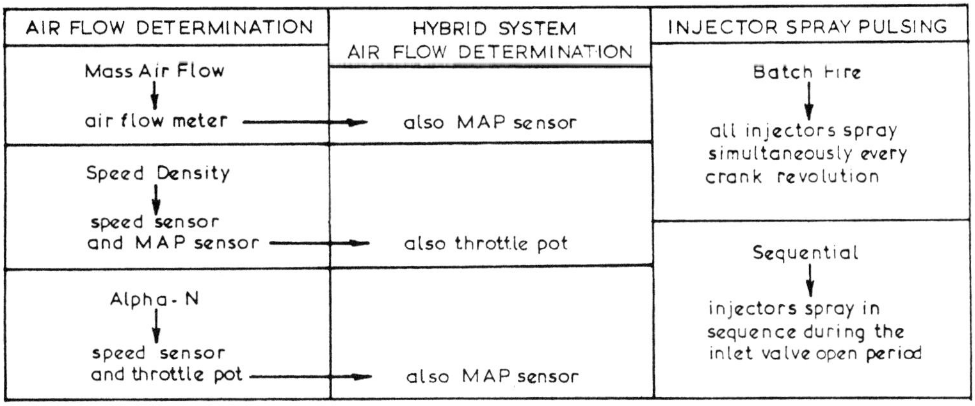

Fig. 18.4. Engine management is differentiated both by how engine air flow is determined and by how the injectors are fired.

car the tailpipe sensor can perform an engine-saving role by overriding the ECU at any time the exhaust indicates a dangerous lean-out, say leaner than 13.2:1.

However, there are good reasons not to rely on the Lambda to tune the fuel injection system for you. At full throttle, for example, the engine may run safe at an air/fuel ratio of 12.7:1, but remember that we are searching for maximum power. Therefore, at some engine speeds we may make more power at 12.5:1, while at other rpm the magic number may be 13.1:1. Turbo engines suffer more exhaust gas contamination, and they benefit from the piston cooling effect afforded by richer mixtures, so many show power increases at A/F ratios down to about 11.8–12.0:1. Also, in sequential systems we lose the benefit of individual cylinder trim, where we adjust air/fuel ratio differently for each cylinder to make maximum hp.

Another ECU input in maximum-effort engines is exhaust gas temperature. With a sensor in every header tube, depending on what we found in testing, the ECU would make programme adjustments to keep exhaust gas temperature (EGT) below a certain danger figure. Thus in one engine the safe EGT may be 1050°C (1922°F), with the turbo being the limiting factor, while in another we may lose valve seats at 780°C (1436°F).

Apart from differentiating engine management systems by looking at the three basic ways of how the ECU determines engine air flow, we also define systems by how they handle fuel injector pulsing duties. As discussed back in Chapter 13, the most basic systems batch-fire the injectors, while the more expensive systems employ sequential injection (*Fig. 18.4*). Some systems are also said to be semi-sequential or group firing, but this is a nonsense as they do not offer any advantage over the ordinary batch-fire arrangement.

WHAT TO LOOK FOR IN AN ENGINE MANAGEMENT SYSTEM

Beyond this we get into the area of very specialised engine management systems where you can virtually specify anything you want, money permitting! The optional features available include: spark and fuel trim to permit ignition timing and mixture adjustments to each individual cylinder; injector phasing to control when, during the inlet valve open period, primary, secondary and tertiary injectors fire; electronic boost control; turbo anti-lag system

The idle air stepper motor controls engine idle speed by increasing air flow in a bypass passage around the throttle plate when the air-conditioner, cooling fan, etc. cuts in. Not all aftermarket ECUs can control this function.

control; traction control; data logging of engine, suspension, transmission and chassis; secondary fuel system control, or water injection control; intercooler spray control.

In road cars, and sometimes in competition cars too, there are some other features we may require or perhaps find highly desirable. These would include: radiator or charge cooling fan control; idle air stepper motor control; closed loop Lambda control; silencer (muffler) bypass control; inlet tract switching control.

While it is obviously very necessary to focus on engine management features that are absolutely vital, we have to be cautious that we do not become so dogmatic in demanding a particular feature that we back ourselves into a corner and ultimately lose anything which we had hoped to gain with that particular engine management system. What can very easily happen is that we are virtually forced to purchase a particular unit because it is the only one with a feature which we feel we must have. The problem could be that that particular EMS demands a huge investment of time, both on the dyno and on the track, to get it set-up, and if that input isn't given it will cause the engine to run worse than a more basic system. Another problem can be that the EMS is just so specialised that there are only a handful of people about who actually understand how to properly programme the thing. If you can't find someone who knows both the engine management system and your engine you may never realise your dreams. In fact, it is more probable that your dreams will quickly become a nightmare! In similar vein, keep in mind that some EMSs are well known and understood in one country or one branch of motor sport but are unknown elsewhere. If you are outside that circle of knowledge then it could be risky to set your focus on such a unit.

Another thing you may need to plan for is upgradeability. If you are in a very competitive league where progress is constant, you will most likely want an EMS backed by a manufacturer who will provide regular inexpensive upgrades and extensive technical back-up. If you are not in such an area of motorsport, keep on the lookout for the 'junk' from the upper echelons of the sport. The EMSs that they are tossing out for peanuts could well be a high-end system that is more than adequate at your level of competition.

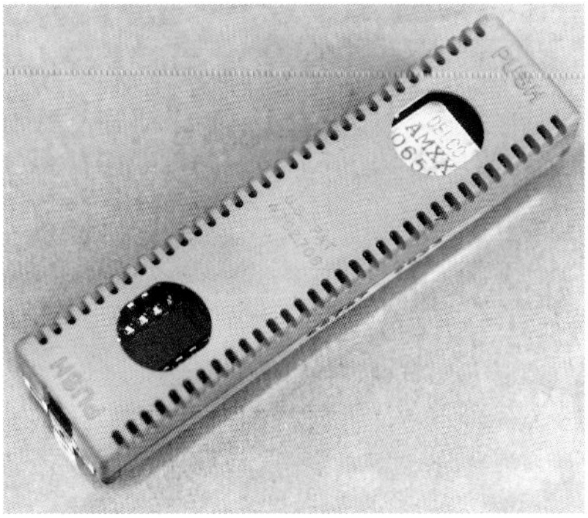

Left: An EPROM chip can be reprogrammed to adjust many parameters affecting engine performance.

Right: The chip plugs into the ECU socket as shown.

THE FACTORY COMPUTER

Very often stock engine management systems are rubbished and then replaced by inferior aftermarket systems from a 'name' manufacturer. There is nothing wrong with getting rid of an ECU which is either impossible to reprogramme or else does not have many of the necessary control features required to allow the engine and car to perform at an acceptable level. However, I question why you would want to toss an easy to programme ECU with everything you need to make street driving as pleasant as possible, and fit an expensive EMS which can't properly control basic things such as the idle air stepper motor or the radiator cooling fan.

Also, when you decide that it is necessary to replace the factory computer, don't automatically think of purchasing an aftermarket unit as you could find a number of good inexpensive ECUs at your local breakers' (wrecking) yard from an earlier model of your car, or another model from the same manufacturer, or even from another brand of car. I know it goes against the grain for some people to do this, but it doesn't make much sense to reject a part on such grounds.

For example, where I operate you can find any number of GM Delco '808' (so called because these are the last digits of the part number) computers for about the same price as 5 litres of Mobil 1. Using Kalmaker software, they can be reprogrammed to work wonderfully in many applications. Most, though, snigger at such an idea, which is great for those in the know as there will be a ready supply of such ECUs and matching wiring harnesses around for next to nothing for a long time. These units may be only 16Kb and not provide sequential injection, but they are considerably more sophisticated than many low spec aftermarket ECUs. Also, because they have a removable EPROM 'chip', it is simple and inexpensive to have one chip burned with a programme suitable for everyday road use on pump fuel, and another chip for weekend competition on an exotic brew.

In the early '90s GM began using a 32Kb ECU which supported semi-sequential injection on 4-cylinder and 6-cylinder engines. These units are more expensive than the Delco 808, but are a steal in comparison with the price of an equivalent aftermarket

ECU. The semi-sequential injection capability isn't a bonus, but the improved computing ability greatly improves idle control and cruise fuel economy in road cars. Also, these units have a high speed serial data link which enables you to read out in real time, using Kalmaker software and a laptop, exactly what is going on with regard to ignition timing, injector pulse width, etc. Actually, the early Delco 808 can be upgraded very simply by fitting the later 32Kb MemCal (GM speak for the EPROM chip).

Soon after this, GM switched to a 64Kb Delco unit which also incorporates auto trans control. They called this a powertrain control module (PCM). It did not support sequential injection either, but this feature came later when GM adopted the 128Kb three plug Delco unit in the mid-'90s. These dropped MAP sensing and instead used mass air-flow sensing, but they can be switched back to MAP sensing.

With Nissans it is sometimes possible to change back to an earlier reprogrammable ECU. The R32 Skyline GT-R could handle an extensive power upgrade with relative ease owing to the stock computer's design. When the next model (the R33) came along, that all ended – the ECU was now a 'no go' zone. However, for about an eighth of the price of the 'brand' Japanese computers you can obtain second-hand units out of R32s. Without any reprogramming these will often provide a 10–15hp gain owing to the later model being set to run rich and retarded to help the CAT control emissions. With a dyno session to optimise fuelling and spark you can expect another 10–12hp, and if the car has a new free-flow exhaust and inlet system the gains are likely to be even higher.

Clearly, when it comes to selecting an engine management system, there are a huge number of factors to consider. Unfortunately, many base their decisions on hardware specifications. For example, looking at the above-mentioned Delco units, it could easily be concluded that the 16Kb 808 was useless even as a boat anchor and that the 128Kb

A Nissan R32 computer is programmable. This one, sourced from a 2-litre 6-cylinder turbo model is destined for use in an R33 Skyline GT-R. When the stock R32 GT-R programme is burned into the chip it will provide an immediate 10hp–15hp improvement. With further tuning on the dyno, another 10hp–12hp will most likely be realised.

hardware would be the obvious choice. However, hardware specifications are only a small part of the equation. I've already mentioned that you require a system that can be easily programmed, but on top of that requirement there are other factors which will influence our decision.

CHOOSING AN AFTERMARKET COMPUTER

Rather than hardware, we should perhaps focus more on software. Unfortunately, engine tuners don't know much about software. If they did they would be in that industry making a lot more money. Consequently, manufacturers of engine management equipment can't appeal to them to make purchasing decisions based on impressive software capability. Rather, they have to talk to us in language we understand, so we are bombarded with hardware sell. How many million calculations it will make per second, how it still functions in adverse environments (heat and vibration), and so forth. However, as engine tuners, we have to question: 'So what if the hardware can make a million calculations per second? If the software is flawed the instructions issued will be useless anyway!' Really, we must never look at the hardware in isolation but rather at the whole package.

Also, we need to remember that the engine management system has to integrate with other areas of the vehicle. The Australian MoTeC system is a good example of the kind of features we would be looking for in a high-end competition EMS. The MoTeC accepts a huge range of sensor inputs and will drive a diverse range of fuel injectors. It has multiple injector and ignition outputs, so can batch-fire, group-fire, sequentially-fire or phase-fire injectors, and it can trigger single coil or multi-coil ignition systems directly from the ECU. Modern race cars rely very heavily on data acquisition and complex LCD dash displays. The MoTeC interfaces with such systems from companies like Pi Research and CIC.

If we were looking at engine management purely from the aspect of engine performance, as in the numbers we see on the dyno, then the Autronic ECU will provide performance equal to, and some would say superior to, MoTeC units. The Autronic has numerous impressive features, such as a full three-dimensional cold-start feature which doesn't simply add a percentage increase to the injector open time but rather operates on a load versus coolant temperature map. The boost pressure can be controlled as a function of coolant temperature (low boost when the engine is cold) and as a function of what gear has been selected (less boost in lower gears to control wheelspin or transmission damage). Programmable outputs can switch such things as the water pump in the water-to-air intercooler system when the inlet air temperature reaches a certain temperature. Like numerous stock type ECUs, the Autronic also has a limp-home mode, and it will flash out fault codes to inform you of problems without having to be connected to a laptop. However, unlike the MoTeC computer there is very little flexibility in terms of sensor configuration, especially in the area of ignition pick-ups where only a few specific ignition triggers can be deciphered. Some tuners, though, consider this as a positive as the Autronic does not suffer signal 'noise' problems like other ECUs designed to function with a broad range of sensor types. Also, for popular vehicles such as the Mitsubishi Lancer Evo and Subaru Impreza WRX, the Autronic can be supplied as a board which slips into the factory ECU case and connects to the stock harness, to retain a totally stock appearance.

Left: The MoTeC engine management system has great flexibility.

Below: An Autronic replacement board for a Subaru Impreza WRX has all the features of the high-spec Autronic SMC unit.

Richard Aubert, the man behind Autronic and previously the software expert at MoTeC, has devoted all his energy to producing and constantly refining a brilliant ECU, and unfortunately documentation isn't slick, or easy to follow. Don't let this put you off the Autronic units, though. I understand Richard is in the process of correcting this matter, but there are alternative sources. A couple of enthusiastic Autronic users, Steve Cox (www.quickfitmotorsport.com.au) and Ray Hall (www.turbofast.com.au), both with a lot of experience wiring and mapping the unit in a wide variety of applications, decided to take matters into their own hands and produce their own manuals.

A mid-range engine management system, with the unlikely brand name EMS (also from Oz) has one interesting feature which adds to its user friendliness. I usually don't like, nor do I recommend units which are programmed from a hand controller. Tuning is too much a fiddly hassle for my liking. Fortunately, the latest EMS brand units give you the option of calibrating the ECU using either their hand controller or a laptop. I would choose the laptop route every time. However, having said that, the hand controller does have one appealing function – it can provide live-data as you drive. As such, it can save you fitting extra gauges that you would otherwise need to obtain to provide you with such information as charge air temperature and the air/fuel ratio. For this latter function to operate, a wide-band Lambda sensor must be fitted to replace the stock exhaust oxygen sensor.

Having the air/fuel ratio meter function built into the hardware is unique and enables the vehicle owner to report back to the tuner any fuelling problems after the system has been initially mapped on the rolling road. Car manufacturers have a power-train team spend months, at a cost of many millions, in an effort to get the programming correct. Obviously, a tuner in a few hours on the dyno can't duplicate every conceivable engine operating condition to achieve similar results. With the EMS computer providing an air/fuel ratio readout, any tuning glitches can be fairly quickly identified as being either a fuelling problem or an ignition advance problem during normal operation on the street.

There are other aspects to a computer being user-friendly. As mentioned, if you want to wire the thing up yourself, or call on the local auto-electrician with no knowledge of this sort of stuff, then you will want extensive totally accurate and easy to understand documentation supplied with the ECU. This is where a mid-range unit such as the Haltech has no peers. The Haltech, as far as I know, is also the only unit capable of operating the idle speed stepper motor in GM and Toyota vehicles. Also, it has pulse modulation boost control and turbo anti-lag control as standard features.

In spite of all the advances which have been made with aftermarket ECUs, we have to realise that none will do all the things that the standard ECU fitted in today's cars will do. In general, the aftermarket ECU is designed mainly with ease of programming and engine management as the priority. So, in a road car, how do we keep the security system functioning and the air-con/climate control working, while at the same time achieving the injector, timing and boost control that our modified engine demands? Basically, the answer is that you can't unless the stock ECU is retained.

Below left: The EMS brand electronic control unit has a number of useful features. In spite of what the label seems to indicate, this is not a fuel only control.

Below right: The Haltech ECU is supplied with excellent documentation to simplify installation.

REPROGRAMMING THE STOCK ECU

The very first option we should always carefully examine is the possibility of reprogramming the stock ECU. In the past it was popular to re-chip the ECU; either swapping the EPROM chip for a chip with a different programme or else by soldering a new chip on to the stock ECU circuit board for those computers which didn't have a plug-in chip. Usually these replacement chips were supplied by mail order, and at huge expense. As such they didn't do much for performance, not to mention engine life and fuel economy, in spite of very impressive advertising claims. The truth is tuning of this sort can only be done on the dyno. Even a dozen precision-assembled race engines of the same specifications, put together by a top engine builder, will require slightly different ECU calibration to deliver optimum performance, so how can a programmer burn a perfect chip for an engine he has never seen? The truth is he can't, he is only guessing.

However, if done properly on a dyno, using good software which enables the tuner to access the manufacturer's original programme codes and correctly modify those codes, so-called chip tuning is very successful and has much in its favour. First of all it ensures that everything in the car will keep functioning as the manufacturer originally intended. Depending on car specification, that can include stuff like the sound system, navigation system, anti-theft devices, climate control, electric seats and mirrors, etc. Also, if the manufacturer's cruise mode and idle mode maps aren't fiddled very much the car should idle well and cruise nicely without surging and flat spots; and it should pass emissions tests.

RUNNING TWO COMPUTERS OR A PIGGY-BACK SYSTEM

What, though, if the stock ECU cannot be reprogrammed either because nobody has been able to crack the programming codes or because the reprogramming options available are either too limited or too complex? There are two alternatives which both retain the stock ECU: either we can run two ECUs together or we can use an interceptor piggy-back ECU arrangement. What path we ultimately take will depend on how far we have gone with our modifications.

Personally, I prefer to use two ECUs as it allows considerably more flexibility. For example, we are not limited as to how far we can go with our modifications, and when we bring the car back to stock, or if we resell it, it is very easy to remove the second ECU and have everything working normally again. When we go with this set-up the new ECU will control only the engine management side of things, i.e. fuel injection, ignition timing, boost, etc. The stock ECU will retain control of all the other functions.

There can be problems, though, with the two ECU systems when there is a connection between engine management and other vehicle operations. Traction control is one problem area, as is electronic throttle control and electronic clutch arrangements. Frequently it is difficult, if not impossible, to achieve the required degree of cross-talk between two ECUs to get these sophisticated systems working satisfactorily, so the only alternative is to opt for an interceptor piggy-back ECU arrangement. These things aren't actually an ECU, nor are they a chip, even though one very capable unit would appear to imply that by its name – Unichip.

The Unichip piggy-back interceptor manufactured in South Africa is only sold through distributors as part of an install and map package.

An interceptor is cut into, or plugged into, the wiring harness to receive and then modify signals prior to the stock ECU receiving them. (Some systems modify the signals going out from the ECU, while others modify a few of the incoming signals as well as some outbound signals.) As a result of receiving input that has been tampered with the stock ECU will then send out a different set of fuelling and ignition advance instructions. And if the stock installation features a pulse modulation boost control valve, that setting can also be modified. Cars with speed limiters and over-boost cut out can also have those functions disabled.

However, the interceptor has its limitations, and with road cars, if it is possible, I prefer to reprogramme the stock ECU or run twin ECUs. What has to be understood is that the interceptor is limited by what the manufacturer has put on the stock maps. It can only replace one set of tuning codes with another bundle that are already available on that particular map. Consequently, an interceptor is only suitable for reasonably moderate engine modifications. Usually ignition timing isn't a worry as most allow ±12–15° from the stock settings, but fuelling is a problem as you will often be limited to not more than a 15–20% increase and it usually isn't possible to change fuel pressure or fit larger injectors as many stock maps don't have sufficient latitude to permit the interceptor to lean the fuelling off the required amount when the engine isn't at wide open throttle and full boost.

However, if there is no alternative but to employ an interceptor, then we have to accept the limitations that go along with that course. This means choosing the best interceptor for the job at hand and then taking care to keep our tuning efforts within the bounds of what the interceptor can comfortably accommodate. Very often the reverse occurs, with the enthusiast modifying the engine and then being disappointed by the final result because the engine spec is way beyond what any interceptor can handle.

PROGRAMMING THE COMPUTER

Regardless of the path we have chosen, the end result will only be as good as the steps we have taken to ensure that our engine management system is correctly

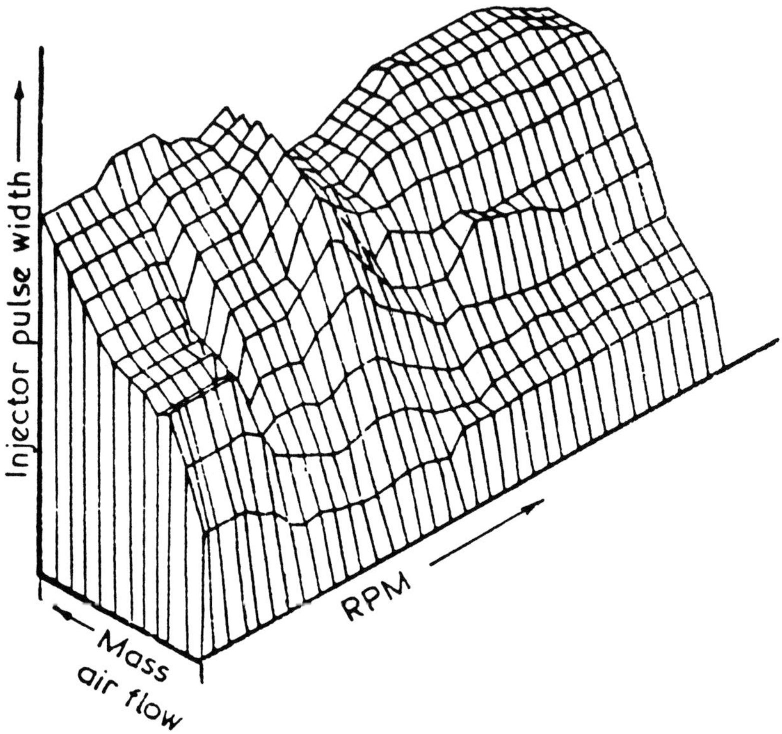

Fig. 18.5. A fuel map with 11 load sites (air flow axis) and 36 engine speed sites (rpm axis); a total of 396 sites.

programmed (*Fig.18.5*), enabling our engine to produce maximum power along with good reliability, and also good fuel economy if that is of importance. In essence we have to be prepared to put in the time running on a dyno to get the right set-up. That might sound simple enough but it is anything but. The big problem is that there are lots of tuners around who don't have a clue about mapping, and equally there are many dynos about that are completely unsuited for mapping.

A new ECU and the mapping of it represent a huge investment for most people, and I'm not just talking in terms of the cash outlay involved. Really, it doesn't matter how much money and effort you have put into the engine and the rest of the car, if the engine management system isn't spot on the whole exercise has been pretty well wasted effort. Therefore, it is imperative that you choose carefully. Don't just think in terms of buying an ECU, or a chip, or an interceptor. Rather plan on buying a package that includes suitable hardware and software, plus a tuner who understands and can get the best out of that hardware and software, and who is working on a dyno suited to mapping.

So often I see enthusiasts pay good money for rubbish results. They buy the ECU they think they need and then try to find someone who can map the thing. I realise that it is very much a catch-22 situation. We are dealing with relatively new and constantly evolving technology here, and every tuner has to learn somewhere. That is fine if he is doing it in his time and at his expense, but when you are paying the bill it is not such a great idea, especially as you will invariably find that the exercise is repeated several times, with you going from one tuner to another before you actually have the car running satisfactorily, let alone the way you know it should.

CHOOSE THE RIGHT DYNO

When it comes to developing engines the engine dyno is always the way to go. However, mapping is different. Generally, it is preferable for mapping to be done on a chassis dyno – a rolling road. This is especially so in the case of turbo engines as underbonnet heat has a large influence on engine tune. Obviously, you will gather from this statement that the bonnet should be closed for final map fine tuning to most closely replicate the actual environment that will exist on the road or track.

However, not just any rolling road is suitable. What we need is a modern, eddy current electronic dyno that allows the operator to hold the roller speed. Many newer dynos are the much cheaper inertia type, with the Dynojet perhaps the most well known. Instead of having an electric brake or a water brake these have a very heavy roller ($2\frac{1}{2}$ tons in the case of the Dynojet) which the tyres accelerate. Calculating the rate at which the roller accelerates, the dyno computer can then do a print-out of the vehicle's hp at the wheels. That's fine if your only concern is determining the engine's power while accelerating at full throttle, as the test is very fast and the likelihood of engine damage is minimal. However, when properly tuning an engine we are not just interested in how it runs from low rpm to maximum revs at wide-open throttle. We want to get it accelerating well without a stumble and without detonation over a broad range of engine speeds and a wide range of throttle openings. Also, we want the engine to be capable of

Any engine management system is only as good as the time and effort put into careful mapping, preferably on an eddy current dyno.

holding maximum engine revs at full throttle for several seconds just as it will be called upon to do lap after lap on a race circuit. (In acceleration mode the engine doesn't get time to heat up fully internally so it is very easy to go overboard with excessive boost and spark advance, or a too lean fuel mixture that looks fine on the dyno but will destroy the engine on the track.)

The eddy current dyno, when switched to closed loop mode, has this ability as an electric brake holds roller rpm regardless of load. Therefore, when mapping, the operator can calibrate the ECU with the engine held at a fixed engine speed. Having dialed in the rpm he wants the dyno to hold, he can then set about tapping in different fuel and ignition settings, and the engine speed will not change – only the torque readout will change, indicating whether the new settings are to the engine's liking. When happy with the result at that rpm point and throttle angle the operator can move on to the next site until the map has been completed.

A similar thing can also be accomplished with a water brake type chassis dyno, but it is more of a fiddle and takes longer. Like the eddy current type, the water brake can be used to map at any throttle opening, but the problem is that as the spark or fuel is altered and power output goes up or down engine revs will rise or fall in response to those changes, moving the engine to a different rpm and load site on the map. The operator then has to adjust the water brake to alter the load on the engine to get back to the map target site. Naturally this takes time and costs you money. And there is always the risk of the operator becoming frustrated at the engine revs constantly jumping about and simply giving up before the best setting has been established. This matter of the engine rpm changing is especially troublesome at the instant high output turbo engines come on boost, and of all engines they are probably the ones we need to be extra careful with to ensure that the mapping is spot on.

The very fine resolution possible on the Autronic engine management system makes it much easier to map on very temperamental high-output turbo engines. Again, the labelling is misleading; it is not a fuel only unit.

MAP RESOLUTION

While we are on the topic of mapping we need to spend some time on the subject of resolution, i.e. the number of rpm and load sites on a map. I am not sure why, but in recent times more individuals are requesting – some are demanding – finer and finer resolutions. Unfortunately, the manufacturers of engine management systems have gone along with this nonsense to such an extent that some maps have hundreds of sites waiting to be filled in. Regardless of what you have perhaps read elsewhere or been told, you need only eight load sites for a naturally aspirated engine, and up to double that number with forced induction. Many competition engines really don't require that number. Anything more is overkill and a total waste of time. For rpm sites some people are fooling around with 100rpm and 200rpm steps. With a competition engine you don't want more than about 11 or 12rpm sites. In a road car, most people stick to sites at every 500rpm step. This is fine but really isn't necessary. There is nothing wrong with 1000rpm steps as soon as you get into the area where the engine spends the least time. For competition engines that means the bottom half of the rev range, and for road cars the top end. Hence, even if the ECU can be taken down to extremely fine resolutions, don't do it, that 'feature' is only there to accommodate those who don't mind wasting their time or think they know more than the best tuners.

For those really into this sort of thing, they really should try the Autronic ECU – it has sensitivity fine enough to allow map sites as close as 1kpa (0.145psi) on the load axis and 1rpm on the other axis! However, all those sites are provided for a very good purpose. Some ECUs demand that all sites be filled in (the Haltech for example) but the Autronic simply makes the provision to enable very precise engine management during those areas of the rpm/load range where the engine suddenly gets very busy. The turbo may rapidly make loads of boost in unison with the variable cam mechanism performing a big cam swing. Consequently, over a very limited rpm range, the engine may require major fuelling and spark advance refinements which may not be adequately met by sites at say, 3500rpm and 4000rpm. In fact, it may be a waste of time having sites at those two points because the big influences are all occurring between 3680rpm and 3930rpm. Therefore, it may be more effective to have a three site cluster in that rpm span, with the closest site either side perhaps 600–800rpm away.

Chapter 19

Building a rugged engine

Engine reliability and integrity are always a major concern with forced induction engines. Some people tend to be extremely optimistic in this area, and thus dismiss any suggestion that forced induction places additional stresses on engine components. When it is pointed out to such individuals that every vehicle manufacturer that has ever added any sort of supercharging has always spent a lot of money strengthening their engines in various ways (in spite of their penny-pinching in other areas), these folk generally shrug this off, muttering that manufacturers are a conservative bunch and as peak cylinder pressures are not much higher anyway what is there to worry about?

Both of these comments, of course, have some validity. However, there is no getting away from the fact that forced induction engines are more prone to failure in certain areas than naturally aspirated engines. And the converse is sometimes also true. In some areas, naturally aspirated engines are more at risk than those which are supercharged. This is because naturally aspirated engines usually have to be spun at much higher rpm when making decent hp. Consequently, con rods in particular tend to be yanked apart.

While it is correct that with forced induction we are usually making considerably more power at lower rpm and thus less likely to break con rods, it has to be recognised that invariably the higher levels of hp made available from smaller engines subjects them to greater mechanical and thermal stress. While the heat and pressure levels in the cylinder are not significantly higher, they do stay closer to the peak number for many more degrees of crankshaft movement. That is how more power is produced – not by moving the peak higher, but by sustaining cylinder pressure at a level closer to the peak for much longer. As a consequence, certain engine components have a considerably shorter 'rest' period between each combustion cycle. Therefore, parts such as the cylinder head, the valves, the head gasket, the pistons and rings and upper cylinder area will be subjected to a greater heat load as there will be less time available in which to shed combustion heat.

In similar fashion the same components will be distorted by combustion pressure for a longer period, so rather than simply 'springing' back to their initial shape they may be more inclined to remain distorted. This then leads to other problems. For example, an out of shape piston creates more friction against the cylinder wall. So, in addition to the extra heat it is already carrying owing to the longer combustion 'soak' time, now we have additional heat being put into it from scuffing against the bore. On the surface, it could appear quite easy to fix this problem by using a more robust piston. Often the reality is that our resorting to stronger, usually heavier, pistons simply introduces a whole new set of problems. If we maintain the same rev limit, the heavier pistons will place considerably higher loads on the piston pin, con rod and con rod bolts in particular, and to a lesser extent also on the crankshaft. Consequently, all these parts may require beefing up as well.

It is usually recognised that quite a number of supercharged engines fail because of detonation. Frequently this is blamed on excessive boost or ignition advance when, in fact, the culprit could just as easily be less than effective piston ring seal. Remembering for a moment that cylinder pressure is very high for considerably longer in a forced induction engine should alert us to the need for a piston ring seal as close to perfect as possible. If it isn't there will be a massive increase in blow-by past the rings into the crankcase and sump. This blow-by gas, along with any oil which it collects along the way, will possibly be routed into the inlet system of street cars if the manufacturer's stock ventilation system is retained. The oil mist will very effectively lower the fuel octane rating, opening up the way for detonation. Also, the blow-by gas, in forcing the piston rings off the cylinder wall, reduces the amount of heat transfer out of the piston crown. The hot piston adds heat to the air/fuel charge as it enters the cylinder, making it more likely to detonate.

This cylinder was badly pitted by aggressive boring ripping chips of cast iron out. Obviously, that was followed by extremely shallow honing. The end result was a 'reconditioned' turbo engine with no power, no ring seal, plus massive oil consumption and detonation problems.

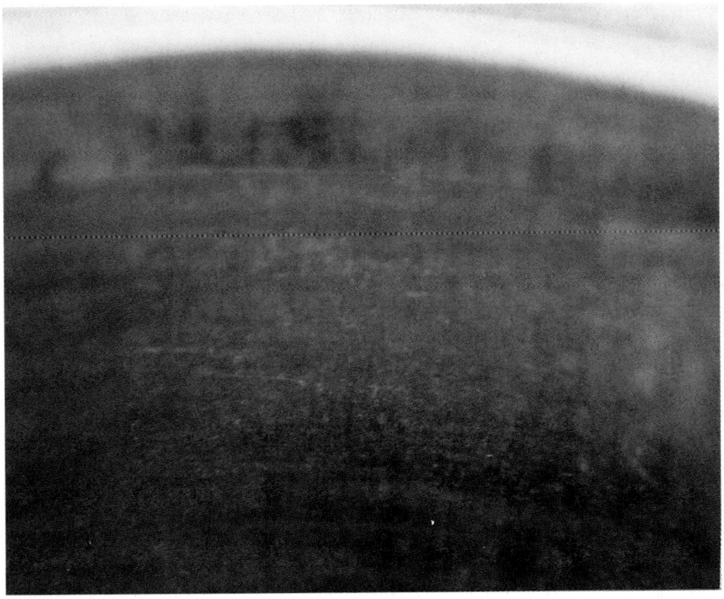

Unfortunately, high performance is often associated with boost pressure, camshaft timing and lift and exhaust sizes. However, many hidden horsepower can be found by proper attention to the bottom end of a motor. An additional consideration, as we have already touched on, is the added reliability we can expect when things are done properly and to an exacting specification.

CYLINDER BLOCK PREPARATION AND CRACK TESTING

Logically, the place to start is the cylinder block, because it is the home for everything else in or on the engine. First, inspect the water passages for rust or scale. If any is present it will be necessary to have the block boiled and cleaned in a chemical bath. This should be done with all the welch plugs (freeze or core plugs) and oil gallery plugs removed.

A mildly dirty block can be steam cleaned or washed with solvent. At the same time, the oilways should be carefully cleaned, using brushes designed for this task. Stud holes, deserve the same treatment, particularly head and main bearing cap stud holes. Finally, high-pressure air must be used to clean and dry off everything. Even brand new (green) blocks should be cleaned as outlined, with particular attention to ensure that no casting sand has been left in the water passages.

With the block perfectly dry, it can be visually inspected for cracks. A competition block should be pressure crack tested; really it is a waste of time and money if this is not done first. You can do a pressure test yourself as only very basic equipment is required. Preferably the head should be bolted to the block (it is acceptable to use an old head gasket for this test if it is not torn) and both head and block should be fully stripped, but with welch plugs in place. All water openings should be blocked off with steel plates. One plate should have an air line fitting installed (or you may prefer a tubeless tyre valve). Connect to this your air line tap and pressure gauge (the use of an old oil pressure gauge is suitable). Now slowly open the air valve and gradually let the air pressure come up to 40psi. Do not let pressure come up too quickly because if a welch plug is not in tight or if there is a weak spot in the block it could blow out and cause you injury.

If everything is holding satisfactorily, again open the air valve and let the pressure slowly build to 50psi. Some people advocate that you should check at 60psi, but this can be quite dangerous if there is a weak spot, and in any event most cracks will become evident at 45–50psi. (Note that heavy duty thick-wall race blocks cast with special chrome moly alloy iron can be tested up to 100psi.) It is really more a matter of how carefully you look for cracks as you slowly work your way over the block (check the head too, and inside the inlet and exhaust ports) than going to very high air pressure.

To make the cracks more clearly visible, some prefer to spray over the entire outside, then the inside, of the block with a penetrating aerosol like WD-40, carefully covering just a small area at a time, while others choose to spray on water with a little added liquid detergent. Either way, any cracks will be shown up by the presence of air bubbles.

If you find a crack do not be in a hurry to discard the block, unless it is obvious that it is not repairable either because the crack is too large or it is in a critical location, as some cracks can be inexpensively repaired. A cracked block that has been repaired may be just as strong as one that was never cracked. In fact, some engines routinely crack in certain areas, so if you are too fussy you may have to check out many second-hand blocks to find one that is not cracked. Then, after you race it a few times you may find that it too has cracked, whereas the block that was initially inspected and subsequently repaired may now be stronger in that particular spot and so not give any more trouble.

With the block crack-tested, take a plug tap and clean the head stud and main bearing cap threads of dirt or burrs. This is important to get a true stud tension reading. Any stud holes that have not been chamfered should be, to prevent threads pulling up; a pulled head stud thread can cause head gasket sealing problems, while a pulled main bearing cap thread can allow a bearing to turn. Finally, check the depth of each hole to be certain that the stud will not bottom.

To avoid future cracking, carefully inspect both the inside and outside of the block and grind away any casting slag. The main area for concern is around the main bearing webs, the sump pan deck and, in Vee motors, around the oil drain back holes and the rest of the valley area.

MAIN BEARING ALIGNMENT

The next operations involve 'squaring' the block. A check must be made to ensure that the main bearing bores are perfectly aligned, as any misalignment will wreck the bearings and possibly even the crank. It will also soak up a lot of power as the frictional losses are greater. I have found new blocks with up to 0.007in misalignment, so do not assume that the alignment is correct. Any misalignment can be corrected by line-boring. When main bearing caps are replaced by heavy-duty items, line-boring will also be necessary. To ensure that alignment is maintained, the main bearing caps should be numbered and have the front position marked. This will assist in fitting each cap in its correct location each time the motor is rebuilt.

With the main bearing bores true, the block should be checked at each corner to guarantee that the distance between the top of the block (the deck) and the crankshaft centreline are identical. If the block is out of true, milling will be required.

CYLINDER BORING

The cylinder bores must be true at 90° to the crank centreline (i.e. across the motor) to keep the frictional losses low and maintain good ring sealing. If the bore is canted slightly to the front or the rear, the piston pin will hammer out, so a check must also be made in line (i.e. at 180°) with the crank. There must not be any steps or taper from the top to the bottom of the bore (*Fig. 19.1*).

Boring and honing should be carried out by a firm with precision equipment able to maintain a tolerance of 0.0003in. Forget about machine shops using small boring machines that bolt on to the block, as the tolerance can be as bad as 0.002in.

On a street engine, the machinist sets up the boring bar centred on the existing cylinder bore. However, there is no guarantee that the existing bore centres are correctly located (indexed) in relation to the crank throws or the combustion chambers in the cylinder head. Therefore, in a race engine I prefer to see a bore index plate bolted to the block as a reference to locate the bore centres correctly. This ensures that all the cylinders are centred over the crank throws, and it also means that cylinder heads can be

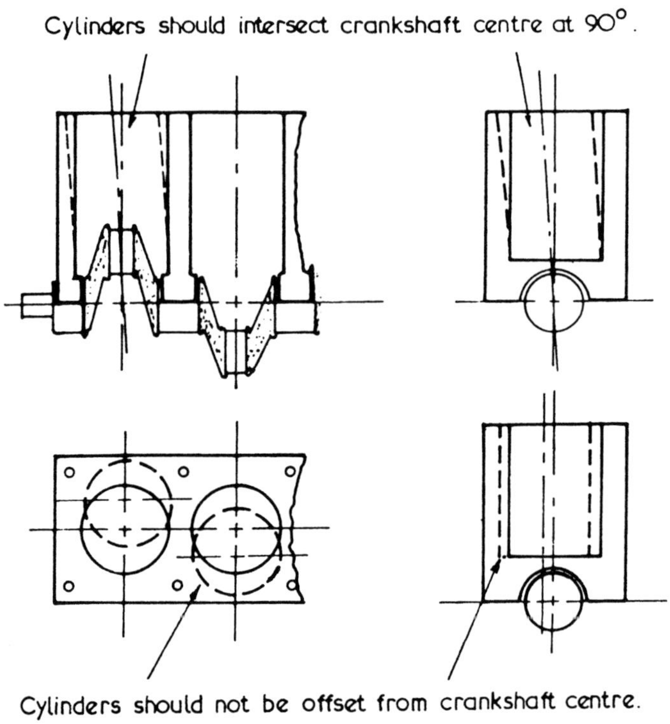

Fig. 19.1. Cylinders must be accurately bored.

swapped between blocks with the certainty that combustion chambers will all be correctly located over the cylinders.

When a race engine is being bored, I like to see only 0.010in taken out with each cut; with a road motor this can be increased to 0.015–0.020in. So, while a 0.030in or 0.040in overbore could be achieved with one cut on a road engine, I would want to see two cuts made on race blocks. This costs more because of the time involved, but the bores will be easier to hone accurately because the boring tool marks will not be so deep and rough.

CYLINDER WALL THICKNESS REQUIREMENTS

There are definite limits as to how far a cylinder can be overbored. If the cylinder walls become too thin they will warp owing to the pressure and heat, which will allow blow-by past the rings, resulting in lost power. Another problem, not often recognised, is that at high hp levels the main bearing webs can actually break away from the bottom of the cylinders.

Normally we would only want to rebore the cylinders to rectify sloppy factory machining or to restore cylinders which were worn; we are not seeking to increase the capacity of the engine. Remember that there is no benefit to be gained if an increase in bore size will induce cylinder warping or block flexing. Both conditions will rob us of power and lead to reliability troubles.

It seems to me that many people do not understand just how little wall thickness there is, even in a stock block that has not been re-bored. Today many cast-iron blocks have a designed cylinder wall thickness of only about 0.110–0.140in when finished with a standard bore diameter. However, during casting, a core can very easily shift 0.040in. This means that the wall thickness can be down to 0.070–0.100in, and we are not talking high-grade alloy steel here; some manufacturers use some pretty ordinary 'soft' cast iron in their blocks. Obviously, with today's race engines making about 120–160hp per cylinder, such thin walls, if overbored, will either crack or, at best, cut the power of the engine owing to inadequate ring seal.

Basically, for good ring seal I reckon that you need a cylinder wall thickness of 0.095–0.105in minimum for an engine expected to make 65–70hp per cylinder. This is most critical at the top half of the stroke because it is the top area of the cylinder where we require good ring seal, first to make compression, then, after ignition, to retain cylinder pressure to drive the piston down and make power. Towards the bottom of the cylinders we may be able to drop approximately 0.020in in wall thickness, but there are many other factors to consider, such as operating rpm, crank stroke, main bearing web strength and whether the engine is ever likely to suffer detonation.

As power levels increase, so does the need for thicker cylinder walls. At 75hp per cylinder I would be looking at a minimum wall of around 0.0110–0.130in; at 85hp the wall should be 0.150–0.180in and over 100hp 0.220–0.250in. These latter two thicknesses for engines at power levels of 85 and over 100hp per cylinder might appear excessive, but if you want a block to seal well and have a service life of up to 5000 race miles at 8500rpm, that is the sort of wall thickness you must be considering. I know that some race teams get away with less wall, but I also know that they are scrapping blocks with less than 300 miles on them. Considering the amount of time and money invested in a race block, this does not make much sense, and I wonder how often they lose an expensive crank because of the main bearing webs breaking away.

HOW TO MEASURE CYLINDER WALL THICKNESS

Checking a block for wall thickness and core shift takes a lot of time and care. However, the reward is that the engine will produce more power and you will not be wasting money on machining a 'thin' block that you will have to scrap after just a few races. I have seen some people trying to figure out cylinder wall thickness in the strangest of ways. Some, apparently, feel that they can simply measure the thickness of the web between adjacent cylinders. Then, with the welch plugs knocked out, they use a bunch of feeler gauge strips to measure the width of the water jacket between those cylinders, then calculate the wall thickness. Unfortunately, they forget that core shift makes such measurements pretty well meaningless.

Some race teams rely on sonic testing to provide reasonably accurate information on metal thickness. The main thing with sonic testing is for the operator to use care and good judgement, but it is also wise to recognise that a sonic test will not detect a porous area in a block or an area of 'dirty' metal. Also, the actual metal thickness may be 0.020in less than the tester is indicating.

Therefore, along with sonic testing it is vital to do a pressure crack test, as discussed earlier, and it is wise to measure the block using basic measuring tools. In fact, this is what race engine builders had to do before cheap sonic testers became available. I do not know of any machinist's callipers that will do the job, so you will probably have to build your own. What you need is a pair of vernier callipers with a set of very long jaws. One jaw has to be the correct size and shape, and with just enough 'kick' in the leg to fit down through water jacket holes in the deck of the block. I've seen some people cut up a pair of 6in or 8in outside callipers and weld the legs to the jaws of their verniers, but I generally find that outside callipers have too much 'bow' in the legs to fit down into the water jacket.

What I prefer is to chop up two pairs of callipers, one an inside calliper and one an outside calliper, then bolt one leg of the outside calliper correctly orientated to one jaw of the vernier. Then I reverse one leg of the inside callipers and bolt it to the other jaw of the vernier. With a bit of grinding, this leg will fit down into the water jacket through holes in the deck of the block, and it generally has enough kick to contact the outside wall of the cylinders. With these legs bolted, rather than welded, to the vernier callipers it is possible to swap them about or to pull them off for grinding to fit down into the water jackets of different blocks.

STRENGTHENING OPEN-DECK BLOCKS

A few engines, though, have problems that go well beyond cylinder wall thickness. Some, for example, have an open-deck design where the top of the cylinder is unsupported or only lightly braced by the top of the block. At low power levels this doesn't usually cause problems, but as the hp and rpm rise so does cylinder and block flexing. The solution for small Hondas is to fit a block guard. This is inserted into the water jacket to fit between the block and cylinder to add rigidity.

At times the solution can be more radical, and expensive. Some Subarus have an open-deck block. Upgrading to a closed-deck block means a big outlay. The K-series Rover 1800 engine has a different problem. Even before we add forced induction, the engine does nasty things like splitting the cylinder liners as we approach 200hp. The solution here is to shrink a steel sleeve around the outside of the liner up toward the top.

In a correctly honed cylinder an even cross-hatch pattern, without waviness or boring tool marks underneath, should be clearly visible.

CYLINDER HONING

The final operation performed on the block is to hone all cylinders to their finished size. During cylinder honing operations, a 2in thick honing plate should be fitted, along with the head gasket, and tensioned to normal head stud tension. This operation is necessary on racing engines to guarantee perfect bores, as the pull of head studs always causes a certain degree of cylinder distortion. When a honing plate is fitted at the specified head tension, the block is distorted during the cylinder finishing procedure. Therefore, the bores will be true when the head is fitted. The main bearing caps, when tensioned, also distort the cylinders, so these should be fitted as well.

I prefer to bore the cylinders to within 0.004in of the finished size, then hone each cylinder to fit a particular piston. In this way I can keep piston clearances to a tolerance of about 0.0003in.

The actual honing procedure varies a little from engine to engine, but a fairly standard method using the preferred Sunnen CK-10 automatic hone is as follows. First, using 220 grit stones, about 0.003in of material is removed to bring the bore to 0.001in from its final size. Then another 0.0005in is taken out with 280 grit stones, and finally 400 grit stones are used to bring the bore to its final diameter. This results in a very accurate bore with a fine finish that requires very little break-in.

The cross-hatch pattern left by the hone on the cylinder walls is critical. It must be just right if the piston rings are to bed-in quickly and have a long life. I prefer a 45° cross-hatch with a finish of 10–12 micro-inches. This type of finish makes it necessary to run the rings in, but they last for a long time and seldom leak. A finish any smoother will not hold enough oil and will allow a glaze to form on the ring face and bore wall. Oil consumption will be a problem, and power will be lost owing to blow-by. A rougher finish will usually eliminate the need to bed in the rings; however, ring life is greatly reduced. Keep in mind, too, that glazing can again be a real problem, but not through lack of lubrication; a rough finish acts like a file on the rings, and the added friction increases their temperature and allows glazing to form.

FINAL BLOCK PREPARATION

The upper and lower lip of each cylinder should be chamfered lightly to remove the sharp edge produced by boring. A smooth-cut half-round file will do the job, but take care not to let it slip and nick the cylinder wall. While you have the file at hand, lightly dress the sharp edges of the main bearing caps and webs.

Two more inspections are necessary for pushrod racing engines running aggressive cams. The camshaft bearing bores must be in line and each tappet bore must be of the correct diameter and perpendicular to the centreline of the camshaft. A tappet tipped off centre may easily dig into the lobe of a racing cam, causing premature wear and/or breakage.

About the only other parts of the block to which we have to give consideration are the main bearing caps and studs. I use aircraft-quality high-tensile studs when these are available. Standard production main bearing studs should not be re-used on racing engines, and the same applies to big-end bolts. Usually, the standard cast-iron main bearing caps fitted to modern engines are acceptable, but some older engines, or very-high-hp race engines, may require steel caps for reliability. In the very hottest motors I

Stock square-shoulder main bearing caps like these are usually very strong. The tapped holes in the centre are to provide more bracing to the main bearing girdle, further enhancing bottom-end strength.

sometimes use a main bearing support saddle. This is a one-piece device that supports all the main bearings, and sometimes takes the place of the individual main bearing caps. It may extend out to the oil pan deck and transfer the bearing load so that it is shared by the outside of the block as well as the main bearing webs.

After all the machining work has been completed, the block should be thoroughly washed with hot, soapy water. Be sure to get all traces of honing grit scrubbed out of the cylinders, using a bristle scrubbing brush. Blow the block dry using compressed air, and spray all the cylinders, tappet bores and bearing bores with a water dispersant such as WD-40.

A main bearing girdle reinforces the block and main bearing caps.

CRANKSHAFT MATERIAL AND HEAT TREATMENT

With the block prepared, we now turn our attention to the crankshaft. The majority of engines use cast nodular iron crankshafts, which generally give good service if sensible rpm limits are observed and the rods and pistons are not too heavy. Some of these come from the factory with rolled fillets, which significantly increases their fatigue resistance, and some may be heat-treated using a process such as Tuftriding. However, as the rpm limit is raised, so the need for a better-quality forged steel crank becomes more apparent.

Forging increases the density of the component, as the metal is literally squeezed into the required shape and compacted, resulting in a stronger core and better fatigue resistance. However, it is really the type of steel and the heat treatment that it receives after being forged that determines the crank's suitability for use in a race engine. Because a crankshaft twists and bends with every power thrust, the steel must have a high tensile strength and high fatigue resistance. High tensile strength – the ability to resist breaking under high loads – is important, but the ability to endure repeated high bending and twisting loads – fatigue resistance – is also important in high-rpm race engines.

In general, the production forged cranks fitted by Japanese and European car makers are very strong and suitable for high-rpm competition use; some will also feature rolled fillets, and a few will be heat-treated by the nitriding process. Not so good, however, are the forged steel crankshafts fitted by some American manufacturers. For example, the steel crank that Chev use in some of their small-block V8s is none too strong and cannot be recommended for use in competition engines; they are forged from fairly poor grade, and often 'dirty', 1053 steel. A nitrided version is also available, and while being a touch stronger they are production-machined without suitable big fillets. In fact, Chev's cast-iron crank, if ground with decent fillets, will last longer when limited to about 7000rpm and less than 500hp.

Another worry with many production V8 cranks is the forging method employed. Typically they are forged 'flat', then the front and rear throws are 'twisted' 90° to give the required crankpin orientation. However, in production the throws will not be twisted exactly 90°, which means that not only the crankpins but also the counterweights will be 2° or 3° out. The raw forging has enough meat in it to enable the crankpins to be machined at 90° and the crank can still be balanced, but the incorrectly indexed counterweights will impose undesirably high bearing loads in a high-rpm competition engine. For this reason 'non-twisted' forged cranks, where all four rod journals are forged in place , are preferred for V8 race engines.

BASIC PREPARATION AND BALANCING

Assuming that the standard crank is suitable, we must have it crack-tested before doing any work on it. If it is free from flaws, carefully and fully demagnetise it (the same applies to rods), as any residual magnetism after a Magnaflux crack test will attract ferrous metal particles to the area, which will soon wipe out the bearings. Next a check should be made to determine its straightness. Cranks can be straightened, but generally it is a waste of time as the pressures of combustion and inertia loads will reverse the straightening process. Because a crankshaft bent by more than 0.002in increases bearing loads, it must be machined undersize by 0.010in, or else discarded.

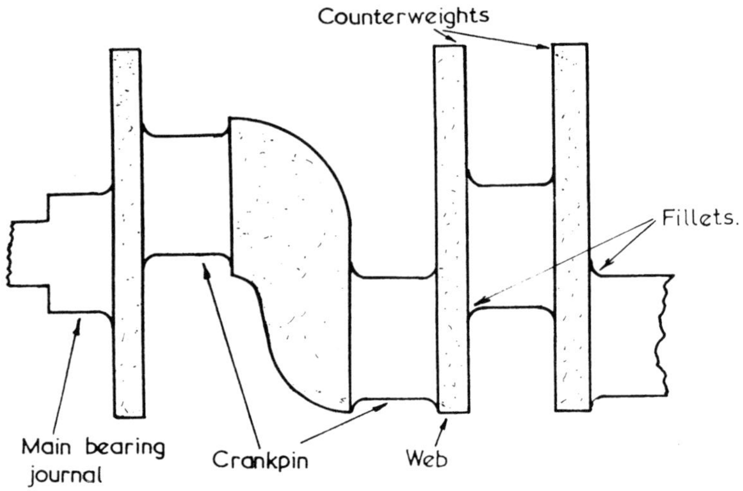

Fig. 19.2. Crankshafts need to be accurately machined with generous fillets.

Next, measure each main bearing and crank pin journal. Remembering that crank journals wear oval, measure the diameter at several angles around the journal. Also measure at each end and in the middle of every journal, as wear can vary along its length. Ovality and taper should be less than 0.0003in in a high-performance or racing crank. If the wear is greater than this, grinding the journals undersize will be necessary. However, grinding weakens the crankshaft, so a crank replacement may be necessary in high-horsepower and high-rpm engines. When the crankshaft is machined, instruct the grinder to pay close attention to maintaining the fillet radii specified by the manufacturer (*Fig. 19.2*); any reduction here will weaken the crankshaft. On the other hand, an increased radius serves to strengthen the shaft; however, this should not be overdone or we can arrive at a situation where the big-end or main bearings are locking on to the fillet.

Any casting slag should be ground off the crank and the area dressed. These slag spots can become a stress raiser or future failure point.

To assist in the smooth operation of the motor and to reduce bearing inertia loading, the crankshaft should be dynamically balanced; this will increase its life through a reduction in the shock loading and vibration that any imbalance would cause. Most manufacturers already do a reasonably good job of balancing their cranks, so unless the engine is for competition use or is going to spend a lot of time running in excess of 6000rpm, it is probably not worth the expense of having the crankshaft re-balanced. An exception to this is the case of V8 engines. If these are fitted with pistons or rods much heavier than standard, the crank has to be re-balanced regardless of whether it is to be used in a road engine or a race engine.

HIGH PERFORMANCE AND RACING CRANKSHAFTS

When the stock crankshaft is unsuitable for our application, then we will have to go upmarket to a high performance or a full racing crank. In the UK, competition cranks are often forged from EN40B steel, while in the USA 4340 and 5140 steel (5140 is inferior to 4340) are preferred. At this point it is wise to realise that just because a crank is said to be 4340 steel, it is no guarantee of its quality or fatigue resistance. Technically, all 4340 should be the same, but it is not. The steel may be 'dirty', containing many impurities, or the alloy may not be up to specification. Also, there is the question of what heat treatment the part has received after the forging and machining operations, as this has a significant influence on the metal's strength and fatigue resistance. Some heat treatment imparts considerably more strength and fatigue resistance, but because there may be greater risk of failure during the process, which may result in an entire batch of cranks being discarded, a crankshaft manufacturer could choose a cheaper and 'safer', albeit inferior, heat treatment. Obviously, such a crank will be less expensive to manufacture, but as it will have less fatigue resistance and a corresponding shorter 'life', it may not be any cheaper in the long run.

Another factor influencing the cost of a racing crank is its weight. Naturally, complex machining costs money; a 20% weight reduction could add considerably to the price of the crankshaft. However, any reduction in reciprocating mass will pay dividends in improving acceleration out of corners, and in reducing bearing loads.

The most expensive, and strongest, cranks are fully machined from a bar (billet) of high-grade steel that has been hammer forged. After being machined, the billet crank is shot-peened and heat treated to further enhance its strength and fatigue resistance.

The cranks of very-high-rpm or endurance race engines should generally be fully counterbalanced. All crankshafts are counterbalanced overall, but not at each crankpin. As a result, the individual throws are subjected to forces that, in combination, tend to make the crank whip and flex. Such increased loading at the bearings can wipe them out at high rpm, and in endurance engines fatigue failure of the crank could occur. The way to overcome these problems is, therefore, to counterbalance each throw. This does not change the overall crank balance, but achieves an internal balance for each throw. Obviously, extra counterweights increase the crank's reciprocating mass and this in turn slows acceleration off the turns. Therefore, in sprint-type engines, where bearings can be replaced frequently and the crank is regularly crack-tested, a fully counterbalanced crank may be a liability.

In the high-rpm engine, crankshaft oilway modifications are undertaken to prevent bearing failure from lack of lubricant. Many feel this means that engines such as American V8s, which have only the upper main bearing grooved, should have the main bearing journals cross-drilled. This is a total waste of money for road engines and it may not even be necessary for race engines. Looking at *Fig. 19.3*, you will see what can be done to improve the quality of lubrication at the big ends while at the same time reducing the amount of oil that is pumped away from the main bearings. With this system, when one crankpin hole is blocked off owing to centrifugal loads on the rod, the other hole feeds oil to the crankpin bearing. When carrying out this modification, great care is necessary to ensure that the new holes break fully into the original oilway.

So far we have only considered the crankshaft from the aspect of engine reliability, but it can also affect overall engine performance. Theoretically, crank throws should be of equal length and correctly indexed (phased). Slight differences in the stroke from cylinder to cylinder will not have a profound effect on performance, but incorrect phasing will knock power. A 4-cylinder engine should have a piston at top dead centre every 180°, a 6-cylinder every 120°, and an 8-cylinder every 90° (180° with a single-plane crank). If any cylinder has a crank throw out of phase by 5° this will have the same

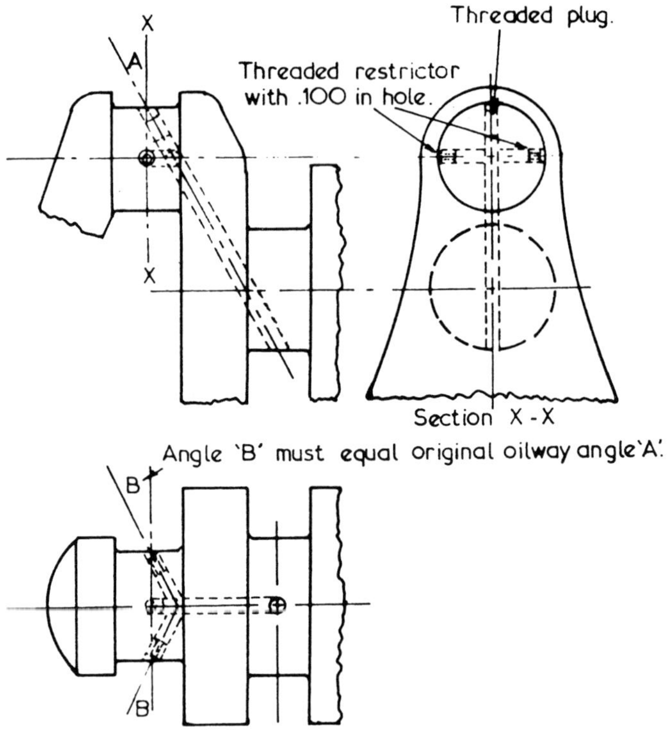

Fig. 19.3. Some engines require crankshaft oilway modifications.

effect on performance as an ignition or camshaft timing error of 5°. An expensive race crank should have perfect phasing, but always check to be sure.

CON ROD STRESSES AND MATERIALS

The connecting rod provides the mechanical link between piston and crankshaft, and is subjected to alternating compression and tensile loads. It is true to say that the connecting rods in an engine have a tougher job to do than any other component, so it is not surprising that a large number of failures in high-performance and racing engines are caused by the rods letting go.

The highest load is reached when the piston is at top dead centre (TDC) on the exhaust stroke. This tension load can range from a few thousand pounds in a small low-rev engine to well over 15,000lb in larger high-rev engines. Interestingly, this maximum load occurs on the non-firing stroke and is caused by the inertia of the reciprocating assembly comprising the small end, piston and pin. At TDC the piston is suddenly stopped, then reversed, and it is this sudden change that produces the high tensile load. On the compression stroke the load is not as high (up to 11,500lb) as compression builds up slowly, and soon after TDC, when combustion has finished, the load changes from tension to a light compression load.

It is not only the fact that stresses are high in a rod; more important, they are applied and reversed each time the engine completes a cycle. This pulsing is much worse than if the loads were constant and applied continually, and it is this that leads to eventual fatigue and failure. A rod has to survive millions of stress cycles during a liftetime so it has to be tough and thoroughly prepared.

Most rods are forged from carbon steel, but they may also be made of cast iron, aluminium or titanium. Cast-iron rods have no place in a modified engine and should not be used. Titanium is prohibitively expensive and is used only in big-budget race engines. Aluminium rods are light and reasonably strong, and if the engine detonates they transfer less of the shock load to the bearings and crankshaft. However, they should be used only in drag race and sprint-type engines where frequent replacement is possible.

Some time ago, Chevrolet released information on connecting rod fatigue testing that they had carried out. Some aluminium rods failed after 150,000 cycles. The heavy-duty small-block rod, polished and shot-peened, survived 1–2 million cycles at the equivalent of 7500rpm, while the Chevy Bow Tie rod survived a minimum of 10 million cycles at 8000rpm. When it is considered that the average race engine is required to stay together for 300–500 miles between rebuilds, the equivalent of 1–1½ million cycles, it is evident that care is needed, first in selecting the correct rod, and second in regular crack testing and replacement.

Connecting rods designed for the most stressful conditions are usually forged or machined from 4340 or E4340 steel. However, as with crankshafts, the type of steel used is no guarantee of a quality rod. Many other factors such a steel purity, accuracy of machining, heat treatment, shot-peening, design and weight also enter the equation.

The standard rod is what we refer to as an 'I' beam rod because if you were to cut through the beam connecting the little and big ends the steel section would look like an 'I'. This type of rod can be very strong, and manufacturers like Oliver, in particular, and Crower have proved that their 'I'-beam rods hold up very well in high-rpm endurance engines.

Possibly the most respected manufacturer of con rods, Carrillo, use an 'H'-beam design, which is a little stronger below the pin boss. However, the real secret of Carrillo's superiority over other rods, at great expense I might add, is their tight control of the steel alloy (E4340) and similar tight adherence to their chosen heat treatment and shot-peening standards.

Many heavy-duty rods are more massive than the standard component, so there are a few things that must be considered during installation. As they are physically heavier, the crankshaft counterweighting will have to be re-balanced. Since the big end of the rod is larger, care must also be taken to verify that it, and the rod bolts, clear the block and camshaft.

HOW TO PREPARE STOCK CON RODS

If you intend to retain the standard con rods, they must be carefully, and individually, hand picked. Rods with any forging irregularities or indentations should be avoided. Bumps standing proud of the rod can be removed, but indentations are stress points, likely to cause failure. The little-end eye should be in the centre with an even thickness of material around it. Again, the big-end should be well formed and symmetrical. Select rods that are reasonably equal in weight; a heavy rod may be excessively weakened if material was ground off to balance it with the lighter rods. If they are not already numbered, number each rod and cap, using a number punch, to ensure that they are never mismatched.

These indentations are stress raisers, indicating that this stock rod should not be selected for use in a high output or high rev engine.

Standard connecting rod preparation must always include a check to ensure that the rod is not twisted or bent, as either condition will wreck the piston and/or big-end bearings. Also necessary is re-sizing to bring the rods to an equal length. At the same time the big-end inside diameter should be checked for size and concentricity with the bearings fitted and bolts torqued to specification. The little-end bush should be honed to give the correct piston pin fit.

The failure point for many rods is at the corner formed by the flat machined for the rod bolt and nut seat. These corners must be radiused to avoid stress concentrations in this area. Every stock con rod will require some modification here (*Fig. 19.4*).

Fig. 19.4. Stock rods may crack through at the corners of the flats machined for the rod bolt and nut seats. To relieve stress concentration in these areas the rod cap is radiused in the corners and the rod is notched. Rods using bolts without nuts do not require relief notching.

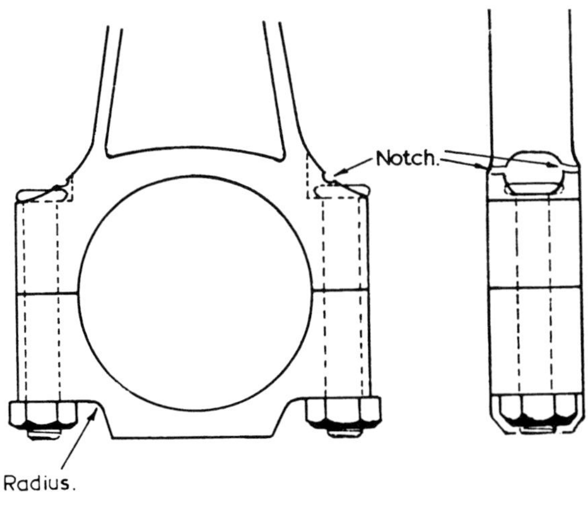

ROD POLISHING AND SHOT-PEENING

The tough skin formed on the con rod by forging gives the rod much of its strength and fatigue resistance, so it should never be polished unless it is followed by shot-peening, to create another work-toughened and compressed skin. It is a waste of time polishing the entire rod, then having it shot-peened. If you look at the shank of a rod you will see along its edges, a rough band where metal appears to have been sawn away. That is where the excess metal, called flash, was squeezed out from between the forging dies when the rod was being made. Later most of the flash is trimmed off, but a bead is left, as you can see. Of course, there is no hard skin along this ridge; in fact, its roughness is a stress raiser, so this ridge should be removed on a sanding belt. Give the entire beam a polish with fine emery cloth, then follow up with buffing and shot-peening.

The spot-peening process has a very useful place in connecting rod preparation. Fatigue failure almost always starts at the surface owing to a combination of two things. First, under most forms of loading, maximum stress is at the surface, and second, surface imperfections are stress raisers. By bombarding the surface with steel shot, the surface layer is compressed and unified. This means that when the component is under load, any tensile stresses at the surface will be reduced and any compressive stresses will be increased. However, because peening can seal over any surface cracks, all crack-testing must be done first. Also, any straightening should be carried out before peening, or this will remove the effect of the peening. After peening, check the component dimensionally, as there will be some growth and, if the peening has been done incorrectly, bending may occur.

CON ROD RATIO

The relationship between con rod length and the crankshaft stroke, called the con rod ratio (con rod length ÷ crank stroke = con rod ratio), is of importance in the high-performance and racing engine. Most rod ratios range from 1.5 to 2:1; and 1.65 to 1.75 is average.

For a long time it was argued that the length of the rod directly affected the performance of an engine. Basically, the idea was that a long rod would improve top-end performance and also reduce piston and cylinder wear, while a short rod would pick-up bottom-end performance, but with increased cylinder wall loads, piston and cylinder wear would increase.

Obviously, a long rod will cause the piston to dwell at TDC longer and to move away from TDC more slowly, so there will be a change in the way combustion of the fuel/air mixture occurs. However, this does not affect the power output of an engine with flat-top, or dish-top pistons. It is really only in nitrous and naturally aspirated high-compression race engines with very high top pistons, which badly upset the combustion process, that we see this influence coming through with an increase in hp in the upper rpm range.

This does not mean that if our engine has flat-top pistons we can forget about connecting rod length. With shorter rods, engine wear is a problem with which we have to contend. While in high-rpm sprint engines this may not be a matter for concern, the increased cylinder wall and piston loadings evident in short-rod engines certainly make it one, as any bowing of the cylinder reduces ring seal and the hp goes off. Hence, I do not like rod ratios below about 1.65:1, and I would prefer it to be around 1.8:1. This is worth thinking about before you fit a long-stroke crank, as the tendency for the piston to rock in the bore is also increased as the rod-length-to-stroke ratio is reduced. Pistons with a reduced compression height and shorter skirt must be used with stroker cranks, otherwise the top would pop out of the cylinder and the skirt would contact the crank counterweights. This aggravates the wear and ring seal problem, as short pistons rock more than those of a standard length.

CON ROD BOLTS AND CORRECT TENSIONING

Of course, standard preparation of the con rods must include a careful selection of high quality con rod bolts. Rod cap bolts have to be designed to withstand a considerable force, to stop the cap from breaking away from the rod. At times, it is necessary to fit bolts of a larger diameter than standard. Whatever the diameter, the tensile strength of the bolts should be 185,000psi minimum. Prior to assembly, carefully inspect every big-end bolt with a loupe (small magnifier) of about 4x, checking for any nicks or imperfections on the surface 'skin' of each bolt. There should not be a square shoulder where the shank meets the bolt head, but rather a generous radius. Note that stock big-end bolts should not be re-used, but high-quality boron steel bolts will generally not require replacement until after several rebuilds.

Frequently, rod bolt failure can be traced back to incorrect tensioning techniques. If the bolts are over-tensioned they may fail at less than their designed rpm limit. On the other hand, when big-end bolts are not sufficiently preloaded – stretched – they can loosen and fail.

There are three aspects to obtaining the correct bolt preload. First, use the specified lubricant and apply it not only to the bolt thread, but also to the thread of the rod (or the nut). Also, lube the underhead area of the bolt (or the base of the nut) where it bears against the con-rod cap. If the bolt manufacturer recommends engine oil and you apply hypoid diff oil, moly cam lube or moly rod bolt lube, the bolts could end up overstretched. All these lubricants will cut the friction between the threads more than engine oil, so a torque wrench reading of 45lb could be stretching the bolt to the equivalent of a 55lb reading. Conversely, if the bolt manufacturer specifies moly bolt lube and you apply engine oil, the bolts will also not be correctly preloaded, and they may loosen in service.

Second, stick to the bolt tension specified by the bolt manufacturer. Do not assume that because standard bolts are torqued to 45lb, the manufacturer of the high-grade bolt has made an error in stipulating a tension of, say, 38lb.

Third, use the correct tensioning procedure and an accurate tension wrench. This involves torquing up the bolts three times. After the first tensioning the bolts are backed off, then they are torqued a second time and backed off. These two operations polish the threads and the bolt head (or nut) and cap contact areas. On the third and final torquing the correct bolt preload will be obtained.

Naturally, the techniques described above will be successful only when either new or within-spec bolts are fitted. If used bolts, which have been stretched beyond their design limit, are installed, bolt failure is inevitable. For this reason all new bolts must be accurately measured using a micrometer, and their length and the position where they are fitted must be recorded. Later, when the engine is dismantled, the length of all rod bolts can be checked and compared with their new length. Any bolt that has stretched more than 0.001in (check with the bolt manufacturer, as this will vary) must be discarded. A bolt that has yielded more than 0.001in has either been subjected to an over-rev or else it has been over-tensioned at the time of installation.

BEARING MATERIALS AND CLEARANCES

Bearings in a powerful supercharged engine, as well as providing a low-friction wearing surface, also have to absorb tremendous shock loads. Therefore, only top-quality trimetal bearings should be used for the mains and big ends, although white metal bearings are satisfactory for carrying the camshaft.

I use only Vandervell lead-indium and Clevite 77 bearings because of their special features, both types being able to stand up to the worst competition loads when correctly installed. These bearings are steel backed, with a copper-lead intermediate layer that gives the bearing good fatigue strength and load-carrying capacity, as well as resistance to hydraulic break-out. The running surface for the Vandervell bearing is a precision overlay of lead-indium, while the Clevite 77 overlay is lead-tin.

Correct bearing clearance is an obvious necessity. Excessive clearance will promote knocking and pounding and allow excessive oil throw off into the cylinder bores; this in turn will cause higher frictional losses in the cylinder and increase oil consumption. Excessive clearance at the big-ends will lead to oil starvation at the main bearings, with resultant bearing failure. Insufficient clearance will cause rapid bearing deterioration as a result of the increased temperatures that come about owing to insufficient oil flow or a thin oil film (Table 19.1). A good rule of thumb is 0.0009in clearance for each inch of shaft diameter for main bearings and 0.0012in clearance per inch of shaft diameter for rod bearings when running 15W-50 full synthetic oil at 100–110°C (212–230°F). With a thinner oil such as 0W-30 bearing clearances can be a little tighter. Con rod side clearance (end-float) also affects bearing lubrication, so this should be checked on every big-end to ensure proper oil control.

Table 19.1. Bearing clearances for trimetal copper lead bearings

Diameter of shaft (in)	Clearance between shaft and bearing	Side clearance
1.5	0.0012–0.0017	0.004–0.006
2.0	0.0015–0.002	0.005–0.007
2.25	0.0018–0.0025	0.005–0.007
2.5	0.0022–0.0027	0.005–0.007
2.75	0.0024–0.0028	0.006–0.008
3.0	0.0025–0.0028	0.007–0.009
3.25	0.0025–0.003	0.007–0.009

Note: In engines where two con rods share a common crankpin (e.g. a V8) multiply the side clearance by 3 for steel rods, and by 4 for aluminium rods.

A slightly different situation exists when an aluminium alloy block is used in a race engine. Because aluminium expands more than cast iron when heated, what were ideal main bearing clearances at room temperature could become excessive at race engine temperature. For example, a clearance of 0.002in cold will increase to at least 0.003in, and more probably 0.004in, when hot. To address this situation the engine should be assembled with tighter main bearing clearances, which will frequently be close to what the manufacturer quotes for stock engines. Therefore, if the race clearance with an iron block is normally set at 0.002in, I would probably be looking at 0.001–0.0013in main bearing clearance when running an alloy block. However, with such tight clearances the oil and water should be pre-heated prior to engine fire-up.

BEARING INSTALLATION PROCEDURE

There are just a few simple rules to follow for correct bearing installation. Accurately measure and record the inside diameter of every main bearing and big-end housing without bearings fitted. Now do the same with the crankshaft main and crankpin journals. Unwrap all the bearings and carefully wash them in clean solvent to remove the protective film, but do not polish them with any sort of abrasive pad as the overlay is very thin and is easily damaged. Do not worry about any little 'peaks' in the overlay – they will soon flatten out when the engine is run. Next, measure the bearing shell thickness and double the measurement as there is a shell on each side of the shaft. The difference between the shaft and housing diameters and the doubled shell measurement is the space left for running clearance. When measuring the bearing shell thickness, be sure to take the measurement well away from the parting line as race bearings typically taper down by 0.0005in and up to 0.0015in over the last ¼in to ½in. This eccentricity is manufactured into race bearings because at very high rpm the connecting rod bore pinches in owing to rod stretch.

Many Clevite 77 and Vandervell bearings are available with three levels of eccentricity, 0.0005in, 0.001in and 0.0015in. Ideally a bearing should show wear over about 65–75% of its surface. If it is showing contact almost up to the parting line, fit bearings with more eccentricity, but when the contact area is too short, bearings with reduced eccentricity are called for.

When you have calculated the running clearances for each bearing you will probably find that some have a little more than you want, while others may be a touch tight. If the engine is a popular unit for competition use, the bearing manufacturer may have special 0.001in undersize and oversize bearings available. Using just one half of such a bearing allows the clearance to be opened up or closed down by just 0.0005in. When mixing a 0.001in undersize shell with a standard size shell the recommended orientation is to fit the shell with the thicker wall in the upper position in a big-end and in the lower position in a main bearing.

After ensuring that all the bearing housings and bearing shells are perfectly clean and dry, fit the shells and check oil hole alignment; any misalignment should be corrected using a small round key file. After filing, carefully dress the steel back of the shell to remove any metal fraze. Next, coat all the bearings with engine oil and fit the crankshaft (do not use a 50–50 mixture of oil and any substance like STP). Fit the main bearing caps in their correct order with arrows facing the front of the block and gradually tighten them down. Before final tightening, the crank should be tapped to each side with a soft hammer in order to line up the bearing caps. Now check the crankshaft end-float; it should be within 0.004–0.006in with a cast-iron block. If the clearance is more than this, fit thicker thrust washers.

When it comes to fitting the big-end bearings, follow the same procedure as for the fitting of the main bearings. There are still a few historic race engines that use lock tabs on the big-end bolts. Throw these away, as they crush and give a false bolt tension reading. Use Loctite on the threads and you will have no problems with loose bolts.

Note that when the bearing shells are pushed into their respective housings they should feel 'springy', and 'snap' into place; this indicates good bearing crush. A lack of crush in new bearings indicates that either the bearing housing is distorted or that bearings with oversize shells are required to match housings that have been bored oversize. With used bearings, a lack of crush usually indicates that the bearings have been hammered by detonation, but may also indicate a distorted bearing housing.

As well as going in for special oil additives, some tuners are advocating the use of products that must be baked on to the bearings at very high temperatures for several hours; they are supposed to provide bearing protection when lubrication is inadequate. I do not recommend such products, as any time a bearing is heated in excess of 200°C (392°F) for more than about an hour it will be ruined.

PISTON DESIGN AND MANUFACTURE

Moving further up in the block, the next components for consideration are the pistons, piston rings and piston pin. First, you must decide whether the standard piston is up to the job. If the motor is in sports tune, it is quite probable that the standard cast and slotted pistons will be satisfactory. Higher states of tune will demand unslotted cast pistons, and in racing tune unslotted forged pistons will be necessary.

Most production engines use cast pistons, as they are easy to produce and shape as required. Some high-performance engines are fitted with high-quality forged or cast pistons right from the factory. By looking inside the piston you will be able to see whether it is cast or forged; cast pistons have quite intricate under-crown shapes around the gudgeon pin boss, while forged pistons are smooth inside and lack intricate struts and braces.

The cast piston, either conventional or hypereutectic, has a relatively low material density. Whereas the ordinary cast piston is made from a low silicon (7–8%) aluminium alloy, the hypereutectic piston is cast from an alloy containing about twice as much silicon (usually 15–20%) and can withstand higher temperature and pressure. The additional silicon makes the piston harder, which increases wear and scuff resistance, and also reduces piston expansion when heated. Therefore, the hypereutectic piston will run tighter piston-to-cylinder-wall clearances, and as it can be cast with intricate underhead struts it can be made reasonably strong and generally lighter than a forged piston. For these reasons hypereutectic pistons are sometimes fitted as standard in high-performance naturally aspirated engines. However, they are not a good choice for forced induction engines as the added silicon leaves them quite brittle and, as such, prone to breaking when subjected to detonation.

By comparison, forged pistons are much denser and consequently have a higher tensile strength. They are capable of withstanding higher pressure and heat loads than the cast piston, and their higher density improves their thermal conductivity to the extent that a piston crown temperature reduction of 38°C (100°F) is usual. Because of these beneficial features, forged pistons are usually the best choice for high hp forced induction engines.

The weak points in many stock pistons are the oil drain/expansion control slots in the oil ring groove.

The worst aspect of many pistons fitted in production to naturally aspirated engines is the slot for oil drainage behind the oil control ring. This usually extends almost from one boss around to the other, on both sides of the piston. Slots weaken the piston considerably, allowing the skirt to break away from the top. In standard production engines this can be a problem, so you can imagine what you are up against if slotted pistons are retained when an engine has some form of supercharging added. Pistons suitable for high-performance use (either cast or forged) do not have any slots for expansion control or oil drain-back. Instead, small oil-drainage holes are drilled right the way around the oil control ring groove. Extra piston clearance and special design take care of expansion.

Racing pistons may be of either the full skirt or the slipper variety, and some may be a cross between both, with a skirt extending to below the pin boss, but cut back to the thrust faces below this point. The slipper designs are lighter, but piston life can be as little as 250–300 miles at sustained high rpm. Also, depending on such things as rod length, crank stroke, piston diameter and piston clearances, piston ring sealing and oil control may not be as good as with a full skirt piston. As well as these problems I sometimes find that I have to use a heavier gudgeon pin to reduce pin tower flex, and the heavier pin tends to offset the weight advantage of the slipper piston. Consequently, I have a preference for the full skirt design. Together with a long taper-wall gudgeon pin (with a length at least 0.7 times the bore diameter), piston life will be anything from three to six times longer than an equivalent slipper type.

PISTON RUNNING CLEARANCE

We tend to think of pistons as being round, but actually the skirt is cam-ground an oval shape, and the piston also tapers from bottom to top; both features being necessary to prevent seizure. In operation the top of the piston is almost twice as hot as the bottom of the skirt, so it expands more. Because of the extra metal around the pin bosses, more

A piston with high ring placement and narrow ring lands is only suitable for very low output forced induction engines.

heat is directed to this area, elongating the piston across the piston pin axis. The piston skirt is ground oval to compensate for this. Most pistons are from 0.005in to 0.012in less in diameter across the pin axis than across the thrust faces. Be careful to measure piston clearance only on the thrust faces, and either at the bottom of the skirt or up near the pin (check this with the piston manufacturer).

Accidentally dropping a piston may damage the skirt and lead to eventual seizure through skirt distortion. Never bang a piston pin out with a hammer and drift; this is a sure way to push the skirt out of shape. If the pin will not push out easily, heat the piston in boiling water or oil, then gently tap it out with the con rod secured in a vice to prevent any pressure at all on the piston.

For better lubrication, and to allow for extra piston expansion, high-performance engines must have more piston-to-cylinder clearance than models off the showroom floor. It is best to consult the piston manufacturer on this, but in general I would be looking at a minimum of 0.002in, and preferably up around 0.003in clearance in road engines running cast pistons with bore sizes in the 80–110mm range. Road engines running forged pistons require about 0.0012–0.0015in clearance per inch of bore, while competition engines typically require about 0.0015–0.0018in per inch of bore. However, some forged pistons (for example, some manufactured by JE) run only 0.003in clearance just like cast pistons. There should be a maximum of 0.0005in difference in clearance between the tightest and sloppiest cylinders. If the tolerance is greater than this, try swapping the pistons around in different cylinders. When it is finally decided which piston goes where, number each inside the skirt using a suitable marking pen. Do not mark the crown as this may be machined later to give the correct deck height.

VALVE-TO-PISTON CLEARANCE

As the majority of engines have heads with inclined or canted valves, some thought must be given to providing adequate valve-to-piston clearance. There should be a minimum of 0.060in vertical clearance, although for push rod engines, which in competition are more likely to suffer valve float, I prefer 0.080in for the inlets and 0.0100in for the exhaust. The cut-out diameter for safety is 0.120in greater than the valve head diameter, but in race engines, where there is no build-up of carbon on the valves or pistons, this can be reduced to 0.050in.

There are two basic ways to go about checking the safe valve working clearance, but both methods involve a deal of work. Both techniques also require that the motor be almost totally assembled. The cam must be installed and accurately timed, and the head and head gasket need to be fitted.

The first way involves the use of a piece of modelling clay pressed down on the top of a piston. After the clay has been smoothed out, spray it and the valves in the cylinder that is being checked with a light coating of WD-40. This will help prevent the clay from sticking to the valves later, when the clearance check is made. Fit the head and head gasket, tension it down and adjust the valve clearances. Now very carefully turn the motor over two full turns to completely open and close both the inlet and exhaust valves. Remove the head and measure the thickness of the clay over both valve cut-outs. Next, cut the clay, using a sharp wet knife, to check the clearance around the periphery of the valve head.

The other method is accurate only if the valve cut-outs are central under the valve, as there is no provision for a check to be made on the clearance between the edge of the valve head and the edge of the cut-out. In this instance light checking springs need to be fitted instead of valve springs. First tension down the head and head gasket as before, and adjust the valve clearances. Next, set-up a dial gauge on the inlet valve and turn the motor over until the valve is fully open. Zero the dial gauge and push the valve open as far as it will go. The dial gauge reading gives the clearance. Measure the exhaust valve clearance similarly.

With the pistons at TDC the piston crown position is measured. This low pressure turbo engine is set-up with the pistons protruding out of the bores to close up the squish clearance to help cruise combustion, economy and throttle response.

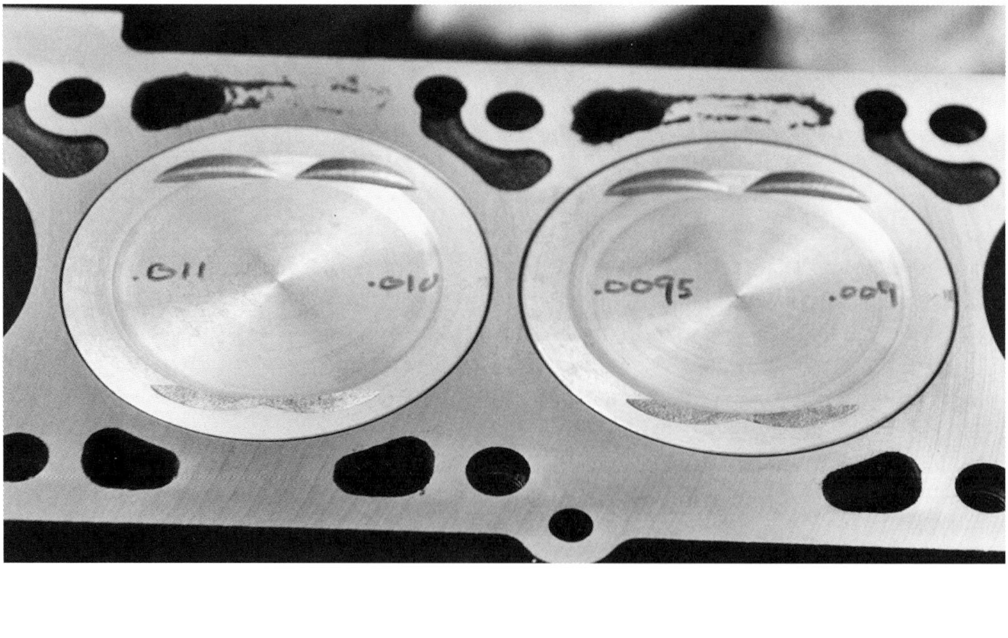

PISTON-TO-CYLINDER HEAD CLEARANCE

To assist in equalising compression and combustion pressures, the crown of every piston must rise to the same point in each cylinder; this is called the deck height of the piston. In a street engine we would want to be running minimum deck clearance to assist cylinder scavenging. A good solid engine with steel rods can run with a minimum of 0.040in clearance between the piston crown and the squish area of the head. Therefore, if the head gasket is 0.030in thick when compressed, we can run a minimum deck clearance of 0.040in – 0.030in = 0.010in. This means that the pistons would all be machined so that they are 0.010in from the top of the block at TDC.

Aluminium connecting rods stretch more at high revs, and aluminium has a greater coefficient of expansion than steel, so we would need at least 0.070in piston-to-head clearance. Engines that have steel rods but suffer block flex or crankshaft whip, will require 0.060in clearance.

PISTON CROWN SHAPE INFLUENCES COMBUSTION

As I mentioned earlier in this book, the top of the piston is a part of the combustion chamber, and those high compression lumps on the top of high compression nitrous pistons can add to, or detract from, the complete combustion of fuel inducted into the engine. They may retard flame travel after ignition and upset complete cylinder scavenging during the exhaust cycle. On the intake stroke, the dome can restrict the critical initial air/fuel flow, then disrupt final mixture homogenisation before ignition.

To improve air flow into the cylinder at low valve lift, the intake valve pocket must be laid back to reduce valve shrouding (*Fig. 19.5*). As the initial flow is along the floor of the port and around the short radius towards the squish area of the head, this modification will assist in cylinder filling. Many tuners worry about removing metal from the piston dome because of the slight drop in compression ratio. However, the gain from improved flow and better combustion will far exceed the slight power losses from the drop in compression ratio.

Fig. 19.5. Piston crown modifications to improve inlet flow and aid combustion.

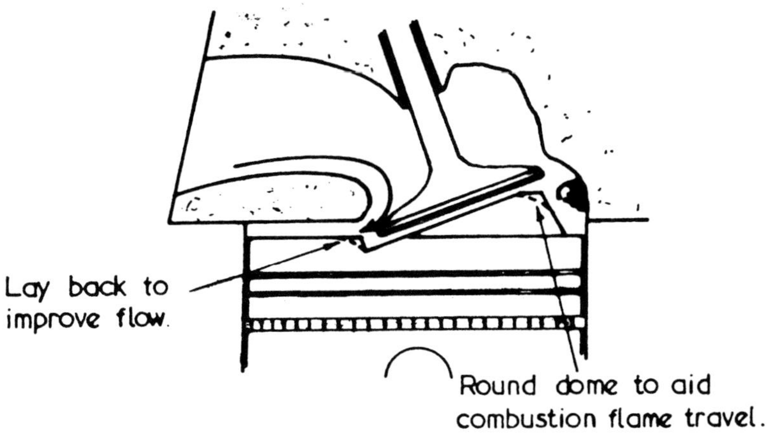

After ignition, we want the flame to travel smoothly across the piston dome and back towards the squish area. Any abrupt edge on the compression lump will disrupt the flame front as it moves across the piston, leaving pockets of mixture that are either unburned or only partially ignited. Rounding and smoothing the top and sides of the dome does much to reduce this problem.

The exhaust valve cut-out must not be laid back, as this is usually not advantageous. The sharpness of the exhaust valve pocket does cause some flame disruption, but exhaust flow and cylinder scavenging are usually superior when the exhaust valve is shrouded by the cut-out.

After the engine has been run for a time, a careful study should be made of piston dome coloration, which is a surprisingly accurate means of investigating combustion completeness. Keep in mind that combustion begins before the piston reaches TDC and continues well past TDC. If at any time something happens to cause the flame to stop, the piston colour will indicate where the dome is causing the cessation.

All engines will show an area of little or no carbon build-up on both the piston and cylinder head. This is the squish (or quench) area where no combustion takes place owing to the closeness of the flat face of the head and piston. Any other areas on the piston dome where there is little or no carbon build-up likewise indicates that the dome is coming too close to the combustion chamber and is retarding the ignition flame at and around TDC.

The solution is to machine the compression domes lightly until the carbon build-up is more even in quantity and colour. Because the flame moves from the spark plug out across the combustion chamber, you will need to keep in mind the direction of flame travel when you come to remove any metal from the dome. Generally, it will not be necessary to machine the entire uncoloured area as it is usually the area immediately in front of and around the beginning of the uncoloured area, relative to the direction of flame travel, where the dome is coming too close to the combustion chamber wall and creating a pocket of unburned fuel. The place of least flame activity is usually around the high side of the piston dome in the area of the intake valve cut-out, close to the cylinder wall. If you can modify the dome to colour evenly in this area, you are well on the way to achieving good combustion.

Engines that are to be supercharged or turbocharged may require a low compression ratio to avoid detonation. To achieve this, dished top pistons should be used for best combustion control. Some feel that flat-top pistons that have a lower than standard compression height for use with stroker cranks can be used, but in street engines this isn't a good idea. If a flat-top piston were used with a deck clearance of, say, 0.150in to reduce the compression ratio to 7:1, the top of the piston would be so far away from the squish area of the head that combustion, at cruise in particular, would be upset by a lack of squish. A dished-top piston has a flat band around the outside of the crown that will come into close contact with the head and provide the required squish for good burning.

PISTON PIN DESIGN, CLEARANCE AND RETENTION

As mentioned previously, I prefer a long piston pin with a tapered bore to cut its weight. However, if we go too far we end up with a pin that is nice and light but which allows the piston to flex excessively. The end result will be cracking in the top of the piston where the pin towers join the crown. The ends of the pin should be chamfered to bear up into round wire circlips and the length should be such that when the pin retainers are fitted, pin end-float is preferably 0–0.001in; more than 0.005in is unacceptable in a race engine.

The piston pin is retained by a variety of means in the production engine, the most common being either an interference press fit in the connecting rod, or by flat circlips. If the latter are used, they should be coated with Loctite. Ensure that the circlip grooves are clean and undamaged so that each circlip can seat securely. The interference press fit (0.0015–0.0025in) is suitable for racing engines, but care must be taken to assemble and remove the piston from the rod without causing any piston damage.

As racing engine speeds are now moving past 8000rpm, many engine builders are experiencing problems with circlips (either flat or round wire). At very high rpm, or if the engine detonates, they may close up, owing to the forces exerted on them, and drop out of the piston into the bore. To overcome this it is necessary to employ another means of pin retention. Some have reverted to interference press fit, but generally I prefer to use double Spirolox. Never re-use old circlips or old Spirolox retainers.

The fit of the pin in the piston exercises more control over piston dimension than is often appreciated. The additional clearance given to a piston by cam grinding is nullified if the piston is unable to expand correctly because of a tight pin. The pin clearance (pin-to-piston and pin-to-rod) in the high-performance and racing engine must be greater to compensate for forces that the standard production engine is unlikely to experience. At high engine speeds, crankshaft whip and con rod distortion can severely load the piston pin, and if it has too tight a fit, piston-to-cylinder scuffing or seizure may result. In extreme instances the pin will push the side out of the piston.

For most engines I set pin-to-rod clearance at 0.0008–0.0015in, and pin-to-piston clearance a little tighter at 0.0005–0.0008in. This will alleviate any binding between rod and pin or between pin and piston that could lead to piston damage. However, no matter how free the pin is, any con rod misalignment could wreck the piston, so have the rods checked to ensure that they are not bent, twisted or offset.

PISTON RINGS

The piston ring, as it slides up and down the cylinder wall, has to function like a bearing, but also has to seal off the combustion chamber to keep the gases from escaping past the piston into the crankcase. Most engines use three rings per piston, but some very low boost high rev engines may get by with only two rings.

The first, or top, ring is called the compression ring; its purpose being to contain the combustion pressure so that maximum power can be obtained. This ring also has the burden of dissipating most of the piston crown heat (about 80%) to the cylinder wall.

The second compression ring has the task of backing up the top ring in sealing off the combustion charge. However, its primary function is to support the top ring in providing another path by which to transfer heat from the piston crown to the cylinder wall. Also, it assists the oil ring in scraping excess oil off the cylinder wall.

The bottom ring is the oil scraper or oil control ring. Its job is to scrape oil from the cylinder wall, but ensure that enough oil remains behind to lubricate the upper rings and assist in sealing.

As indicated in Table 19.2, the oil control rings cause a good deal of engine friction, and we all know that friction costs hp. Therefore, many manufacturers are able to supply oil rings in three grades of tension. Standard tension rings are those mostly fitted to ordinary road engines by car makers, but low-tension oil rings are now being fitted to some production cars and are the preferred option in competition engines as they usually provide oil control almost as good as standard tension rings, but with reduced parasitic losses. High-tension oil rings are rarely used, but may be necessary in some engines running at very high rpm with poor rod length-to-stroke ratios.

Table 19.2. Comparison of piston ring function and friction

	Pressure sealing	Oil control	Frictional losses
Top ring	78%	5%*	30%
Second ring	22%	45%*	20%
Oil ring		70%	50%

*Top and second rings only assume such high levels of oil control when oil control rings are worn excessively or are sticking.

What we have to keep in mind is that in most forced induction engines the component that is going to cop the greatest abuse is in all probability the pistons. Apart from maintaining the correct fit and clearances, the way we can best help the pistons is by keeping them cool. There are basically four areas we have to look at to achieve this, but with a single exception the solution to this task revolves around what we do with the piston rings.

Right at the beginning of this chapter we looked at why the pistons in a forced induction engine get a lot more heat into them. Now we have to find a solution to move that heat out of the crown of the piston, otherwise it will soften like putty. While a row of oiling jets screwed into the main oil gallery and squirting a continuous flow of oil up into the piston crown is a considerable help in this regard, we have to see it as being just part of the answer.

Fig. 19.6. Piston ring placement is critical in high output forced induction engines.

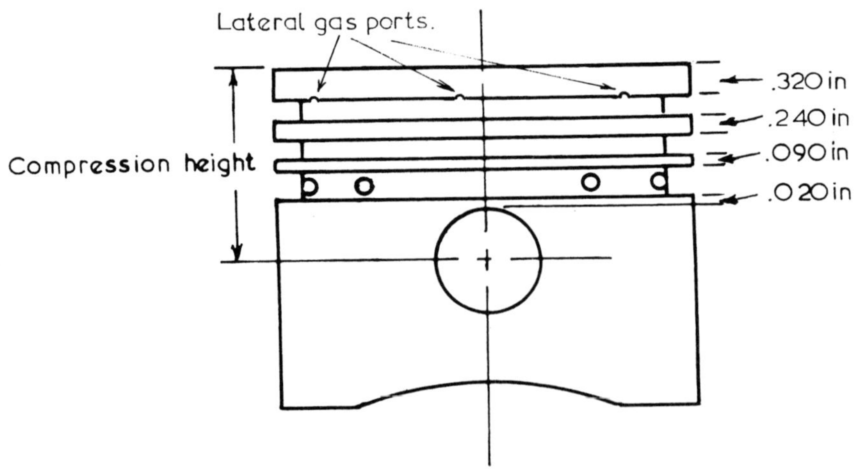

PISTON RING PLACEMENT

The very first thing we need to adjust is the actual placement of the rings on the piston. As shown in *Fig. 19.6*, we have to move the rings further down the piston and make the top ring land much thicker. This means that the top ring should be 8.0–9.5mm from the crown and the second ring 6.0mm from the top ring, with the oil control ring a further 2.5–3.0mm down. The reason we do this is that as we move further down from the crown of the piston it is cooler, and so the piston material is stronger and better able to properly support the top ring which is being levered by tremendous pressure.

The material under the top ring bends, either because it is very soft through being too close to the top of the piston where it is the hottest, or it is weak owing to insufficient material thickness to resist bending. When this happens, a chain reaction is set off. Initially the top ring tilts at an angle to the cylinder wall. This marginally reduces its seal against the wall, but the major impact is that the heat transfer path from the piston crown via the top ring becomes narrower. With more heat being retained in the piston following each combustion cycle the piston becomes hotter, and softer.

In some cases the material under the top ring will bend and then jam the second ring in its groove. In this state the second ring is unable to provide any back-up for the top ring in moving heat out of the piston. Consequently the piston seizes or collapses and the engine blows. Alternatively the overheated piston will add more heat to the inlet charge causing the engine to detonate, and then it blows.

PISTON RING SEALING

Now lets take a look at *Fig. 19.7* and see what goes on when the compression ring seals off the combustion chamber. Many feel that it is the ring's inherent radial tension that holds it against the cylinder wall. Radial tension does help, but it is gas pressure behind the back of the ring that forces its face to seal against the wall.

A detrimental phenomenon that may occur in a high-rpm engine to reduce the ring's sealing ability is called ring float or flutter. As the piston approaches TDC it is slowed

Fig. 19.7. Only the top ring is shown to illustrate how combustion pressure behind the ring forces it against the cylinder wall to effect a good seal. When ring mass causes it to swap sides in the groove, gas pressure is unable to get behind the ring.

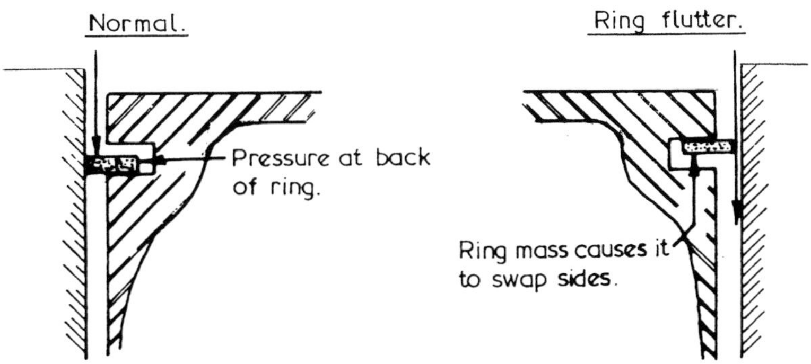

down by the connecting rod, but the rings try to keep on moving, and if they have enough weight each will break contact with the lower side of its groove and bang into the top of the groove. When this happens, the ring seals off the gas pressure in the combustion chamber so that it cannot get behind the back of the ring and force it against the bore wall. Any gas pressure that may have been behind the ring quickly leaks into the crankcase, and combustion pressure forces the ring to collapse inwardly. Immediately the ring breaks contact with the bore wall, the combustion gases blow-by into the crankcase.

Radial tension in the ring is unable to prevent blow-by caused by ring flutter. However, a certain degree of radial tension is necessary for good sealing, otherwise the pressure at the back of the ring would only be equal to the pressure trying to force the ring off the cylinder wall. This would allow blow-by, and it is this that we normally see occurring when the rings are old and have lost their tension.

Ring flutter allows blow-by, but it can also wreck engines. When the ring loses contact with the bore wall it is unable to transfer heat from the piston to the water jacket. As we have already discussed, the end result can be melted pistons or severe detonation.

PISTON RING WIDTH

The narrower a ring for a given radial depth, the higher the rpm at which ring flutter commences. This has led many tuners to believe that they should use very narrow rings in all competition engines. However, narrow rings have a disadvantage: as we have noted, the compression rings are the primary heat transfer path for the pistons. Reducing the ring width will slow heat transfer, so the piston may overheat and soften. The top ring land may distort, which will result in the hp falling off, or worse the piston will melt.

The other negative aspect of narrow rings is their reduced service life. This may not be a concern in sprint engines if you have the time and cash to replace rings regularly. However, in an endurance engine, rapid wear of the top ring could see a serious drop off in power before the race concludes, hence my recommendation to keep right away from narrow rings unless they are necessary.

For the majority of high-performance road and competition engines running up to about 8000rpm I recommend either $^1/_{16}$in, or 1.5mm, top and second compression rings. Engines running up to 9500rpm would run a 0.043in, or 1.2mm, top ring and either a $^1/_{16}$in, or 1.5mm, second compression ring. Above 10,500rpm I would usually go down to a 1mm top ring and a 1.2mm second ring.

PISTON GAS PORTS

Apart from reducing the weight of the top ring, by reducing its width, there is something else we can do to help keep away from ring flutter in competition engines. By adding a series of lateral gas ports above the top ring we are able to maintain gas pressure behind the ring, even if the ring does swap sides as the piston moves past TDC. In effect, these gas ports help an ordinary piston ring work like an 'L' section Dykes ring by ensuring that there is always an open path for gas pressure to get in behind the ring and push it out against the bore wall (*Fig. 19.8*).

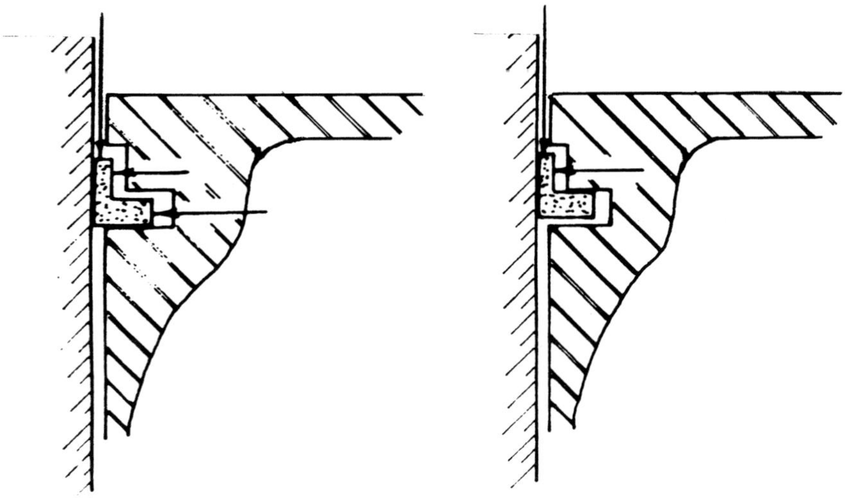

Fig. 19.8. Dykes ring design ensures that combustion pressure always has an open path to get behind the ring and push it against the cylinder.

In earlier times, builders of drag race engines routinely drilled many gas ports from the crown of the piston down through to the back of the ring groove. Unfortunately, this practice exposed the piston and ring to a lot more heat, so component life was dramatically shortened. Lateral gas ports are nowhere near as aggressive. Even so, I recommend that no larger than a 2mm bit be used to mill a half round port to the full depth of the ring groove. The number of these ports will vary according to piston diameter. Below 83mm we should not exceed 6 ports, up to 93mm this would increase to a maximum of 7 and up to 105mm and 115mm we increase to no more than 8 ports and 9 ports respectively.

PISTON RING LIFTING

This brings us to another problem that many tuners do not appear to understand. As well as the top ring being thrown up to the top of the ring groove as the piston rapidly reverses its direction of travel at TDC, it can also be pushed up by pressure building up between the top ring and the second ring. What happens is that gas escaping past the face of the top ring and through the gap between the ends of the top ring may not get past the second ring and into the crankcase as quickly as gas is building up in this area. This pressure will then work in opposition to combustion pressure which is pushing the top ring down onto the bottom face of the ring groove to effect a gas seal. Consequently, the ring breaks full contact with the bottom face of the ring groove and combustion pressure then escapes under the top ring. That in turn increases the volume and pressure of gas trapped between both rings, further reducing the effectiveness of the seal between the lower surface of the top ring and the bottom face of the ring groove.

Clearly, this scenario is going to cost us hp because gas pressure which should be driving the crankshaft is simply leaking into the sump. However, that isn't the end of the story. That gas escaping down under the top ring then proceeds to give up some of its heat to the lower part of the piston. With this part of the piston now much hotter, it is unable to act as a heat sink and pull some heat out of the top of the piston. Even more serious is the fact that the top-ring and piston are not making good contact for quite some time, and as this is the piston's primary heat rejection path the piston overheats, and in serious cases fails. Before that point is reached, the engine will be down on power as the hot piston will heat the inlet mixture, and to keep out of detonation the boost and spark advance will have to be reduced.

There are two things we can do to keep gas pressure from lifting the top ring. First, we must ensure that the second ring has sufficient end gap. Everyone frets about losing gas pressure through the ring gaps and into the sump. Therefore, we have worked to bring the ring gaps down smaller, but we need to keep in mind that there is a limit to how far we can reasonably go. A supercharged engine with thick, rigid cylinder walls should have a second ring gap at least as wide as the top ring. If the cylinders are prone to flexing a little, then you need more gap in the second ring.

The second thing we can do is increase the space between the two rings to provide a larger reservoir for gas to escape into. In a supercharged engine we gain a little extra volume anyway as we use a thicker ring land, but in addition to that we can machine a gas balance groove into the ring land. This groove may appear insignificant, but it adds considerable volume to the area and, as such, reduces any tendency for pressure to build up and lift the top ring.

Never do we want the pressure to equalise above and below the top ring. If that happens we only have the weight of the ring to keep it sealing against the bottom of the ring groove, which isn't anything like sufficient. In fact, this is part of the reason why when we add vacuum to the crankcase our horsepower rises. We cover this in more detail in the following chapter, but basically what we find is that when we keep crankcase pressure below zero by means of a scavenge pump drawing blow-by gas out, then we gain hp.

Even though naturally aspirated Formula 1 engines run only one compression ring, they still employ a pressure balancing groove below the top ring to prevent ring lifting. Note the oil drain-back holes break-out below the bottom oil rail to improve oil control and reduce parasitic losses.

Part of this increase is because there is less friction. The low pressure environment means that we don't have a couple of litres of oil clinging to the crankshaft or suspended as droplets in the crankcase area for the crank and rod big-ends to smash into and thus cause drag. Also, the vacuum literally pulls oil out of the cylinders. Therefore, pistons and oil rings draw off less power scraping oil out of the bores. Because the oil rings have less work to do we may be able to drop the ring tension right off and cut parasitic losses even more.

The other portion of the power increase is owing to the bigger pressure differential existing above and below the top ring, increasing the effectiveness of the seal between the face of the ring and the cylinder, and also between the bottom of the ring groove and the lower edge of the ring. A better seal means a bigger push on the crankshaft, and it ensures best possible heat transfer out of the piston to the water jacket.

ZERO GAP RINGS

At this point some readers may be questioning how the above information can be accurate when zero gap second compression rings are held up by some race engine builders as being good for power. Clearly, with no gap, the second ring's escape route for blow-by gas is quite restricted and this must be causing the top ring to lift. However, to my mind the merit of zero gap second rings relates more to how they help those engines operating at extremely rich air/fuel ratios achieve some semblance of engine life. These rich mixtures may be necessary to cool the engine internally, or they may be the product of crude injection systems forced on us by race regulations.

When we have an engine running very rich methanol or nitro mixtures, cylinder lubrication is marginal. Added to that, a huge volume of fuel can escape past the rings into the sump, severely diluting the lubricating oil and exposing the engine to wrecked big-end bearings and cam lobes/tappets in particular. When zero gap second rings are fitted they appear to do a better job of stemming the flow of this liquid fuel down past the rings. The rings and cylinder bore then bed-in much better and so effect a superior seal. Also, because the lube oil is not diluted to such a degree, the engine makes more power after a couple of hours operation than an engine with conventional rings owing to the cam lobes being closer to their new shape and the engine bearings being healthier. As well as keeping parasitic losses lower, bearings in good shape help to subdue crank harmonics, and this in turn improves the accuracy of the valve timing events and ignition firing events.

Consequently, while it may appear on the surface that zero gap second rings offer many benefits, such is not the case. Outside the areas I have just mentioned, i.e. when extremely rich alcohol or nitro mixtures are present, they are not recommended.

RING END GAP

This in no way implies that we can simply unwrap a new set of rings and stick them straight onto pistons without first checking the end gaps. Actually, the converse is true; we have to be extremely careful to fit each ring perfectly square in its bore and then measure the end gap and increase it if necessary by filing the ring ends. The minimum gap per inch of bore for blower and turbo engines is 0.006in. This is suitable also for light nitrous doses, but a competition nitrous engine will want this increased to 0.0065in per inch of bore. Oil ring rails must be at least 0.004in per inch of bore diameter, but a larger gap won't hurt.

Some people will, no doubt, question such large ring gaps, but let me assure you that when the engine is actually running at full power the top ring gap will close up to around 50% owing to the heat put into it. If the engine gets hot enough to detonate, the gaps will go close to closing right up. One way you can check if you have sufficient ring gap is by inspecting your old rings. If the ring ends are polished the ends are touching, therefore increase the gap by 0.002in. If you don't, you could break the rings, severely score the bores or ruin the pistons!

In practice, most blow-by is because of bore distortion and leakage past the ring faces rather than through the gaps, so be careful when it comes to setting the end gaps. Also, keep in mind that once sonic velocity is achieved through the ring gap, the flow volume will not rise with an increase in combustion pressure or boost pressure. Therefore, as we tune our engine to higher performance levels, blow-by through the ring gaps will not increase. Overall blow-by, though, will worsen, but this is because of more bore distortion and increased leakage past the ring faces.

Some people will point to superior leak down test figures as evidence that zero gap second rings and smaller gaps, especially on the second rings, is the route to less blow-by and thus more hp. Let me assure you that, in spite of what you may have previously been led to believe, there is virtually no relationship between extremely low leak down numbers and increased hp.

GROOVE CLEARANCE

These days, ring breakage is rare and can usually be attributed to worn piston ring grooves that allow the ring to flop around and break, or excessive taper of the bore, causing radial flutter and subsequent ring breakage. If new rings are fitted to a worn bore, and a ridge is present at the top of the bore, the new rings will bang into the ridge and break. Excessive piston-to-bore clearance can lead to ring failure owing to the piston rocking and twisting the rings as it passes TDC.

Generally, a racing engine should be set-up with very tight groove clearances of 0.001–0.0015in. Certainly, there should not be more than 0.002in side clearance between the ring and ring groove. A road motor should, ideally, have the same groove clearance as a racing unit, but up to 0.0035in clearance is acceptable (*Fig. 19.9*).

RING DESIGN AND MATERIAL

In the past, compression ring material was usually plain grey cast iron, either plain, chrome-plated or moly-filled. However, most high-performance top rings are now ductile nodular cast iron, which is much more durable and heat resistant, and less prone to detonation cracking than plain grey iron. There is nothing wrong, though, with using plain cast iron in the second ring position.

For the top ring I prefer Sealed Power/Speed Pro moly-filled nodular cast iron rings. Nodular iron has almost three times the strength of conventional grey cast iron. It is ductile rather than brittle and can be bent without breaking. Molybdenum belongs to the same chemical family as chrome; however, it has a lower coefficient of friction and high resistance to abrasion. Its thermal conductivity is several times greater than that of either

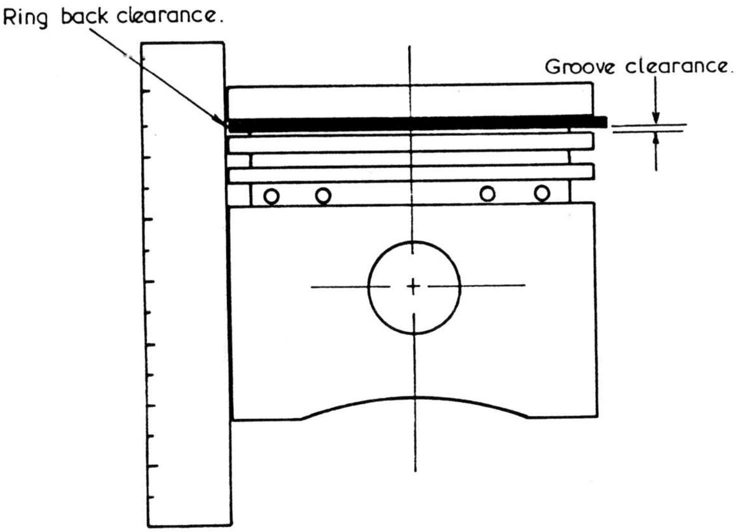

Fig. 19.9. Before rings are fitted, the back clearance and groove clearance should be checked. With used pistons, carbon may build up in the groove, causing ring jamming and breakage. Excessively worn grooves allow the rings to twist and flop about, unable to seal effectively and likely to break.

cast iron or chrome-plated cast iron, and its porosity acts as an oil reservoir, reducing scuffing and cylinder wear.

For oil control I like the multi-piece pressure-back type. The idea is to keep frictional losses to the very minimum while at the same time maintaining adequate oil control. Therefore, only low-tension-type oil control rings should be used for race engines. The Sealed Power/Speed Pro low-tension oil ring is excellent in this type of application.

With oil control rings, the tendency has been to go to very narrow three-piece rings and, surprisingly, no attempt is being made to use the space made available to move the gudgeon pin higher, which raises the question of why bother with narrow oil rings? If there is no problem with sufficient space between the piston ring and the top of the piston I prefer to run with a $^3/_{16}$in, or 4mm, oil ring pack. Such a wide groove allows for four good size oil drain-back holes on both sides of the piston. Along with a full piston skirt with a square bottom edge to scrape oil off the cylinders, such good oil drain-back ensures that the top ring will seal effectively rather than 'aquaplaning' over a deep film of oil. A good top ring seal also means more hp, and less oil in the combustion chamber reduces the likelihood of power-sapping trace detonation. Only in very limited circumstances would I consider going down to a $^1/_8$in, or 3mm, oil ring, and then I would insist on large 3.2mm oil drain holes that actually broke out of the bottom of the oil ring groove by around 1.2mm so as to drain oil from below the lower rail.

HOW TO FIT RINGS

When rings are being fitted, care is needed to avoid fitting them the wrong way up and to prevent damage by incorrect installation. Compression rings can be permanently twisted if they are fitted in the groove at one end and gradually screwed around until the entire

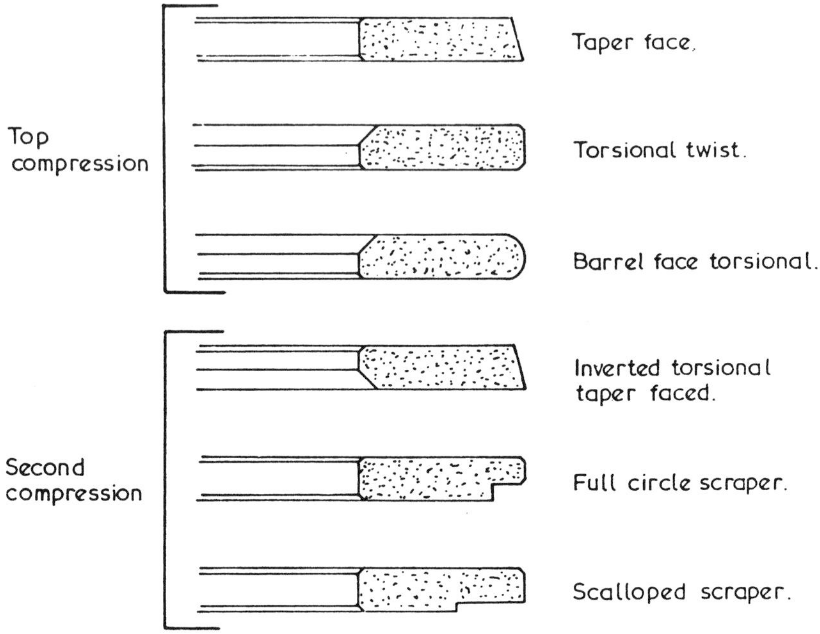

Fig. 19.10. Rings must be fitted the correct way up to effect a good seal and maintain proper oil control.

ring is in place. Instead, they should be expanded sufficiently to fit over the piston, then allowed to drop into the groove. Special expander tools are available for this purpose, but I prefer two 0.015in feeler blades placed between the ring and piston. The blades provide a bearing surface and prevent the ring from digging into the piston.

Taper-face rings are marked 'TOP', or have a dimple mark, and should be fitted that side upwards, providing the greater diameter, or contacting edge, on the lower face of the ring (*Fig. 19.10*).

The 'dimple' near the ring end indicates that this side of the piston ring faces up.

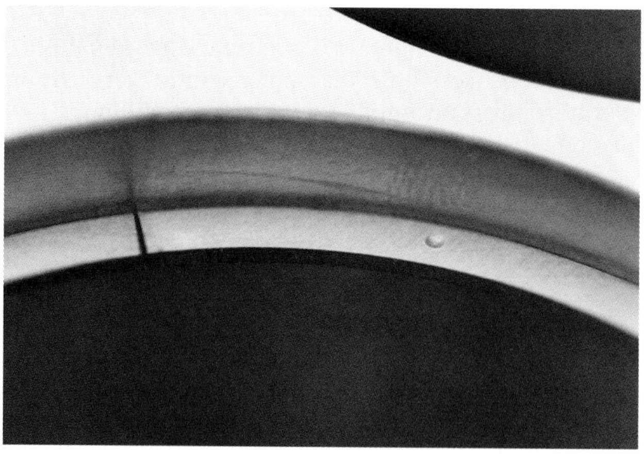

The unbalanced section, or torsional twist ring, is fitted with the inner chamfered edge uppermost. This causes a slight dish in the ring face and provides the same characteristics as the taper-face ring, in that the lower edge makes a high-pressure contact with the cylinder wall. Whereas the taper-face ring eventually wears parallel, the torsional twist ring retains its characteristics.

The 'Multo-Seal' type second compression ring is fitted with the serrations in the ring face downwards.

I have always accurately positioned the ring ends in order to reduce gas and oil leakage through the ring gap, even though some maintain that this is unnecessary in a race engine. It has been proved that the rate of ring rotation to crank speed is about 1:1,000. In other words, in an engine operating at 7000rpm the compression rings will rotate at 7rpm, hence the idea that it is a waste of time to stagger the gaps. However, I still maintain that the ends of compression rings should be staggered 180°, and oil rails should have their ends staggered at least 90°.

Before fitting the assembled pistons in the engine, dip each entire piston in engine oil and oil the walls of each cylinder. Some feel that the oil should be fortified with some additive like STP; some even paint straight STP onto the cylinder walls and pistons. This is wrong. Piston rings are not designed to cut through an additive like this; the end result could be glazed rings and bores.

BEDDING-IN RINGS

The initial bed-in of rings is achieved by giving the car a full throttle burst for a few seconds, followed by snapping the throttle shut and coasting for a few more seconds. This should be repeated at least 12–15 times with the engine at normal operating temperature. Accelerate the vehicle in top gear from the slowest speed it will pull in that gear.

By giving the engine full power, the high gas pressures force the rings out against the bore wall. Snapping the throttle shut causes a vacuum in the cylinder, which draws up extra oil. This, and a low engine speed, minimises the risk of glazing and allows the ring face and cylinder wall to cool.

After the rings are initially bedded-in, the engine can be operated at up to 80% of its rev/power potential, but constantly vary the speed. If this is not done, the rings may still glaze. After about half an hour of this on the dyno or the race circuit, the rings can be considered as being run-in. A road engine with chrome rings should be run in for around 200 miles, preferably in one session. Again, constantly vary the engine speed and include heavy acceleration for short bursts as for initial bedding-in. Avoid constant high speed until the engine has done 500–700 miles.

I prefer to break in my engines on the dyno, carefully controlling the engine speed and load to ensure long engine life and maximum power. This takes two to three hours for most types of engine. When the engine cools, the tappets are adjusted and the head is re-tensioned. Then, following a warm-up to bring the water and oil up to temperature, full load power tests are made to determine ignition advance, fuelling, etc. By the end of the tests the engine is fully run-in, having spent at least four hours running on the dyno (Table 19.3). Some feel that it is preferable to run an engine on the dyno from some external power source. Theoretically, this should smooth out the differences in fit and surface finish, but this solves only part of the problem. Internal temperatures influence the fit and shape of parts, and to further complicate matters, clearances will alter with mechanical loads as well as with temperatures. It follows, then, that the engine must be run-in under its own power, initially very lightly loaded, then progressively loaded to produce the required running surface.

Table 19.3. Standard break-in procedure on an engine dyno

Rpm	Engine load (torque lb/ft)	Time (mins)
3500	40	10
4000	50	30
4500	70	30
5000	85	30
5500	100	30

Note: This is the procedure used for engines of 1,600–2,000cc. Other engines require a different run-in load according to their displacement and torque output.

TESTING RING SEAL

When full power testing is being carried out, a blow-by gauge should be connected to the crankcase, which will give a quick overall indication of how well the rings are sealing. With the amount of scavenge I usually run on the dry-sump pump, I expect to see the blow-by gauge reading less than 1–1.5cfm. If the power is down and the gauge is reading higher than this, I can be pretty certain that there is a ring seal problem somewhere.

An alternative is to block off all the engine breathers except one, then connect a 'T' into the open breather hose. Connect one branch of the 'T' to a 'U'-tube water manometer, and to the other branch, which is vented to the atmosphere, connect a drilled restrictor plug. If the engine is dry sumped, I would try an 8mm vent hole. The idea is to use a restrictor small enough to produce a pressure reading of around 2in of water in a carefully built engine with excellent ring seal; then on another occasion it is very easy to quickly check on the ring seal. If, for example, the engine is down on power and the water gauge is showing a blow-by pressure of 9in with the same-size breather restrictor as previously used, you know that the ring seal is not good.

Obviously, for this latter method to be of any value, a leak-down test must be done first. Then, having established that all cylinders have an average leak-down of, say, 3%, you can be reasonably sure that you have excellent ring seal. Therefore, your initial manometer 'calibration test' to establish what size of breather restrictor is required to produce a blow-by pressure of 2in of water will have real relevance.

The leak-down test is a much less accurate indicator of ring seal, so I prefer to use it only after I see an abnormal blow-by reading, to isolate which cylinder(s) is/are causing the high blow-by reading. Obviously, a static, comparatively low pressure test like this cannot accurately reflect what the ring seal is actually like when the engine is operating; it can only point to where a problem may exist. Note that this test will also show up leaky valve seats, so you need to listen carefully for air leakage to determine in which area of the engine the problem lies.

The actual leak-down test is fairly straightforward. After bringing the engine up to temperature, pull all the spark plugs, then bring No. 1 piston up to TDC on the compression stroke, connect the leak-down tester in the spark plug hole and connect the compressor air line to the tester. The leak-down tester will then display a percentage air loss reading. Ideally we want only 2–3% leakage, but 4% is acceptable. If it exceeded 5% I would look for problems in that cylinder. Every piston is brought up to TDC and tested in turn for leakage.

Really, any serious competitor should do a leak test at the race track as soon as a practice session is concluded, and likewise at the end of competition for the weekend. If the leak-down shows up a problem, look for it and fix it before your next race. Some foolishly haul a race car with a tired engine hundreds of miles to a race meet and spend the weekend frustrated because they are off the pace, or they even blow the engine because it so badly needs attention. What they should have done was to save their money for meals, accommodation, fuel, etc., and spent it to give the engine a light hone and a set of rings, and touched up the valve seats and perhaps replaced the valve springs and bearings.

HARMONIC BALANCER

The pulley or harmonic balancer attached to the nose of the crankshaft has its part to play in the durability of an engine, and the surrounding environment. Many four-cylinder engines use a pressed metal drive pulley, spot welded or riveted to a hub. At higher engine speeds, i.e. 6000rpm-plus, a pulley of this construction is liable to fly apart, wrecking anything in its path. To avoid this, a cast or machined component should be used.

The harmonic balancer is fitted to dampen crankshaft torsional vibrations that could, if not controlled, wreck the crankshaft. At times the balancer and pulley may be in one piece, or a separate pulley may be bolted to the harmonic balancer. Again, if this pulley is a pressed metal component, scrap it to avoid problems.

The actual harmonic balancer is made of three parts bonded together. A rubber belt is bonded between the outer inertia ring and the hub. Failure is normally caused by the inertia ring losing its bond with the rubber. If this occurs at high engine speed you can be assured of spectacular and expensive results. To prevent this, the harmonic dampener must be carefully inspected on a regular basis. If you race, do this after each race meeting. To assist in your inspection you should mark a distinct, common line between the hub and inertia ring. If the two marks move out of line, it indicates that the ring is moving on the hub, so scrap it. Under no circumstances should you try to repair the balancer by bolting the inertia ring to the hub. This will wreck your crankshaft just as surely as running 2000rpm over the limit. The inertia ring must be able to move, within limits, to dampen crankshaft vibration.

There is little that can be done to improve the reliability of the harmonic balancer except to check it for run-out before fitting. The other modification necessary, if a fully counterbalanced crank is fitted, is to machine away the counterweight cast as part of the hub.

Before you fit the harmonic balancer (or pulley) to the engine, paint it black. After fitting, accurately find TDC and file a good groove into it, to align with the TDC mark on the engine timing case. Paint the groove with silver or white paint. If you also paint black the area around the timing marks on the engine timing case, and the TDC and ignition timing marks or pointer silver, the ignition timing will be much easier to adjust later on. To adjust the timing accurately you must be able to see the timing marks easily, so anything that you can do to make them stand out must help.

FLYWHEEL

The flywheel is attached to the other end of the crankshaft. To ensure that it remains attached, two good dowels ($^3/_8$in) and retaining bolts that have been coated with Loctite and correctly torqued provide good insurance. To cut costs, manufacturers seldom fit dowels, so it will be necessary to machine the flywheel and crank to fit these.

Over the years there has been a fair amount of controversy with regard to flywheel lightening. Some claimed that a lightweight flywheel improved acceleration through reduced inertia, others stated that the only benefit was better crankshaft life because the twisting load on the end of the crank was reduced, and there was less risk of flywheel explosion owing to the inertia load on the centre of the flywheel being lessened. At this time we can state and prove that any reduction in an engine's reciprocating mass, or rotating mass, will improve acceleration. Therefore, lightweight pistons, rods, cranks, clutches and flywheels all benefit performance.

A very light flywheel is undesirable for a road vehicle as it can produce a lumpy engine idle and low speed surging. The reason manufacturers use a relatively heavy flywheel in the first place is to absorb the uneven torsional impulses coming through the crankshaft and keep the engine turning smoothly at low engine revs. A fairly hot engine with a relatively heavy flywheel is much more pleasant to drive on the road than one with an overly light one.

If you decide to use a lighter flywheel it is not always necessary to buy a new lightweight version or even have your old flywheel turned down. Some manufacturers have, over the years, built many engines of various capacities around the same basic design. Therefore, the flywheel fitted in production to the 1000cc version may also fit the 1500cc derivative and save possibly 5–7lb; if you have the 1500cc engine in mind for modification, you can use the lighter flywheel from the smaller engine.

Lightening should only be done near the outside of the flywheel. Removing metal from the centre is fatal and likely to lead to failure. When metal is removed from the outside it makes life easier on the centre of the flywheel as the inertia loads are reduced. This is important, as some flywheels are none too strong. To prevent distortion from heat, do not machine away so much of the clutch side of the flywheel that there is only a thin ribbon of metal left, just wide enough to provide a seat for the clutch plate friction material. If the clutch is subjected to severe usage, that thin ribbon may distort, owing to localised heating, and allow clutch slip.

CYLINDER BLOCK DECK PREPARATION

Of course, engine ruggedness and reliability extends to what we do in the area immediately above the engine block, too. The deck of the block needs to be perfectly flat and completely free of any foreign material such as rust, grit, or the remains of the head gasket and sealants before we fit the cylinder head. The way to get rid of such rubbish is by very deliberate use of a flat scraper. If the block has dowels or studs still in place, be extremely careful to cut in close to the base of these. Personally, I do a better job using a bootmakers knife in this area. When undertaking a clean-up of the deck, take care that dislodged material doesn't find its way into oilways, the sump, etc. Do not, under any circumstances, use emery paper or wet-and-dry paper, as abrasive particles will undoubtedly end up in the cylinders, and in push-rod engines you can add the cam and lifters, too. Don't think that compressed air will help – most likely it will spread the grit deeper into the engine.

HEAD GASKET SELECTION

Obviously, all this preparation is intended to provide a nice flat bed for the head and head gasket to seal against. In a supercharged race engine the head gasket has to withstand pressure of up to 1800psi, so obviously we may require something better than the stock gasket if the engine began life naturally aspirated. However, it isn't supercharging as such that gives head gaskets a hard time. The real culprit is detonation, so if we keep the engine free of detonation, blown head gaskets generally will not be a problem.

Copper asbestos gaskets are definitely out, as the thin annealed copper sheet is just too soft to hold up. On the other hand, stainless steel sheet gaskets are too hard to deform into minute surface imperfections, so they also give trouble. Steel/copper/asbestos head gaskets at times work better than stock composite gaskets. This type has a mild steel face and cylinder fire ring, with copper sheet on the other face, and asbestos in between. Some high-performance composite gaskets use a special cylinder fire ring which is better able to withstand detonation than a sheet mild steel or sheet stainless steel fire ring. A round wire steel ring is fine with cast-iron heads but is too aggressive with aluminium, so a pre-flattened steel ring gasket or a round wire hard copper ring gasket may work better.

After this we get into more specialised sealing methods, sometimes in conjunction with a stock type composite head gasket and sometimes with a full copper head gasket. The latter may be obtained in thickness much greater then the traditional 0.035–0.045in. Around 0.060in is common, but up to 0.095in is possible with special heat treatment. Such thick gaskets act as a compression ratio lowering device as well as a seal. In addition to a flat gasket, some type of 'O' ring may be fitted around each bore to exert a greater compressive load on the gasket. With the softer stock-type composite gaskets, the first choice is the stainless steel 'W' ring followed by the copper 'O' ring, but the mild steel round wire 'O' ring may also be suitable. However, the latter is more likely to be used with a solid copper type gasket. The usual procedure with steel 'O' rings is to machine the block to accept 0.041in mild steel wire, such that it stands proud of the deck about 0.006–0.008in. As such, it exerts a considerable compressive force against the gasket and head.

HEAD GASKET CLAMPING FORCE

Apart from detonation, the other big destroyer of head gaskets is insufficient clamping force between the head and block, allowing the gasket fire ring to flap about in response to the violent fire storm going on in the combustion chamber. We have already discussed the need for a perfectly clean and flat deck surface on both the head and block, and later we will cover correct head tension and tensioning technique. That leaves us with a mechanical clamping problem caused by poor design of the block and/or head and also insufficient and/or inefficient clamping devices – read, head studs or bolts – around each cylinder bore.

The most common cure for blown head gaskets is to fit head studs if the manufacturer is relying on bolts to clamp the head to the block. Studs clamp better than bolts and they cause less distortion and pulling of the block deck (which is none too solid in some engines), so that sealing is improved. When initially fitting the studs, torque them to 30lb/ft. Remember that when studs are fitted to Vee-style engines it will often be impossible to remove the heads, owing to a lack of clearance, unless the engine is first lifted out of the vehicle.

With some engines the factory has deleted bolts where they were too difficult to fit in a normal assembly line situation. However, for a high-performance or race engine, we may be able to devise a method of adding those missing bolts and it is amazing how just four additional $^{5}/_{16}$in NC bolts tensioned to 18–20ft/lb cures the sealing problems of powerful big-block Chevys.

After this we have to consider the integrity of the deck of the cylinder head and block. If we find the deck of the head is so thin and lacking in internal bracing that it is bowing under pressure and not clamping tight against the head gasket, we have to come up with a solution. If an aftermarket head is available it may have a thicker, stronger deck. When there is no alternative but to stick with the stock head, investigate what you can do to strengthen it. This may involve tapping some holes into the water jacket in the top of the head and screwing long $^{1}/_{4}$in NC bolts into those holes to push up against the water jacket side of the deck in order to reinforce and brace it against bowing.

If the trouble is the deck of the block bowing, we may be able to reinforce it by converting to a dry deck arrangement. All water passages and oil return passages in the block have to be plugged with threaded plugs, and any holes that cannot be plugged must be welded closed. This serves to strengthen and brace the deck. The deck is then ground perfectly flat. Naturally, when a dry deck is used, an external water tube must be devised to carry coolant from the block into the head, and if a wet sump lube system is used, an oil return from the head to the crankcase will be required.

Another way to reinforce the block deck is by means of an 8–12mm thick steel plate screwed and glued, using a high tech sealant, to the deck. After bonding and curing is complete, the deck plate is surface ground and the cylinder bores honed. The additional deck height also allows for the use of longer con rods and/or more crank stroke. However, we have to use a piston design which keeps the top ring from travelling into the deck plate.

HEAD BOLTS AND TENSIONING

Even if an engine does not have a history of head gasket sealing problems, we can easily bring problems on ourselves. For some time now a number of manufacturers have used stretch-to-yield type head bolts. These are a use-once and then throw away item. If reused they will continue to stretch and so not provide the clamping force necessary for the head gasket to do its job. Consequently, if you stick with this type of stock bolt, follow the factory recommendation and use them only once. The same applies if the factory has what appears as a non-conventional head tensioning method. If they say warm the engine and apply additional bolt tension, then do it.

Now that we have sorted out the head gasket/clamping situation, it is time to fit the head. The head bolts should have their threads buffed and oiled (also oil under the head

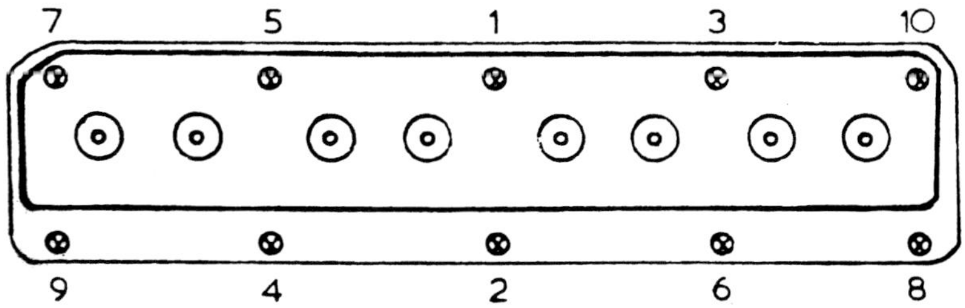

Fig. 19.11. To avoid head warpage and head gasket damage, the manufacturer's tensioning sequence should be followed. Always tension the head in about four stages, and when removing the head reverse the tensioning sequence, after first allowing the engine to cool.

of the bolt) and be tightened in the sequence recommended by the manufacturer. A typical sequence is shown in *Fig. 19.11*. This sequence must be reversed when the bolts are being loosened to remove the head. I always tension the head in four or five progressive steps, to prevent head warpage. Therefore, if the head tension is 80 lb/ft you would take it down to 30, 50, 60, 70 then 80. After the gasket has had time to set (usually 10–15 minutes), go over the bolts again. That's about it if the head is alloy. If the head is cast iron, the motor should be started and brought to normal operating temperature and the bolts re-tensioned. Final gasket set should take place by the time you have done 300 miles, when the head (alloy or cast iron) should be tensioned again.

Something to keep in mind when tensioning head studs or bolts on an engine that has short studs along one side of the head and longer studs down the other side is that long studs stretch more than short ones. Therefore, the head gasket will not be clamped as tightly down one side of the head. To overcome this, tension on the longer studs should be increased by approximately 10%. For example, if the recommended tension is 65lb/ft, torque the short studs to this figure, and the long studs to 72lb/ft.

CYLINDER HEAD HEAT TREATMENT FAILURE

Many head gasket failures, however, are not because of inappropriate head gaskets being fitted, incorrect tensioning or a block deck that is not perfectly flat. The problem, if it is an aluminium alloy head, could be that the head has been annealed when it was previously overheated and blew a head gasket. What happens is that the hot gases destroy the heat treatment and the head goes soft. Consequently, it will no longer hold a head gasket and it has probably lost its crush on the valve seats as well. In this state the head also changes shape; it will be shorter and narrower. Hence a head that does not easily drop down on the block dowels or is jamming on the head studs is probably annealed. If you use head bolts rather than studs and you find trouble getting the bolts started into the block, it means the same thing; the head has gone soft.

This can be checked with a Brinell or Rockwell hardness tester. On the Brinell scale, a good head will read in the 95-plus range, and anything under 75 is too soft. When testing on the Rockwell B scale, this translates to a hardness in the 48 to 60 range being ideal, while anything under 38 is soft. Obviously, the place to test is in an area where the gasket has been blown. Do not be content with a hardness test in just a couple of places.

A casting flaw could throw the reading way off, so over an area of about a square inch test in at least 10–12 spots.

If the head is soft and you have a lot of money invested in it, it may be worthwhile to have it heat-treated afresh. This is not cheap and it may not be possible in heads where the cam runs directly on the aluminium. Ask around to find what specialists can do, or if it is a head from an after-market supplier, get them to give you a price. Basically, the seats and guides are knocked out and it is then heated in an oven at 520°C (968°F) for around five hours (some manufacturers specify up to 20 hours). Then it is quenched in 80°C (176°F) water and tempered (reheated) at 200°C (392°F) for another four hours (up to 20 hours), then allowed to cool slowly in the air. Following this, the head will be machined for oversize guides and seats and the deck will have to be resurfaced. This all costs a lot, but if the head is a good one with lots of fancy port and chamber work, you will not want to throw it away. Note also that any time a head is welded close to a head gasket fire ring line or valve seat it will also require a fresh heat treatment.

Chapter 20

The lubrication system

In the past the lubrication system was simply that – a means of pumping lubricating oil between vital engine components to prevent metal to metal contact, and premature wear or seizure. Today's engines, however, demand much more of the lubrication system, as the lubricating fluid is called upon to perform a bigger role in cooling the bearings, pistons and valve springs. This is especially true of forced induction engines as there are clear limits on how much heat we can move out of the piston crown to the water jacket via the piston rings. The only alternative to this route is to use the lubricating oil, literally squirting it up into the top of the piston to transfer heat from the piston to the oil.

However, in performing the dual roles of lubricating and cooling, the oil ends up in large quantities in places where we don't want it. The reason we don't want it in certain places is because this increases parasitic losses – costs us horsepower and impedes engine acceleration. And we don't want it in certain areas because while there the oil is whipped up, becomes aerated and consequently unable to carry out so effectively its lubricating and cooling activities on its next pass through the engine. So, when we think of engine lubrication we have to consider well beyond just the lubricating oil in isolation. We have to look at the entire delivery system, the oil return system and the oil cooling system. Also, we want to examine ways of enabling the oil to do its work more efficiently while winning back hp previously eroded by parasitic losses.

Obviously, stock or lightly modified engines will work quite well without any modification of the lubricating system. And, as manufacturers take parasitic losses ever more seriously in their quest for higher hp and reduced fuel consumption, the opportunity for us to make gains in this area in road cars is becoming smaller. Therefore, for many readers their prime concern will be the lubricating oil itself. They will be wanting to choose an oil that minimises engine wear and provides a power gain by cutting friction and other parasitic losses.

Apart from its lubricating and cooling roles the engine oil has a number of other basic tasks that we most probably don't even think about. These are: provide a seal between piston rings and cylinder, permit easy engine starting from cold, keep the engine clean of

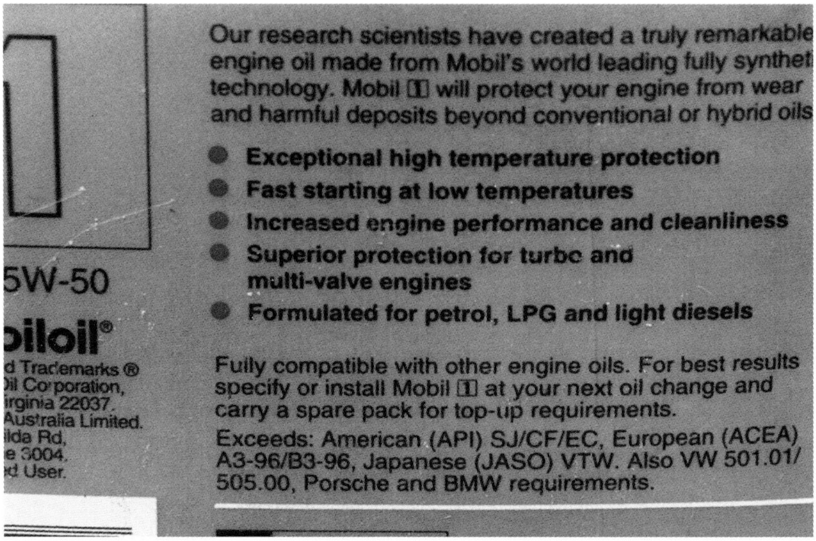

The oil packaging usually lists some of the tests which the oil has passed.

sludge, minimise combustion chamber deposits, not 'poison' the Lambda probe and catalytic converter, provide rust and corrosion protection, and resist foaming. To indicate just how well an oil performs in these areas, various test organisations around the world assign engine oil for street use a series of letters and/or numbers to tell us what tests the oil has passed. While there is a question mark over the validity of such tests, they do give the average consumer some idea of how well a particular oil does its job.

OIL TEST STANDARDS

For example, the American Petroleum Institute (API) assigns petrol engine oils a service rating such as SD, SE, SF, etc. Currently the toughest API rating is SJ, and to carry that rating an oil must pass a series of tests with the engine running on an engine dyno. One test monitors the oil's ability to prevent valve train wear in an ohc 2.3-litre American Ford. Like its 2-litre European cousin, this engine is notorious for valve train problems; in its early life being subject to a recall for abnormal cam lobe wear. Zinc is a great lifter/lobe anti-scuff agent, but unfortunately it does nasty things to Lambda probes and CATs and it had to be removed from the oil additive package. A substitute has yet to be found, so all street oils now struggle with this test. (Race oil and some motorcycle oil formulations are superior in this respect. They don't have to meet car manufacturer's demands regarding Lambda sensor and CAT life, so zinc can still be added by the oil companies.) Another test monitors the oil's capacity to prevent wear and resist oxidation in a 350 cubic inch Oldsmobile run for 64 hours with the oil at 149°C (300°F) sump temperature. Poor quality oils finish up as tar by the end of this test, but the better oils, usually full synthetics, will still look like oil at the finish of the 64 hours. In fact, some oil companies now duplicate this test and extend it up to 128 hours to show how well their synthetic oils hold up. Consequently, while the majority of oils available may achieve an SJ rating, literally just scraping through to gain a pass, others will easily pass and a few will far exceed the test standard.

It is these latter oils in which we should be interested. Each will have its strengths and weaknesses. Others may not be outstanding in any particular area, but may be close to the top in a number of categories. Generally, it would be just such an oil that I would choose. However, for a special situation I may also select an oil with that particular attribute which I deemed to be absolutely essential.

VISCOSITY AND HORSEPOWER

For example, my usual oil of choice for road engines is Mobil 1, 5W-50 grade viscosity. It provides a very high level of engine protection first and foremost. There are one or two other oils about which are similar, or even marginally superior in this regard, but they are not so readily available and they are considerably more expensive. Also, these are down on power and show increased fuel consumption. The horsepower king, while still providing adequate engine wear protection is the 0W-40 grade of Mobil 1. As you can see in Table 20.1, when tested against a large number of oils, including some with lower viscosities, it was a clear winner. In comparison with Mobil 1 5W-50 it has slight fuel economy benefits, too, in the order of 1–1.5% under highway cruise conditions. Therefore, when it is maximum hp that I am after 0W-40 tri-synthetic is the grade I choose. However, it allows slightly more engine wear than 5W-50 and if the power advantage is to be maintained it has to be changed more frequently as it thickens up more rapidly because of a significantly higher evaporation rate.

In recent times the idea has been that low viscosity oil, being thinner, will always help an engine make more hp. However, viscosity is only half of the picture – friction modifiers are the other half. You would think that a light oil would always guarantee a hp benefit but, as you can see when you compare the two Castrol products, that isn't the case. The thicker 10W-60 Castrol R synthetic made more maximum power than most of the oils tested, while its thinner sibling was close to the bottom in overall hp. Clearly, if we are looking to gain hp from a particular type of oil, then there is more to it than simply choosing one with a very low viscosity.

Table 20.1. Synthetic Oil Horsepower Comparison

Rpm	Mobil 1	Shell Helix Ultra	Castrol R	Amsoil	Mobil 1	*Royal Purple	BP Visco	Castrol SLX
	5W-50	5W-40	10W-60	10W-40	0W-40		5W-40	0W-30
3000	143.2	144.8	140.5	141.2	142.9	140.7	139.8	138.6
4000	202.9	201.3	203.6	203.2	202.4	202.5	202.3	201.2
5000	263.1	262.5	262.4	262.5	266.6	262.8	261.5	260.8
5700	290.4	290.0	293.3	290.7	297.4	294.0	289.9	289.7
6200	304.3	303.5	306.5	305.6	311.2	308.7	303.6	305.1
7000	280.7	278.9	282.7	281.4	288.1	284.5	279.1	282.4

*Royal Purple does not state a viscosity rating, but lab testing shows it to be in the 0W-20 to 0W-30 range.

Note: Test engine was a slightly modified Subaru Liberty RS turbo. Oil and water temperatures were controlled at 95–97ºC (203–207°F) and 85–87ºC (185–189°F) respectively.

UNDERSTANDING VISCOSITY AND VISCOSITY IMPROVERS

The viscosity of the oil, as shown on its packaging, relates to the oil's ability to flow through a graduated hole at a specific test temperature: –18°C (0°F) oil temperature for the winter rating (that's the W-number), and 100°C (212°F) for the SAE rating. A viscous (high-viscosity) oil leaves a physically thicker oil film under any condition than a non-viscous (low-viscosity) oil. This fact influences both engine wear and blow-by. Obviously, if an oil film is so thin that it cannot keep moving parts from contacting one another, wear results. Realising that gases in the combustion chamber are at a high pressure following ignition and that the oil film between rings, piston and cylinder wall is assisting in sealing, we know that the oil film must be viscous enough to withstand this pressure. However, if too heavy an oil is selected, it will fail to get between closely fitting parts in sufficient quantity to lubricate or conduct away heat. Tests on viscous oils in the 60 and 70 grades have given poor results because of temperature build-up at the bearings and the inability to lubricate in tight areas. Added to this, these oils have been found to hold heat rather than dissipate it, and to trap air, so de-aeration is a problem.

All oils become thinner when hot, and thicker when cold, but they do not all react to temperature changes in the same way. For example, two 40 grade oils that possess the same viscosity at 100°C (212°F) may have radically different viscosities at higher and lower temperatures. The oil that thins the most at high temperatures will also thicken the most at low temperatures, so obviously it is the oil that shows the least change in viscosity that is the most desirable one to use. Such an oil is said to have a high viscosity index, while the one most affected is regarded as a 'low VI' oil. An oil with a high VI is most desirable for racing engines, providing that the high rating has not been obtained by using additives.

Mobil 1 synthetic is recommended for all engines. 5W-50 is best suited for general use, while 15W-50 is for highly stressed competition engines.

VI improvers, referred to as polymers, have enabled the development of multi-grade or multi-viscosity oils. Polymers are molecules that expand and thicken the oil as the temperature increases, and reduce in size and thin the oil as it cools. Therefore, an oil such as 20W-50 will have the characteristics of 20W oil at low temperatures and of 50 SAE at high temperatures.

Large advances have been made in polymer chemistry, but these VI improvers are still not completely shear-stable at high loads. During shock periods they tend to lose their structure momentarily and collapse, so a 20W-50 oil would only be offering the protection of 20W oil under those conditions. This loss of structure is particularly likely to occur at heavily loaded areas such as the bearings and the tappets. For this very reason you will find that most camshaft grinders recommend that their cams be run-in using a straight 30 or 40 grade oil.

BLENDED AND SYNTHETIC OILS

However, adding polymers or seeking out special mineral oil base stocks is not the only way of producing an oil with a high VI. Another route is to blend synthetic base stock with mineral oil base stock to produce a blended synthetic. Such an oil thus combines mineral oil, synthetic and polymers to produce a high-grade, wide-viscosity oil suitable for use in both high-performance road and competition engines.

Taking the process a step further, we have so-called 100% synthetic oils. In reality they are not 100% synthetic, as a considerable portion of engine oil is taken up by an 'additive package' that is blended into the base stock. However, if the base stock is fully synthetic, without any mineral or vegetable oil blended in, it is said to be 100% synthetic. An example is Mobil 1, which is described as not containing polymer VI improvers; the wide range of 0W-40 and 5W-50 being made possible by special refining processes.

While mineral oils are produced from petroleum crude oil after the distillation process has removed a fair portion of substances like tars, petrol, diesel, kerosene, etc., synthetics go considerably further in the refining process until the very basic components of the crude are separated. These base components are then combined into new man-made molecular structures to produce synthetic base fluid. Some synthetics may use components derived only from petroleum crude, but others may combine molecules from a variety of carbon sources: coal, natural gas, animal fat and bone marrow, etc.

Synthetic oils offer many benefits in high-performance engines over even the very best mineral and vegetable oils. Because they are so thin when cold they lubricate better when an engine is first started. Then, as the engine and oil come up to temperature, they do not thin out like mineral oil, thereby reducing cam/lifter wear, bearing/crank wear and piston ring/cylinder bore wear. Also, synthetics are more oxidation-resistant, so they are less stressed by high operating temperatures.

OIL ADDITIVE PACKAGE

Earlier it was mentioned that a considerable part of engine oil is, in fact, an additive package. Beside polymers, oils also contain a number of other compounds. Some work as anti-oxidants, some as detergents, some as foam suppressants and others are anti-wear additives that in total may constitute 20% of the oil volume.

Oxidation occurs owing to the presence of oxygen and combustion by-products in the oil and owing to the temperature of the oil itself. As the oil temperature increases past 100°C (212°F), the rate of oxidation accelerates and the anti-oxidant becomes less effective.

The by-product of oxidation and combustion is known as sludge. To keep this in suspension, rather than allow a build-up inside the engine, detergent is added. Racing oils, however, may contain very little detergent, and for this reason they should only be used in road vehicles when they carry an API rating such as SH, SJ, etc.

To counter aeration, a foam suppressant is added, which will prevent the formation of foam under the conditions for which the oil is designed to be used, but it cannot eliminate the foam resulting from the presence of water in the oil or from air sucked into the lubrication system either through a fractured pick-up or because of oil surge that has allowed the oil pump to pump air.

Obviously, this additive package eventually wears out, and when the anti-wear compounds are depleted engine wear accelerates rapidly; when the anti-oxidants and detergents break down, sludge begins to build up in the engine, and so forth. This situation should be avoided, and it can be if the oil is replaced while these various compounds are still working effectively.

HOW OFTEN THE OIL SHOULD BE CHANGED

How often the oil should be changed will largely depend on the air temperature and driving conditions. Stop-start driving and cold weather under 10°C (50°F) are hardest on oil, while long distances between stops with little or no engine idling in warm weather – 20°C (70°F) – is very easy on today's oils. In the first instance I would be changing Mobil 1 every 4000 miles or four months, whichever came first. In the latter situation Mobil 1 would provide good protection up to 12,000 miles or six months. However, I would change the filter at the midway point.

In race engines, how often you change the oil will probably depend on your budget more than anything else. However, there is no good reason to throw away synthetic oil after every race meeting. Obviously, if you have just done a 24-hour event you would toss out the oil, but for most engines there is probably no need to replace the oil any more frequently then when the engine is stripped for a rebuild. Really, when you get your oil tested for the presence of metal and contaminants, the report should state how the viscosity has changed and if the additive package has been depleted.

One major concern with oil in competition engines is fuel contamination. Rich mixtures wash off into the sump and even just a little methanol does nasty things to the oil. To get around this the oil can be drained after each competition meeting and gently heated at 90–100°C (195–212°F) to boil off the fuel. In fact, some barbecues see more use cooking fuel and water out of oil than cooking meat!

OIL TEMPERATURE AND OIL COOLERS

No matter how good your oil is, it must be maintained at the correct temperature to lubricate effectively. A good deal of engine wear takes place because engines are operated with cold oil and/or water. An engine should never be driven hard until the oil reaches 70°C (158°F), while the ideal operating temperature is 95–110°C (203–230°F).

If sump oil temperatures regularly exceed 120ºC, an oil cooler will help engine bearings and oil life.

It can go as high as 130°C (266°F) for short periods, but oil breakdown and excessive oxidation will take place above this temperature. In a race engine, bearing failure is a possibility any time the oil temperature goes past 130°C (266°F), and generally hp is lost over 110°C (230°F).

Years ago, an early modification was to fit an oil cooler, but with advances in oil technology this is often no longer necessary for street engines, or even for sprint type competition engines. An exception would be when we are limited in what we can do to bring water cooling temperatures down without a large outlay of money and/or major re-engineering of the plumbing system. In such situations the simple addition of an oil cooler could help to get heat out of the engine to the extent that no change to the cooling system is needed.

Occasionally, at the finish of big events, I have heard podium finishers express their amazement at standing there despite a problem in the oil cooling system having sent oil temperature sky high for much of the race. Since an oil company often features in the list of sponsors, it is sometimes difficult to determine if such tales of high oil temperatures are free advertising or for real. However, I know of a race car completing several hundred miles of an endurance event with the oil cooler bypassed after accident damage. Temperature in the oil tank went as high as 170°C (338°F), but on strip down engine components did not appear to be in any worse condition than if temperatures had been normal. That was on full synthetic; 15W-50 Mobil 1. I have also seen full throttle 24-hour track endurance tests by manufacturers with the same mineral oil in their sump as their new vehicles in the showroom. During such tests, sump oil temperatures of 150°C (302°F) and higher, up to 180°C (356°F), for the entire test have been recorded. Yet subsequent engine teardown revealed that such high temperatures were not a problem. This is not an endorsement of very high oil temperatures, but they do show that, at least for short periods, a new engine, or one still with very good piston ring seal, and within

Oil cooler take-off blocks must be cast or CNC machined without 90º internal passages. Also to be avoided are 90º bends in the oil line fittings.

The bearings were ruined and the crank journals ended up worn like this, all because of a dirty oil cooler.

spec bearing clearances, can operate at 120–130°C (248–266°F) sump oil temperatures. I would only recommend an oil cooler if under hard driving conditions the oil is regularly in excess of 120°C (248°F). I base that recommendation on the fact that I have seen a number of highly stressed race engines lose bearings when the oil went past 130°C (266°F).

When an oil cooler is necessary there are several 'musts' which should be kept in mind. As already noted, cold oil wrecks engines and deteriorates quickly, so in road vehicles either an oil thermostat will be required, or the oil cooler should have a shutter or blind fitted which can be adjusted to suit driving conditions and the ambient temperature. The oil lines connecting the cooler into the system should be at least ½in bore, preferably ⅝–¾in, so that there's no flow restriction. Also, if a take-off sandwich block fitting between the oil filter pad and oil filter is used, it must be of a non-restrictive design. Many such adapters are simply drilled with passages intersecting at 90°, severely restricting oil flow. The adaptors you should select will have oil passages which are cast or CNC machined into the alloy block. They should be finished to look as though they were 'ported' to maximise oil flow. Oil coolers are impossible to clean out, so do not purchase a used oil cooler. It could be full of dirt or bearing material. Also, if your engine or turbo suffers bearing failure it is always prudent to replace the oil cooler, thereby avoiding possible damage from pumping rubbish into your rebuilt engine and turbo.

After fitting an oil cooler, continue to monitor the engine oil temperature. While high oil temperatures may lead to inadequate lubrication, and subsequent failure of bearings, tappets and cam lobes, oil cooled below about 85°C (185°F) is no good either. In a competition engine, vast amounts of money are expended to pick-up a few additional hp, but many fail to realise just how much power cold oil is costing them. Basically, you want the oil at 95–105°C (203–221°F); dropping it down to 85°C (185°F) will cost at least 1% hp unless you are running 0W-20 or 0W-30 oil.

OIL FILTERS

The next requirement of an efficient lubrication system is adequately filtered oil. Any solid material in the oil will act as an abrasive, wearing away at bearings, crank journals, tappets, cam lobes, etc. The oil filter must remove the majority of this type of material from the oil. Very few stock filtering systems are totally effective, so because long engine life is at least dependent on clean oil, steps must be taken to install an effective filter.

The term 'Micron Rating' is often misused in filtration technology. The important figure is the one that indicates the smallest particles totally removed by the filter. For example, a filter may have to stop all particles over 40 microns from passing through, but 5% of particles of 20–30 microns size are allowed through. A system suitable for an expensive engine should remove all particles over 5 microns, but this ideal is not possible at present.

A micron is one millionth of a metre – a human hair is about 50 microns thick. The larger the particles circulating in an engine's oil, the more likelihood there will be of abrasive wear. The challenge facing filter manufacturers is to stop all of these abrasive particles with a filter that has an acceptable service life. You can do your part by changing the filter regularly, as specified by the manufacturer, and by re-installing the

If possible, choose a larger capacity oil filter. If there is insufficient space in the stock location, consider installing a remote mount.

new filter correctly. If the type fitted to your engine has a replaceable element, be sure to remove all traces of sludge from inside the canister before fitting a new element.

A race or rally engine with a lot of money invested in it requires a higher level of filtration than that offered by over-the-counter spin-on or replacement-element filters. The standard type of oil filter may have a bypass in case it becomes clogged with sludge and other material; this is done for the benefit of the motorist who does not bother to have his filter changed regularly, and also to prevent the filter element from rupturing when the oil is cold and viscous. Once the bypass opens, as a result of excessive pressure, unfiltered oil is allowed to circulate through the engine. It can easily be seen that the area where the bypass is located is where much of the sediment material from the oil collects. When the bypass opens this rubbish is flushed through the engine (*Fig. 20.1*).

Fig. 20.1. Bypass type oil filters are not recommended.

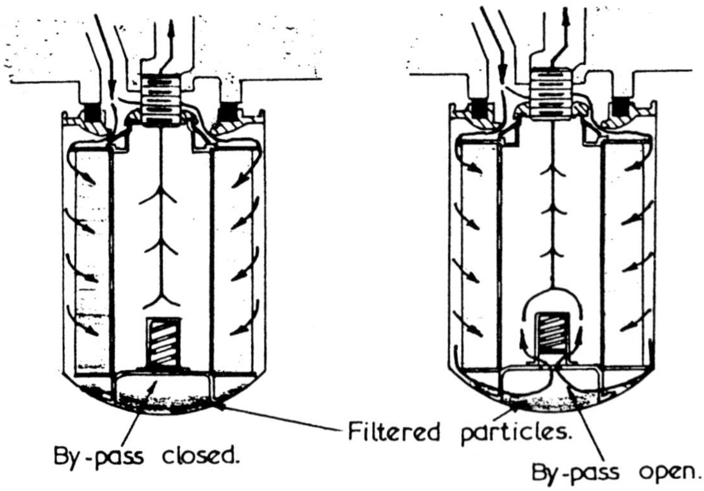

A racing oil filter must not contain any bypass, and consequently must be burst-proof at oil pressure up to 200psi. It must also be fine enough to stop abrasive particles from circulating with the oil without reducing oil flow to a dangerous level, particularly during cold starting.

OIL PUMP AND PRESSURE

As engine tune is increased and bearing clearances become greater, the need to investigate the flow potential of an engine's oiling system grows in importance. In lower stages of tune the standard oil pump, if in good condition (check the manufacturer's specifications for gear clearances, etc.) will usually be acceptable. It may be possible to increase the oil pressure and oil flow a small amount by fitting a stronger relief valve spring or by fitting a spacer under the existing spring. Once the engine reaches semi-race tune, a higher-capacity oil pump, which uses wider gears, will most likely become necessary to keep sufficient oil passing through the engine. It may also be found necessary to change the oil pump drive if the standard drive is not able to cope with the additional oil pump load.

Some tuners, in an endeavour to supply oil in sufficient quantities to the bearings, increase oil pressure to 100–120psi. This should never be necessary if the lubrication system is properly modified. In fact, if an 8000rpm race engine is experiencing any failure at 60–70psi, it usually indicates that work is required to open up oilways and/or restrict bleed-off to the cam and lifters. A competition engine operating above 8500rpm may benefit if the oil pressure is raised to 80–85psi. However, keep in mind that the power required to pump oil through an engine is considerable, so do not increase the power drain to ancillary equipment by unnecessarily increasing the oil pressure.

PREVENTING OIL SURGE

Oil surge can be a serious problem, and potentially hard to overcome. Any vehicle with a wet sump (i.e. an oil reservoir in the sump) will be prone to oil surge unless the

The stock sump baffle needs to be extended right across the oil reservoir area, and a substantial turned down lip should be added around the dipstick and oil pick-up holes.

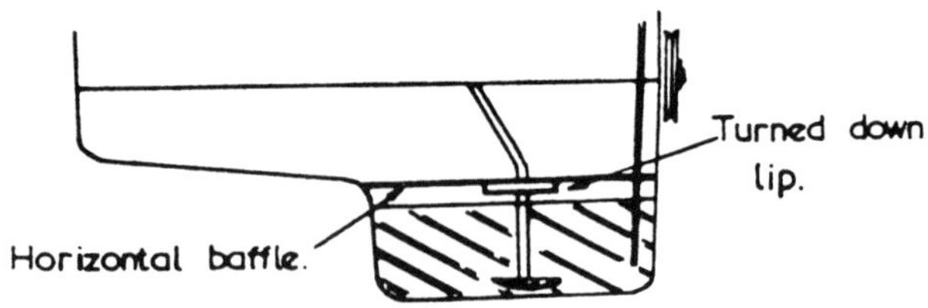

Fig. 20.2. A basic wet sump anti-surge baffle.

manufacturer has taken preventative measures during production. Oil surge is caused by braking, acceleration or cornering forces surging the oil away from the oil pick-up, allowing air to be pumped to the bearings. If allowed to go unchecked, bearing failure becomes imminent.

To reduce surge, the trend is to fit vertical baffles in the sump, but these have little effect as the oil will rise over them during hard acceleration or braking. Instead, a flat, horizontal baffle should be fitted in the sump about ¼–½in above the full oil level, first checking that at this height the rods will not hit the baffle. The baffle should cover the entire oil reservoir area in the sump, and have a hole cut in it just large enough for the oil pick-up to fit through. Around the hole a ½–¾in turned-down lip is required, to discourage oil from surging up through the hole. This hole will also take care of oil drain-back into the reservoir. Do not forget to also cut a hole for the dipstick (*Fig. 20.2*).

The height of the oil pick-up above the floor of the sump must also be adjusted. If it is closer than ¼in, oil flow into the engine will be restricted; if more than ⅜in, the engine will gulp air with G forces around 0.8. Remember, when measuring, to include the thickness of the sump gasket.

In engines subjected to the full range of G forces – left and right turns and acceleration and braking – the pick-up should be located right in the middle of the sump's oil reservoir. However, in a drag race engine it would be offset towards the rear of the sump, and in a speedway machine running anticlockwise it should be off to the right side.

To ensure that the baffle is working and to keep a check on the oil pressure, a reliable capillary-type (not electric) oil pressure gauge is required. A sudden drop to zero will

This oil pressure sender switch operates at 30psi and is useful when assessing sump baffles and sump oil fill levels.

indicate surge or a low oil level; a small pressure decrease indicates bearing failure or a blocked oil filter.

When testing the effectiveness of sump baffling, gauge readings can be missed, particularly surge readings induced by cornering forces. For this reason I also fit an electric oil pressure switch, adjusted to operate at 30–45psi. I wire this switch to a 3in stop/tail light mounted on the dash and fitted with a 21-watt lamp. If the light flashes on, low oil pressure is indicated. (A vehicle driven at night will also need a change-over switch to connect with a 5-watt lamp in the light, otherwise you could be momentarily blinded if the 21-watt lamp flashes on in the dark.) If you have data acquisition, then use this to keep a check for surge at least during sump development, and whenever you compete at a new track.

DRY SUMP SYSTEM

The dry sump type lubrication system avoids many of the pitfalls of the wet sump arrangement, so is the preferred lubrication system in competition engines. *Fig. 20.3* shows a typical dry sump system. The oil is scavenged from the engine and is piped to an external oil tank, usually passing through the oil cooler en route. The oil pump will have at least two scavenge stages drawing oil and blow-by (plus air, if the engine is unsealed) back to the oil tank. The oil tank is designed to not only store the oil, but also

Fig. 20.3. A dry sump lubrication system with one pressure and two scavenge stages. The components illustrated are: 1. scavenge/pressure oil pump; 2. dry sump oil pan; 3. full flow oil filter; 4. oil tank; 5. breather catch tank; 6. oil cooler; 7. aircraft screen filter; 8. scavenge lines with inlet screen filters; 9. tank return lines; 10. suction line; 11. pressure line; 12. breather lines.

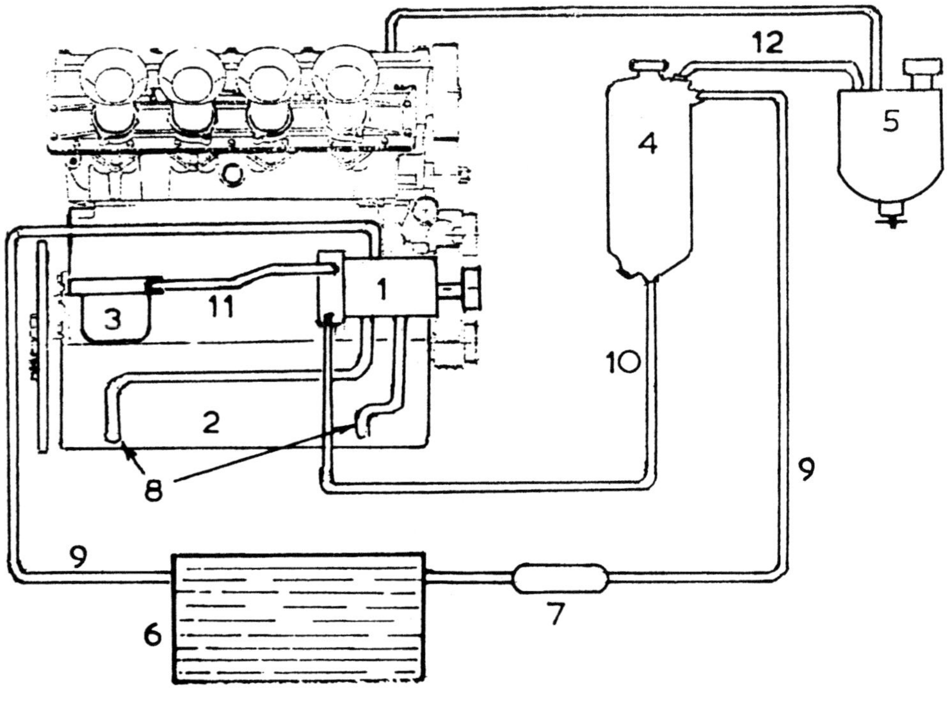

to de-aerate it effectively, for if aerated oil were pumped back into the engine, bearing failure could result. The pressure stage of the pump then draws this de-aerated oil from the oil tank and pumps it through the oil filter and into the engine.

The beneficial aspects of this type of system are obvious. The absence of oil surge helps to increase engine life and reduces the risk of bearing failure. The vehicle can be lower without the risk of sump damage, as a shallow oil pan without a reservoir is used. Oil leakage from the engine is reduced as the scavenge stages reduce pressure, caused by a blow-by, within the crankcase. Depending on the scavenge pumps used, crankcase pressure should be less than a couple of inches of water at maximum engine speed and load in a race engine with good piston ring/cylinder seal. As I will explain later, this reduced crankcase pressure contributes to a power rise of the order of 3–4%, in spite of the extra hp required to drive a big dry sump pump.

The disadvantages of the dry sump system mainly revolve around the expense and additional weight, but a dry sump system can also give many problems if not correctly engineered. The main areas where I see deficiencies are in the design and mounting of the oil tank, and the size and routing of the pressure and scavenge lines.

OIL TANK

The size of the oil tank, to some extent, will be dependent on the size of the engine and the quantity of oil being pumped into it, the effectiveness of the scraper/scavenge system in rapidly returning oil to the tank, the effectiveness of the tank in quickly de-aerating the stored oil, and engine oil consumption and race distance. Note that the oil tank must not be filled more than two-thirds full.

A tall round oil tank with the oil line entering the top of the tank at a tangent works best (*Fig. 20.4*). With the oil entering at a tangent it will tend to swirl around the tank wall, which helps the air separate out. To further aid de-aeration, a horizontal baffle

Fig. 20.4. A dry sump oil tank.

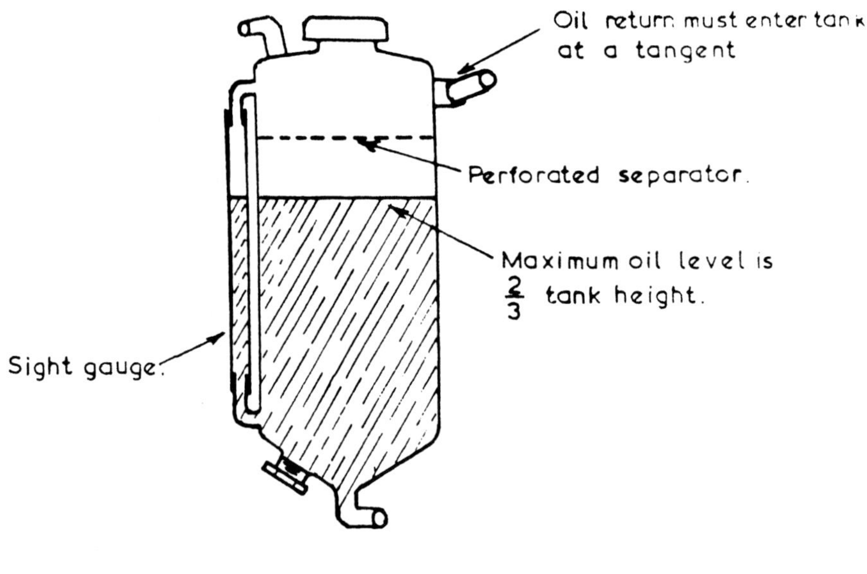

perforated with 4–5mm holes should be located about midway between the top of the oil and the oil inlet. Then, by keeping the height-to-diameter ratio of the tank greater than 2:1 – preferably around 2.5:1 – the oil will give up more air, reducing the air content from as much as 7% down to a round 2–3%. Also, by keeping the tank tall, and with the pump suction line drawing from the bottom of the tank's cone-shaped floor, the pump will always be drawing oil into the engine regardless of the G forces being encountered.

Mounting the oil tank in the engine compartment as close to the engine as possible cuts down the length of the various oil lines, and thereby reduces flow losses. However, flow losses also occur when hoses are too small in diameter or are bent too tightly. The minimum radius of bends should be 2in; if you do not know what that looks like, take a length of 4in exhaust pipe. That has a 2in radius, so no bend should be tighter than that. The actual diameter of hoses will depend on oil flow rates through the engine and whether they are scavenge or pressure lines. A typical 2-litre race engine will have an oil flow rate of around 15 litres per minute, while a 6-litre V8 will be double that. So, while a 2-litre could run ½in bore pressure hoses and ⅝in bore scavenge and suction hoses, bigger V6s and V8s would need ⅝in pressure lines and ¾in scavenge and suction lines.

ENGINE BREATHERS AND CRANKCASE VACUUM

With a dry sump system installed, large engine breathers are not required. With small engines, a ⅜in or ½in hose from the cam cover to the breather catch tank is ample, while for bigger V6 and V8 engines you should run two hoses. There are two schools of thought here. Some tuners advocate running an 'open' breather (one that vents directly to the catch tank), while others prefer a 'closed' system (one which incorporates a check valve between the engine and the catch tank). Personally, I prefer a closed system as this allows the scavenge pumps to run down the crankcase pressure very low, perhaps below zero when there is minimal blow-by. With zero crankcase pressure, hp rises a little and oil leaks around the crank and cam seals, for example, dry up. If the engine runs sour during a race and blow-by increases crankcase pressure beyond what the scavenge stages can handle, the breather pipe flapper valve opens to vent excess blow-by back to the catch tank. With an open breather, pressure in the crankcase will never drop down to zero, because if the scavenge stages have excess pulling capacity beyond handling blow-by they draw air into the engine via the breather pipe. In the closed system this cannot happen, because the check valve allows only one-way flow in the breather pipe.

In recent years the lubrication system has been carefully scrutinised in the search for more power. It should be obvious that a crankshaft rotating at 8000rpm and banging into the gallons of oil draining from the camshaft, lifters and rockers, etc., will be experiencing a certain degree of drag. To get some idea of what we are getting at here, stick your hand out into the airstream sometime at 100mph and feel the effect. This drag consumes a lot of power and results in oil frothing (aeration) and heating, owing to the friction involved and energy being expended. 'V'-type push rod engines are the worst offenders in this respect, as without modification they dump all the oil from the cam and lifters on to the crankshaft.

You learned during school physics lessons that fast-moving air creates a low-pressure zone, which is why aeroplanes fly, etc. A crankshaft spinning at 8000rpm is also moving the air at quite a speed, and this too creates a low-pressure region in the area in which the crank and rods are moving. The rest of the crankcase may be experiencing a pressure of

2psi, depending on the effectiveness of the breathing system on your engine. Let us assume that there is a difference in pressure of 5psi between the high-pressure and low-pressure zones in this engine. What will tend to happen is that the oil pouring through the engine bearings will continue to cling to the crankshaft, because the centrifugal force of the crank is unable to overcome the pressure differential within the crankcase. Of course, part of this mass of oil must continually be released, but in a big V8 there may be three pints of oil clinging to the crank, and being carried around by it to cause fluid drag additional to that being caused by oil drain-back from the cam and lifters.

An effective dry sump system can help with this problem in the following way. If large scavenge pumps are used, the air pressure in the sump pan can be brought to below zero or close to equal the low-pressure zone surrounding the crankshaft. This will allow the force of gravity to have its way and pull a considerable amount of oil away from the crank. Basically, this explains some of the power increase shown in dry sump engines.

The other reason for the rise in hp is improved piston ring to cylinder wall seal. Whenever we increase the pressure differential existing above and below the top compression ring, we get a superior ring to cylinder wall seal. This comes about because the pressure above the ring improves the seal between the lower edge of the ring and the bottom of the ring groove. If we achieve zero leakage here, more combustion pressure builds behind the ring forcing it hard against the bore to effect a more gas-tight seal. Obviously, with added combustion pressure trapped above the piston a bigger shove is transferred to the crank, which we record as increased hp.

There are limits to the amount of vacuum desirable in the crankcase. First, it costs us

The pan evacuator system uses exhaust pulses to draw a vacuum on the crankcase. A good system will pull up to 36in water (2.65in Hg) vacuum, which is useful but a lot less than that possible with a vacuum pump or a dry sump scavenge pump. The pan evacuator cannot be used on turbo engines. The fitting into the exhaust pipe is an anti-backfire valve.

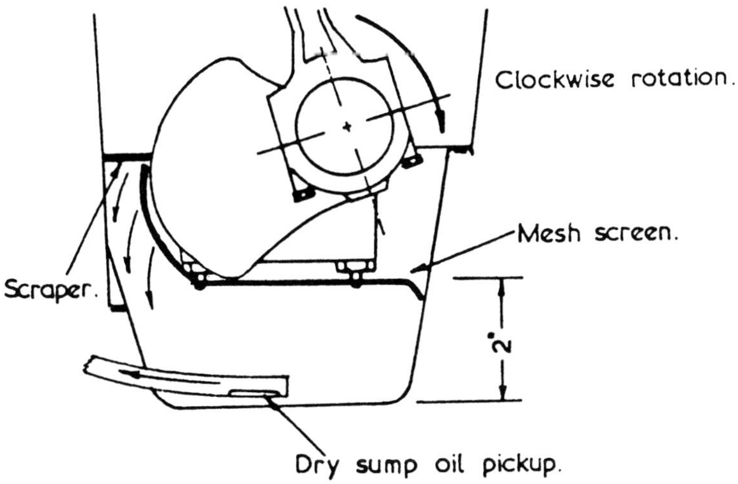

Fig. 20.5. An oil screen and scraper.

power to drive a scavenge pump harder, so we reach a point of diminishing returns. Also, we can run into quite a serious lubrication problem. The high vacuum can pull so much oil away that piston pins seize unless we pressure lube them using special rods with an oil passage from the big-end to the little-end. Exhaust valves also run dry and seize. Some have experienced serious cylinder lube problems, and scavenge pump reliability problems. Consequently, my recommendation is not to exceed depressions of 10–12in for street engines. For competition engines, start out at 14–15in vacuum and carefully inspect the engine before going further. Personally, I don't think there is any gain above about 18in, but I understand some big money teams have been finding hp in the 22–25in area.

SCREENS AND SCRAPERS

The dry sump system cannot get all the oil away from the crank, and for this reason we need a system of screens and scrapers to assist (*Fig. 20.5*). The screen should be of rigid 16–20 gauge steel mesh running the full length of the crankshaft and positioned, as shown, in close proximity with it. The idea is that the screen will prevent oil from splashing back onto the crank after it has been flung to the floor and sides of the oil pan; the screen effectively dissipates much of the oil's energy on the way through, so there is less energy available for splash-back when the oil hits the pan.

A rigid 22 gauge steel scraper positioned along the sides of the oil pan deck will scrape oil from the crank and rods if machined to extend to within just a few thousandths of an inch of these.

To improve the oil drain off from the scraper, the side of the oil pan should be extended out an inch if possible. With these modifications you can expect a further 3–4% power increase over and above that possible with dry sumping alone.

Do not despair if you cannot afford a dry sump system, because the screen and scraper set-up can be adapted to work almost as effectively with a wet sump. There are two ways to approach the problem. If ground clearance is not a consideration, deepen the sump to give 1½in clearance between the screen and the horizontal anti-surge baffle. For improved oil drain from the scraper, move out the side of the sump 1in, the same as for the dry sump oil pan.

If ground clearance is a problem, the only alternative is to widen the sump to lower the oil level. Again there should be 1½–2in clearance between the screen and the horizontal baffle.

OIL FILL LEVEL

Having considered the foregoing information, it should be obvious that the sump should never be overfilled, otherwise you are sure to have the crank and rods dipping into the oil every revolution. If that is not bad enough, you will end up with a situation where too much oil is being thrown up the cylinders; the rings have to drag it off the walls and that consumes power. If the rings cannot cope with this additional oil load you will have combustion chamber and plug oiling problems, which will affect combustion and run the engine into detonation.

Personally, I keep the oil level well below the full mark on the dipstick. First, I check for oil surge with ½ litre less. Then, if there is no surge problem, I drop another ¼ litre and so on down to a maximum of 1 litre below full. With a complex baffling system, involving several swinging trap doors and a swinging pick-up, I have been able to drop to a 50% oil fill. This is dangerous territory, though, and the hp gains are minimal. Lowered by 1 litre, some engines have shown maximum hp gains of around 2½%, but a little over 1½% is more usual. At cruise, though, the gains are much greater, and this pays a nice fuel economy dividend.

CRANKCASE AIR PASSAGES

We may have an opportunity to go much further in our endeavours to reduce crankcase windage losses. For example, in the upper echelons of motorsport, engines are built with the crankcase divided into separate compartments in order to limit air movement and

To reduce parasitic losses, and numerous other problems, never exceed the 'max' oil level mark. If possible, run closer to the 'min' mark.

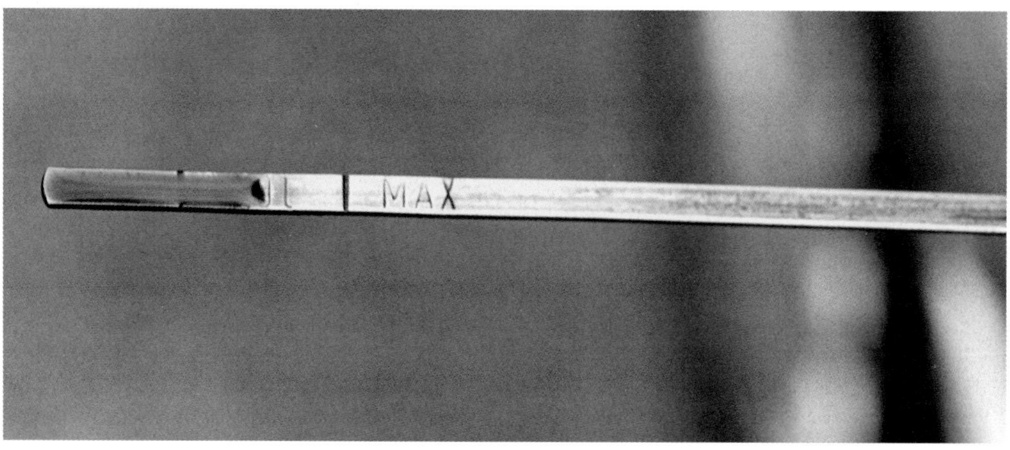

subsequent frictional losses caused when a piston rapidly descends in its cylinder. Obviously, if an engine has an individual cylinder capacity of 500cc, each time a piston descends it has to push 500cc of air out of the way. If it is only a 4 cylinder engine working at 7000rpm, we are pushing the equivalent of 7000 litres of air back and forth in the crankcase every minute! That costs us power even if we have a slight vacuum in there. Remember, normal air pressure is 14.7psi, so if we apply a vacuum of 15in, about 7.4psi, we are still moving a lot of air (about the equivalent of 3480 litres). When we divide the crankcase into compartments we arrange things such that the air being displaced as the No. 1 piston descends moves next door to fill the void created as the No. 2 piston rises. In this manner, air movement is restricted to a smaller area, windage losses decrease and hp rises.

Car engine manufacturers have done something similar. In the late '80s I noticed that some engines began appearing with a large passageway between the main bearing webs and the bottom of the cylinder bores. Some tuners suspected this was to save engine weight. However, those who were observant also noted that at the same time these passages appeared the very same engines were outfitted with windage trays which wrapped up close to the crankshaft. While these pressed metal trays did a reasonable job of pulling oil off the crank and rods, and prevented the hurricane being generated by the spinning crank from drawing oil up out of the sump, they created a secondary windage problem. There was now no easy path for the air to shuttle between cylinders as the pistons travelled up and down. The solution was to add these holes to interconnect the bottom of the cylinders. The resultant windage losses produced hp gains in the region of 2%, and in a few engines it was over 2½%.

The windage tray and scraper strip oil off the crank and rods and prevent a low pressure area around the spinning crank from drawing oil up out of the sump.

These passages between the cylinders and main bearings are supposed to allow the air below the pistons to easily shuttle back and forth as the pistons go up and down. Before being cleaned up they partly block flow and introduce stress raisers into the block.

Using a grinder, all corners are radiused to help prevent main bearing web cracking. Also, air flow is improved, cutting parasitic losses.

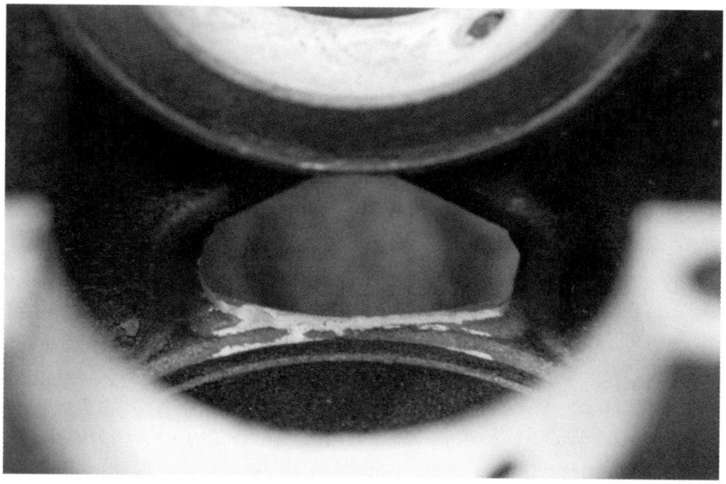

In some blocks these holes are machined, but generally they are cast. Either way they need some help to properly do their job, and at the same time we can partly get rid of a potential failure point in the block. We have to view these passages in a similar light to that of the inlet ports. They are expected to move the same gas volumes, but in considerably less time. Cast passages are usually quite large. However, they are often partially blocked by casting slag. This stuff has to be ground out to improve flow capacity and, more important, to get rid of a stress raiser which could end in the main bearing web breaking away from the bottom of the block. Machined passages are better in this respect. However, as they are quite small be sure to radius both ends of the hole to help flow.

WET SUMP BREATHERS

To vent crankcase pressure, most engines in standard tune rely on a single $^1/_2$in or $^5/_8$in hose connected to the air filter or inlet manifold. Any wet sump competition engine should have at least a 1in hose connected into each rocker cover and another 1in hose connected into the block. Take care, when connecting breathers into the motor, that oil is not going to be splashed up the breather pipe by a rocker arm or a con rod. If such is the case, steps will have to be taken to fit a suitable deflector at the breather inlet.

The problem with these breather systems is that they fall far short of what we actually want to help us make more hp, and improve engine reliability. The stock breather system is required by law on road cars to cut pollution of the air we all breathe. We aren't allowed to simply dump blow-by gases into the atmosphere as is done on competition engines. Therefore, this gas, along with any oil mist that the oil/air separator in the cam cover has failed to remove, is recirculated into the inlet tract. Clearly, this gas occupies volume and so reduces the power potential of the engine. The oil mist, though, can wreck engines by lowering the fuel octane and setting off detonation.

CRANKCASE VACUUM PUMP

How can we get around this and still do our part in not ruining the environment? In the past I have used the belt-driven smog pump that was fitted on some engines back in the '70s and '80s to pump air into the exhaust in order to lower emissions. However, rather than have the pump draw in atmospheric air, I connected it to suck blow-by gas out of an oil/air separator tank which I connected to the crankcase and cam cover. These pumps are quite reliable if not spun too fast and are protected from exhaust back-fires by a check valve. Rather than pump the blow-by gas into the exhaust manifold, I connected a single stainless steel pipe into the exhaust at the exhaust header collector. Obviously, in turbo engines these gases must be recirculated after the turbo, either into the dump pipe or just prior to the CAT.

The main problem with these smog pumps, properly called an AIR or Air Injection Reactor pump, is that they are big (5in dia x 3–4in long) and didn't pull a really good vacuum on the crankcase. However, in more recent times Moroso have marketed two dedicated vacuum pumps which are very effective. The big 4-vane 22641 pump will pull over 15in vacuum while the smaller 3-vane 22640 pump will draw up to about 12in. The 3-vane is a better street pump, but is also useful in endurance events where you are required to run a wet sump. The 4-vane race pump will wear out fairly rapidly if spun faster than about 4000rpm. The life of both pumps is enhanced if some oil mist is allowed to flow through them. This means that it is best to locate the oil/air separator after the pump, and in road cars with a connection into the exhaust the back-fire check valve is located after the separator. Street engines routinely show a 3–4% hp rise at 10–12in vacuum; over double that which I record with smog pumps.

Chapter 21

The cooling system

Your engine may be air-cooled or water-cooled, but in both cases you rely directly or indirectly on a flow of air to stabilise the temperature of the cylinder head and the cylinders. The cooling system of every internal combustion engine performs a vital function: the dissipation of heat in order to maintain normal engine operation.

The heat engine relies on the conversion of fuel into heat, and then into mechanical energy to produce power at the crankshaft. Only about one-third of this heat is converted into power; another third is eliminated through the cooling system. This alone should alert you to the load placed on the cooling system. If our engine is generating 300hp we are also calling on the cooling system to get rid of the equivalent of 300hp of heat energy. Now, if our stock engine produces 150hp, the manufacturer obviously installed a cooling system capable of handling that heat load and not much more, so it would be foolish of us to think that the stock arrangement could cope with 300hp.

AIR-COOLING

If your engine is air-cooled, there is not a great deal that you can do to increase the engine-cooling capacity to cope with higher power outputs. Therefore, it is essential to ensure that the cooling system provided by the vehicle's manufacturer is operating at 100% capacity.

Heat radiation from the cooling fins is retarded by the presence of oil and mud, so ensure that the fins are clear. Fins that are silver-coloured can have their radiating capacity improved by a coat of matt black paint.

Air-cooled cars rely on a fan to circulate cooling air over the head and cylinder. Think carefully before you decide to modify a cooling system of this type, as there are several arrangements in service from the various manufacturers. Without a proper understanding of the system, any modification could spell disaster for the engine. It may be possible to raise fan speed to increase air flow over the engine, but on the other hand the fan may already be moving all the air it is capable of moving and any increase in engine rpm or fan speed will only result in a marginal flow improvement. Frequently, increased cooling can be best cared for by thinking more in terms of massively increasing the size of oil coolers, and perhaps upping oil flow in the engine – there is, especially, a need for more flow volume up into the pistons and across the cylinder heads.

When Porsche added turbos to their 630hp 917 racer, raising the output to 1000hp and then later to 1100hp, they successfully increased fan speed from a maximum of 7400rpm up to 9000rpm, a 22% increase. However, cooling air flow rose from 2400 litres/sec to 3100 litres/sec, indicating that the big fan on the 5-litre flat-12 (5.4-litre on later versions) was more efficient at this higher speed. Also, you should note that owing to the much higher hp, full throttle was used less frequently and for significantly shorter periods. As a result, air flow per hp was dropped from 3.8 litres/sec in the naturally aspirated engine to 2.8 litres/sec in the 1100hp turbo engine without any problems with overheating.

MONITOR CYLINDER HEAD TEMPERATURE

Before you even think about modifying a fan-air-cooled engine, be sure to fit a cylinder head temperature gauge; it could save you a lot of money. The liquid surrounding the combustion chamber of a water-cooled motor tends to act as a heat sink and stabilises the combustion chamber temperature. An air-cooled head has very little in reserve to take away a sudden increase in heat, so the combustion chamber and piston temperature could take a sudden and unexpected rise, perhaps large enough to melt the pistons. For this reason the head temperature must be continually monitored when you are engaging in high-speed motoring.

What is the maximum safe head temperature? That depends on where you measure it and on the quality of the pistons. If pistons suitable for competition are being used, and the temperature is being recorded from a thermocouple washer at the base of the spark plug, I would suggest throttling back once the temperature exceeds 220ºC (428°F). If the thermocouple is attached elsewhere on the head, I would advise that the same action be taken at a 200ºC (392°F) read-out.

LIQUID-COOLED ENGINE, CORROSION AND FREEZE PROTECTION

Cooling by water or liquid is usually thought to be more straightforward than by air, and to a degree this is so. For instance, it is easy to accommodate a large power increase from a liquid-cooled motor and avoid overheating; you simply use a larger radiator. Other aspects of water cooling are, however, not so elementary.

The two major deterrents to proper heat transfer from the combustion chamber and cylinder to the cooling medium (usually water) are deposits and air in the cooling system.

Metallic oxides (e.g. iron oxide, or rust, and aluminium oxide) are formed in the water passages. A deposit 12-thousands of an inch thick will cut heat transfer by up to 40%. In order to maintain optimum heat transfer within the engine, the cooling system should be chemically cleaned in a bath when the engine is stripped down, and when operating should contain an inhibitor that will keep water jacket surfaces clean and free of deposits.

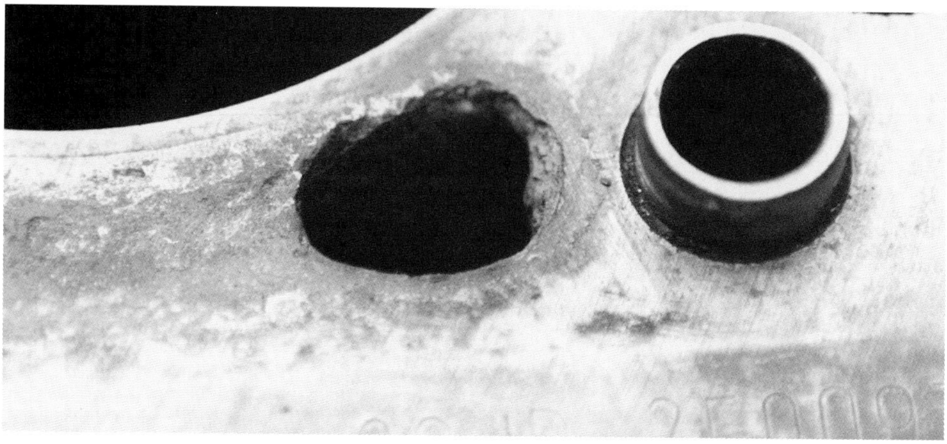

This engine blew a head gasket because of corrosion in water passages. The glycol coolant looked all right, but through a lack of cooling system maintenance had, in fact, turned to acid.

There are two basic types of inhibitors: chromates and non-chromates. Sodium chromate and potassium dichromate are two of the best and most commonly used water-cooling system corrosion inhibitors. These chemicals are toxic, so handle them carefully. Non-chromate inhibitors (borates, nitrates, nitrites) provide protection in either water only or water and permanent anti-freeze systems. Chromates must not be used in systems protected by anti-freeze.

When freeze protection is required, a permanent-type anti-freeze must be used. If a solution of 30% by volume anti-freeze is used, additional inhibitor protection will not be required, but solutions of less than 30% concentration do not provide sufficient corrosion protection, so a non-chromate-type inhibitor should be added. Concentrations over 50% adversely affect heat transfer and should not be used in high-performance vehicles. Note that I do not suggest the use of even low concentrations of anti-freeze in competition engines because of reduced heat transfer.

Ethylene glycol base anti-freeze is recommended for use in all high-performance engines needing freeze protection. Methyl alcohol base anti-freeze should not be used because of its effect on water pump seals and radiator hoses, and because of its low boiling point.

Also, anti-freeze containing cooling system sealer additives must not be used because of the possibility of the sealer plugging the radiator core tubes and various other areas in the cooling system. In fact, a sealer or stop-leak of any description is not recommended; except in an emergency to get you home or to finish the race. Then, as soon as possible, it should be flushed out of the cooling system by a cooling specialist using high-pressure air and water.

Petroleum-derived products such as soluble oil, often used as a water pump lubricant and rust inhibitor, should never be used in the cooling system. A 2% concentration of soluble oil will raise the cylinder head deck temperature by around 10%, owing to the reduced heat transfer efficiency of the coolant. One popular radiator stop-leak contains a high proportion of soluble oil, which is another reason for keeping clear of radiator sealers. When added to water, soluble oil turns it a milky colour.

To prevent corrosion and mineral build-up in water jackets, a mixture of distilled water and inhibitor, or a mixture of distilled water and anti-freeze, should be used, and should be regularly replaced as recommended.

ELIMINATING AIR BUBBLES

The presence of air bubbles in the coolant reduces its heat transfer capacity and the efficiency of the water pump. Air can be sucked into the system through a leaking hose or gasket, and gas bubbles can form in the system owing to localised boiling or water pump cavitation.

In the first instance, air can be kept out by ensuring that the system is free of air or water leaks, and by maintaining the coolant at the correct level. However, a good deal more is involved. At the very least, take care to bleed the cooling system carefully whenever it is disturbed in any way to replace a hose, etc. You must ensure that there are no pockets of air trapped in the block and head, or, in the case of road cars, in the interior heater. A few engines will self-bleed fairly satisfactorily, while others may have to be bled through a factory-fitted bleed nipple. Some, however, require the addition of a bleeder nipple high up on the block right under the deck and/or a bleeder in a high point in the head.

After letting trapped air escape through the bleeders, do not assume that the cooling system is now free of air. With a race engine running a restrictor, start the engine and throttle it up and down for a couple of minutes. The vibration and water surge should get most air bubbles moving to the high spots, so open the bleeders to release any air and top up the radiator.

A road car is a bit different as we also have to worry about getting air out of the heater and from under the thermostat. What we need to do, after initially bleeding the system, is to take the car for a run for about 15 minutes with the heater full on. The thermostat will open in this time and the combination of engine vibration and water surge in the passages should get any trapped air moving either into the high point in the head or into the radiator. At the end of the run, with the engine still running, carefully open the top bleeder to let any air escape. With that done, shut off the engine and allow it to cool. When down to ambient, check the coolant level and top up if necessary.

Even after all these steps, the cooling system will probably contain some pockets of trapped air that will only be eliminated by evacuating the system using a vacuum pump. People accept the fact that refrigeration and air-conditioning systems must be evacuated prior to being re-gassed, and in reality race car cooling systems deserve the same treatment. What is required is to pump all the air out of the cooling system and bring the vacuum down to around 25in Hg. After turning the vacuum pump off, the vacuum should hold at that figure; if it does not, there is a leak somewhere that has to be eliminated. With all leaks rectified, the system can then be completely filled with coolant, free of any air pockets.

THE PRESSURISED COOLING SYSTEM

In the second instance, gas bubbles or steam pockets can be prevented by pressurising the system to the degree necessary to prevent the coolant boiling. Many wonder why it is that we pressurise the cooling system. The boiling point of water is 100ºC/212°F (at sea level), so why is it necessary to increase the boiling point by pressuring the system when most cars operate at around 90ºC (194°F) and most competition vehicles operate at an even lower temperature – usually 70–75ºC (158–167°F)?

First, the system is pressurised to prevent boiling after the engine is turned off. Once the coolant stops circulating, its temperature climbs rapidly from its normal 90ºC (194°F) to something like 110ºC (230°F), way past the boiling point of water at sea level pressure (14.7psi). If the water is boiled each time the engine was stopped, a considerable amount of coolant would be lost, and if the system was not re-bled an air pocket would form.

Second, regardless of what the temperature gauge is reading, the temperature is very high in the water passage around the combustion chamber. Remember that the temperature gauge is only giving a reading of the circulating water temperature, not the temperature of the water around the exhaust valve seat, where the temperature is well above the boiling point of water. To prevent the water around the combustion chamber from boiling and forming a steam pocket, the cooling system has to be pressurised. If the coolant were allowed to boil here, localised heating of the metal would occur, creating thermal stress points that would lead to cracking of the metal.

By pressurising the system using a 14psi radiator cap, the boiling point is raised to approximately 125ºC (257°F) at sea level. As well as preventing boiling when the engine is switched off, the radiator pressure cap also serves to stop gas bubbles in a number of other situations. For example, a road engine may be given a lot of throttle at low engine speed, which will give rise to rapid heating of the combustion chamber, exhaust port and valve area. At low engine rpm the water pump is turning slowly, so water flow is limited and the water pump will not be creating any pressure in the block and head. It is only the radiator cap that stops this sort of localised boiling.

A similar sort of thing occurs when a race car pits. The top of the engine is extremely hot because of all the full-throttle driving out on the circuit, but now, with the engine idling while adjustments are carried out, tyres changed, etc., the engine would boil without a pressure radiator cap.

WATER PUMP CONSIDERATIONS

When water pump speed increases to the peak efficiency speed of 3000–5000rpm, it is not the radiator cap that stops boiling but water pressure created by the water pump. Even many race engine tuners do not seem to understand that, regardless of the radiator cap pressure, a water pump spinning at maximum efficiency rpm will produce a pressure head of around 30–40psi in the engine block and head when water flow out of the head is limited by a thermostat or restrictor plate. This pressure packs coolant around the top of the cylinders and around the combustion chambers to carry away combustion heat and stop an insulating blanket of gas bubbles forming in these areas.

In the third situation, gas bubbles from water pump cavitation have to be tackled on two fronts: water pump design and water pump rotation speed. Even the best of pumps cavitate if you spin them fast enough, but some stock pumps begin to form bubbles at relatively low rpm because of both poor design and cost-cutting manufacturing techniques. In the past, water pump impellers were cast with a closed back, and clearance between the pump body and impeller was kept tight to ensure maximum flow volume. However, in modern times, stock pump impellers tend to be pressed from light metal, which results in an open-back impeller often spaced well out from the pump body. This design will not only pump less volume, which means reduced block pressure, but it will also begin to bubble the water at lower pump speeds.

Race pumps are available for some of the more popular engines; look for one that has a good close-fitting, closed-back, cast impeller. Some so-called race pumps merely feature a stronger bearing and large-diameter shaft, but with a poorly designed body and impeller, flow volume and cavitation resistance may be no better than the stock pump.

If no good race pump is available, you will have to modify the stock pump. It may be possible to modify a closed-back cast impeller from another pump; this could involve searching out old engines in scrapyards. An alternative is to close up the back of an open impeller by welding a backing plate onto it. Then, when the impeller is

Below left: An open back impeller is inefficient, has poor pumping capacity and a strong tendency to cavitate at only low rpm. In fact, what appears to be rust on the impeller is actually erosion owing to cavitation.

Below right: A closed back impeller is much more efficient and far less prone to cavitate. The four holes promote coolant flow to cool and lubricate the seal.

pressed back on to the pump shaft, press it on further to reduce the pump-body-to-impeller-vane clearance.

With that sorted out, we now have to turn our attention to pump speed. In a well-designed pump, water flow will increase proportionally to pump rpm up to about 5000rpm. Between 5000rpm and 7000rpm water flow continues to increase, but the rate of increase begins to taper off rapidly as the water starts to bounce around off the impeller vanes and pump body. So, past a pump speed of about 5000rpm it is costing more hp to move proportionally less water. In a 650hp race engine, a water pump spinning at 5000rpm will be consuming around 15hp; increase the speed to 7000rpm and the parasitic loss increases to 25hp for only a 15–20% increase in water flow.

As pump speed increases, the pump will begin to cavitate, at which point not only does a lot of air get introduced into the system, but flow rates will also plunge by 20–25%. Apart from excessive pump speed and poor pump and impeller design, water temperature and block pressure also influence at what point cavitation will set in. As the average system temperature increases past about 85ºC (185°F), the pump rpm at which cavitation begins decreases. Likewise with block pressure; less block pressure means that cavitation will occur at a lower pump speed. In a race engine this means that we must limit water pump speed and control engine block water pressure, and not concern ourselves only with water temperature.

First, I would look at pump speed. While most pumps tend to work at peak efficiency at 5000rpm, they do vary, so check with the pump manufacturer or do some flow testing yourself. If the pump peaks at 5000rpm and the engine produces maximum hp at 7750rpm, I would fit a big pulley to the pump to underdrive it to 65% of engine speed. Therefore, at 8500rpm, the engine redline, the pump will be turning at around 5500rpm, and down at 6000rpm it will be running at 3900rpm. Basically, I am aiming to keep the pump in its maximum efficiency zone of 3000–5000rpm.

CONTROLLING COOLANT FLOW RATES

Next, I would look at restrictor size, to control the rate of water flow through the engine and radiator, and to get water pressure in the block up around 30psi. Let us say that a restrictor with a 16mm hole is fitted and the block pressure is where I want it, but the water temperature is too cold. The options are to go to a smaller restrictor, to a lower pump speed or a combination of both. If the testing was being carried out in hot weather, I would go for less pump speed to reduce parasitic losses, but I would keep an eye on block pressure, as I would not want it to drop off. If the testing was during cold conditions, a smaller restrictor would be fitted. If that brought the water temperature up, but there was a steep rise in block pressure to over 35psi, I would probably try a slightly slower pump speed combined with an intermediate-size restrictor hole.

However, what if the engine was too hot on the first test with a 16mm restrictor? If going to a 19mm hole did not bring the temperature down and we were losing block pressure, we would have three options: a larger radiator, more pump speed or a combination of both. Assuming that it was a hot day test, I would go for more pump speed with a limit of around 6200rpm. However, if the temperature was way up on a cold day, more pump speed might not get you out of trouble when the weather warmed up. Therefore, a larger radiator might be the way to go.

Some, of course, would just go for bigger and bigger restrictors, or discard the restrictor altogether. This is not a good move; certainly the temperature gauge may come down to where you want it, but consider what is happening inside the engine. With no outlet restriction, water pressure in the block and head will be right down. With lowered water pressure, steam pockets will form in the hottest areas of the engine. The combustion chamber, exhaust valve and piston crown will then overheat, driving the engine into detonation. At this point hp will fall, and if the situation continues the engine will be destroyed.

An old wives' tale states that if you discard the thermostat or the restrictor the engine is damaged because water is flowing through it too quickly to draw off excess heat, or it is flowing through the radiator too fast to give up its heat. This is not so. What causes the engine damage is insufficient water pressure to pack the coolant in tight around hot spots in the engine. Then, any water that hits these hot spots dances about like water droplets on a sizzling barbecue plate without drawing off much heat. As the water boils off, the size of the hot spot grows as a bigger and bigger steam pocket forms.

Really, what we want is reasonably rapid water flow through an engine, as this tends to reduce the incidence of stagnant high-temperature pools. Also, the rapid flow will scrub off gas bubbles as they appear in the hottest parts of the engine before they have a chance to congregate into an impenetrable steam pocket. In fact, the solution to cooling problems is not so much a matter of moving more water through an engine, as moving less more rapidly. This will pay large dividends in more hp and better engine reliability.

COOLANT FLOW PATH

What we have to be thinking about is the accurate delivery of coolant at higher velocity to critical areas where hot spots are most likely to form – around combustion chambers, exhaust valves, spark plugs and the tops of the cylinders. Straight away you can see that we have a problem with the way water travels through most stock engines. Typically, the water pump forces all the coolant into the front of the engine block. Then, in theory at any rate, the water courses towards the back of the block, taking up heat as it goes. On reaching the rear of the block, the water moves up into the head through the large water passages interconnecting the cylinder block and the head. From here the coolant flows towards the front of the head, taking up heat from the combustion chambers and exhaust ports as it goes. The water then exits the engine through the thermostat or restrictor and enters the top of the radiator to be cooled.

However, the theory falls down because, to prevent steam pockets around the tops of the cylinders, we need a series of vent holes, usually 0.060–0.125in in diameter, drilled in the top of the block. While these holes are primarily to vent air and steam that would otherwise be trapped in the top of the block, they also allow the free passage of coolant from the block into the head. Also, as the spark plug bosses are often very close to the surface of the head, it is usual to find a series of holes in the block that direct water flow from the block up around the spark plugs. It should be obvious that all of this coolant flow straight from the block to the head reduces the volume and slows down the flow velocity of coolant past the far-end cylinders and combustion chambers. The result is that the rear cylinders and combustion chambers run too hot, while those of the front run too cool. Therefore, to keep the back cylinders out of detonation we have to run the engine with a less than ideal compression ratio and/or boost and reduced spark advance. All this costs us hp and engine efficiency.

Of course, if we had the time and resources to experiment, we could run each cylinder at a different compression ratio, higher at the front and progressively lower towards the back cylinder. If the race regulations permitted sequential fuel injection and ignition trim control we could trim the quantity of fuel injected into each cylinder, running the hot cylinders richer and cool ones leaner, and likewise with the ignition, we could increase advance on the cold cylinders and decrease it on the hot ones. All this effort would unleash additional hp and improve engine reliability, but the rewards are nowhere near as great as would be achieved by correcting the coolant flow problems.

It should be obvious that as the combustion process takes place in the combustion chambers, which are located in the head, it is the head where we have to focus the cooling effort. Yes, the block does require some cooling, but because by comparison the cylinders have such a huge surface area to give up their heat to the coolant, cooling difficulties in this area are minor. Consequently in a race engine we want to move a lot more water through the head, to take excess heat away from that area. However, if we follow the conventional coolant flow path this also means circulating more water through the block, but as the cylinders are already overcooled this would not be a good idea.

The solution is to change the coolant flow path. Rather than all water flow from the water pump being directed into the block, then up into the head, why not take the bulk of the flow straight from the pump to the head? Such a move reduces parasitic hp losses because we do not have to push bigger and bigger volumes of water through the engine, and frictional losses reduce as we allow cylinder wall temperatures to rise. Therefore, in a road race engine we would usually direct 65–80% of pump volume straight into the back of the cylinder head, and I understand that some tuners have gone to 90% without problems. Of course, in a drag engine the block often gets no flow at all; the block has water in it and the water passages still match those in the head, but all flow goes into the back of the head. Even the stock V8 Jaguar has only half the pump output going into the front of the block, while the other half goes directly into the back of the heads.

One problem area with V6 and V8 engines is the manner in which coolant from the right and left heads comes together in the inlet manifold water passages before exiting through the thermostat or restrictor. Because flow is not the same in both banks – typically there is a 2–3% flow difference – water from the head with more flow will tend to reverse flow into the other head. This not only restricts coolant flow out of that head and causes it to overheat, but because there are two columns of water crashing head on into each other in the inlet manifold passage, there is a lot of flow turbulence. A few hp can, therefore, be wasted driving the water pump harder to force sufficient water volume through the engine to keep it cool. In a race engine the solution to this is to forget about bringing the water out through the inlet manifold, but instead to put a water outlet in the front of each head, if possible positioning it off-centre to favour flow along the exhaust side of the heads. Then, coolant flow from the two heads can be brought together without turbulence through a pipe fabricated in the same style as two branches of an exhaust header blending together at the collector. The coolant flow restrictor or thermostat must be located in the system after the two pipes join.

Regardless of whether we follow the traditional pattern of cooling (with all of the coolant going into the block, then up into the head), or use a split system (with coolant diverging at the pump to both the block and head), we must concern ourselves with the size and number of coolant holes that interconnect the block and head. Obviously, it is only at the rear of the block where we want volume flow of coolant going into the head. All other holes must only be as large as necessary to vent air and steam from under the deck of the block into the head, or to allow metered coolant flow from the block to a specific area in the head. Therefore, we may have flow from the block to purge gas bubbles from certain areas such as from around the bosses for the head studs; we may have flow up around the spark plugs and the exhaust ports. Note here that we are talking metered flow, not bulk flow. For example, in the small-block Chev, a single 0.090in hole drilled below the centre exhaust ports feeds enough coolant to lower the cylinder head deck temperature in that region by 70ºC (158°F)!

However, what do we regularly find in production blocks and heads? Frequently these coolant transfer holes are around $^1/_2$in in diameter, and at times larger. Usually they are so large not for the sake of coolant transfer, but to make it easier to get the core sand out at the end of the casting process. Naturally, big holes in the top of the block and correspondingly big holes in the head mean that a lot of coolant entering the front of the block will flow straight up into the front of the head, and only a small volume will course to the back of the block to cool and purge steam from around the end cylinders. Therefore, the rear cylinders will overheat and there will be reduced coolant transfer in this area from block to head to purge gas bubbles, cool spark plugs, etc.

HEAD GASKET COOLANT HOLES

Manufacturers attempt to rectify this undesirable situation by punching small holes in the head gasket where they desire only limited coolant transfer, and they punch larger holes where they want more flow. This works well enough on the road, but at the much higher flow rates in a competition engine there is usually an excessive amount of coolant transfer between the block and head, particularly at the front of the engine. Remember, too, that manufacturers have to make allowances for poor maintenance over a long period, so the holes may be larger than necessary to provide adequate flow even when partly blocked by rust and scale. Of course, we do not see this sort of problem in a competition engine, so the transfer metering holes can be smaller.

It is at this point that we come up against a problem in that the manufacturers of heavy-duty head gaskets generally punch their gaskets with the same hole size (or larger) as the original manufacturer. There are a number of ways to get around this problem. Preferably, if there are large round holes in the block, tap them out and screw in soft aluminium pipe plugs. If the holes are smaller, gently knock in a short length of aluminium rod; obviously these plugs must not protrude above the block deck. After fitting them, drill the appropriate-size metering holes, which will usually be in the range of 0.060in to 0.125in. Note that in competition engines using a restrictor in place of a thermostat, the bypass hole is left completely blocked.

Where the transfer holes are an irregular shape they can be cleaned with a file or die grinder, then plugged with a silicone sealant such as Silastic. Coolant metering is taken care of by a length of aluminium rod, pre-drilled to the required size, being inserted into the silicone sealant. I prefer to run a drill through the sealant after it has cured, then force in the drilled rod, but some like to hold the pre-drilled rod in place, then pump sealant in around it to plug the core hole.

Clearly, these small metering holes in the block must match the holes in the head gasket to allow the venting of air and the flow of coolant into the head. Some tuners just

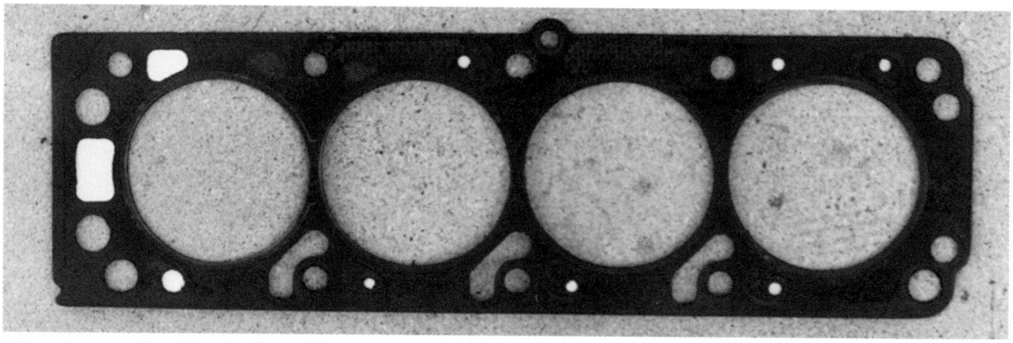

The six small white holes allow only limited coolant flow from the block into the head, thereby ensuring that the majority of flow is through the three large white holes towards the rear of the block.

punch large holes in the head gasket, but this is a job that will have to be repeated every time the head gasket is replaced. What is preferable, if solid plugs have been inserted into the block deck, is to use the head gasket as a template and drill each plug to match with the corresponding hole in the gasket. Naturally, where the deck has been plugged with sealant and a pre-drilled rod has been inserted, it may be necessary to punch a matching hole in the head gasket.

Because we want the bulk of coolant circulation from the block to the head to take place through the large holes at the rear of the engine, we do not want them restricted in any way. With this in mind it is a good idea to smooth them out and match them up using a die grinder or a file. In a conventional flow system this encourages coolant flow right to the back of the engine, rather than flow short-cutting through transfer holes towards the front. With a split-flow system a good flow of coolant in this region is also

The coolant hole on the right is blocked by casting slag. This must be ground out to encourage coolant flow from the back of the block up over the far end combustion chamber and exhaust port.

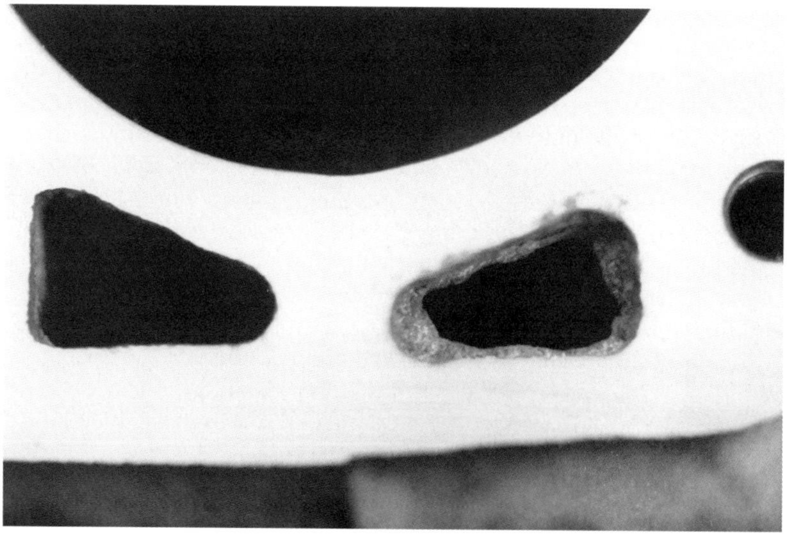

important; first, to cool the top of the rear cylinder adequately and, second, so that the majority of the hot water from the block goes into the head at this point to mix with cooler water entering the back of the head, which has come straight from the water pump. This serves to balance engine temperature front to rear.

COOLING HIGH-OUTPUT ENGINES

While the modifications we have just discussed will provide adequate cooling in the majority of engines, we have to accept that as we push forced induction engines to very high power levels it may be advisable to scrap the stock type cooling system. This means changing to an external competition water pump and completely altering the coolant flow path through the engine block and cylinder head. The aim being that coolant flow should be equally divided between cylinders, and then after cooling just that single cylinder the coolant is returned to the radiator.

How this is achieved will vary from engine to engine. I will describe the basic concept for an in-line engine with four to six cylinders, so if you have a Vee or flat engine, follow the same pattern, dealing with each bank as a separate engine. The simplest arrangement is to introduce all coolant flow from the water pump into one side of the block. When the cylinders are siamesed we would have only a single inlet in the centre of the block. However, as I will explain later, if the cylinders are not joined and we are unable to seal the water passages between them, we will require two or three inlets respectively for 4-cylinder, and 5- or 6-cylinder engines. The main thing we want to do is get maximum water movement up close to the top of the block, and we want to equalise cylinder to cylinder temperatures as much as possible. Generally we would knock out existing welch plugs and tap the holes to accept suitable water fittings. Obviously, because we have got rid of the stock water pump we have to cover that hole in the front of the block. Also, it will be necessary to fill the large coolant flow holes in the deck at the other end of the block.

Next, we have to decide on a suitable flow path across the head. Note that we are not talking about coolant entering the head at one end and then exiting at the opposite end, after

Take a few moments to study both blocks. The one on the left is stock while the other shows what BMW did to make it suitable for 1,000hp during the Formula 1 turbo era. The main change is the addition of coolant passages at the ends of the block and the elimination of six small steam holes between cylinders.

having flowed over every combustion chamber. No, what we want in a high-output race engine is for the bulk of the coolant to be divided up and enter the head along one side then flow across the head, exiting at several points into a water manifold on the opposite side.

Some race engines actually split water flow with about 70–80% going into the head and 20–30% going into the block. Using this more complex arrangement, which is really only necessary when we exceed about 350hp per litre, a water manifold is attached to both sides of the head. An obvious advantage of such an arrangement is that if in a 4-cylinder engine we have a water manifold with four outlets, the size of those outlets can be adjusted to equalise the temperature of each combustion chamber. Also, the individual ports of the water manifold do not have to terminate where they attach to the side of the cylinder head. Rather, what we can do is extend each port right into the water jacket such that the coolant is blasted into the main hot sport; usually around the exhaust valve seats. From there the water will course over the combustion chamber, then exit the head into the exhaust water manifold.

AN ALTERNATIVE COOLING METHOD

When we opt for the simpler arrangement of flowing all water into the block, and from there into the head, we are not able to exercise such precise control of flow across each combustion chamber. Of course, if we have an engine management system offering individual cylinder fuel and spark trim, small temperature differences between combustion chambers can be compensated for, but in an endurance race engine where fuel consumption is a concern, injecting additional fuel into a 'hot' cylinder isn't a satisfactory solution.

Fig. 21.1. Revised coolant flow path for a high-output engine.

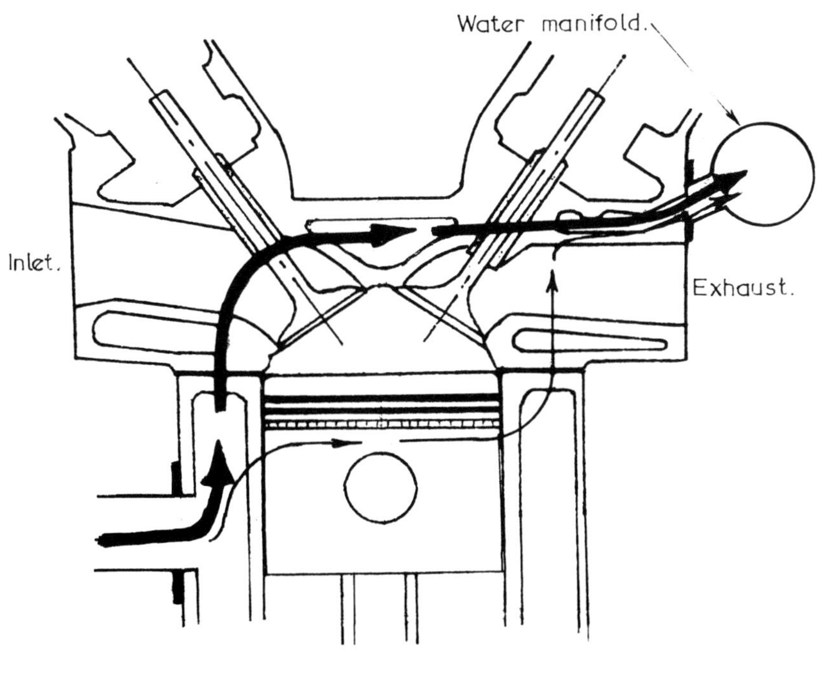

What we have to decide early on is, on what side of the head do we want the water to exit? If we decide it should exit on the exhaust side, then most flow (around 65–80%) should travel from the block into the head along the inlet side. This will ensure that, for the most part, water will move from the block, straight up into the head, turn 90º and flow over the combustion chambers, then past the exhaust ports whereupon it enters the water manifold and returns to the radiator (*Fig. 21.1*).

To achieve that flow path we may not have to do much more than punch bigger holes in the head gasket along the inlet side. In all likelihood, coolant holes in the block and head will already be larger than necessary to achieve the flow rates required, and we will have to balance flow across each combustion chamber by adjusting coolant hole sizes along the inlet side of the head gasket. This may sound simple enough but usually involves long dyno sessions with temperature probes stuck down onto the exhaust side of all combustion chambers, testing head gaskets with various combinations of different hole sizes to even out the temperatures. We can do the same sort of thing with track testing, but that is even more tedious.

Some tuners choose, if possible, to have the coolant exit on the exhaust side of the head. It is frequently more convenient to locate the water manifold on that side, but usually the reasoning is that inlet ports should be kept as cool as possible to maximise charge density, and flowing coolant from the exhaust side, across the combustion chambers, and out past the inlet ports gets the inlet hotter. While there is merit in that argument there are other issues which may, depending on the particular engine and its hp level, be a more pressing consideration. Another view is that it is at times possible to achieve more control on the direction and quality at flow around the exhaust valve seats by having the water enter the head on the exhaust side. Then, because the exhaust valves will be cooler, they will add less heat to the inlet charge. Also, being cooler, they are less likely to cause detonation after combustion has begun.

However, regardless of whether we settle on the coolant exiting on the exhaust or the inlet side, that decision will have a bearing on what side of the block the coolant should enter. The arrangement illustrated in *Fig. 21.1* with the inlet and outlet on the opposite side is correct; but we could still run into problems if the cylinders were non-siamesed. For example, if the engine had four cylinders and a single water inlet in the centre of the block, there would be minimal water flow around the outside of the end cylinders – most flow would be through the three gaps between the cylinders. If we had two water inlets located at about a quarter distance and three-quarters distance, flow around the cylinders would be evened up, but there is a better way. Depending on where openings are located, it is often possible to block the gaps between cylinders with silicone sealant. We don't want to convert the block to being siamesed. Rather, leave the top of the gap open down to about 25% of engine stroke, and close the gap from there to the bottom of the cylinder. This forces flow around the end cylinders, and the small gap helps cooling around the top of the bore. Distortion of the bore is reduced, ring seal is better and hp rises slightly.

Usually it is possible to block those gaps, even if you have to add a plastic extension hose (use 4–6mm id) to your silicone gun. What, though, if you can't? Well, as pointed out, two inlets would be the way to go (three inlets for 5- and 6-cylinder blocks). If we were to put additional inlets in the opposite side, we would simply compound cooling problems around the end cylinders.

With a block that has siamesed cylinders we can improve cooling and reduce bore distortion, if the siamesed walls are sufficiently thick, by adding a small coolant hole in

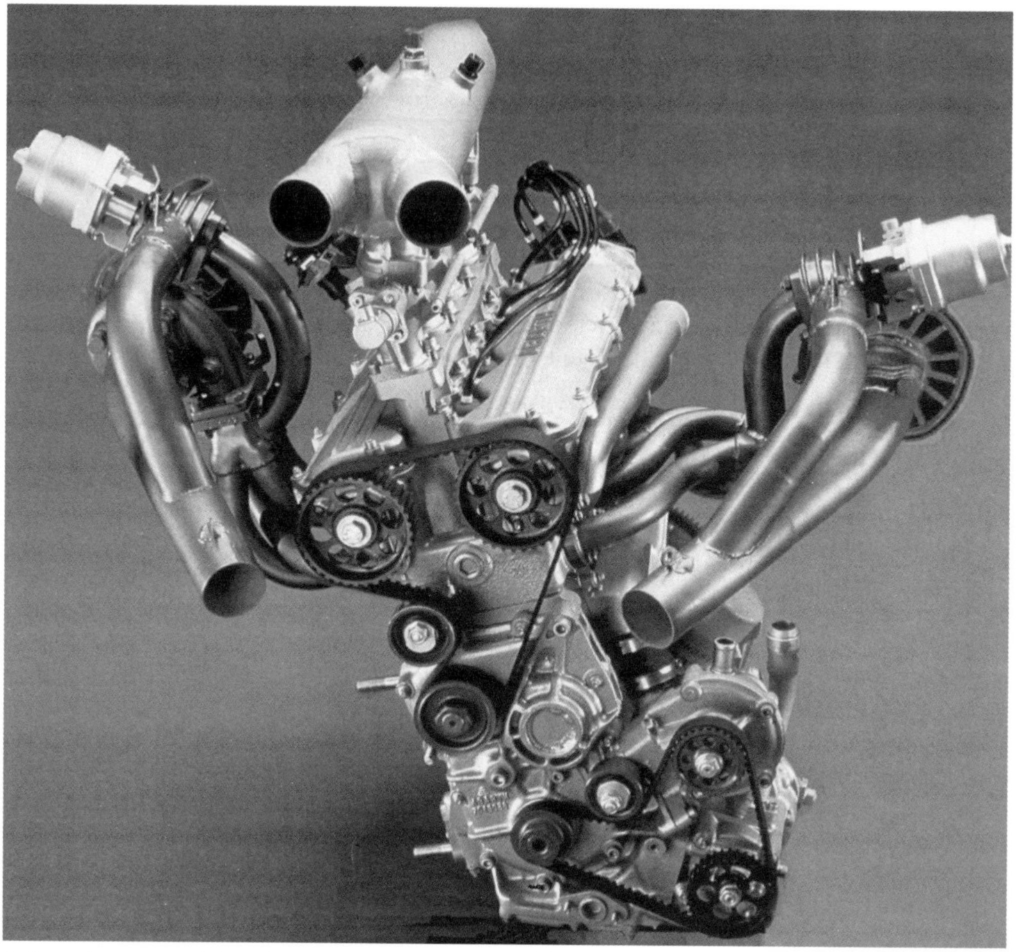

A Lancia sequential turbo competition engine. Note the water manifold along the left side of the head.

between cylinders. This is precision work, so the block must be set-up on a mill. Obviously, if the hole goes off course the block becomes scrap! Basically, the web joining the cylinders has to be around 6mm or more wider. If it is 6mm you will be restricted to a 2mm hole. The idea is to set the mill up in order to leave a minimum wall of 2mm either side of the hole. First, you drill bigger access holes through one side of the block. Later, these will be tapped and plugged with small pipe plugs. The coolant hole will be located as high as possible up under the deck of the block. The idea of this is to get rid of any steam that may become trapped there, as well as move a little coolant through that area which is the hottest part of the cylinder.

RADIATOR MAINTENANCE AND DESIGN

The actual heat exchange between the coolant and air takes place at the radiator. Bugs and debris that restrict air flow through the core should be cleaned out to maintain cooling efficiency. Cleaning out balls of rubber thrown up off the track is always a problem. The easiest method seems to be to soak the radiator, after plugging all openings, for a day or two in a tray of solvent. Then, direct compressed air at the back of the radiator. Any balls that will not budge can be carefully picked out. A large tray of solvent is both a health and fire hazard, so be very careful. Any fins which are bent must be straightened, as bent fins present as much impediment to air flow as trapped rubbish. The radiating efficiency will be improved further if the radiator is regularly repainted with matt black paint. This will also serve to increase the life of the core, by reducing the effects of external corrosion.

However, when it comes to a competition radiator many more factors are involved, such as whether it will be constructed of copper or aluminium, how many rows of coolant tubes it will have, what fin count will be most suitable, etc.

Track debris and bent fins both restrict air flow through the radiator. Debris should be picked out and bent fins should be straightened using flat-blade tweezers.

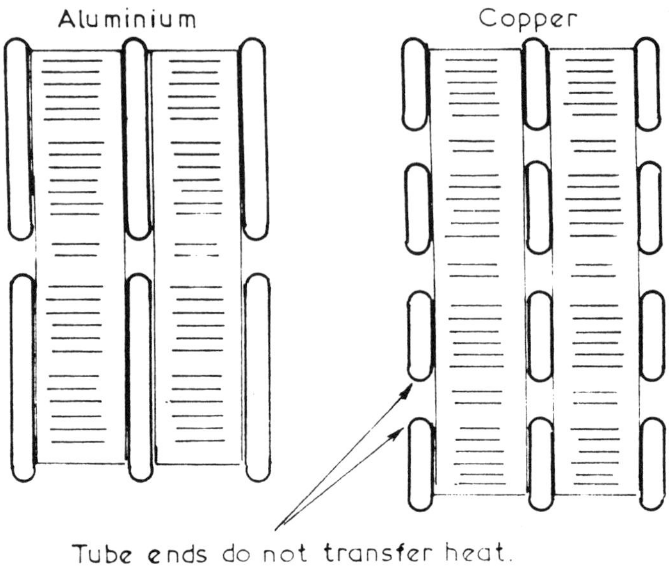

Fig. 21.2. Aluminium radiator tubes are twice as wide as copper tubes.

For many people, the decision between a copper or aluminium radiator is based on possible weight savings, but this is really a bit of a nonsense as most weight in a radiator consists of the volume of the coolant it contains! Copper conducts heat better, but in a crash aluminium will show more resistance to damage; also it is less likely to be damaged by stones or debris thrown up off the track by other competitors.

Really, the biggest difference between the two is their relative strengths. Copper is fairly weak, so to avoid coolant tubes exploding from water pressure, the tubes must be kept quite narrow. Aluminium on the other hand is much stronger, so the tubes can be twice as wide. At first glance this may not appear to be such an advantage, but if you take a look at *Fig. 21.2* you will understand where the gain lies. The water flow characteristics of a four-row copper radiator are pretty much the same as a two-row aluminium core. However, when we consider heat transfer area, the aluminium tubes easily win to the tune of up to 20% gain. The fins, which transfer heat to the air passing through the radiator, only contact the flat sides of the tubes. The rounded ends of the tubes have no finning, so there is no heat transfer of any significance from this area. Therefore, with only half the number of tube ends, the aluminium radiator will give up more heat to the fins.

Another big difference between the radiators installed in road cars is with regard to their construction philosophy. Radiator tubes are much more expensive than fins, so road cars tend to skimp on the number of tubes and attempt to compensate with a very high fin count (*Fig. 21.3*). At low speeds this presents no problem, but at race track speed all those closely spaced fins massively reduce air flow through the radiator. Even with good ducting, which should force the air entering the nose cone to flow through the radiator, a stand-off results: a high-pressure column of air builds up in front of the radiator, and though some trickles through there is insufficient flow to take up heat rapidly enough from the radiator fins. It is a bit like pouring oil down a funnel. The rate of flow is

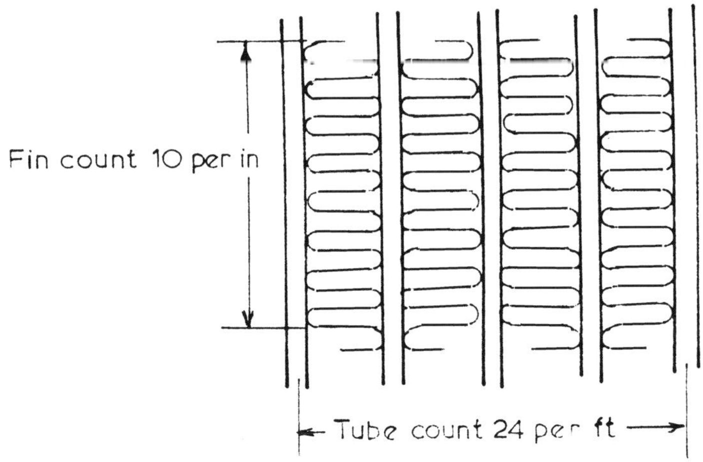

Fig. 21.3. Radiator tube and fin density affect radiator performance.

controlled by the size of the hole in the bottom of the funnel; the size of the hole at the top and the amount you pour into the funnel, even to the point of overflowing, really does not change the rate of flow through the funnel by very much.

The only way around this is to construct a radiator with many more tubes – but not side by side, for unless we can make the radiator wider this just adds to the air flow problems. What we usually need are more tubes placed one behind the other. Typically, a road car will have a two-row, or perhaps a three-row, copper radiator, while a race car will maybe have a radiator that is four or five rows deep, and to assist air flow the tubes may be spaced a little bit further apart across the front of the radiator. So, whereas a road radiator may have the tubes closely bunched at, say, 29 tubes per foot with a fin count as tight as 17 per inch, a competition radiator may space out the tubes to 22 per foot with a fin count opened up to 10–12 per inch, or even as low as 8 fins per inch in dirt-track cars where there is inadequate protection to prevent dirt getting in and blocking air flow.

If you have a close look at the fins you will better understand why a high fin count is so bad for good air flow through the radiator. Rather than having a smooth flat face, the fins are very finely louvred across their surface. These tiny louvres create turbulence to increase their radiating efficiency, and as they are packed more closely together it is this turbulence across their surface that cuts air flow through the radiator.

When it comes to correcting a cooling problem, racers often go about it the wrong way; they adopt the road car radiator mentality of closely bunched tubes and high fin counts. When that fails to work they may exacerbate the problem by moving to an even higher fin count! Because the overheating just gets worse they drive the water pump faster, which in turn increases parasitic losses at best, or else runs the pump into cavitation. Alternatively, they may throw out the restrictor, so water pressure in the block and head goes down (remember that we want about 30psi here), so big steam pockets develop, the engine detonates and loses power at best, or perhaps lunches a head gasket or piston.

Really, what we want is a large number of radiator tubes but a low fin count. This costs more money, but you may save a great deal in the long run. If the radiator brings the temperature down too far, run the pump slower, but fit a restrictor with a smaller

Louvres on the fins increase air turbulence across the surface of the fins, thus raising their heat-shedding capacity.

orifice to get block pressure back up to 30psi. Less pump rpm will cut parasitic losses, and reduced air flow resistance through the radiator may see a speed gain on fast straights.

DOUBLE-PASS RADIATOR

Another way we can increase the heat dumping ability of the radiator is to build it as a double-pass unit (*Fig. 21.4*). As you can see when a splitter plate is welded in one tank, the effect is to force all the coolant through half the number of tubes, but over twice the distance as the water flows from one side to the other side across the top half of the radiator, and then back again through the lower half. Providing the water pump is up to the task, a two-pass radiator will dump 15–20% more heat than an otherwise identical conventional single-pass radiator.

Fig. 21.4. A double pass radiator halves the number of tubes and doubles their length.

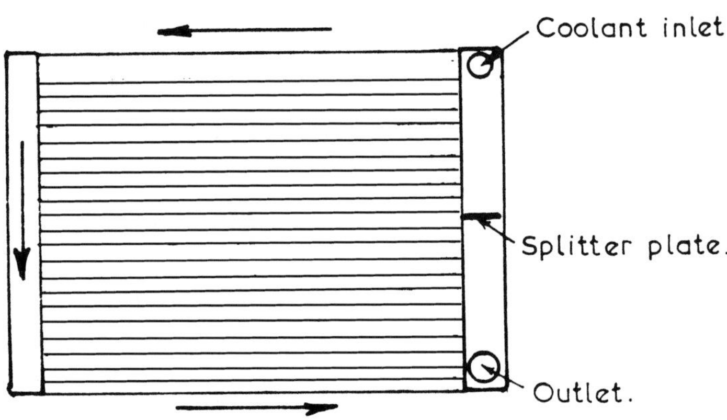

This happens in spite of the flow rate through the tubes dropping 25–30% owing to increased friction and water turbulence. What then accounts for the increase in efficiency? Obviously the water travels twice as far, giving up heat as it goes. However, the real gain comes as a result of increased water turbulence in the tubes. At low flow rates the water clinging to the tube walls pretty well stalls and, in effect, becomes an insulator which reduces how much heat is shed by the main body of coolant flowing through the centre of the tube. When the flow rate is pushed up by effectively halving the number of tubes, water turbulence increases and this has the effect of breaking up and scrubbing away this stagnant insulating blanket clinging to the tube walls. This improves the heat transfer rate from the coolant to the radiator.

REGULATING COOLANT TEMPERATURE USING A THERMOSTAT

A high-temperature thermostat, which maintains the water temperature at 88–95ºC (190–203°F), is usually fitted to production cars, but that is not the temperature for best power. The 88ºC (190°F) thermostat is fine if you want good heater efficiency in winter, and the higher temperature may help cruise fuel economy and help the oil boil off contaminants, but apart from that there is no reason for its use.

Some feel that high engine coolant temperature in the range of 90–100ºC (194–212°F) is good for power. The theory is that anything lower just takes away heat energy, which we should be using to push the pistons down and produce more hp. To an extent this is true at low engine speeds and small throttle openings, but generally in a full-throttle high-rpm dyno test, a production engine will make more power at coolant temperatures in the 70–80ºC (158–176°F) range. Only dyno testing will confirm what is ideal for a particular engine, but in my experience I rarely see an engine that makes more

The thermostat on the right is a bypass type. When it opens, the 'hat' closes the engine bypass port, ensuring that all coolant actually passes through the radiator. Never replace this type with an ordinary thermostat.

power running higher than 84ºC (183°F) or less than 68ºC (154°F). I have seen engines gain 2–3% hp when the temperature was dropped from 90°C (194°F) to 70ºC (158°F); conversely I have seen gains when the temperature was pushed up from 65°C (150°F) to 80ºC (176°F). Formula 1 engines are now operating at 105–110ºC (221–230°F), but remember they have had huge sums expended on them to keep the entire inlet tract at not much above ambient temperature. Also, they have a horrible slow-burn combustion chamber, basically the four holes in the piston crown for valve clearance plus a 'flame bowl' under the spark plug, demanding ridiculous amounts of ignition advance approaching 50º before TDC.

The water temperature in a road car can be lowered by fitting a cooler thermostat, providing that the radiator is large enough. Whenever a thermostat is replaced, use the correct type. Many modern engines use a dual-function bypass-type thermostat. When open to allow water flow out of the head into the radiator, this type of thermostat also closes off the engine bypass port. If an ordinary thermostat is fitted, this port will remain open, thereby allowing a large volume of very hot water to circulate in a closed loop within the engine. This may lead to engine overheating because of the volume of hot water from the bypass mixing with coolant from the radiator and significantly raising its temperature as it flows into the block.

RADIATOR COOLING FAN

The fan fitted to most cars consumes a good deal of power, which could otherwise be used beneficially to improve performance or reduce fuel consumption. Generally, a road speed of 25mph and above will provide sufficient moving air to cool the engine. Therefore, a fan will not normally be required in competition vehicles.

If a fan is required it would be of benefit to fit an electric unit, which can be an engine-saver on rally cars. You never know when you are likely to be bogged or when you are going to come up against a greasy half-mile climb in a rally; either occurrence would probably cause coolant boiling in the absence of a cooling fan.

Liquid-cooled vehicles with rear-mounted or side-mounted radiators will almost certainly require a larger cooling fan and also a separate expansion tank connected to the radiator. Some engines have a tendency to overheat if the radiator header tank is not completely full in order to give proper water flow down through the radiator core; an expansion tank ensures that the top tank is always full.

If a front-mounted radiator is installed on such cars, the fan can be removed and you can be sure of maintaining the correct coolant temperature under competition conditions. This is a simple modification on the Mini, particularly if of the Clubman body design.

Whenever an electric fan is fitted, it should be mounted behind the radiator to avoid blocking off air flow into the radiator. Testing has shown that a fan mounted in front of the radiator has about 80% the efficiency of one mounted behind the radiator. Also, remember that the fan may have a current draw of around 10 amps, so be sure that the alternator can handle the additional load; modern race engines place a heavy demand on the electrical system, so do not just assume that everything will be fine. I have seen engines shut down when the fan was switched on because the ECU could not handle the electrical surge. I have also seen engines lose a lot of power and come close to a meltdown because the fuel pump slowed and fuel pressure at the injectors dropped.

Some drivers run the electric fan all the time to get sufficient cooling, but this often just indicates that a bigger radiator with better air flow is needed. An exception is for race cars running with a grill closed off to improve the car's aerodynamics. Others use a thermostat control to switch on the fan at a pre-set temperature. Personally, I feel that this just complicates things. I prefer to see the driver operate a simple switch to bring the fan into operation only when it is required. Then, if he is given instructions such as 'fan on on long boggy hills', 'fan on under yellow', fan on into the pits', etc., he will not have to worry about studying the temperature gauge in such situations and make a decision as to when the fan should go on.

HELPING AIR FLOW THROUGH THE RADIATOR

Another situation when an electric fan will provide an improvement in cooling is when air flow through the radiator is poor owing to excessive pressure build-up in the engine bay. Really, we need to work to exhaust air out of the engine bay to get air pressure down as low as possible behind the radiator. However, while a solution is being found an electric fan may provide a short-term fix. Reducing air flow under the front of the car will help, as will opening holes into the wheel tub area. At times the problem can be traced back to an inappropriate bonnet air scoop, which may be providing good air flow into the engine air intake, or cooling air flow over the turbocharger, but which is 'over stuffing' the engine bay with air.

A duct forces all air entering the below bumper opening through the intercooler, a/c condenser and radiator. A secondary role is for the duct to steer air entering the small above-bumper bonnet opening across to the right headlight and into the engine air intake.

Of course, every effort must be made to ensure that all the air entering at the front of the car actually passes through the radiator; we do not want any air bypassing it by going over, under or around it. This means blocking off any openings around the radiator, or else constructing a duct from the front of the car that seals up against the radiator's edges.

THE BENEFITS

To some, no doubt, all of this discussion and concern over cooling may appear to be over the top. However, let me remind those folk that during the Formula 1 turbo era the major problem confronting teams, even allowing for their vast resources, both monetary and technical, frequently related to engine cooling. While readers may not be pushing the boundaries to the same extent as the F1 crowd, the fact remains that when we generate very high hp levels out of relatively small mass produced stock based road engines, we are moving a considerable distance away from what the manufacturer's powertrain engineers ever envisaged, or made provision for. Consequently, if we are to attain such power outputs with relative reliability we have to spend much in the way of time and resources to take care of engine cooling.

The other aspect is, if we are ever going to attain big hp, then to some extent that outcome will be attainable only when we equalise, as much as practicable, combustion chamber and cylinder temperatures. Until that is achieved, there will always be at least one cylinder giving less than 100%. And if we don't have individual cylinder engine management capabilities, the situation is even more depressing, as the cylinder which is the detonator will dictate just how much boost, fuel and spark we give all the other cylinders.

Chapter 22

Making the factory forced induction engine better

The world's motor manufacturers undeniably build some brilliant cars. However, all car enthusiasts are individuals, and I might add that, in truth, some are engine enthusiasts rather than car enthusiasts. Whatever, as individuals we have our own ideas of what attributes are important, and what is absolutely essential in a car. Conversely, there are other areas of the car, while being a prime concern for the masses, may not even rate as necessary for you. Consequently, what modifications one individual may rate as essential in his quest for a better car may meet with scorn from another.

Some of these differences, of course, relate to the end use of the vehicle. If the car is your home for 500 miles each week, then it obviously cannot be devoid of creature comforts. It will have to be reasonably quiet at normal road speeds. The suspension must have some compliance, not jarring your body over every little join in the surface. The engine will have to be reliable and not be too thirsty, nor require a diet of special fuel. Finally, the car will have to be fun to drive quickly, without being irritating or frustrating at cruise or in slow traffic. In effect, what you want is a car that provides the thrills and pleasure of something like a Porsche 911 or 928 at an affordable price. Plus, just as importantly, rather than being a product of your wealth, it is a creation of your personal skills and imagination.

This chapter is about creating an engine for just such a car. It is not about building a narrow focus engine for a race car, or for a semi-race car that is only driven for a few miles on twisty roads at the weekend. With those ideas in mind we must first carefully evaluate our engine to determine where we can obtain the best performance increase for the least amount of money.

ASSESSING WHERE MODIFICATION SHOULD BEGIN

Most factory forced induction engines have quite efficient 4-valve cylinder heads. Consequently, you are not going to find big performance gains from simple cylinder head modifications. Also, most enthusiasts want to keep away from opening up the

Most factory forced induction engines are fitted with 4-valve cylinder heads that flow very well in stock form.

engine until it is actually necessary – when the engine needs an overhaul or all other avenues to gain performance have been exhausted. There are exceptions, such as GM's family of blower engines based on the original Buick V-6. Being a 2-valve pushrod engine it responds very well to basic cylinder head work, but even so it is unlikely that this would be the first place you would go looking for more hp.

The same applies to camshafts. Turbo road cars will occasionally respond to revised cam lobe profiles. However, this would be a modification that you would get into only after all other performance options had been fully exploited. Well before that move, providing the engine was a twin cam unit, we would lengthen the cam overlap period by retarding the exhaust cam and advancing the inlet cam. However, factory blower engines are quite different in this respect, and very early on we may resort to cams with a sportier lobe profile. This is especially the case with an engine like the V-6 Buick. It has only one cam, and replacements are cheap. With 2-valve heads it doesn't breathe especially well, so help from more cam duration will certainly be desirable. However, a hotter cam will have an adverse effect on fuel consumption around town. Also, we will have to choose quite carefully as a cam in combination with other breathing modifications could tip us over budget if the computer can't cope, or the injectors are unable to flow sufficient fuel.

This gets us back into traditional, basic, modification territory – the inlet and exhaust system. In the past engines were noisier, and once under way, even at quite low speeds, cars were subject to the additional noise from air leaks around doors and glass. Consequently, vehicle manufacturers did not have to be so paranoid about silencing noise from the exhaust and inlet pipes. As build quality improved, consumers got fussier and near silence sold cars. On top of this, governments imposed ever more stringent noise regulations. These pressures have forced car makers, producing cars for the masses, to develop very quiet induction and exhaust arrangements. As a result, many performance engines have been very effectively nobbled.

MODIFYING THE NISSAN SYLVIA

Enthusiasts generally do not have the same disdain for noise and in many lands, once a vehicle is registered, noise laws become more liberal, thereby opening the way for excellent power increases to be won quite easily. As an example, take the Nissan Sylvia S15 (200SX) turbo. The basic car is quite good, but even with six well-spaced gears it is handicapped by its weight and only 197hp from the 4-cylinder 2-litre engine (there is a 248hp version available in some markets). Also, unlike its big brother Skyline GT-R, the intercooler is quite modest and, as well as being unable to keep inlet charge temperatures down, it creates quite a deal of flow restriction. Obviously, a larger intercooler is an expensive addition, but it was reckoned that, if the desired 50% power increase was to be attained while maintaining perfect road manners (with no computer remapping), this would have to be part of the inlet modification package.

So, an oversize intercooler was constructed with a core measuring 620mm x 280mm x 80mm – about as big as can be easily fitted in the snout of the Sylvia. Upstream of the intercooler the stock airbox and air filter was ditched in favour of a big K&N cone filter clamped to a bellmouth attached to the stock air snorkel. In turn the K&N was boxed in using aluminium sheet to ensure that only cool air at ambient temperature was drawn into the turbo.

On the exhaust side, the entire system aft of the turbo was replaced. Stainless steel tube 3in in diameter was chosen for the dump pipe section from the turbo to the high-flowing catalytic converter. Because the CAT puts a lot of heat into the exhaust gas, increasing its volume, it was found that 3½in pipe was required for the rest of the system, which ended with a stainless steel resonator. This was a straight-through unit with a one-piece perforated tube wrapped in stainless steel shavings.

The Sylvia 200SX is strangled by this tiny and tortuous air box duct.

Table 22.1 shows the result of testing on a eddy current wheel dyno. At the wheels the stock engine produced a maximum of 169.2hp at 6200rpm, which was a bit higher than expected. With all the inlet and exhaust modifications, this rose to 253.6hp at 6950rpm, still on 98RON unleaded. Not quite the hoped for 50% increase, but pretty close. Very impressive were the low speed hp gains which began just past 3000rpm. Even though boost wasn't fiddled with, boost pressure rose from 9psi stock to just over 13psi owing to the free flowing inlet and exhaust. However, as testimony to the intercooler's increased charge cooling capacity, the knock sensor did not detect any detonation.

Table 22.1. Nissan Sylvia 200SX dyno testing

	Horsepower	
Rpm	Stock	Modified
3000	93.9	93.7
3500	114.0	128.7
4000	124.1	139.5
4500	135.4	179.7
5000	146.2	206.5
5500	166.7	228.2
6000	168.2	239.1
6250	169.2	243.8
6500	165.6	249.4
6750	158.2	252.1
7000		253.6
7500		244.1

On the road the engine was, as expected, every bit as tractable as the stock unit. In fact, it seemed to be much more responsive, and easily out-accelerated other 2-litre turbo cars, and many well over 2 litres. With so much extra power on tap, and so easily accessible, it was speculated that fuel economy would nosedive. Such wasn't the case. In fact, cruising on motorways, there was an improvement of just under 8%.

SUBARU IMPREZA WRX MODIFICATIONS

Some other turbo engines do not provide such easy power gains. At times the stock turbo is simply too restrictive. This was the affliction of the Subaru Impreza WRX from the MY97 model, when Subaru switched from a Mitsubishi TDO5 turbo to a smaller TDO4 turbo in an endeavour to reduce lag and increase the car's appeal to a larger group of prospective buyers. Consequently, some looking for more performance change to the larger IHI turbo fitted to the more powerful Sti version. However, with stock gear ratios and taller diff ratios this produces a good deal of lag (the Sti has closer ratios and very low ratio diffs). Others, though, revert to the earlier stock turbo which gives a good power increase without so much lag. Unfortunately, this isn't a straight swap, and when the necessary machining and other modification costs are added in it becomes an extremely expensive conversion for the modest performance gain. For many, this means that they must work within the constraints imposed by the small turbo.

Table 22.2. Subaru Impreza WRX (MY99) dyno testing

				Horsepower			
Rpm	Test 1	Test 2	Test 3	Test 4	Test 5	Test 6	Test 7
3000	69.8	73.7	75.5	75.0	79.8	82.5	92.7
3500	97.2	100.6	106.1	105.7	112.6	117.3	126.2
4000	124.7	131.5	138.7	140.2	142.5	148.1	154.0
4500	141.3	147.5	158.2	160.1	167.6	171.9	173.8
5000	147.8	155.9	166.3	170.0	176.3	178.4	179.8
5300	150.2	160.9	171.6	173.7	182.4	183.3	183.7
5700	146.3	158.6	175.0	177.2	183.0	184.4	186.8
6000	142.2	156.1	172.3	173.6	180.4	180.9	182.0

Test 1 – Stock with 95RON unleaded fuel.
Test 2 – K&N pod filter with cold air supply, 13psi boost and intercooler water spray.
Test 3 – As above plus 3in exhaust.
Test 4 – As for Test 2 plus 4in competition exhaust.
Test 5 – As above but with twin 3in competition exhaust after CAT.
Test 6 – As above with 98RON unleaded fuel.
Test 7 – As above with reprogrammed ECU.

Again, the obvious power gain areas would be the inlet and exhaust. Also, reprogramming of the ECU would be investigated to see if any additional hp could be found there. Table 22.2 sets out the results. After getting a set of baseline figures for the

When Subaru downsized to a TD04 turbo, performance gains became more difficult to realise, particularly if the smallish stock intercooler was retained.

stock engine, the first change was to rig up a full intercooler water spray to help compensate for the inefficient turbo. Following that, we got rid of the restrictive airbox. In its place a spun aluminium intake trumpet was fitted in the inlet snorkel and flow meter, and a big K&N pod filter was clamped to this bellmouth. The filter was boxed in using aluminium sheet to ensure air at ambient temperature was supplied to the turbo at all times. With improved inlet flow, maximum boost rose almost $^1/_2$psi to match what the factory claims to be the standard boost figure. Overall, there was only a small hp increase, but the improved breathing would become more apparent when the bigger free flow exhausts were tested.

The first exhaust tested was an off-the-shelf mandrel-bent 3in unit. The dump pipe wasn't anything fancy, having only a short tongue about 100mm long to divide turbine and wastegate exhaust flow right at the turbo exit. The exhaust was quite noisy at idle and low speed, but was acceptably loud at cruise and under full power. The maximum power increase was around 10% right near the top of the power band. However, the small turbo was really stopping the engine, with boost falling away once past 4500rpm.

In Test 4 a 4in mandrel-bent exhaust was fitted. Although rated as a 4in system it was actually 3in between the turbo and the CAT, as the 4in dump pipe had a 3in pipe welded inside to carry turbine flow, while wastegate flow was handled by the remaining space. This system didn't really provide any appreciable gain over the 3in system. It was supplied as a street exhaust, but there was no way it could be used on the street – the noise meter reading showed that it would have only just scraped through at the race circuit!

The exhaust in the next test was also supposed to be a street exhaust. At idle and cruise the noise level was borderline acceptable, but under full power acceleration it was just as loud as the previous 4in exhaust. In fact, it had the same style of dump pipe, with a 3in pipe within a 4in pipe, and the CAT was the same brand and was located in the same position. However, after the CAT the system divided into two 3in pipes, each with their own resonator. The system appeared to be way over the top, something to appeal to loud boy-racer types. What a surprise, then, when it produced such a good hp increase. It is rare to come across an engine under about 400–450hp that will benefit from an exhaust of this size.

There was no way this exhaust could qualify as being suitable for the street. Apart from the noise factor, ground clearance was poor. Consequently, the system was only left in place to determine what sort of power levels could be attained with the stock turbo and intercooler and the rest of the car set-up as it would be for weekend club days. With that aim the next test was to change to 98 RON unleaded. (At the time of these tests 98RON was not available outside major cities, so it was not wise to tune a street car for anything over 95/96RON.) Surprisingly, considering the high inlet temperature at higher rpm, the extra octane made little difference to top-end power. However, at lower engine speeds, where the turbo was capable of pumping full boost, hp did rise.

The Subaru ECU has learning ability but it is fairly slow to respond to fuel octane changes. Also, the ECU obviously can only make changes within limits set by the factory programme. Therefore, for the final test, the ECU was remapped for both spark advance and injector pulse width. This produced good gains at lower rpm but, unfortunately, the top end was not going to respond – the small turbo was still having the final say!

CONSIDER AN ENGINE SWAP

Another route to possible big power gains in rear-wheel-drive cars is an engine transplant. Quite a number of Nissans, for example, feature a variety of engine sizes and body styles, but a closer look reveals that what, on the surface, appear to be quite different cars are indeed very similar underneath. Such is the situation with the S13 Sylvia. This particular car was purchased because it looked good, it had a turbo and a 5-speed gearbox, and it was cheap. The problem was that the engine was past its 'use by' date. In his haste to snare a bargain, the buyer had failed to notice just how far down on performance this tired engine was, and by then it was too late. He checked out the cost of reconditioning the stock 1.8-litre 4-cylinder engine (CA18DET); it was more than he had paid for the car! The only option was to fit a second-hand engine from a late model low-mileage wreck. The 2-litre 4-cylinder SR20DET engine would have been his first choice, but demand was high so there weren't many to choose from and their prices were way beyond what he could afford. During this search it was noted that the heavier 2-litre 6-cylinder RB20DET engine was plentiful and only around half the price. Apparently it had become an orphan – enthusiasts all wanted the bigger 2.5-litre RB25DET. A few quick measurements indicated that there was ample room under the bonnet to accommodate the longer 6-cylinder engine. Another factor was that Nissan quoted 215hp for the six verses 197hp for the four. Also the six is renowned for its smoothness and ability to rev.

The engine mated up to everything without any fuss. Slight engine mount modifications were required. The main tasks were building a new exhaust and changing the wiring loom and ECU. The front struts and brakes were a straight swap from the 6-cylinder car. However, even though the new engine was running sweetly, performance was no better than from the tired old 1.8-litre unit. The air/fuel ratios were checked and found to be all right. The next check was for turbo boost pressure as the hi-tech ceramic turbine wheel is susceptible to damage. The pressure at the inlet manifold was only just over 9psi. This was extremely low, but before pulling the turbo off to inspect it for damage, boost was tested at the turbo outlet. The gauge now showed almost 13psi, so where were we losing nearly 4psi? All joints in the intercooler ducting were tested for leaks – none could be found. The intercooler was then pressure tested in a water bath, but there were no bubbles, so obviously there were no cracks allowing boost to escape. All this meant the small stock intercooler was causing a huge pressure drop and, no doubt, massive restriction to flow.

To find out just how much, the car was run on the wheel dyno with the factory cooler in place. Table 22.3 shows the result – a maximum of 154hp at the wheels at 6500rpm, around 30–40hp less than it should have been. A used intercooler with a core measuring 600 x 250 x 60mm was modified to fit in the front of the car. Power now rose to just over 213hp at 6800rpm and boost at the inlet manifold now as 13½psi. However, the engine appeared to be down on power under 4000rpm so it was decided that we would move the cams to try to obtain more low speed grunt. Unfortunately, nothing we tried produced a gain in that area. In fact, we either had no change or a slight loss of 2–3hp. However, when the valve overlap period was increased by 7° (inlet cam advanced 3°, exhaust cam retarded 4°) we picked up around 8–17hp in the upper rpm area without any loss at the bottom end.

Table 22.3. Nissan Sylvia 6-cylinder conversion dyno test

Rpm	Horsepower		
	Test 1	Test 2	Test 3
3000	49.4	50.6	52.1
3500	74.0	73.7	74.2
4000	97.7	106.5	108.3
4500	106.1	144.4	148.7
5000	119.3	163.3	171.0
5500	134.5	182.1	192.4
6000	146.9	207.8	216.5
6500	154.3	212.0	224.6
6750	147.8	213.4	226.3
7000	143.8	209.2	223.4
7250		191.9	206.6

Test 1 – Stock S13 intercooler.
Test 2 – 600 x 250 x 60mm intercooler.
Test 3 – As above with inlet cam advanced 3° and exhaust cam retarded 4°.

It was suggested that the bigger turbo from the 2.5-litre engine should be tried, but that was beyond the budget. At any rate, the bigger turbo usually only helps low speed power if the boost is raised. The major benefit is that at the top end it will add 15–20hp with the stock cams. Actually, the stock turbo is up to the task but it is let down by the exhaust manifold. A set of tube headers should always be a high priority for a single turbo 6-cylinder engine. The cast manifold kills exhaust flow and softens the pulses which are so necessary to help the turbine quickly spool up. These factors also tend to lessen possible gains from moving the cams, and if bigger cams are installed hp will often be down everywhere.

Obviously, not all intercoolers were created equal. While the Nissan Skyline GT-R intercooler pictured, and that fitted to the Pulsar GTi-R are both quite efficient and flow well, those Nissan fit to some other models are not up to the task.

The larger RB25 turbo would have raised the RB20's power output.

MODIFYING A BLOWN ENGINE

With factory blower cars it takes a lot more work to achieve the sort of power increases that can be fairly easily picked up from simple changes to turbo engines. In Table 22.4 you can see the effort required to wring a 40% power increase from the supercharged 3.8-litre Buick V-6 factory fitted in the rear wheel drive Holden Commodore. The stock engine spins an Eaton M90 blower at a drive ratio of 1.68 times engine rpm to produce maximum boost of 7psi without intercooling. The cylinder heads are only 2-valve, but they do flow quite well in spite of relatively small valve sizes of 1.8in inlet and 1.52in exhaust. The inlet manifold plenum, with its nicely bellmouthed runners, no doubt assists overall flow on the inlet side. However, on the exhaust side the situation deteriorates. The section up to the two catalytic converters doesn't look too bad, but at only 2in the pipe size appeared to be too small. After the CATs, the two pipes remain at 2in right up through the first silencers (mufflers). Then just before the rear axle the two pipes are savagely deformed to clear the suspension and are then brought together into one pipe with a single silencer right at the rear of the car.

The first modification would be to get rid of the entire stock exhaust. In its place a full high-performance twin $2^{1}/_{2}$in mandrel-bent system designed for the 5.7-litre all alloy Chev V8 version of the Commodore was fitted. This system had bigger CATs, two stainless steel silencers in front of the rear axle and a pair of small resonators. Just after the CATs the two exhausts were siamesed over a length of 10in as this has been shown to help power in the V-8 to the tune of 4.5–6% over the entire rev range. Naturally, the V-8 headers could not be used, so a custom set of headers, using $1^{7}/_{8}$in tube 30in long for

Table 22.4. Buick 3.8 V-6 dyno test

	Horsepower			
Rpm	Test 1	Test 2	Test 3	Test 4
2000	95.4	106.3	99.1	103.4
2500	117.5	131.3	124.7	132.8
3000	126.7	143.3	148.0	155.5
3500	135.0	151.4	157.6	168.9
4000	141.9	162.5	174.8	189.3
4500	147.3	170.8	186.9	210.0
5000	153.3	175.0	197.2	211.1
5250	144.5	168.6	198.8	207.9
5500	132.1	158.4	184.6	195.2
6000			166.1	175.8

Test 1 – Stock with 98RON fuel.
Test 2 – As above, plus modified airbox, 1⅞in x 30in headers and twin 2½in exhaust.
Test 3 – As above, plus sports cam and reprogrammed ECU, only the CAT back section of exhaust from Test 2 retained.
Test 4 – As above, plus water injection and reprogrammed ECU, 91RON fuel.

each primary, were borrowed from a Buick race engine. The power increase was quite good, with gains of well over 20hp at the top end of the rev range, and even at 2000rpm power was up just under 11hp.

The next change was to replace the very mild factory cam with something that would help engine breathing at higher rpm. The cam chosen has a duration of 218° at 0.050in lift on both the inlet and exhaust lobes, and valve lift of 0.469in. It was ground with a lobe separation angle of 110°. The problem with a cam change of this magnitude is that the ECU has to be remapped to get the ignition timing and fuelling back on track, so there is a good deal of effort involved. As expected, bottom-end power fell away a little and there was an increase of almost 25hp in peak hp. While the ECU was being reprogrammed, the rev limit was moved up 500rpm to 6000rpm.

As the engine had barely topped 200hp at the wheels, there was doubt as to whether this cam really needed the custom tube headers and big bore CATs. Consequently, these were replaced by the stock cast headers, 2in engine pipes and stock 1.4-litre CATs. From the catalytic converters back, the high-performance twin 2½in V8 system was retained. The results are what you see in Test 3 (Table 22.4). The power hardly changed. The biggest loss was 2.7hp at 5000rpm. So, with the money saved on the exhaust pieces, it was decided to take modifications a step further elsewhere.

There were a number of options. Some choose the fairly simple route of installing 2:1 pulleys to drive the blower faster (12,000rpm is the maximum recommended blower speed for this M90) and give more boost. In reality, going past the stock 6–7psi (it varies, many cars read closer to 6psi than the factory quoted 7psi) is pretty well pointless unless an intercooler is fitted. That wasn't an option on this car. First, it had to travel big distances at night over roads where frontal contact with straying animals was always a possibility. Apart from the expense of replacing the intercooler after such altercations, a

The Commodore's air box is a reasonable size, but the air intake is a disaster.

major concern was that the vehicle should remain driveable. A smashed headlight, a busted grill and bent air-conditioner condenser is inconvenient, but it won't stop a car. A broken intercooler will, and even if it could be easily bypassed, the radiator is often wrecked owing to the intercooler being pushed back onto it. Second, from an engineering viewpoint it would have been exceedingly difficult as the blower is mounted in the Vee, bolted directly to the inlet manifold.

An alternative would be to add toluol injection into the blower, but with a waste spark type ignition this is a worry. Far better in this situation to stick with water injection. On this particular car boost increased by over 1psi, to almost 7½psi, because of the water closing up clearances in the blower. As the owner wanted to switch back to using 91RON fuel, it was decided that this was about the limit, so the stock 1.68:1 drive pulleys were retained.

The first water injection system consisted of a brass spray bar with two tiny holes discharging finely atomised fluid into the blower inlet. This stopped detonation at some engine speeds and throttle positions, but in other speed/load areas hp fell away when the engine detonated and the knock sensor instructed the ECU to pull back ignition advance. Unhappily, when water discharge rates were increased to prevent detonation in those places power dropped off at the other rpm owing to too much water stalling combustion. Adding more spark advance didn't appear to have much effect.

To get around the problem it was decided to run with a rising rate regulator and fuel injection pump based system. The brass spray bar would come into operation at approximately 3psi boost (this varied depending on inlet charge temperature). Then, as boost increased, the adjustable rising rate pressure regulator increased water flow rates more in line with what was actually required, and the very high pressures (over 100psi) ensured good atomisation. A Bosch fuel injection pump pushed the water (with a little soluble oil added to prevent corrosion in the system) through the spray bar. Test 4 (Table 22.4) shows the very satisfactory result with good hp increases at all engine speeds. Particularly useful were the bottom-end improvements which wiped out the losses caused by the camshaft change. Considering that these figures were obtained on 91RON fuel (the previous tests were using 98RON) it is even more remarkable.

Chapter 23

Putting theory to the test

It is very easy when planning modifications to be lulled into believing that the end result will always match the figures we read on the calculator screen after punching in a string of numbers. The reality is that many times they won't, and especially so if we have had to introduce a number of compromises along the way to get around unplanned-for problems. For example, having tested your choice of turbo on the car you may have concluded that all is well. The car drives all right, it responds well when the throttle is floored, it feels powerful, nothing fell off or broke, etc. However, if we are to be truly methodical and we want to turn this into a real learning experience that may enable us to move on to another level in the future, we need to check that we are getting the degree of performance we had worked out on paper, and prove from instrument readings that all is indeed all right. This cold, hard analysis will probably give you some surprises and disappointments, but at the end of it you will be better equipped to properly analyse the true performance gains and determine if and where improvements could be made.

Really, to avoid possible engine damage we should do at least two instrument checks before the engine is given anything more than some low boost (50–60%) running. This was covered in greater detail in earlier chapters, but you need to understand that before you progress to the next stage of testing your choice of turbo, you must ensure that the fuelling is correct and that the boost pressure/spark advance relationship is also correct.

AIR/FUEL RATIO AND DETONATION TESTS

First, check the air/fuel ratio. This would be tested at cruise, 1psi boost, 25% boost, 50% boost, 85% boost, and finally at full boost. Obviously, if at any level the fuelling was lean, you would discontinue testing and rectify the problem. During the A/F ratio check you would also have a knock (detonation) indicator connected, and if at any time during the fuelling test the indicator signalled detonation was occurring, testing would be suspended. If both instruments showed a lean condition and knock simultaneously, then obviously getting the A/F ratio right could see the end of the detonation. With all fuelling

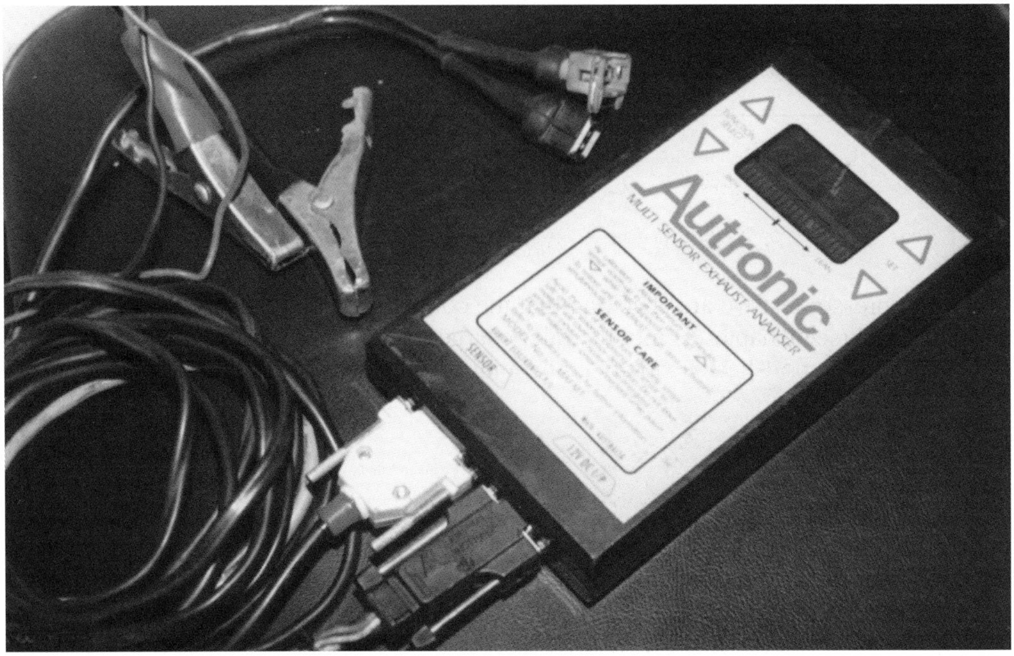

An Autronic multi-sensor exhaust analyser will accurately measure A/F ratios from as rich as 9.4:1 to as lean as 36.0:1.

With care, a digital voltmeter connected to the Lambda sensor will give a rough idea of whether the mixture is safely rich or dangerously lean. If you have no alternative to this method, any reading under about 930–950 millivolts at full throttle should be considered lean for a forced induction engine, i.e. approximately 13:1.

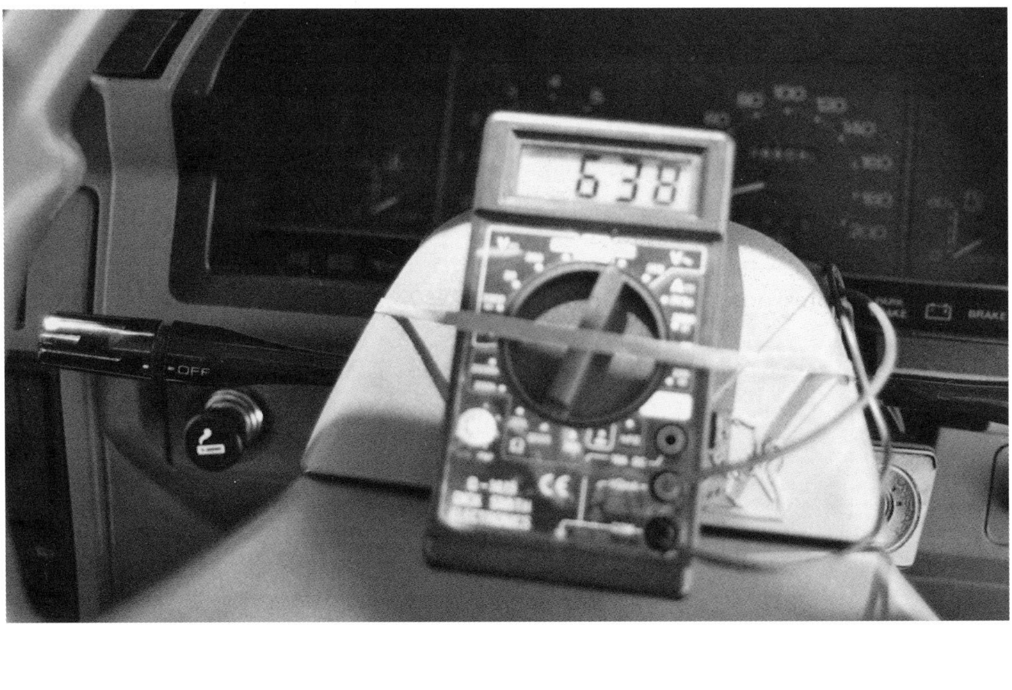

and knock activity corrected, several high load runs would be made holding maximum rpm and boost for around 20 seconds. This duration ensures that the fuel system is up to the task, and that there is no detonation activity at much higher internal heat loadings.

The next tests are to evaluate our choice of turbo compressor and turbine. I prefer to carry out these tests separately, but if you can watch several instruments and keep the car on the black stuff at the same time, they can be combined.

COMPRESSOR EFFICIENCY TEST

In the compressor test we are checking that the compressor is not putting more heat than we calculated (see Chapter 5) into the inlet charge. If there is a big discrepancy between calculated and actual heat levels it could indicate that we made the wrong choice of compressor. Consequently, to provide the airflow demands of our engine the compressor is working in a low efficiency area of the compressor map.

Compressor efficiency should be measured in the area of the power band that is most important to you. For a competition engine that is probably maximum hp revs and down at 60% of redline rpm. For a road car the mid-range is much more important, so here we would test at 50% and 75% maximum rpm. During the test we have to hold maximum boost steady for about 10 seconds, and we have to repeat the test for accuracy. For a road car, three checks at both engine speeds is the minimum, but for a competition engine you want double that. Obviously, we don't want to cook the brakes holding steady both revs and maximum boost for 10 seconds while we get a temperature reading at the compressor outlet (we are not checking intercooler efficiency, so it has to be the compressor outlet temperature). Consequently, we do these tests in lower gears, usually second or third for the high rpm tests and third or fourth for the lower rpm tests.

You will recall from Chapter 6 that our guinea pig 2-litre engine was to run at 11.4psi maximum boost and maximum rpm of 7200. Therefore, the tests would be at 3500rpm (approximately 50%) and 5500rpm (approximately 75%). In addition to the compressor outlet temperature, we have to record the inlet temperature (the inlet should be outside the engine bay drawing in air at ambient temperature). The boost pressure must also be checked on a known accurate gauge (accessory gauges are notoriously inaccurate). For the tests we would prepare a chart similar to Table 23.1.

Table 23.1. Compressor efficiency tests

75% rpm = 5,500 gear = third
50% rpm = 3,500 gear = fourth
Calculated compressor outlet temperature rise °C = 74.2*

Test	rpm	Boost (psi)	Inlet °C	Outlet °C	Rise °C
1	75%	11.8	31	108	77
2	75%	11.8	31	109	78
3	75%	11.8	31.5	109	77.5
4	50%	11.5	32	107	75
5	50%	11.5	31.5	108	76.5
6	50%	11.5	31.5	109	77.5

*Calculated at median inlet temperature of 31.5°C and 11.4psi boost.

Clearly, when we compare the test results the outlet temperatures are somewhat higher than calculated by about 5%. This can be explained to some extent in that during the 75% rpm tests the turbo was overboosting by 0.4psi owing to the internal wastegate being too small to bypass sufficient exhaust flow. When that is taken into account the temperature rise is very close to what theory said it should be. The higher temperatures at 50% revs can also be accounted for owing to the compressor being less efficient at the lower air flow rate.

TURBINE PERFORMANCE TEST

The next test is to check our choice of turbine housing. In general, mildly tuned street engines should reach full boost on wide-open throttle at the engine's peak torque rpm. Fitting a turbine housing any smaller than this usually doesn't help improve low speed performance, and it will as a rule result in high rpm power losses owing to excessive exhaust back pressure killing exhaust flow out of the engine. Ideally, the exhaust back pressure, when measured at the turbine inlet and with a 2ft long open dyno exhaust connected to the outlet, should not be more than about 25% higher than the maximum boost pressure at peak hp. Therefore, in this engine, with maximum boost of 11.8psi, exhaust back pressure should not exceed 14.8psi at maximum power revs with an open exhaust. As we are testing with a full street exhaust, we can add 1psi for a total of 15.8psi. Adding 1psi is, in fact, being quite generous. Any good high-performance exhaust should have less than 1psi back pressure. Even the very quiet exhausts fitted by car makers to their performance models rarely exceed 2psi, and quite frequently when the rear silencer (muffler) is replaced by a straight-through resonator, pressure falls to 1psi.

With pure performance engines a higher power output is desired. In this instance the turbine housing selected will often not achieve peak boost until around 500–700rpm higher than a more mildly tuned engine's peak torque revs. A turbine housing of this A/R ratio considerably cuts exhaust back pressure. Consequently, at maximum power rpm exhaust pressure at the turbine inlet will be about the same as maximum boost pressure.

This car, though, didn't fall into the latter category. So, we were expecting that at maximum hp revs, which we estimated to be close to 6000rpm, exhaust back pressure should not exceed 15.8psi. We were a bit disappointed, then, when the pressure recorded at 6000rpm was just over 17psi, and at maximum revs it rose to a fraction under 20psi. However, the next tests to check on intercooler efficiency revealed that boost at the compressor outlet, as opposed to the pressure at the inlet manifold, was in fact 13psi. Hence an exhaust back pressure reading of a little over 17psi was quite good. Some tuners are a good deal more liberal in this area, reckoning that back pressure in the range of 2–2.5 times boost pressure is satisfactory.

The early Porsche 911 turbo is a good guide to work from in this regard. It produced peak hp at 5,500rpm. At that point it made 11.6psi boost and showed an exhaust back pressure of 16.3psi. This was achieved with the stock, very quiet, exhaust in place. A further handicap was that this model was fitted with an air-injection system to clean up exhaust emissions. The air being injected into the exhaust ports added more gas volume, and any combustion taking place here added more exhaust pressure. Incidentally, this helped the turbo spool up more rapidly at lower rpm when exhaust flow was lower, but it added to exhaust back pressure at higher revs (Table 23.2). Note the figures at 3000rpm where the 911 produced maximum torque. At that point turbo boost has peaked (it does slowly creep higher but basically the peak is at 3,000rpm) indicating good turbine size selection by the factory. Also at the torque peak, exhaust back pressure is approximately 70% of the boost pressure, another indicator of good matching in a performance application.

Table 23.2. Porsche 911 exhaust turbine back pressure

Rpm	Boost pressure (psi)	Exhaust back pressure (psi)
1500	1.5	1.1
2000	4.0	2.6
2500	9.6	6.0
3000	10.9	7.6
3500	11.0	8.7
4000	11.0	10.3
4500	11.2	13.0
5000	11.4	14.7
5500	11.6	16.3
6000	11.6	17.2

When developing a forced induction package, car makers perform a multitude of tests to establish that the outcome matches what calculations had predicted. When Porsche developed the intercooled 911 over two decades ago they didn't have the hi-tech data gathering equipment that's now available. They had to rely more on the sort of basic test instruments tuners use today.

INTERCOOLER EFFICIENCY TESTS

When it comes to intercooler efficiency there are two tests – one for pressure drop caused by flow restriction, and the other to determine the intercooler's ability to reduce charge temperature. Basically, the pressure loss should not exceed 10%, but less is better if cooling efficiency is not compromised. This engine was recording 13psi at the compressor outlet and 11.8psi at the inlet manifold, so it was in the range of 'acceptable' rather than 'good'. However, this would be conditional on its cooling ability, which was expected to be above 80% efficiency.

As in the compressor efficiency test, intercooler efficiency was checked at 75% maximum engine speed. In a competition car we would test at the bottom of the power band, at the top of the power band and about midway. During the test we have to hold maximum boost for about 10 seconds, holding the car on the brakes in second or third gear. For consistency we really need to repeat this at least three times, on each run recording the ambient air temperature, the intercooler inlet air temperature and the intercooler outlet air temperature. If we were checking the insulating efficiency of heat shielding and reflective blankets added to inlet ducting, we would also have temperature probes at the compressor outlet and at the inlet manifold. These figures would be recorded in a chart similar to Table 23.3.

Table 23.3. Intercooler efficiency tests

75% rpm = 5,500 gear = third

Test	Boost (psi)	Ambient °C	Inlet °C	Discharge °C	Drop °C
1	11.8	31	109	47.5	61.5
2	11.8	31	110	49	61
3	11.8	31.5	111	49.5	61.5

CALCULATING INTERCOOLER EFFICIENCY

Remembering that we are comparing the temperature drop due to the heat removed by the intercooler with the temperature of the charge air leaving the compressor, we calculate intercooler efficiency using this formula:

$$\text{Intercooler efficiency \%} = \frac{\text{cooler inlet temp} - \text{cooler discharge temp}}{\text{cooler inlet temp} - \text{ambient temp}} \times 100$$

$$= \frac{100 - 49}{100 - 31} \times 100$$

$$= \frac{61}{79} \times 100$$

$$= 77.2\%$$

Therefore, with an efficiency of 77.2%, the intercooler is reasonably efficient. However, in view of the pressure drop across the core it is obvious that its flow efficiency relative to its cooling efficiency is not good. Generally, I would expect an intercooler to be approaching an 85% cooling efficiency for a 10% pressure restriction. Down at 77% efficiency the pressure drop should not have exceeded about 8½%.

CHECK FOR PRESSURE LEAKS AND DUCTWORK PROBLEMS

At this stage we should not conclude that the pressure loss is all owing to the intercooler core not flowing efficiently. The first check should be to ensure there are no pressure leaks anywhere. This means checking every join in the ducting, all manifold and throttle body gaskets, clearance between the throttle spindle and throttle body allowing air to escape, every hose connected to the inlet system anywhere between the compressor outlet and the cylinder head for cracks or leaky connections, leaks around the fuel injectors. (Remember the 'O' rings sealing between the injectors and manifold eventually harden and crack, or if they are fitted dry they can be stretched out of their groove and unable to effect a seal.) Before fitting, the intercooler should have been checked for cracks or porous welds. If it hasn't been, now isn't too late.

When all those checks come up clear, we next need to examine our duct work from the turbo to the cooler, and then to the inlet manifold. Are all the bends really necessary? Do they need to be so sharp, etc.? Also, check that all joins are smooth internally. At times there are flow-destroying steps at every join where a flexible sleeve links two sections of rigid metal ducting. These steps should be eliminated. Only after all of these inspections, and after appropriate correction, should we conclude that the intercooler has a flow problem.

The very same applies with regard to its cooling efficiency. With an air-to-air unit we must reassess how effectively we are ducting cooling air flow to the intercooler, and then ensure that it isn't allowed to flow around the intercooler rather than through the heat-radiating fins. If at the same time we have engine overheating maladies we should suspect that we have an air flow problem into the nose of the car, or else we have too much pressure behind the radiator which is either blocking or drastically slowing air speed through the intercooler and radiator. Short of testing in a wind tunnel, the best ways to check for either problem are by means of a simple water manometer and/or using a dyno fan and water spray or smoke stream as described earlier in Chapter 10.

WATER-TO-AIR INTERCOOLER CHECKS

If we have a water-to-air cooler, then we should first ensure that water flow is sufficient and that the water radiator is the correct size. Begin by checking that water entering the charge cooler is at the very same temperature as the ambient air. If it isn't, then we have to check that cooling air flow over the water radiator is adequate, that water circulation rates are rapid enough and that reservoir volume is sufficient. When all these areas are in order, suspicions should be aroused as to the size of the water radiator and the adequacy of air flow across its fins. As in the case of air-to-air units, we must correct any tendency for air to flow around the outside of this cooler rather than through it.

Finally, there is the problem of air in the water cooling system. This acts as an insulating blanket and prevents cooling water from coming into contact with the cooling fins whose job it is to draw off charge heat in the charge cooler proper. This is the more common difficulty. However, air can also be trapped in the water radiator and thereby

Water-to-air intercooling can be deceptively inefficient, often owing to air in the water cooling system or inadequate water flow rates. Only temperature tests will reveal the true story.

prevent the water from giving up its heat to the tubes and fins. Only thoughtful location of bleeders and very careful bleeding may be enough to purge most of the air out of the system. Vibration will assist in dislodging air bubbles, so after driving the car rebleed the system a couple of times. Really difficult systems may require being evacuated to quite low vacuum levels in much the same manner as refrigeration units and air-conditioners prior to being regassed.

OTHER PERFORMANCE THIEVES

Obviously, the same attention to detail would be expended on any other system which was holding the engine back from reaching its true potential. This could be in areas such as the fuel system, ignition system, lubrication system, cooling system, etc. The same principles apply to cars being supercharged by a blower or nitrous oxide injection. However, these have additional areas which may be the weak link in the chain. For example, the blower drive arrangement may be far from ideal allowing belt slipping, or energy may be wasted with misaligned pulleys, and so forth. Unfortunately, many who set out modifying an engine never realise their dreams. When things do not work out as planned immediately, they very quickly abandon the project and initiate another plan, perhaps repeating the same basic mistakes.

Chapter 24

Other considerations

In earlier chapters we have looked into methods of increasing engine performance by means of forced induction. Now we want to consider what impact any proposed modifications will have on the vehicle itself. And, because this is a two way arrangement – an engine has to work with the rest of the vehicle – we have to recognise what modifications are feasible for the engine. Also, there are safety and legal issues. Those areas are beyond the scope of this book, but obviously any modifications we make must harmonise with local laws so that we can continue to operate our cars on public roads. Safety is of even greater importance, which means that no modification should compromise your ability to arrive home safely in your car at the close of each day. It goes without saying that when we plan major power upgrades there must be similar improvements in relation to the performance of the suspension, tyres and brakes. (These issues are covered in detail in my book *Modern Engine Tuning*, published by Haynes.)

POWER BRAKE ISSUES

The first matter concerns only those cars which left the factory with a naturally aspirated engine and power brakes. If we add a blower or turbo we need to make sure we have a suitable one-way valve installed in the vacuum line which goes from the inlet manifold to the brake booster. The brake servo will have a one-way valve in the fitting where the vacuum hose connects, and while they may hold up at low boost levels it must be recognised that ordinary valves were designed to hold back only normal atmospheric pressure. If the valve fails, you will no longer have any power assistance – the brake pedal is going to go hard and it is going to take much, much longer to stop! Consequently, try to obtain a valve from a turbo car to fit your brake servo, or else retain your stock valve and install a high pressure one-way valve somewhere in the vacuum line. While you are at it, get rid of the stock spring clips securing both ends of the vacuum hose. They open under pressure, allowing supercharger boost to leak, or worse the vacuum hose may blow-off. In their place use Jubilee hose clamps.

Serious power demands equally serious brakes, suspension and tyres.

At high boost levels this one-way valve in the brake booster vacuum line can fail, and the spring clamp can allow the hose to blow off.

TRANSMISSION TORQUE LIMITS

The next matter also primarily concerns naturally aspirated cars, although it may involve some supercharged cars as well. We have to remember that the standard clutch and transmission were designed with a particular torque limit in mind. Some manufacturers have a reputation for using components that just scrape through at standard torque levels. When we add some form of forced induction, or we increase traction by means of grippier tyres, add a limited slip diff or lock the diff, then we put a lot more strain on the clutch and transmission. Some people choose to ignore such considerations. They build in the power level they want, they beef up the clutch and accept that transmission failures are going to become a familiar experience. That is, until they realise that (a) it's not fun being regularly stuck under the car replacing broken parts (b) they run out of money or (c) the local breakers' yard exhausts its stock of gearboxes, axles and diffs!

To my way of thinking it is much wiser to avoid such problems. The solution may be to limit how frequently these components are subjected to drag starts and power shifting. Some choose to follow the course chosen by many vehicle manufacturers. They simply cap modifications to the engine at a level that will not overload the transmission. Others view either course as too much of a compromise and are willing to invest in an engine management system that limits boost at certain revs, and/or in specific gears, in order to keep torque levels in check. The ultimate solution may be to fit a tougher bullet-proof transmission. Perhaps components from another production car are suitably robust and can be modified to fit. For a road car this is usually wiser than going for a tough competition-type transmission. The latter are often described as being able to handle the torque of a truck. Frequently that is simply a polite way of saying it sounds and feels like a truck transmission – the shift is heavy and slow, it clunks between gear changes and it's noisy!

GEARBOX RATIOS AND GEARING

High on my list of priorities for any car are suitable gearbox ratios. A car with a set of nicely spaced cogs is always much more enjoyable to drive. This isn't such a concern with a blower engine, but for turbo cars it is so important that my advice is that you should buy a car with a good set of gears, or else be prepared to spend a lot of money for a custom gearset. Other than that, be willing to modify the engine to suit the stock ratios. This will limit what you can do in the way of engine or turbo upgrading. However, it will ensure that your car is more fun to drive on a twisting road, and ordinary secondary roads.

The gearbox and diff are really just torque multipliers. Hence an overall gear ratio (i.e. gearbox ratio x diff ratio) of 3:1 multiplies engine torque by three, giving three times as much twisting force at the wheels as compared with that measured at the flywheel. When we lower the gearing to say 4:1 we increase the twisting force by $33\frac{1}{3}\%$ and enhance acceleration significantly. Looking at Table 24.1 you will note just how much more rapidly a Subaru Impreza WRX accelerated in 3rd gear and 4th from various initial speeds. In comparison with 3rd or 4th, the car is really struggling to accelerate in top gear. This lethargy is partly because of high gearing in top gear (intended to provide gentle on the ears cruising) and partly owing to sluggishness because of the turbo. When we install a gearset from the Sti V-3 model the engine revs much harder in top gear, but note how much harder the car now accelerates (Table 24.2). It would take a lot of work on the engine to achieve that level of acceleration with the stock gearing, and at the conclusion of the exercise you would not be able to accelerate the car in top gear at anything less than 80mph.

Table 24.1. Subaru Impreza WRX roll-on acceleration

	Acceleration		
Overall gear ratio	35–55mph	50–70mph	60–80mph
3rd gear – 5.34:1	3.4 sec	3.6 sec	3.9 sec
4th gear – 3.78:1	6.1 sec	5.4 sec	5.8 sec
5th gear – 2.89:1	11.2 sec	9.8 sec	8.7 sec

Tyre size 205/50 x 16

Table 24.2. Subaru Impreza WRX/Sti gearset acceleration

	Acceleration		
Overall gear ratio	35–55mph	50–70mph	60–80mph
3rd gear – 6.88:1	2.5 sec	3.2 sec	3.8 sec
4th gear – 5.11:1	3.6 sec	3.8 sec	4.1 sec
5th gear – 3.69:1	6.3 sec	5.6 sec	6.1 sec

Tyre size 205/50 x 16

The same kind of principle applies when it comes to accelerating out of corners. If the engine is highly tuned, the car needs a nice close set of gear ratios. To help you understand this, think about a gearbox with a big gap between 3rd and 4th. In 3rd the gearing may be too low, so the car will be making a stack of boost which will badly upset traction and handling, particularly if the road is wet. However, because 4th gear is too big of a step up, the engine will bog out of the corner and acceleration will be leisurely to say the least. There are two ways around this problem. Either we modify the engine to provide a broad power band and almost zero turbo lag, or else we install a close ratio gearset.

WORKING OUT ACCEPTABLE GEAR STEPS

Ideally, in a road car, we want five close ratio gears for performance, plus a big overdrive top gear for effortless cruising on motorways – that means six speeds in total. Obviously, many cars still only have a five-speed gearbox, but we can get by with five nicely spaced gears. Top gear won't be as tall, to keep the gaps manageable between the lower gears, so the car will not be as quiet at cruise. To give you some idea of what gear steps are acceptable take a look at Table 24.3, where the gearsets of a number of performance cars are listed. They are all six-speed gearboxes; the manufacturers who built these cars intended that all provide high levels of performance and driving satisfaction. You can see quite clearly that once past 3rd gear the gaps between gears are all less than 24%. The gear steps are calculated using the formula:

$$\text{step \%} = \frac{\text{lower gear ration} - \text{higher gear ratio}}{\text{lower gear ratio}} \times 100$$

When we get to checking the five-speed gearsets in Table 24.4, it is obvious that some manufacturers make it quite difficult for tuners by introducing very big steps between gears. The Subaru Liberty RS, like its more recent cousin the Impreza WRX, has an unacceptably large gap, over 29%, in the vital 3rd to 4th step. The others, though, are all a little under 24%. This is what we want, anything wider than 25% is too big unless the car has a big engine relative to its weight. The 4th to 5th step isn't so important for a road car. Most drivers want a fairly quiet cruise so are prepared to accept less than magical acceleration. In such instances a step of 22–24% is acceptable. On the track, that will usually be too wide, with something less than 20% providing improved performance, depending on track speed and car aerodynamics. A car with the aerodynamics of a brick and high maximum track speeds calls for a smaller step, under 18%, while a slippery car on a slow track can get by with a bigger gap over 22%.

Table 24.3. Six-speed gearbox comparison

	Audi S3 (1.8-litre)		Nissan Sylvia (2-litre)		Subaru WRX Sti (2-litre)		Nissan SkylineGT-R (2.6-litre)		Porsche 911 (3.6-litre)	
	gear ratio	gear step	gear ratio	gear step	gear ratio	gear step	gear ratio	gear step	gear ratio	gear step
1st	3.42	+38.3%	3.63	+39.3%	3.64	+34.9%	3.83	+38.4%	3.82	+43.7%
2nd	2.11	+32.2%	2.20	+30.0%	2.37	+25.7%	2.36	+28.4%	2.15	+27.4%
3rd	1.43	+23.8%	1.54	+21.4%	1.76	+23.3%	1.69	+22.5%	1.56	+22.4%
4th	1.09	+20.2%	1.21	+17.4%	1.35	+21.5%	1.31	+23.7%	1.21	+19.8%
5th	0.87	+17.2%	1.00	+23.0%	1.06	+20.8%	1.00	+21.0%	0.97	+22.7%
6th	0.72		0.77		0.84		0.79		0.75	
Axle ratio	4.2:1		3.69:1		3.90:1		3.27:1		3.44:1	

Table 24.4. Five-speed gearbox comparison

	Audi TT (1.8-litre)		Mitsubishi Lancer Evo (2-litre)		Nissan Sylvia (2-litre)		Subaru Liberty RS (2-litre)		Nissan Skyline GT-R (2.6-litre)	
	gear ratio	gear step	gear ratio	gear step	gear ratio	gear step	gear ratio	gear step	gear ratio	gear step
1st	3.30	+41.2%	2.79	+30.1%	3.32	+42.8%	3.55	+45.1%	3.21	+40.1%
2nd	1.94	+32.6%	1.95	+26.2%	1.90	+31.1%	1.95	+29.7%	1.93	+32.4%
3rd	1.31	+21.4%	1.44	+23.6%	1.31	+23.7%	1.37	+29.2%	1.30	+23.2%
4th	1.03	+18.4%	1.10	+24.5%	1.00	+16.0%	0.97	+19.6%	1.00	+24.8%
5th	0.84		0.83		0.84		0.78		0.75	
Axle ratio	3.94:1		4.53:1		3.69:1		3.90:1		4.11:1	

Note: Alternative Mitsubishi Lancer Evo gearsets are: 1st 2.93; 2nd 1.95; 3rd 1.41; 4th 1.03; 5th 0.72. Or 1st 2.78; 2nd 1.95; 3rd 1.41; 4th 1.03; 5th 0.76.

The Mitsubishi Lancer Evo engine has always been noted for good bottom-end performance.

The Subaru Liberty B4 has sequential turbos and a nice set of gears, but still suffers serious turbo lag.

PROPER GEAR RATIOS CAN ONLY HELP SO MUCH

However, we should not conclude that a lack of urge from the engine can be totally corrected by fiddling with the gearing. Getting the gearing right will help, but it cannot fully compensate for an engine suffering big doses of turbo lag, or some other mismatch that is destroying mid-range power. Take, for example, the two factory hot rods from Mitsubishi and Subaru. Each version of the Mitsubishi Lancer Evo has been praised for its low-speed grunt, with good power available all the way from just below 3000rpm right to the 7800rpm rev limiter. The Sti version of the Impreza WRX is by comparison a late riser, needing 1000rpm more on the tacho before it gets past a shuffle. When Subaru added a close ratio six-speed gearbox in 2001 things really didn't improve. In fact, Mitsubishi did just the opposite on the Evo VII and fitted a set of wide ratio cogs in their five-speed box to give a much less rowdy cruise.

As Table 24.5 shows, in roll-on acceleration the Evo will eat the Sti in any gear, and this in spite of its slightly taller gearing and one size larger diameter tyres (235/45 x 17 vs. 225/45 x 17). While both cars are advertised as having the same maximum hp (276hp), in truth the Evo is about 10% over that and the Sti is 15% over at 6500rpm. However, in the 2500–5000rpm band the Evo has a clear power advantage thanks to its hi-tech twin-scroll turbo with lightweight titanium innards and variable nozzle turbine. Combined with a long stroke engine, high compression ratio, conservative cams and port sizes this all adds up to a much more flexible power package. So, an Evo rolling out of a corner at 60mph in 4th gear will reach 80mph in 4.3 seconds. Even dropping back two gears the Sti will take 17% longer.

Table 24.5. Subaru Impreza WRX Sti and Mitsubishi Lancer Evo VII comparison

	Acceleration		
	35–55mph	50–70mph	60–80mph
Mitsubishi Lancer Evo VII			
3rd gear – 6.39:1	3.3 sec	3.5 sec	3.7 sec
4th gear – 4.67:1	5.6 sec	4.8 sec	4.3 sec
5th gear – 3.26:1	–	9.3 sec	7.7 sec
Tyre size – 235/45 x 17			
Subaru Impreza WRX Sti			
3rd gear – 6.86:1	4.8 sec	4.4 sec	4.3 sec
4th gear – 5.27:1	6.5 sec	5.4 sec	5.0 sec
6th gear – 3.28:1	–	12.8 sec	10.9 sec
Tyre size – 225/45 x 17			

In the 276hp Liberty B4, Subaru fitted sequential turbos to wake up bottom-end and mid-range performance. This strategy didn't work either. The bottom-end isn't too bad, but there just isn't any mid-range. Between 3000rpm and 4000rpm the engine falls into a black hole, marking the end of the primary turbos usefulness. Then, just before 5000rpm, the secondary turbo begins contributing a good wallop that continues all the way past 7000rpm. As such the engine is very peaky, with an effective power band spanning just 2500rpm. At 3300lb the Liberty is no lightweight either, consequently roll-on acceleration is poor in most everyday driving situations (Table 24.6). In fact, because the engine seems so choked in the mid-range, forward thrust feels much worse than the times recorded on the stop watch. Consequently, when swapping between an Sti and a B4 the Sti actually feels faster even though its times are considerably slower.

Table 24.6. Subaru Liberty B4 roll-on acceleration

	Acceleration		
Overall gear ratio	35–55mph	50–70mph	60–80mph
3rd – 5.77:1	4.1 sec	4.3 sec	4.7 sec
4th – 4.31:1	6.3 sec	6.4 sec	6.4 sec
5th – 3.29:1	9.9 sec	9.0 sec	8.2 sec

Tyre size – 215/45 x 17

Appendix

Table of useful equivalents

1 inch = 25.4mm	1mm = 0.03937in
1 cubic inch = 16.387cc	1 litre = 61.024cu in or 1000cc
1 horsepower = 0.7457 kilowatts or 1.0139 PS	1 kilowatt = 1.341hp 1 PS = 0.9863 horsepower
1 pound foot torque = 1.3558 Newton metres or 0.13824 kg/m	1 Newton metre = 0.7376 pound foot 1 kg/m = 7.2336 pound foot
1 pound inch torque = 0.11298 Newton metres	
1 psi = 6.89476 kilopascals or 68.95 millibars or 2.0345 inches of Mercury (Hg) or 27.67 inches of water	1 kPa = 0.14504psi 1 Bar = 14.5038psi or 100 kPa
°F = (9/5 x °C) +32	°C = (°F –32) x 5/9
1 gallon (Imperial) = 160 fluid oz or 4.546 litres	1 fluid oz (Imperial) = 28.4cc
1 gallon (US) = 128 fluid oz or 3.785 litres	1 fluid oz (US) = 29.57cc

Index

Other books from Haynes Publishing

Four-Stroke Performance Tuning (2nd Edition)
by A. Graham Bell
ISBN 1 85960 435 8

Two-Stroke Performance Tuning (2nd Edition)
by A. Graham Bell
ISBN 1 85960 619 9

Modern Engine Tuning
by A. Graham Bell
ISBN 1 85960 866 3

Engine Management
Optimising carburettors, fuel injection and ignition systems
Haynes High-Performance Tuning Series
by Dave Walker
ISBN 1 85960 835 3

Tuning the A-Series Engine (3rd Edition)
The definitive manual on tuning for performance and economy
by David Vizard
ISBN 1 85960 620 2

Tuning Rover V8 Engines
How to get the best performance for road and competition use
by David Hardcastle
ISBN 0 85429 933 5

Rebuilding and Tuning Ford's Kent Crossflow Engine
by Peter and Valerie Wallage
ISBN 1 85010 938 9

Rebuilding and Tuning Ford's CVH Engine
by Peter Wallage
ISBN 1 85960 007 7

For more information please contact:
Customer Services Department,
Haynes Publishing,
Sparkford, Yeovil,
Somerset BA22 7JJ England

Tel: 01963 442030 Fax: 01963 440001
Int. tel: +44 1963 442030
Int. fax: +44 1963 440001

E-mail: sales@haynes-manuals.co.uk
Web site: www.haynes.co.uk